Gene Delivery
to Mammalian Cells

METHODS IN MOLECULAR BIOLOGY™

John M. Walker, SERIES EDITOR

METHODS IN MOLECULAR BIOLOGY™

Gene Delivery to Mammalian Cells

Volume 2: Viral Gene Transfer Techniques

Edited by

William C. Heiser

Bio-Rad Laboratories

Hercules, CA

Humana Press ✳ **Totowa, New Jersey**

© 2004 Humana Press Inc.
999 Riverview Drive, Suite 208
Totowa, New Jersey 07512

www.humanapress.com

This publication is printed on acid-free paper. ∞
ANSI Z39.48-1984 (American Standards Institute)

Permanence of Paper for Printed Library Materials.

Cover illustration: Foreground graphic from Fig. 1A,B in Chapter 14 (Volume 1) "Delivery of DNA to Skin by Particle Bombardment," by Shixia Wang, Swati Joshi, and Shan Lu. Background graphic from Fig. 3 in Chapter 33 (Volume 2) "Retrovirus-Mediated Gene Transfer to Tumors: Utilizing the Replicative Power of Viruses to Achieve Highly Efficient Tumor Transduction In Vivo," by Christopher R. Logg and Noriyuki Kasahara.

Production Editor: Robin B. Weisberg.
Cover design by Patricia F. Cleary.

For additional copies, pricing for bulk purchases, and/or information about other Humana titles, contact Humana at the above address or at any of the following numbers: Tel.: 973-256-1699; Fax: 973-256-8341; E-mail: humana@humanapr.com; or visit our Website: www.humanapress.com

Printed in the United States of America. 10 9 8 7 6 5 4 3 2 1

1-59259-650-9 (e-ISBN)

Library of Congress Cataloging in Publication Data
Gene delivery to mammalian cells / edited by William C. Heiser.
    p. ; cm. -- (Methods in molecular biology ; 245-246)
  Includes bibliographical references and indexes.
  Contents: v. 1. Nonviral gene transfer techniques -- v. 2. Viral gene transfer techniques.
  ISBN 1-58829-086-7 (v. 1 : alk. paper) -- ISBN 1-58829-095-6 (v. 2 : alk. paper)
    ISSN 1064-3745
    1. Transfection. 2. Animal cell biotechnology.   I. Heiser, William C. II. Series: Methods in molecular biology (Totowa, N.J.) ; 245-246.
    [DNLM: 1. Cloning, Molecular. 2. Gene Transfer Techniques. 3. Cells, Cultured--cytology. 4. Gene Targeting--methods. 5. Mammals--genetics.   QH 442.2 G327 2004]
  QH448.4.G42 2004
    571.9'64819--dc21

                                                                                    2003006980

Preface

The efficiency of delivering DNA into mammalian cells has increased tremendously since DEAE dextran was first shown to be capable of enhancing transfer of RNA into mammalian cells in culture. Not only have other chemical methods been developed and refined, but also very efficient physical and viral delivery methods have been established. The technique of introducing DNA into cells has developed from transfecting tissue culture cells to delivering DNA to specific cell types and organs in vivo. Moreover, two important areas of biology—assessment of gene function and gene therapy—require successful DNA delivery to cells, driving the practical need to increase the efficiency and efficacy of gene transfer both in vitro and in vivo.

These two volumes of the *Methods in Molecular Biology*™ series, *Gene Delivery to Mammalian Cells,* are designed as a compendium of those techniques that have proven most useful in the expanding field of gene transfer in mammalian cells. It is intended that these volumes will provide a thorough background on chemical, physical, and viral methods of gene delivery, a synopsis of the myriad techniques currently available to introduce genes into mammalian cells, as well as a practical guide on how to accomplish this. It is my expectation that it will be useful to the novice in the field as well as to the scientist with expertise in gene delivery.

Volume 1: Nonviral Gene Transfer Techniques discusses delivery of DNA into cells by nonviral means, specifically chemical and physical methods. *Volume 2: Viral Gene Transfer Techniques* details procedures for delivering genes into cells using viral vectors. Each volume is divided into sections; each section begins with a chapter that provides an overview of the basis behind the delivery system(s) described in that section. The succeeding chapters provide detailed protocols for using these techniques to deliver genes to cells in vitro and in vivo. Many of these techniques have only been in practice for a few years and are still being refined and updated. Some are being used not only in basic science, but also in gene therapy applications.

I wish to express my thanks to all of the authors who made *Gene Delivery to Mammalian Cells: Volume 1: Nonviral Gene Transfer Techniques* and *Volume 2: Viral Gene Transfer Techniques* possible. I would especially like to thank those who contributed the overview chapter to each section. They provided invaluable discussions, suggestions, and assistance on organizing those sections. I would particularly like to mention Joanne Douglas, Tom Daly, and Bill Goins for their suggestions on topics and authors, Dexi Liu and Shan Lu for their helpful discussions, and Mark Jaroszeski for his suggestions on organizing the entire editing process.

William C. Heiser

Contents

Contents of the Companion Volume
Volume 1: Nonviral Gene Transfer Techniques

Contributors

RAMON ALEMANY • *Gene Therapy Program, Institut Català d'Oncologia, Barcelona, Spain*

JOSEPH M. ALISKY • *Marshfield Clinic Research Foundation, Marshfield, WI*

QING BAI • *Department of Molecular Genetics and Biochemistry, University of Pittsburgh School of Medicine, Pittsburgh PA*

ED A. BURTON • *Department of Molecular Genetics and Biochemistry, University of Pittsburgh School of Medicine, Pittsburgh PA*

BAOHONG CAO • *Department of Orthopedic Surgery, University of Pittsburgh School of Medicine, Pittsburgh, PA*

MANEL CASCALLÓ • *Gene Therapy Program, Institut Català d'Oncologia, Barcelona, Spain*

SI-YI CHEN • *Department of Molecular and Human Genetics, Baylor College of Medicine, Houston, TX*

DANNY CHU • *Division of Cardiothoracic Surgery, Department of Surgery, University of California San Diego Medical Center, San Diego, CA*

ROY COLLACO • *Department of Biochemistry and Molecular Biology, Medical College of Ohio, Toledo, OH*

J. PATRICK CONDREAY • *Department of Gene Expression and Protein Biochemistry, GlaxoSmithKline Research and Development, Research Triangle Park, NC*

SHEILA CONNELLY • *Advanced Vision Therapies Inc., Rockville, MD*

DAVID T. CURIEL • *Division of Human Gene Therapy, Departments of Medicine, Pathology and Surgery, and the Gene Therapy Center, University of Alabama, Birmingham, AL*

MICHAEL A. CURRAN • *Department of Molecular and Cell Biology, University of California, Berkeley, CA*

THOMAS M. DALY • *Department of Pathology, University of Alabama at Birmingham, Birmingham, AL*

BEVERLY L. DAVIDSON • *Program in Gene Therapy, Departments of Internal Medicine, Neurology, Physiology & Biophysics, University of Iowa College of Medicine, Iowa City, IA*

JOANNE T. DOUGLAS • *Division of Human Gene Therapy, Departments of Medicine, Pathology, and Surgery and the Gene Therapy Center, University of Alabama at Birmingham, Birmingham, AL*

SEYYED MEHDY ELAHI • *Institut de Recherches en Biotechnologie, Montréal, Québec, Canada*

MICHAEL E. EPPERLY • *Department of Radiation Oncology, University of Pittsburgh Cancer Institute, Pittsburgh, PA*

WENDY FELLOWS • *Department of Neurological Surgery, University of Pittsburgh School of Medicine, Pittsburgh, PA*

DAVID J. FINK • *Departments of Molecular Genetics and Biochemistry, and Neurology, University of Pittsburgh School of Medicine, Pittsburgh PA*

DAVID GAGNON • *Institut de Recherches en Biotechnologie, Montréal, Québec, Canada*

JOSEPH C. GLORIOSO • *Department of Molecular Genetics and Biochemistry, University of Pittsburgh School of Medicine, Pittsburgh PA*

JULIE P. GOFF • *Department of Radiation Oncology, University of Pittsburgh Cancer Institute, Pittsburgh, PA*

WILLIAM F. GOINS • *Department of Molecular Genetics and Biochemistry, University of Pittsburgh School of Medicine, Pittsburgh PA*

JAMES R. GOSS • *Department of Neurology, University of Pittsburgh School of Medicine, and GRECC, VA Medical Center, Pittsburgh, PA*

JOEL S. GREENBERGER • *Department of Radiation Oncology, University of Pittsburgh Cancer Institute, Pittsburgh, PA*

MIRTA S. GRIFMAN • *Laboratory of Genetics, The Salk Institute for Biological Studies, La Jolla, CA*

CHRISTINE L. HALBERT • *Molecular Medicine, Fred Hutchinson Cancer Research Center, Seattle, WA*

ROLAND W. HERZOG • *Department of Pediatrics, University of Pennsylvania Medical Center and The Children's Hospital of Philadelphia, Philadelphia, PA*

FRANK HOOVER • *Department of Oncology, Gene Therapy Program, Haukeland Hospital, Bergen, Norway*

JOHNNY HUARD • *Department of Orthopedic Surgery, University of Pittsburgh School of Medicine, Pittsburgh, PA*

TAL KAFRI • *Gene Therapy Center, University of North Carolina, Chapel Hill, NC*

GUNILLA B. KARLSSON • *Microbiology and Tumor Biology Center, Karolinska Institutet, Stockholm, Sweden, and Department of Vaccine Research, Swedish Institute for Infectious Disease Control, Solna, Sweden*

BRIAN A. KAROLEWSKI • *Department of Pathobiology and Center for Comparative Medical Genetics, School of Veterinary Medicine, University of Pennsylvania, and Division of Neurology, Children's Hospital of Philadelphia, Philadelphia, PA*

NORIYUKI KASAHARA • *Department of Medicine, School of Medicine, University of California, Los Angeles, CA*

DOUGLAS KONDZIOLKA • *Departments of Neurological Surgery and Radiation Oncology, University of Pittsburgh School of Medicine, Pittsburgh, PA*

THOMAS A. KOST • *Department of Gene Expression and Protein Biochemistry, GlaxoSmithKline Research and Development, Research Triangle Park, NC*

VICTOR KRASNYKH • *Division of Human Gene Therapy, Departments of Medicine, Pathology and Surgery, and the Gene Therapy Center, University of Alabama at Birmingham, Birmingham, AL*

DAVID M. KRISKY • *Department of Molecular Genetics and Biochemistry, University of Pittsburgh School of Medicine, Pittsburgh PA*

PETER LILJESTRÖM • *Microbiology and Tumor Biology Center, Karolinska Institutet, Stockholm, Sweden, and Department of Vaccine Research, Swedish Institute for Infectious Disease Control, Solna, Sweden*

CHRISTOPHER R. LOGG • *Department of Medicine, School of Medicine, University of California, Los Angeles, CA*

L. DADE LUNSFORD • *Departments of Neurological Surgery, Radiation Oncology, and Radiology, University of Pittsburgh School of Medicine, Pittsburgh, PA*

BERNARD MASSIE • *Institut de Recherches en Biotechnologie, Montréal, Québec, Canada*

MARINA MATA • *Department of Neurology, University of Pittsburgh School of Medicine, and GRECC, VA Medical Center, Pittsburgh, PA*

CHRISTINE MECH • *Genetic Therapy Inc., Gaithersburg, MD*

RAYMOND V. MERRIHEW • *Department of Assay Development and Compound Profiling, GlaxoSmithKline, Research Triangle Park, NC*

A. DUSTY MILLER • *Molecular Medicine, Fred Hutchinson Cancer Research Center, Seattle, WA*

HIROYUKI MIYOSHI • *Subteam for Manipulation of Cell Fate, BioResource Center, RIKEN Tsukuba Institute, Tsukuba, Ibaraki, Japan*

ATSUSHI NATSUME • *Department of Neurology, University of Pittsburgh School of Medicine, Pittsburgh, PA*

AJAY NIRANJAN • *Department of Neurological Surgery, University of Pittsburgh School of Medicine, Pittsburgh, PA*

GARRY P. NOLAN • *Department of Microbiology and Immunology, Baxter Laboratory in Genetic Pharmacology, Stanford University, Stanford, CA*

MIROSLAVA OGORELKOVA • *Institut de Recherches en Biotechnologie, Montréal, Québec, Canada*

MASAFUMI ONODERA • *Department of Hematology, Institute of Clinical Medicine, University of Tsukuba, Tsukuba, Ibaraki, Japan*

ALI OZUER • *Department of Molecular Genetics and Biochemistry, University of Pittsburgh School of Medicine, Pittsburgh, PA*

MARCO A. PASSINI • *Department of Pathobiology and Center for Comparative Medical Genetics, School of Veterinary Medicine, University of Pennsylvania, and Division of Neurology, Children's Hospital of Philadelphia, Philadelphia, PA*

SELVARANGAN PONNAZHAGAN • *Department of Pathology, University of Alabama, Birmingham, AL*

PAUL N. REYNOLDS • *Royal Adelaide Hospital Chest Clinic and Department of Medicine, University of Adelaide, Adelaide, South Australia, Australia*

ROLAND SCHROERS • *Department of Internal Medicine, University of Göttingen, Göttingen, Germany*

ANDREW SMITH • *Department of Biochemistry and Molecular Biology, Medical College of Ohio, Toledo, OH*

NIKUNJ SOMIA • *Institute of Human Genetics, University of Minnesota, Minneapolis, MN*

BILL SPOHN • *Departments of Molecular and Cellular Oncology, University of Texas, M.D. Anderson Cancer Center, Houston, TX*

ARUN SRIVASTAVA • *Department of Microbiology and Immunology, Indiana University School of Medicine, Indianapolis, IN*

CHRISTOPHER C. SULLIVAN • *Division of Cardiothoracic Surgery, Department of Surgery, University of California San Diego Medical Center, San Diego, CA*

MASAYO TAKAHASHI • *Department of Experimental Therapeutics, Transitional Research Center, Kyoto University Hospital, Kyoto, Japan*

PATRICIA A. THISTLETHWAITE • *Division of Cardiothoracic Surgery, Department of Surgery, University of California San Diego Medical Center, San Diego, CA*

BRYAN W. TILLMAN • *Division of Human Gene Therapy, Departments of Medicine, Pathology and Surgery, and the Gene Therapy Center, University of Alabama at Birmingham, Birmingham, AL*

LAURA TIMARES • *Departments of Dermatology, Cell Biology, and Pathology, and the Gene Therapy Center, University of Alabama at Birmingham, Birmingham, AL*

JAMES P. TREMPE • *Department of Biochemistry and Molecular Biology, Medical College of Ohio, Toledo, OH*

DEBORAH J. WATSON • *Department of Pathobiology and Center for Comparative Molecular Genetics, School of Veterinary Medicine, University of Pennsylvania, and Department of Neurology, The Children's Hospital of Philadelphia, Philadelphia, PA*

JAMES B. WECHUCK • *Department of Molecular Genetics and Biochemistry, University of Pittsburgh School of Medicine, Pittsburgh, PA*

DANIEL J. WEISS • *Pulmonary and Critical Care, Vermont Lung Center, University of Vermont College of Medicine, Burlington, VT*

MATTHEW D. WEITZMAN • *Laboratory of Genetics, The Salk Institute for Biological Studies, La Jolla, CA*

DARREN WOLFE • *Department of Molecular Genetics and Biochemistry, University of Pittsburgh School of Medicine, Pittsburgh PA*

JOHN H. WOLFE • *Department of Pathobiology and Center for Comparative Medical Genetics, School of Veterinary Medicine, University of Pennsylvania, and Division of Neurology, Children's Hospital of Philadelphia, Philadelphia, PA*

ROLAND WOLKOWICZ • *Department of Microbiology and Immunology, Baxter Laboratory in Genetic Pharmacology, Stanford University, Stanford, CA*

I

DELIVERY USING ADENOVIRUSES

1

Adenovirus-Mediated Gene Delivery

An Overview

Joanne T. Douglas

1. Introduction

Adenoviruses, which were first isolated in the 1950s, have been developed as gene-delivery vehicles, or vectors, since the early 1980s *(1)*. The adenoviruses constitute the *Adenoviridae* family, which is divided into two genera: the *Aviadenovirus* genus infects only birds, whereas the *Mastadenovirus* genus contains viruses that infect a range of mammalian species. Human adenoviruses are classified into six subgroups based on the percentage of guanine and cytosine in the DNA molecules and the ability to agglutinate red blood cells. They are further subdivided into more than 50 serotypes, primarily on the basis of neutralization assays (reviewed in **ref. 2**).

The majority of recombinant adenoviral vectors are based on human adenovirus serotypes 2 (Ad2) and 5 (Ad5) of subgroup C. These serotypes cause a mild respiratory disease in humans and are nononcogenic. These safety features, coupled with the fact that adenovirus-based vaccines have been administered to humans without ill effects, have favored the development of adenoviral vectors for in vivo gene-therapy applications *(3)*. The safety of recombinant adenoviral vectors is also enhanced by deletion of the E1 region of the genome, which renders the vectors replication-deficient and capable of propagation only in specially designed complementing cell lines. Other advantages of recombinant adenoviral vectors derived from serotypes 2 and 5 include the ability of the vectors

From: *Methods in Molecular Biology, vol. 246:*
Gene Delivery to Mammalian Cells: Vol. 2: Viral Gene Transfer Techniques
Edited by: W. C. Heiser © Humana Press Inc., Totowa, NJ

to be purified to high titers (up to 10^{13} virus particles per mL), which means that it is practical to employ them in vivo. Adenoviral vectors also possess the important attribute of stability in the bloodstream, which means that they can potentially be employed for gene delivery following intravenous administration. Adenoviruses can infect both dividing and postmitotic cells, and have evolved an extremely efficient mechanism for delivery of their genome to the nucleus. The genome remains extrachromosomal, which minimizes the risk of insertional mutagenesis. So-called "first-generation" E1-deleted Ad2 and Ad5 vectors can accommodate up to 7.5 kb of foreign DNA, and the capacity of the vectors can be expanded by additional deletions of the viral genes. These characteristics of Ad2 and Ad5 vectors have spawned considerable interest in their exploitation as gene delivery vehicles, which, in turn, has led to the development of a range of techniques by which their genomes can be manipulated and recombinant vectors generated with relative ease.

However, recombinant Ad2- and Ad5-based vectors also suffer from a number of disadvantages. These vectors possess the tropism of the parent viruses, which can infect all cells that possess the appropriate surface receptors, which precludes the targeting of specific cell types. Conversely, some cell types that represent important targets for gene transfer express only low levels of the cellular receptors, which leads to inefficient infection. Another major disadvantage of Ad2- and Ad5-based vectors in vivo is the elicitation of an innate and an acquired immune response. Considerable attention has therefore been focused on strategies to overcome these limitations, thereby permitting the full potential of adenoviral vectors to be realized.

This overview chapter will review the biology of adenoviruses and adenoviral vectors, discuss the applications of adenovirus-mediated gene delivery and describe the strategies that are being developed to address the limitations of adenoviral vectors. For more detailed coverage of these topics, the reader is referred to **ref. 4**.

2. Structure of Adenoviruses

2.1. Capsid Structure

Adenoviruses possess a nonenveloped icosahedral protein shell or capsid of 70–100 nm in diameter surrounding an inner DNA-containing core (**ref. 2** and references therein). The 20 facets of the capsid are comprised of 12 copies of the trimeric hexon protein, which is the most abundant component of the virion and performs a structural role. Each vertex of the capsid is composed of a pentameric penton base protein in association with a trimeric fiber protein that projects from the viral surface and ends with a globular knob domain. The fiber and penton base both play important roles in the initial steps of the virus-cell inter-

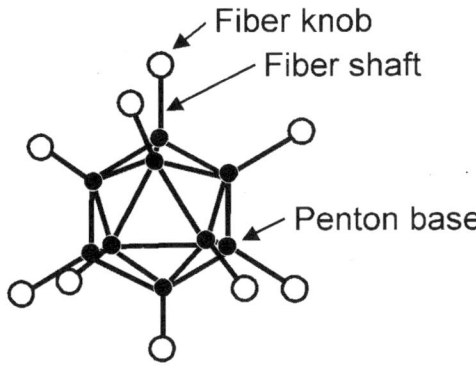

Fig. 1. Schematic diagram of Ad5 virion. The double-stranded DNA genome is packaged within an icosahedral protein capsid. The major structural protein of the capsid is the hexon. Penton capsomers, formed by association of the penton base and fiber, are localized at each of the 12 vertices of the Ad capsid.

action during infection. A number of minor polypeptides are involved in stabilization of the capsid, whereas two additional polypeptides bridge between the capsid and core components of the virion. The capsid structure is depicted schematically in **Fig. 1.**

2.2. Genome Organization

The core of the adenoviral particle contains the viral genome, a linear, double-stranded DNA molecule approx 36 kb in length (**ref. 2** and references therein). The genome is highly condensed and associated with two basic proteins that organize the DNA into a nucleosome-like structure. The *cis*-acting origins of replication of the viral DNA are located in the first 50 base pairs (bp) of the 100- to 140-bp inverted terminal repeat sequences (ITRs) located at each end of the genome. The ITRs play an important role in replication of the DNA. A terminal protein is covalently attached to each of the 5′ termini of the DNA and serves as a primer for DNA replication. The left end of the genome also includes a *cis*-acting packaging signal that directs the interaction of the viral DNA with its encapsidating proteins.

The adenoviral genome is shown schematically in **Fig. 2.** By convention, it is drawn with the immediate early transcription unit (E1A) at the left end, adjacent to the packaging signal. In addition, there are four early transcription units (E1B, E2, E3, and E4); two delayed early units (IX and IVa2); and one late unit (major late), which is processed to give five families of late mRNAs (L1 to L5), all of which are transcribed by RNA polymerase II. Transcription of each of the adenovirus genes leads to multiple mRNAs.

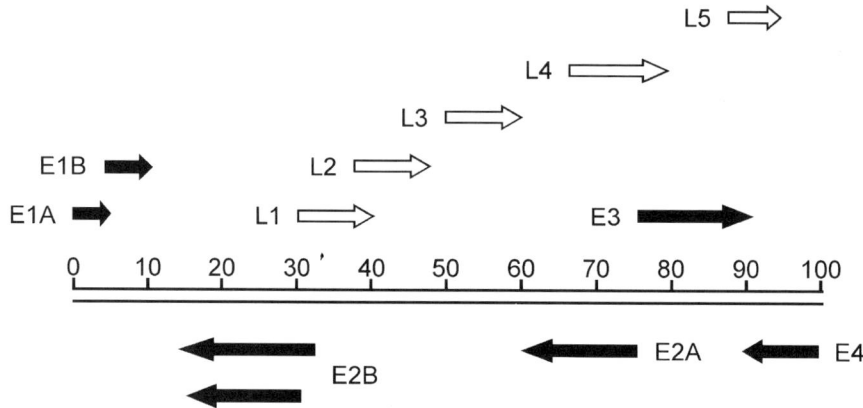

Fig. 2. Schematic diagram of the structure of the Ad5 genome. The Ad5 genome is approx 36 kb long, divided into 100 map units. The direction of transcription is indicated by arrows. Closed arrows represent early transcripts; open arrows represent late transcripts.

3. The Biology of Adenoviral Infection

The rational design of adenoviral vectors is based on an understanding of the infectious cycle of the parental viruses (**ref. 2** and references therein). The replication cycle is conventionally divided into two phases separated by the onset of viral DNA replication. The early phase starts as soon as the virus interacts with the host cell: entry into the cell and transport of the viral genome to the nucleus, followed by the transcription and translation of early viral genes. These events modulate the functions of the host cell to facilitate the replication of the virus DNA and the transcription and translation of the late genes. In permissive cells, the early phase takes 5–6 h, after which time viral DNA replication is first detected. The late phase begins concomitantly with the onset of DNA replication, and involves the expression of the late viral genes, leading to the assembly in the nucleus of the structural proteins and the maturation of infectious viruses. The host cells lyse to release progeny virions about 20–24 h postinfection.

The entry of adenoviruses into susceptible cells requires two distinct, sequential steps—binding and internalization—each mediated by the interaction of a specific capsid protein with a cellular receptor (**Fig. 3**). The initial high-affinity binding of Ad2 and Ad5 to the primary cellular receptor *(5,6)*, designated CAR (for coxsackievirus and adenovirus receptor), occurs via the globular knob domain of the fiber capsid protein *(7,8)*. CAR appears to function purely as a docking site for the virus on the cell surface: the cytoplasmic and

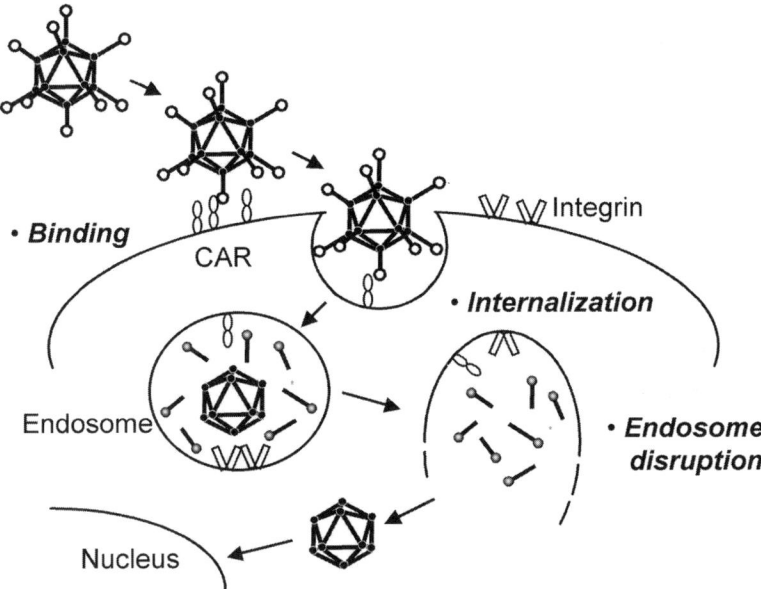

Fig. 3. The pathway of adenoviral infection. The entry of Ad into susceptible cells involves two distinct, sequential steps. The initial high-affinity binding of Ad5 to the primary cellular receptor, CAR, occurs via the globular knob domain of the trimeric fiber capsid protein. Subsequent internalization of the virus by receptor-mediated endocytosis is potentiated by the interaction of Arg-Gly-Asp (RGD) peptide sequences in the penton base protein with secondary host-cell receptors, integrins $\alpha_V\beta3$ and $\alpha_V\beta_5$. The virion then escapes from the endosome and localizes to the nuclear pore, whereupon its genome is translocated to the nucleus.

transmembrane domains of the molecule are not essential for adenoviral infection *(9,10)*. Subsequent internalization of the virus by receptor-mediated endocytosis is potentiated by the interaction of Arg-Gly-Asp (RGD) peptide sequences in the penton base protein *(11)* with secondary host-cell receptors, integrins $\alpha_V\beta3$ and $\alpha_V\beta_5$ *(12)*. The virion then escapes from the endosome, the capsid is disrupted, and the virus is transported to the nuclear membrane. The genome then passages through the nuclear pore into the nucleus, where the primary transcription events are initiated.

Expression of the adenoviral genes is temporally regulated *(2)*. E1A is the first transcription unit to be expressed after the adenoviral chromosome enters the nucleus of an infected cell; its expression requires only cellular proteins. The E1A proteins activate transcription from the other adenoviral early regions and induce the host cell to enter the S phase of the cell cycle. The E1B gene encodes two proteins (E1B 19K and E1B 55K) that inhibit apoptosis and further modulate cellular metabolism to render the cell more susceptible to viral replication. The E2 transcription unit encodes three proteins involved in viral DNA replication: DNA polymerase (Pol), preterminal protein (pTP), and DNA binding protein (DBP). The E3 region encodes multiple proteins designed to inhibit pathways of cell death induced by the host innate and cellular immune response to the infected cell. The E3 proteins are dispensable for the replication of adenoviruses in tissue culture. The E4 gene products perform a range of functions, with distinct proteins playing roles in viral DNA replication, viral mRNA transport and splicing, shut-off of host protein synthesis, and regulation of apoptosis.

The expression of the early adenoviral genes sets the stage for replication of the viral DNA. Replication of the adenoviral DNA starts at the origins of replication in the ITRs at either end of the chromosome, with the terminal protein serving as a primer. The expression of the late adenoviral genes commences with the onset of DNA replication. The late gene products are expressed after processing a 20 kb transcript from the major late promoter, which is attenuated during transcription of the early genes. This primary transcript undergoes multiple splicing events to generate five families of late mRNAs encoding proteins that are part of the viral capsid or are involved in the encapsidation and assembly of viral particles in the host-cell nucleus. Encapsidation of the viral DNA is directed by the packaging signal at the left end of the chromosome. This process is accompanied by alterations in the nuclear infrastructure and the permeabilization of the nuclear membrane, facilitating the egress of the progeny viruses into the cytoplasm. The plasma membrane subsequently disintegrates and the progeny are released from the cell.

4. Adenoviral Vectors

4.1. Design and Construction of Adenoviral Vectors

The most widely used adenoviral vectors for gene delivery are the so-called "first-generation" replication-deficient vectors, in which the E1 region of the genome is deleted *(1,3)*. Deletion of the E1 region, while retaining the ITR and packaging signal, is designed to prevent expression of the E2 genes and thus block viral DNA replication and the synthesis of late structural proteins. E1-deleted Ad vectors are therefore propagated in complementing human cell lines that provide the E1 proteins in *trans*. In order to provide additional cloning space in the vector, the E3 region, which is not necessary for viral replication in culture, is also commonly deleted. Because adenoviruses can encapsidate DNA ranging from 75 to 105% of the length of the wild-type viral genome, these modifications allow up to 7.5 kb of foreign DNA to be accommodated.

A number of different approaches have been used to construct Ad vectors with the E1 region substituted with the transgene of interest (reviewed in **ref. *13***). The classical method employs homologous recombination in an E1-complementing human cell line between two DNA molecules, one carrying sequences mapping to the left end of the Ad genome and the gene of interest, and one carrying the Ad genome with the left end deleted but retaining some sequences that partially overlap the 3′ end of the first molecule. This second molecule can be either a linearized partial viral genome purified from virions (*14*) or a plasmid (*15,16*). This technique suffers from the inefficiency of homologous recombination in mammalian cells, and the need for purification of individual viral plaques, which means it is both labor-intensive and time-consuming. Another big disadvantage is that if no recombinants are generated, the researcher is unable to determine whether the problem is technical or biological.

The past few years have seen the development of new methods to facilitate the generation of E1-substituted Ad vectors by constructing the recombinant vector genome prior to transfection of the E1-complementing mammalian cells, thereby avoiding multiple rounds of plaque purification. One approach that has found widespread use exploits the highly efficient homologous recombination machinery in bacteria to generate a recombinant Ad vector by homologous recombination in *Escherichia coli* between a large plasmid containing most of the Ad genome and a small shuttle plasmid containing the expression cassette flanked by sequences homologous to the region to be targeted in the viral genome (*17–19*). The recombinant Ad genome is then linearized by restriction digestion and used to transfect E1-complementing mammalian cells to produce viral particles and propagate the vector.

4.2. Production and Purification of Adenoviral Vectors

The original E1-complementing cell line, designated 293, was generated by transformation of human embryonic kidney cells with sheared Ad5 DNA (*20*). The cells constitutively express the left 11% of the Ad5 genome and can be used to produce E1-deleted vectors at high titers of up to 10^{13} particles per mL. However, a disadvantage of the 293 cell line is that it allows the emergence of replication-competent adenovirus (RCA) as a result of homologous recombination between the host-cell genome and the vector (*21*). This has led to strategies to avoid RCA by creating rationally designed E1-complementing helper cell lines with minimal or no homologous sequences between the transfected E1 DNA and E1-deleted vector (*22,23*).

The classical method for purification of Ad vectors is cesium chloride density-gradient ultracentrifugation. This is an efficient technique that can yield highly purified viral particles, although it is time-consuming, rather expensive, and is not amenable to large-scale purification of Ad vectors. More recently, Ad vectors have been purified by column chromatography using resins originally developed for protein purification *(24)*. Anion-exchange chromatography is commonly used in an initial purification step, followed by immobilized metal-affinity chromatography or reversed-phase high-performance liquid chromatography (RP-HPLC) as the second step. Column chromatography offers the ability to rapidly purify large amounts of virus to a highly pure state without compromising the viability of the viral particles.

After purification, the concentration of the Ad vector is determined by physical and/or biological methods *(25)*. The most common physical method for calculating the number of viral particles is to disrupt the particles with sodium dodecyl sulfate (SDS) and determine the optical absorbance of the virion DNA at 260 nm, using the conversion factor 1.1×10^{12} particles per absorbance unit *(26)*. Biological methods involve the infection of cells in culture followed by the determination of infectious Ad vector particles, either by counting visible plaques in a monolayer of cells that support replication of the vector, or by histochemical or immunohistochemical staining of cells to detect expression of a viral structural protein or a reporter gene delivered by the vector. The biological titer of the vector is then expressed in terms of plaque-forming units (PFU), infectious units (IU), or transducing units (TU).

5. Applications of Adenoviral Vectors

First generation, E1-deleted Ad vectors can mediate high, albeit transient, levels of expression of the transgene in mammalian cells, resulting in yields of the recombinant protein of up to 30% of total cellular protein. The expressed proteins are subject to the full range of complex posttranslational modifications that might be necessary for their appropriate folding and function. Recombinant viral and mammalian proteins are therefore identical to the native proteins, thereby avoiding the disadvantages associated with expression of these proteins in prokaryotes, lower eukaryotes, and insect cells.

Based on these favorable characteristics, E1-deleted Ad vectors have been employed for expression of recombinant proteins in cultured mammalian cells in vitro *(1)*. Because Ad vectors can infect a range of dividing and nondividing mammalian cells, they have also been widely used in gene-transfer experiments and gene-therapy applications in vitro and in vivo, both in preclinical studies in animal models and in clinical trials in human patients *(3,4)*. However, a number of limitations of first-generation Ad vectors have been identified in the course of these studies.

6. Limitations of Adenoviral Vectors and Strategies to Improve the Vectors

The use of first-generation Ad vectors in vivo is associated with the induction of both an innate and an acquired immune response (reviewed in **refs.**

27,28). Studies in mice and primates have indicated that within the first few hours of administration of Ad vectors by the intravenous route, the viral capsid proteins trigger an acute inflammatory response characterized by the rapid release of inflammatory cytokines, including interleukin-6 (IL-6) and IL-8, and the recruitment of immune effector cells, such as neutrophils, into the liver. This acute-phase toxicity does not require expression of viral genes but is dependent on the dose of vector: minimal toxicity has been shown to result from administration of low doses of E1-deleted vectors to mice.

Over the next 24–96 h, toxicity associated with first-generation Ad vectors results from an acquired cellular immune response. Although E1-deleted Ad vectors are in theory replication-defective, in practice many cells possess E1-like proteins that can activate the E2 genes, leading to viral DNA replication and the expression of the late structural proteins. It has also become clear that the E1-dependence of E2, E3, and E4 gene transcription can be circumvented at high multiplicities of infection. Newly synthesized Ad peptides displayed on the surface of infected cells are recognized and destroyed by cytotoxic T lymphocyte and natural killer (NK) cell-mediated responses. In many cases the expressed transgene product has also been shown to be immunogenic. As a consequence of the elimination of infected cells by the cellular immune response, transgene expression mediated by first-generation Ad vectors in vivo is only transient, lasting 2–3 wk.

In addition to cellular immunity, a humoral immune response is generated to the Ad vector. This leads to a reduction in Ad-mediated gene delivery upon repeat vector administration. Moreover, even the initial vector dose may be inefficient in human patients who possess neutralizing antibodies to the commonly used Ad2 or Ad5 vectors, as a result of prior exposure to the parental viruses.

In those instances where the goal of gene delivery by a first-generation Ad vector is the elimination of the infected cell, for example in cancer gene therapy, the induction of a cytodestructive immune response is beneficial. However, in many cases the eradication of the infected cell would be a serious problem. In an attempt to reduce immunogenicity, subsequent generations of Ad vectors have been designed to be defective for multiple viral genes, in addition to E1. The removal of genes encoding proteins essential for DNA replication (the DNA binding protein, DNA polymerase, and terminal protein), or key regulatory functions (the E4 proteins) has led to vectors that in some studies have been reported to be less immunogenic than first-generation E1-deleted vectors and to mediate longer-term gene expression. However, in other studies these vectors have shown minimal or no advantage over first-generation vectors. The production of these multiply deleted vectors has necessitated the construction of novel complementing cell lines that provide the missing function in *trans*.

The strategy of deleting regions of the viral genome has met its ultimate realization with the so-called "gutted vectors," which retain only the ITRs and packaging signals. The gutted vectors can accommodate up to 36 kb of foreign DNA, and can therefore carry large cDNAs together with appropriate regulatory elements. Production of these vectors requires the use of helper viruses, from which the gutted vectors must be separated and purified, a process that has been simplified by the development of packaging-defective helper viruses. Compared to first-generation vectors, gutted vectors have shown reduced immunogenicity and more persistent gene expression in vivo. Nonetheless, gutted Ad vectors do not integrate into the host-cell genome, which means that the transgene would not be transmitted to the progeny of dividing cells.

Because the acute inflammatory response is directly related to the vector dose, toxicity could be reduced by lowering the number of viral particles necessary for a given level of gene transfer. In this regard, a paucity of the primary Ad receptor on various cell types, including primary cancer cells, airway epithelium, and mature skeletal muscle, has been shown to be associated with a poor efficiency of gene delivery. Thus, a number of strategies have been employed to retarget the Ad vector to a more abundant receptor, resulting in more efficient gene delivery and hence permitting the use of a lower dose of vector. To date, these targeting approaches have either employed bispecific molecules directed against both a viral capsid protein, most commonly the knob domain of the fiber protein, and an alternative cellular receptor, or have involved the direct modification of a capsid protein to allow the recognition of an alternative receptor (reviewed in **ref. 29**). In addition to facilitating more efficient gene transfer, such targeted Ad vectors can confer the benefit of specificity, by enabling selective gene delivery to the target cells. This is particularly important when the therapeutic gene encodes a product that might be beneficial when expressed in the desired target cell but prove toxic to normal cells. Ad vectors with modified fibers have also been shown to afford the ability to mediate efficient gene delivery in the presence of neutralizing antibodies directed against the unmodified fiber proteins.

7. Summary

Ad vectors possess a number of features that have favored their widespread employment both in vitro and in vivo. In fact, the use of Ad vectors is increasing as technologies to facilitate their construction are being developed and refined. First generation, E1-deleted Ad vectors have been shown to be associated with limitations that are being addressed by rational strategies based on the biology of the virus. These advances should allow the realization of the full potential of Ad vectors for gene delivery.

Acknowledgments

Research in the author's laboratory is supported by grants DOUGLA00I0 from the Cystic Fibrosis Foundation, BCTR0100406 from the Susan G. Komen

Breast Cancer Foundation, NIH 1 R03 AR46864, and a grant from the Muscular Dystrophy Association.

References

1. Berkner, K. L. (1988) Development of adenovirus vectors for the expression of heterologous genes. *Biotechniques* **6,** 616–629.
2. Shenk, T. (1996) *Adenoviridae:* The viruses and their replication, in *Fields Virology,* (Fields, B. N., Knipe, D. M., Howley, P. M., Chanock, R. M., Melnick, J. L., Monath, T. P., et al., eds.), Lippincott-Raven Publishers, Philadelphia, PA, pp. 2111–2148.
3. Kovesdi, I., Brough, D. E., Bruder, J. T., and Wickham, T. J. (1997) Adenoviral vectors for gene transfer. *Curr. Opin. Biotechnol.* **8,** 583–589.
4. Curiel, D. T. and Douglas, J. T. (2002) *Adenoviral Vectors for Gene Therapy.* Academic Press, New York, NY.
5. Bergelson, J. M., Cunningham, J. A., Droguett, G., Kurt-Jones, E. A., Krithivas, A., Hong, J. S., et al. (1997) Isolation of a common receptor for coxsackie B viruses and adenoviruses 2 and 5. *Science* **275,** 1320–1323.
6. Tomko, R. P., Xu, R., and Philipson, L. (1997) HCAR and MCAR: the human and mouse cellular receptors for subgroup C adenoviruses and group B coxsackieviruses. *Proc. Natl. Acad. Sci. USA* **94,** 3352–3356.
7. Henry, L. J., Xia, D., Wilke, M. E., Deisenhofer, J., and Gerard, R. D. (1994) Characterization of the knob domain of the adenovirus type 5 fiber protein expressed in *Escherichia coli. J. Virol.* **68,** 5239–5246.
8. Louis, N., Fender, P., Barge, A., Kitts, P., and Chroboczek, J. (1994) Cell-binding domain of adenovirus serotype 2 fiber. *J. Virol.* **68,** 4104–4106.
9. Leon, R. P., Hedlund, T., Meech, S. J., Li, S., Schaack, J., Hunger, S. P., Duke, R. C., and DeGregori, J. (1998) Adenoviral-mediated gene transfer in lymphocytes. *Proc. Natl. Acad. Sci. USA* **95,** 13159–13164.
10. Wang, X. and Bergelson, J. M. (1999) Coxsackievirus and adenovirus receptor cytoplasmic and transmembrane domains are not essential for coxsackievirus and adenovirus infection. *J. Virol.* **73,** 2559–2562.
11. Bai, M., Harfe, B., and Freimuth, P. (1993) Mutations that alter an Arg-Gly-Asp (RGD) sequence in the adenovirus type 2 penton base protein abolish its cell-rounding activity and delay virus reproduction in flat cells. *J. Virol.* **67,** 5198–5205.
12. Wickham, T. J., Mathias, P., Cheresh, D. A., and Nemerow, G. R. (1993) Integrins alpha v beta 3 and alpha v beta 5 promote adenovirus internalization but not virus attachment. *Cell* **73,** 309–319.
13. Danthinne, X. and Imperiale, M. J. (2000) Production of first generation adenovirus vectors: a review. *Gene Ther.* **7,** 1707–1714.
14. Chinnadurai, G., Chinnadurai, S., and Brusca, J. (1979) Physical mapping of a large-plaque mutation of adenovirus type 2. *J. Virol.* **32,** 623–628.
15. McGrory, W. J., Bautista, D. S., and Graham, F. L. (1988) A simple technique for the rescue of early region I mutations into infectious human adenovirus type 5. *Virology* **163,** 614–617.

16. Bett, A. J., Haddara, W., Prevec, L., and Graham, F. L. (1994) An efficient and flexible system for construction of adenovirus vectors with insertions or deletions in early regions 1 and 3. *Proc. Natl. Acad. Sci. USA* **91,** 8802–8806.

17. Chartier, C., Degryse, E., Gantzer, M., Dieterle, A., Pavirani, A., and Mehtali, M. (1996) Efficient generation of recombinant adenovirus vectors by homologous recombination in Escherichia coli. *J. Virol.* **70,** 4805–4810.

18. Crouzet, J., Naudin, L., Orsini, C., Vigne, E., Ferrero, L., Le Roux, A., et al. (1997) Recombinational construction in Escherichia coli of infectious adenoviral genomes. *Proc. Natl. Acad. Sci. USA* **94,** 1414–1419.

19. He, T. C., Zhou, S., da Costa, L. T., Yu, J., Kinzler, K. W., and Vogelstein, B. (1998) A simplified system for generating recombinant adenoviruses. *Proc. Natl. Acad. Sci. USA* **95,** 2509–2514.

20. Graham, F. L., Smiley, J., Russell, W. C., and Nairn, R. (1977) Characteristics of a human cell line transformed by DNA from human adenovirus type 5. *J. Gen. Virol.* **36,** 59–74.

21. Lochmuller, H., Jani, A., Huard, J., Prescott, S., Simoneau, M., Massie, B., Karpati, G., and Acsadi, G. (1994) Emergence of early region 1-containing replication-competent adenovirus in stocks of replication-defective adenovirus recombinants (delta E1 + delta E3) during multiple passages in 293 cells. *Hum. Gene Ther.* **5,** 1485–1491.

22. Fallaux, F. J., Bout, A., van der Velde, I., van den Wollenberg, D. J., Hehir, K. M., Keegan, J., et al. (1998) New helper cells and matched early region 1-deleted adenovirus vectors prevent generation of replication-competent adenoviruses. *Hum. Gene Ther.* **9,** 1909–1917.

23. Gao, G. P., Engdahl, R. K., and Wilson, J. M. (2000) A cell line for high-yield production of E1-deleted adenovirus vectors without the emergence of replication-competent virus. *Hum. Gene Ther.* **11,** 213–219.

24. Shabram, P., Vellekamp, G., and Scandella, C. (2002) Purification of adenovirus, in *Adenoviral Vectors for Gene Therapy* (Curiel, D. T. and Douglas, J. T., eds.), Academic Press, New York, NY, pp. 167–204.

25. Mittereder, N., March, K. L., and Trapnell, B. C. (1996) Evaluation of the concentration and bioactivity of adenovirus vectors for gene therapy. *J. Virol.* **70,** 7498–7509.

26. Maizel, J. V., Jr., White, D. O., and Scharff, M. D. (1968) The polypeptides of adenovirus. I. Evidence for multiple protein components in the virion and a comparison of types 2, 7A, and 12. *Virology* **36,** 115–125.

27. Brenner, M. (1999) Gene transfer by adenovectors. *Blood* **94,** 3965–3967.

28. Young, L. S. and Mautner, V. (2001) The promise and potential hazards of adenovirus gene therapy. *Gut* **48,** 733–736.

29. Barnett, B. G., Crews, C. J., and Douglas, J. T. (2002) Targeted adenoviral vectors. *Biochim. Biophys. Acta* **1575,** 1–14.

2

DNA Delivery to Cells in Culture

Generation of Adenoviral Libraries for High-Throughput Functional Screening

Miroslava Ogorelkova, Seyyed Mehdy Elahi, David Gagnon, and Bernard Massie

1. Introduction

In functional genomics, the use of expression libraries of DNA variants in combination with potent screening techniques is a powerful tool for gene discovery. They allow study of gene and protein function, generation of peptide variants with novel properties, as well as identification of functional short DNA and RNA motifs. In proteomics, generation of large expression libraries of protein variants with random substitutions ("directed evolution") and further screening for novel or improved functions has been commonly used for isolation of proteins with novel characteristics, for improving enzymes, for rapid isolation of antibodies, and for functional protein studies (reviewed in **refs. *1–3*)**. Most commonly, peptide libraries are expressed and screened in prokaryotic systems. Such systems have the advantage of rapid and simple generation of clones expressing single variants, allow high diversity (up to 10^{11}), and can be combined with phage- or cell-surface display technique *(2)*. The main disadvantage of bacterial systems is the absence of posttranslational modifications and native folding of many mammalian proteins, leading to limited applications, particularly when enzyme–substrate-, protein–protein, or protein–RNA interactions are to be studied.

Currently, libraries for screening in mammalian cells have been generated using plasmids, retroviral vectors, or Epstein-Barr virus (EBV)-based vectors

From: *Methods in Molecular Biology, vol. 246:*
Gene Delivery to Mammalian Cells: Vol. 2: Viral Gene Transfer Techniques
Edited by: W. C. Heiser © Humana Press Inc., Totowa, NJ

(4,5). Scoring for new functions or phenotypes upon transfection or transduction led to the identification of genetic suppressor elements, genes involved in growth inhibition and apoptosis, and novel oncogenes. However, more ubiquitous use of plasmid- or retrovirus-based libraries is limited mainly by the range of cells that can be used for efficient transfection or transduction.

Over the past two decades, several generations of replication-deficient, recombinant adenoviruses (AdVs) have been generated and commonly used in gene therapy and functional studies (reviewed in Chapter 1, Part I of this volume; *6*). Thus far, adenoviral vectors have been used for cloning and delivery of single genes. Here we propose the use of AdVs for generating libraries by positive selection of recombinants. When transient expression is sufficient for functional assessment, adenoviral libraries have several advantages over plasmid- or retroviral-based libraries: (1) they are suitable for a rapid functional screening in a broad range of differentiated, dividing, and postmitotic mammalian cells; (2) small stocks of single adenoviral clones with high titers can be easily obtained; and (3) a high level of uniform transient gene expression can be achieved within 24–48 h postinfection, facilitating the screening process. However, the construction of adenoviral libraries without selection based on inhibited growth of nonrecombinants is fairly inefficient. Even if recombinant clones can be selected using reporter genes, i.e., fluorescent proteins or *LacZ*, in a library of several hundred to several thousand clones there will be a large number of parental viruses that have to be eliminated by time-consuming plaque-purification techniques. Thus, an ideal method for construction of adenoviral libraries would ensure that: (1) a very large number of clones are generated following single transfection, and (2) only recombinant viruses are selected. Among the wide variety of methods used for the construction of recombinant AdVs, several allow generation of recombinants without any parental virus background *(7–14)*. However, the number of viral clones generated is, at best, lower than 50 per µg of viral DNA, which is insufficient for generating libraries with high diversity.

Recently, we have designed a new type of AdV devoid of the viral protease gene *(PS)* *(15)*. In a wild-type AdV, the viral protease is expressed during the late phase of infection and is essential for production of mature viral particles *(16)*. Therefore a *PS*-deficient AdV (Ad5-ΔPS) is capable of a single round of replication but cannot form infectious particles unless propagated in a cell line engineered to express *PS*. We explored this feature to establish a positive selection method based on ectopic co-expression of the *PS* and a gene of interest in the E1 region upon recombination with the Ad5-ΔPS parental genome *(17)*. With this method, the parental virus can be eliminated after one round of purification because only the recombinant AdVs that have rescued *PS* can lead to productive infection. Furthermore, we applied the positive selection for rapid generation of adenoviral libraries, as illustrated in **Fig. 1**. The DNA inserts are initially cloned in an Ad5 transfer vector designed to co-express *PS* together with the transgene from two unrelated promotors (**Fig. 1**). The choice of a weak enhancerless promoter for *PS* expression is based on the contention that a low level of *PS* is both sufficient and desirable, and also to minimize promoter interference between the expression cassettes. At the genetic level, the diversity can be generated using a broad range of well-established techniques such as error-prone polymerase chain reaction (PCR), DNA shuffling, random deletions, or cloning cDNAs from various libraries using standard protocols. The method of choice will depend on the particular application and will not be discussed in this chapter.

Because the expression of the gene of interest is not necessary, and could be potentially deleterious during the recombinant AdV generation and production, the various inserts in the library are expressed under the control of a regulated promoter such as the tetracycline inducible (tTA-responsive) promoter *(18–20)*. Using this system, the recombinant clones are generated and propagated in 293 cells, whereas the functional screening is performed in a cell line expressing a tetracycline-regulated promotor trans-activator (tTA) or by co-infection with a recombinant virus expressing tTA *(19)*. Alternatively, a strong constitutive promoter such as CMV5 can be used in a configuration, ensuring its repression in specific cell lines expressing a repressor binding to its cognate operator sequence downstream of the start site. We have recently constructed such an inducible system using the *cymene* operon of *Pseudomonas putida* F1 in which the expression of the genes is regulated by a 28kD repressor molecule (CymR) that binds operator sequences downstream of the start site. CymR is in a DNA-binding configuration only in the absence of cymene or cumate, the effector molecules. Thus, when the cumate operator sequences (CuO) was cloned downstream of the start site of the CMV5 promoter, the expression of the transgene was substantially repressed in 293-CymR cells in absence the of cumate *(21)*. This inducible system permits the construction of AdVs in a cell line in which the transgene expression is minimal (293-CymR), whereas testing of the transgene function can be done by simple transduction of a wide variety of cell lines and primary cells.

The plasmid library is introduced to the cells by transfection following infection with Ad5-ΔPS. Individual clones from the resulting pooled viral stock are isolated as single viral plaques. They are further amplified in small volume stocks and are ready to be used in screening assays. This method allows generation of large adenoviral libraries of several hundred to several thousand clones. We also give an example for high-throughput screening of the amplified adenoviral clones. The screening is designed to determine cell growth and/or viability and is suitable for identification of clones expressing inserts with either cell-

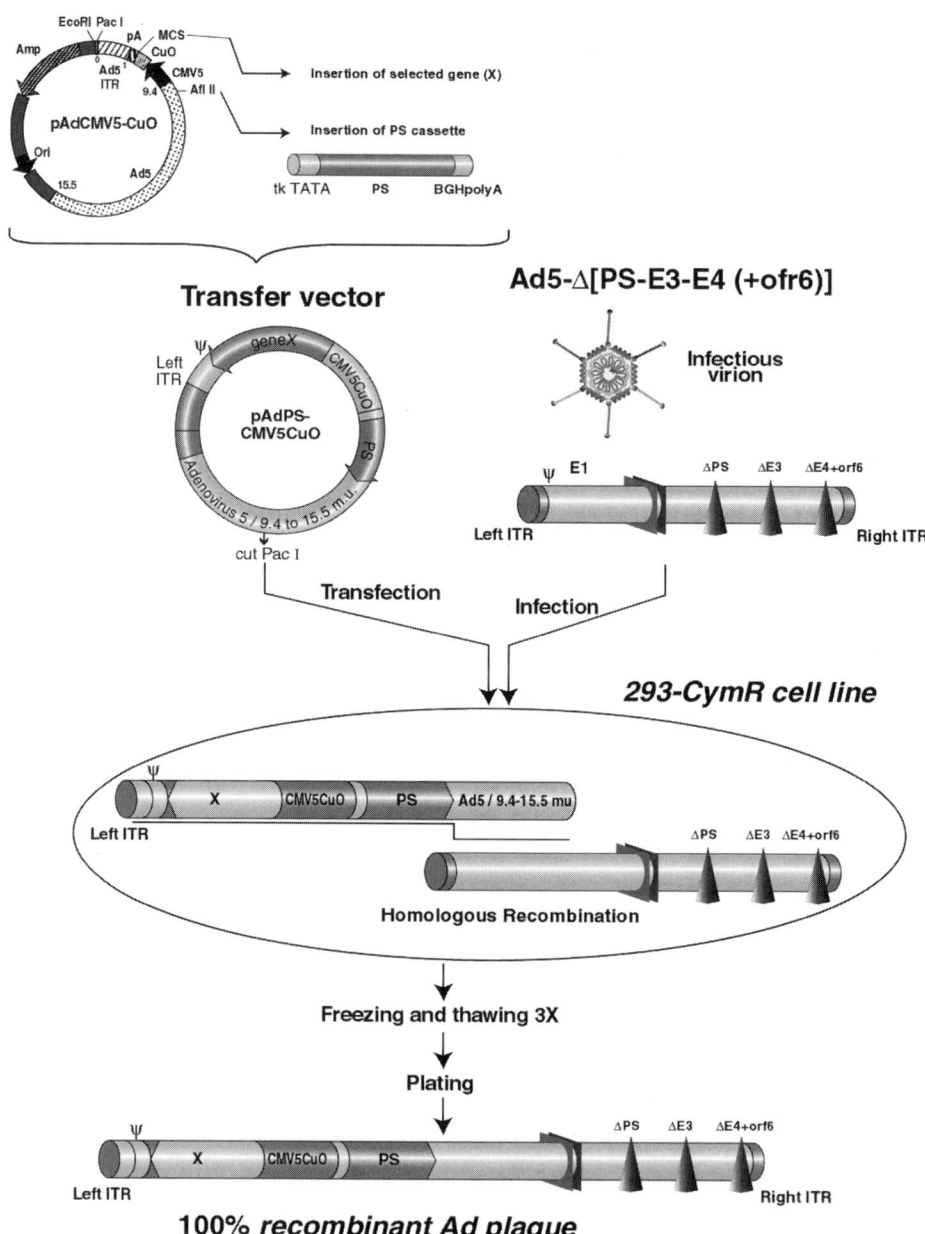

Fig. 1. Schematic representation of the steps required to generate adenoviral library by positive selection of recombinants. The adenoviral transfer vector pAdPS-CMV5CuO used for cloning of DNA variants in the MCS (*Bgl*II/*Not*I/*EcoR*V) carries the left adenoviral ITR and 9.4–15.5 map unit segment of the Ad5 genome for homologous recombi-

cycle dysregulation or pro- or anti-apoptotic properties. Because adenoviral genes in E1, E3, and E4 regions have been shown to interfere with cell-cycle regulation and/or apotosis, the AdVs used for the functional studies are preferably deleted of all of these genes except the E4 orf6. The E1 genes are complemented in 293 cells; the E4 orf6 alone is sufficient for normal growth of E4-deleted virus; and E3 is dispensable for growth in cell culture (reviewed in **ref. 6**). Such AdVs can thus be readily grown in 293-derived cell lines. For functional studies, nonpermissive cells are infected with individual library clones followed by delayed infection with a virus expressing the green fluorescent protein (GFP). Quantification of the resulting GFP expression for each of the co-infections is used as an indication for cell growth or viability. This protocol can include an additional step of apoptosis induction if needed. The assay is performed in 96-well plates and GFP intensity is measured using a FluorImager™ instrument and ImageQuant™ software. A similar assay using GFP as a reporter gene can also readily be developed, for example, to map promoter elements or screen for optimal gene expression in specific cell lines using chimeric promoters assembled with various TATA boxes and enhancer motifs. For other applications, specific assays can further be developed for screening of desired phenotypes.

2. Materials

1. The plasmid pAdPS-CMV5CuO (**Fig. 1**) used to generate libraries is a modified version of pAdCMV5 *(19)*. In addition to the viral left ITR and an Ad5 region allowing homologous recombination in E1 region, the plasmid contains an expression cassette in which the gene of interest (X) is under the control of a cumate-regulated promoter *(21)* and a PS cassette under the control of a tk TATA minimal promotor.
2. Ad5-Δ[PS-E3-E4(+orf6)]: adenovirus type 5 deleted in the *PS* gene as well as in the E3 and E4 regions, except E4 orf6, derived from Ad5-ΔPS *(15)*, propagated in 293-PS cell line and used as nonpurified viral stock.
3. AdCMV5-GFP *(20)*: recombinant adenovirus type 5 with deleted E1 and E3 regions ectopically expressing GFP in the E1 region propagated in 293 cells.

---

nation with the backbone virus Ad5-Δ[PS-E3-E4(+orf6)]. This plasmid carries two cassettes for ectopic co-expression in the E1 region upon recombination: (1) a cumate-inducible cassette for expression of the gene of interest, and (2) a protease cassette controlled by a tk minimal promotor. Recombinant viruses are generated in 293 cells upon infection with Ad5-Δ[PS-E3-E4(+orf6)] providing the backbone ΔPS viral genome and transfection with linearized transfer vectors. The resulting Δ(E1, E3, E4+orf6) recombinants have rescued PS in the E1 region and can replicate in E1-complementing 293 cells, in contrast to the parental virus.

4. 293 (CRL 1573, ATCC, Manassas, VA), HeLa , and the 293-derived cell line, 293-PS, expressing viral PS *(15)* and 293-CymR, expressing the repressor of the cymene operon *(21)*, are all cultured in Dulbecco`s modified Eagle's medium (DMEM) supplemented with heat-inactivated fetal bovine serum (FBS, HyClone, Logan, UT), 2 m*M* L-glutamine (Wisent Inc., St-Bruno, QC, Canada), 1% antibiotic/antimycotic solution (Wisent Inc.). Use 5% FBS for 293 cells and 10% FBS for all other cell lines. Prepare heat-inactivated FBS by heating the serum at 56°C for 30 min to inactivate complement. For all cells lines, cell growth is at 37°C in a humidified incubator with 5% CO_2 (referred to as standard conditions in the text).

5. Phosphate-buffered saline (PBS) (Wisent Inc.).

6. Trypsin-ethylenediaminetetraacetic acid (EDTA) solution (Wisent Inc.).

7. Stock of 5% Seaplaque GTG agarose (BMA, Rockland, ME). Add 5 g of Seaplaque GTG agarose per 100 mL of PBS. Autoclave for 25–30 min to sterilize. Prepare 10-mL aliquots and store at 4°C. Approximately 1 h prior to preparation of overlay, melt the agarose, add warm complete DMEM to 1% final concentration of Seaplaque agarose, and keep at 42°C until use.

8. Luria-Bertani (LB) medium and LB agar plates prepared according to standard protocols *(22)*: 10 g of tryptose phosphate, 5 g of yeast extract, 5 g of NaCl, and 15 g of agar for LB agar per 1 L water; sterilize by autoclaving for 20 min.

9. Twenty-five kDa linear polyethylenimine (PEI) polymer, obtained from Polysciences (Warrington, PA) and prepared according to protocols *(23)*: stock solution (1 mg/mL) prepared in water and neutralized with HCl. Sterilize the stock solution by filtration using 0.22-µm filter and prepare aliquots of 1 mL. Store at −80°C.

10. QIAGEN Plasmid Maxi Kit (Qiagen Inc., Valencia, CA).

11. TOP10 One Shot™ Chemically Competent Cells (Invitrogen Inc., Carlsbad, CA).

12. *Pac* I restriction endonuclease (New England BioLabs Inc., Beverly, MA).

13. Ampicillin (Sigma, St. Louis, MO): Prepare 100 mg/mL stock in water, aliquot 1 mL samples, and store at −20°C.

14. FluorImager™ instrument and ImageQuant™ software (Molecular Dynamics, Sunnyvale, CA).

3. Methods

3.1. Generation of Ad Library Using Positive Selection

3.1.1. Preparation of Plasmid Library for Transfection

1. Perform cloning of the DNA variants in the multiple cloning site of the adenoviral transfer vector (pAdPS-CMV5CuO, *Bgl*II/*Not*I/*EcoR*V) followed

by transformation of TOP10 One Shot™ Chemically Competent Cells following the manfacturer's instructions.

2. Inoculate the transformation mixture into 250 mL of LB containing 100 µg/mL of ampicillin and grow overnight at 37°C shaking at 300 rpm (*see* **Note 1**).

3. Extract the plasmid DNA using a QIAGEN Plasmid Maxi Kit according to the manufacturer's instructions.

4. Linearize 5–6 µg (per 60-mm dish) of the plasmid variants with *Pac*I , which cuts at the 5′ end of the left ITR (**Fig. 1**), by adding the 2 U of *Pac*I/µg of DNA in total volume of 50 µL of water containing 5 µL of NEB buffer 1 and 100 µg/mL of bovine serum albumin (BSA) (*see* **Note 2**). After overnight digestion, purify the linearized plasmids using standard phenol/chloroform/isoamyl alcohol protocol *(22)*:

   a. Add 150 µL of water to 50 µL of *Pac*I digested DNA.
   b. Add an equal volume (200 µL) of phenol/chloroform/isoamyl alcohol (25:24:1), and mix by vortexing.
   c. Centrifuge for 1 min at room temperature. Transfer the top (aqueous) phase to a new microcentrifuge tube.
   d. Add 200 µL of chloroform and mix by vortexing.
   e. Repeat **step 4c**.
   f. Add 2 volumes of 100% of ethanol and 1/10 volume of 3 *M* sodium actate, pH 5.2, to the DNA solution, vortex, and incubate at −70°C for 20 min.
   g. Centrifuge for 10 min and wash the pellet with 400 µL of 70% ethanol.
   h. Dry the pellet briefly and resuspend in 20 µL of sterile TE buffer.

3.1.2. Infection/Transfection Protocol for Generating Adenoviral Library

1. Plate 293-CymR cells in 60-mm tissue-culture dishes at 5×10^5 cells per dish (*see* **Note 3**). Incubate overnight at standard conditions.

2. Remove the medium and infect the cells with 1 mL of fresh complete DMEM containing Ad5-Δ[PS-E3-E4(+orf6)] at multiplicities of infection (MOI) of 1–5 plaque forming units (PFUs).

3. Incubate the plates under standard conditions on a rocking platform for 5 h.

4. Replace the infectious mixture with 3 mL of fresh complete DMEM and let the cells recover for 30 min at standard conditions.

5. For each dish, prepare transfection mixture as follows:

   a. To a 1.5-mL microcentrifuge tube add 300 µL of DMEM without serum, 4.5 µg of linearized plasmid library (from **Subheading 3.1.1.**), and 6 µg of PEI *(22)*.
   b. Mix well and incubate for 10 min at room temperature.

6. Add the transfection mixture to the Ad5-Δ[PS-E3-E4(+orf6)] infected cells.

7. Incubate the cells at standard conditions for 3–4 d.
8. Collect the cells together with the medium and subject them to three freeze/thaw cycles to release the viral particles from the cells. The resulting viral mini-stock represents the initial adenoviral library (*see* **Note 4**). Store the mini-stock at -20 to $-80°C$.

3.1.3. Determination of the Approximate Viral Titer of the Library by Plaque Assay

1. One day prior to plaque assay, plate 293-CymR cells at 2.5×10^5 cells/well into six wells of a 6-well tissue-culture dish.
2. On the day of the plaque assay, make 1:10 serial dilutions (10^{-1} to 10^{-6}) of the viral library mini-stock, each dilution in 1 mL of complete DMEM.
3. Remove the medium from the cells and infect with 1 mL of the appropriate viral dilution per well.
4. Incubate the plates under standard conditions on a rocking platform overnight.
5. Remove the infectious medium and cover the cells in each well with an overlay of 2.5 mL of complete DMEM containing melted Seaplaque agarose at a final concentration of 1%. Leave the dishes at room temperature on a level surface for 5–15 min in order to allow the overlay to solidify.
6. Incubate the dishes under standard conditions for 1 wk. At that time, add a second overlay of 1 mL of fresh DMEM containing 1% Seaplaque agarose to each well.
7. Two weeks postinfection, count the number of plaques per well and multiply by the dilution factor to obtain the approximate virus titer (PFU/mL). Use wells with 10–100 PFUs to estimate the titer.

3.1.4. Plating the Library for Harvesting Individual Viral Clones

1. Plate 293-CymR cells at 1.5×10^6 cells/100-mm tissue-culture dish. The number of dishes depends on the expected library diversity (*see* **Note 3**). Incubate at standard conditions overnight.
2. For each plate, prepare 3 mL of the original viral mini-stock at 50 PFU/mL diluted in complete DMEM.
3. Remove the medium from the cells and infect with 3 mL of the diluted viral mini-stock.
4. Incubate the plates under standard conditions on a rocking platform overnight.
5. Remove the infectious mixture and cover the cells in each dish with an overlay of 10 mL of complete DMEM containing melted Seaplaque agarose at a final concentration of 1%. Leave the dishes at room temperature on a level surface for 5–15 min in order to allow the overlay to solidify.

6. Incubate the dishes under standard conditions for 1 wk. At that time, add a second overlay of 5 mL of fresh DMEM containing 1% Seaplaque agarose to each dish.

7. Viral plaques representing single recombinant clones will appear within 1–2 wk postinfection. At that time, pipet 200 µL of DMEM/well into 96-well plates (*see* **Note 5**).

8. Pick up the viral plaques by piercing the agarose with a 200 µL pipet tip by gently aspirating with a P200 Pipetteman. Transfer the agarose plug to the well and pipet up and down three times. Change the tip after each manipulation.

9. Incubate the plates with the eluted plaques at standard conditions overnight.

10. Store the plate at −20 to −80°C.

3.1.5. Amplification of Individual Viral Clones

1. Plate 293-CymR cells in 96-well plate at 1×10^5 cells per well in 150 µL of complete DMEM. Infect with 50 µL of the eluted plaques.

2. Incubate the plates under standard conditions on a rocking platform for several days until complete cytopathic effect is present in all wells.

3. Subject the plates to three freeze/thaw cycles to release the viral particles from the cells.

4. Store the plate at −20 to −80° C.

5. The amplified viral clones can be further used to infect cells for functional screening assays. At that stage, the titers of the mini-stocks are typically greater than 10^9 PFUs/mL.

3.2. High-Throughput Library Screening for Clones Affecting Cell Viability

The screening assay is designed to measure GFP intensity as an indicator of cell viability. As a basic principle, the infection with individual viral clones is followed by a second delayed infection with AdCMV5-GFP delivering the GFP indicator gene. The assay can be modified for two applications: (1) to identify expressed DNA variants with a pro-apoptotic or cytotoxic phenotype—as a result of reduced cell viability, the synthesis and accumulation of GFP reporter will be impaired; (2) to identify clones with protective or anti-apoptotic properties. In this case, cells can be treated with agents affecting cell viability prior to infection. Increased viability will lead to higher GFP intensity.

1. Inoculate HeLa cells in 96-well plates at 1×10^5 cells per well in 100 µL of complete DMEM (*see* **Note 6**).

2. Infect with 10 µL of the amplified viral clones (*see* **Note 7**) and bring the volume to 200 µL by adding 90 µL of complete DMEM.

3. Incubate the plates under standard conditions on a rocking platform for 12–16 h.
4. Infect with AdCMV5-GFP at MOI of 100 PFU by diluting the viral stock to 1×10^9 PFU/mL and adding 10 µL per well of cells (*see* **Note 8**).
5. Scan the plate on a FluorImager™ for background fluorescence (*see* **Note 9**). Incubate the plate under standard conditions on a rocking platform for 24 h.
6. Scan the plate on a FluorImager to quantify the fluorescence signal. If necessary, incubate the cells for another 24 h and scan again.

3.3. Plaque Purification of Positive Viral Clones

This step is necessary if the positive clones are intended for further applications such as protein production, precise functional studies, or gene therapy. This step assures that the viral clones are homogenous before they are amplified on large scale. One additional round of purification will completely eliminate any possibility of contamination with parental virus or the presence of more than one clone in the preparation.

1. Plate 293-CymR cells at 2.5×10^5 cells per well in 6-well tissue-culture dishes and incubate under standard conditions overnight. Prepare one dish per clone to be purified.
2. The following day, prepare 1:10 serial dilutions (10^{-1} to 10^{-6}) of each eluted plaque (from **Subheading 3.1.4.**) in complete DMEM (*see* **Note 10**).
3. Remove the medium from the cells and infect with 1 mL of virus dilution per well.
4. Incubate the plates under standard conditions on a rocking platform for 2–4 h.
5. Remove the infectious mixture and cover the cells in each well with an overlay of 2.5 mL of complete DMEM containing melted Seaplaque agarose at a final concentration of 1%. Leave the dishes at room temperature on a level surface for 5–15 min in order to allow the overlay to solidify.
6. Incubate the dishes under standard conditions for 1 wk. At that time, add a second overlay of 1 mL of fresh DMEM containing 1% Seaplaque agarose to each well.
7. Two weeks postinfection, collect isolated positive plaques as described in **Subheading 3.1.4., steps 8–10**.

4. Notes

1. To test cloning and transformation efficiency, plate 50 µL of the transformation mixture on LB agar plates containing 100 µg/mL of ampicillin.

2. Other restriction enzymes with even fewer cutting sites, such as *Sce*I, can be used instead of *Pac*I to ensure representation of every insert in the library.
3. The number of plates prepared for infection depends on three parameters. The first parameter is the number of generated independent viral variants, which is dependent on the recombination efficiency. Under optimal conditions, one recombination event per 10^3 cells is expected. The second parameter is library diversity; for example, if the library contains 1000 different DNA variants and the recombination efficiency is optimal, a minimum of 1 $\times$ 10^6 cells have to be initially infected/transfected to ensure the representation of all of the 1000 independent adenoviral clones. The third parameter is the characteristics of the transgenes expressed in the library. Some transgenes are expected to be moderately cytotoxic, pro-apoptotic, or might interfere with the AdV replication. The corresponding viral clones will have impaired growth and, consequently, they will be less represented in the library unless controlled by tight inducible promotors.
4. Along with the pool of recombinant AdV variants expressing PS, the parental virus (Ad5-ΔPS) is also present in the mini-stock because its growth is maintained by trans-complementation. One round of plaque purification (*see* **Subheading 3.1.4.**) is sufficient to eliminate the parental virus and to obtain pure individual viral clones.
5. Ninety-six well plates are convenient because they are compatible for use with multi-channel pipets. This facilitates the next steps of viral amplification and high-throughput functional screening.
6. If the assay is designed for examining cell cycle dysregulation rather than cytotoxicity or anti-apoptotic activity, then 2×10^4 cells per well should be used instead.
7. For more precise results, this assay can be done in triplicate, infecting with 5 μL, 10 μL, and 20 μL of the amplified clones.
8. Given the high amount of GFP released in crude lysates of AdCMV5-GFP-infected 293 cells, CsCl-purified AdCMV5-GFP *(24)* should be used to minimize the background GFP that would otherwise be introduced in the wells with viral inocula from crude lysates.
9. For scanning and quantification of the fluorescence intensity, follow FluorImager™ and ImageQuant™ manufacturer's instructions.
10. In most cases, dilutions of 10^{-3} and 10^{-4} are optimal to obtain a few isolated plaques.

Acknowledgments

We thank Alaka Mullick and Maureen O'Conner-McCourt for critical reading of this manuscript. This work was supported in part by a collaborative research agreement with Q-Biogene (*www.Qbiogene.com*). This is a NRC publication #44845.

References

1. Encell, L. P., Landis, D. M., and Loeb, L. A. (1999) Improving enzymes for cancer gene therapy. *Nat. Biotechnol.* **17**, 143–147.
2. Li, M. (2000) Applications of display technology in protein analysis. *Nat. Biotechnol.* **18**, 1251–1256.
3. Schmidt-Dannert C. (2001) Directed evolution of single proteins, metabolic pathways, and viruses. *Biochemistry* **40**, 13125–13136.
4. Carstens, C. P., Gallo, J. C., Maher, V. M., McCormick, J. J., and Fahl, W. E. (1995) A system utilizing Epstein-Barr virus-based expression vectors for the functional cloning of human fibroblast growth regulators. *Gene* **164**, 195–202.
5. Gudkov, A. V., Roninson, I. B., Brown, R., Kimchi, A., Cohen, O., Kissil, J., et al. (1999) Functional approaches to gene isolation in mammalian cells. *Science* **285**, 299 (Technical Comments).
6. Oualikene, W. and Massie, B. (2000) Adenoviral vectors in functional genomics, in *Cell Engineering*, vol. 2 (Al Rubeai, M., ed.), Kluwer Academic Publishers, Dordrecht, The Netherlands, pp. 80–154.
7. Aoki, K., Barker, C., Danthinne, X., Imperiale, M. J., and Nabel, G. J. (1999) Efficient generation of recombinant adenoviral vectors by Cre-lox recombination in vitro. *Mol. Med.* **4**, 224–231.
8. Bett, A. J., Haddara, W., Prevec, L., and Graham, F. L. (1994) An efficient and flexible system for construction of adenovirus vectors with insertions or deletions in early regions 1 and 3. *Proc. Natl. Acad. Sci. USA* **91**, 8802–8806.
9. Chartier, C., Degryse, E., Gantzer, M., Dieterle, A., Pavirani, A., and Mehtali, M. (1996) Efficient generation of recombinant adenovirus vectors by homologous recombination in *Escherichia coli. J. Virol.* **70**, 4805–4810.
10. Crouzet, J., Naudin, L., Orsini, C., Vigne, E., Ferrero, L., Le Roux, A., et al. (1997) Recombinational construction in *Escherichia coli* of infectious adenoviral genomes. *Proc. Natl. Acad. Sci. USA* **94**, 1414–1419.
11. Ghosh-Choudhury, G., Haj-Ahmad, Y., Brinkley, P., Rudy, J., and Graham, F. L. (1986) Human adenovirus cloning vectors based on infectious bacterial plasmids. *Gene* **50**, 161–171.
12. He, T. C., Zhou, S., da Costa, L. T., Yu, J., Kinzler, K.W., and Vogelstein, B. (1998) A simplified system for generating recombinant adenoviruses. *Proc. Natl. Acad. Sci. USA* **95**, 2509–2514.
13. Ketner, G., Spencer, F., Tugendreich, S., Connelly, C., and Hieter, P. (1994) Efficient manipulation of the human adenovirus genome as an infectious yeast artificial chromosome clone. *Proc. Natl. Acad. Sci. USA* **91**, 6186–6190.
14. Mizuguchi, H. and Kay, M. A. (1998) Efficient. construction of a recombinant adenovirus vector by an improved in vitro ligation method. *Hum. Gene Ther.* **9**, 2577–2583.

15. Oualikene, W., Lamoureux, L., Weber, J. M., and Massie, B. (2000) Protease-deleted adenovirus vectors and complementing cell lines: potential applications of single-round replication mutants for vaccination and gene therapy. *Hum. Gene Ther.* **11,** 1341–1353.

16. Greber, U. F., Webster, P., Weber, J., and Helenius, A. (1996) The role of the adenovirus protease on virus entry into cells. *EMBO J.* **15,** 1766–1777.

17. Elahi, S. M., Oualikene, W., Naghdi, L., O'Connor-McCourt, M., and Massie, B. (2002) Adenovirus-based libraries: efficient generation of recombinant adenoviruses by positive selection with the adenovirus protease. *Gene Ther.* **9,** 1238–1246.

18. Mosser, D. D., Caron, A. W., Bourget, L., Jolicoeur, P., and Massie, B. (1997) Use of a dicistronic expression cassette encoding the green fluorescent protein for the screening and selection of cells expressing inducible gene products. *Biotechniques* **22,** 150–154, 156, 158–161.

19. Massie, B., Couture, F., Lamoureux, L., Mosser, D. D., Guilbault, C., Jolicoeur, P., et al. (1998) Inducible overexpression of a toxic protein by an adenovirus vector with a tetracycline-regulatable expression cassette. *J. Virol.* **72,** 2289–2296.

20. Massie, B., Mosser, D. D., Koutromanis, M., Vitté-Mony I., Lamoureux, L., Couture, F., et al. (1998) New adenovirus vectors for protein production and gene transfer. *Cytotechnology* **28,** 53–64.

21. Mullick, A. and Massie, B. A cumate-inducible system for regulated expression in mammalian cells (patent application filed 04/02).

22. Sambrook, J., Fritsch, E. F., and Maniatis, T. (1989) *Molecular Cloning: A Laboratory Manual.* Cold Spring Harbor Laboratory Press, Cold Spring Harbor, NY.

23. Durocher, Y., Perret, S., and Kamen, A. (2002) High-level and high-throughput recombinant protein production by transient transfection of suspension-growing 293-EBNA cells. *Nucleic Acid Res.* **30,** 1–9.

24. Tollefson, A. E., Terry, H. W., and Wold, W. S. M. (1999) Preparation and titration of CsCl-banded adenovirus stock, in *Adenovirus Methods and Protocols* (Wold, W. S. M., ed.), Humana Press, Totowa, NJ, pp. 1–9.

3

Adenovirus-Mediated Gene Delivery to Skeletal Muscle

Joanne T. Douglas

1. Introduction

Adenoviral vectors can be employed for gene delivery to skeletal muscle, both ex vivo and in vivo. Although the realization of the full potential of adenoviral vectors awaits the development of methods to allow safe and efficient targeted gene delivery to mature skeletal muscle upon intravenous vector administration *(1)*, the current generation of vectors has nonetheless found utility in preclinical studies of gene therapy and in gene-transfer experiments designed to study muscle biology. Features of adenoviral vectors that have favored their use for gene delivery to skeletal muscle include the ability to infect both actively dividing and terminally differentiated cells, as well as their large insert capacity. Gutted adenoviral vectors are capable of carrying the large dystrophin gene together with regulatory sequences, and are therefore appropriate vehicles for gene-replacement therapy for Duchenne muscular dystrophy. In addition to their suitability for in vivo gene-therapy applications, adenoviral vectors have been used ex vivo to transfer genes to myoblasts prior to myoblast transplantation into muscle.

Adenoviral vectors also suffer from a number of limitations as vehicles for gene delivery to the muscle. On the one hand, there are generally recognized issues of short-term gene expression and the induction of both innate and acquired immune responses in vivo, which are being addressed by improvements to the vectors, as discussed in the overview chapter. However, adenovirus-mediated gene transfer to the muscle suffers from the unique problem of a matu-

From: *Methods in Molecular Biology, vol. 246:*
Gene Delivery to Mammalian Cells: Vol. 2: Viral Gene Transfer Techniques
Edited by: W. C. Heiser © Humana Press Inc., Totowa, NJ

ration-dependent decrease in the efficiency of infection (reviewed in **ref.** *2*). Thus, although adenoviral vectors can efficiently infect neonatal muscle in vivo, mature skeletal muscle is only poorly transduced. This finding correlates with in vitro studies that demonstrate that adenoviral infection of mononucleate myoblasts is much more efficient than infection of myotubes (cylindrical, multinucleate muscle cells generated by fusion of myoblasts) or isolated muscle fibers.

A number of potential barriers to efficient adenoviral infection of mature skeletal muscle have been suggested. In this regard, the primary cellular receptor for adenovirus serotype 5, the coxsackievirus and adenovirus receptor (CAR), is barely detectable in adult (60-d-old) murine skeletal muscle, although low levels are present at in 3- and 10-d-old mice *(3)*. This downregulation of CAR expression correlates with the known developmental decrease in susceptibility to adenoviral infection. Muscle-specific overexpression of CAR in transgenic mice has been shown to render the mature muscle more susceptible to adenoviral infection *(4)*. This suggests that tropism-modified adenoviral vectors capable of CAR-independent infection of muscle would be able to accomplish more efficient gene transfer than vectors with wild-type fibers *(1,5)*. However, an additional hurdle is introduced by the evidence that the basal lamina presents a physical barrier to efficient infection of mature muscle by adenoviral vectors *(6)*. To date, no strategies to overcome this obstacle have been proven to be practical and efficacious in vivo.

In summary, currently available adenoviral vectors are capable of efficient gene delivery to myoblasts ex vivo and to immature skeletal muscle upon direct intramuscular injection in vivo. Accordingly, adenoviral vectors have found widespread application in these delivery contexts. Much effort is currently being expended to generate an adenoviral vector capable of efficient gene delivery to mature skeletal muscle upon systemic administration.

In this chapter, we describe methods for adenovirus-mediated gene delivery to myoblasts, myotubes, and myofibers in vitro, and for intramuscular injection in vivo. Although methods are described for gene transfer by current adenoviral vectors, only minimal modifications will be necessary for future, more efficient vectors.

2. Materials

2.1. Sources of Cells and Reagents

1. Mice (C57BL/6, C57BL/10, *mdx* etc.) are available from The Jackson Laboratory (Bar Harbor, ME).
2. The murine C2C12 myoblast cell line (**ref.** *7*; ATTC Number CRL-1772) is available from the American Type Culture Collection (ATCC), Manassas, VA.

3. Normal human skeletal muscle cells (Clonetics™) and the recommended growth medium are available from BioWhittaker, Inc. (Walkersville, MD).
4. Dulbecco's modified Eagle's medium (DMEM), Opti-MEM, chick embryo extract, glutamine, sodium bicarbonate, penicillin/streptomycin, fetal calf serum (FCS), horse serum, Dulbecco's phosphate-buffered saline (PBS), Hank's balanced salt solution (HBSS), dispase, and trypsin-ethylenediaminetetraacetic aicd (EDTA) (GIBCO™ brand) are available from Invitrogen (Carlsbad, CA).
5. Tissue-culture plates precoated with collagen type 1 or Matrigel (BD Bio-Coat™)are available from BD Biosciences (Bedford, MA).
6. Collagenase type I and type XI are available from Sigma (St. Louis, MO).

2.2. Adenoviral Vectors

1. Purify adenoviral vectors by cesium chloride density-gradient ultracentrifugation or column chromatography.
2. Dialyze at 4°C to remove cesium chloride if necessary.
3. Determine the concentration of the vectors by physical and/or biological methods *(8)*.
4. Store adenoviral vectors in 10-µL aliquots at −80°C.

2.3. Solutions and Culture Medium

1. Growth medium for C2C12 myoblasts: DMEM with 4 m*M* L-glutamine supplemented with 10% FCS and 1% penicillin/streptomycin and adjusted to contain 1.5 g/L sodium bicarbonate and 4.5 g/L glucose. Cells must not be allowed to become confluent because this will deplete the myoblastic population in the culture.
2. Growth medium for primary murine myoblasts: DMEM with 2 m*M* L-glutamine supplemented with 10% FCS, 10% horse serum, 0.5% chick embryo extract, and 1% penicillin/streptomycin.
3. Differentiation medium for C2C12 myoblasts: DMEM with 4 m*M* L-glutamine supplemented with 10% horse serum and 1% penicillin/streptomycin and adjusted to contain 1.5 g/L sodium bicarbonate and 4.5 g/L glucose.
4. Differentiation medium for primary murine myoblasts: DMEM with 2 m*M* L-glutamine supplemented with 1% FCS, 1% horse serum, and 1% penicillin/streptomycin.
5. Growth medium for myofibers: DMEM with 2 m*M* L-glutamine supplemented with 10% horse serum, 1% chick embryo extract, and 1% penicillin/streptomycin.
6. Solutions for preparation of myoblasts by enzymatic digestion of muscle: 0.2% (w/v) collagenase type XI in HBSS; dispase (2.4 U/1 mL HBSS); 0.1% (w/v) trypsin-EDTA in HBSS. Prepare fresh.

7. Adenoviral vector dialysis buffer: 10 mM HEPES, 1 mM MgCl$_2$ containing 10% glycerol.

3. Methods

3.1. Adenovirus-Mediated Gene Delivery to Myoblasts, Myotubes, and Myofibers In Vitro

3.1.1. Preparation of Primary Myoblast Cultures

Although the commercially available murine myoblast cell line C2C12 is suitable for many purposes, it is often necessary to isolate primary murine myoblasts from skeletal muscle *(9,10)*.

1. Sacrifice mice by a standard painless technique.
2. Immediately remove the forelimb and hindlimb muscles, dissect the bones, and rinse the muscle in PBS.
3. Chop the muscle into a coarse slurry using scalpels and transfer to a 15-mL conical tube.
4. Add 0.2% (w/v) collagenase type XI in HBSS and dissociate the muscle tissue by enzymatic digestion at 37°C for 1 h, with continuous gentle rocking of the conical tube.
5. Centrifuge at 2000g for 5 min. Collect the cells and incubate in dispase (2.4 U/1 mL HBSS) for 45 min at 37°C. Then incubate the cells for 30 min at 37°C in 0.1% (w/v) trypsin-EDTA in HBSS.
6. After the enzymatic dissociation, centrifuge and resuspend the isolated muscle cells in growth medium.
7. Preplate the cells on collagen-coated 6-well plates and incubate at 37°C in 5% CO$_2$. After about 1 h, decant the supernatant containing nonadherent cells from the well and replate in a fresh collagen-coated well (*see* **Note 1**).
8. Subculture myoblasts before they become confluent, to avoid differentiation into myotubes.

3.1.2. Differentiation of Myoblasts into Myotubes

1. Fusion of myoblasts into myotubes is induced by changing from myoblast growth medium to differentiation medium. Cylindrical multinucleate myotubes will form within 5 d.

3.1.3. Isolation of Myofibers

Single myofibers can be isolated from the extensor digitorum longus (EDL) muscle of mature mice *(11)* or from the soleus and gastrocnemius muscles (en bloc) of immature mice *(6)* (*see* **Note 2**).

1. Sacrifice mice by a standard painless technique.
2. Immediately remove the muscle by dissection and rinse in PBS.

3. Transfer the muscle to a 50-mm diameter × 18-mm deep plastic Petri dish. Add 1 mL 0.2% (w/v) type I collagenase in DMEM and dissociate the muscle tissue by enzymatic digestion at 37°C for 2 h, with continuous gentle rocking.
4. Transfer the muscle to a 50 mm × 18 mm plastic Petri dish using a wide-mouth Pasteur pipet (*see* **Note 3**).
5. Under a stereo dissecting microscope, liberate single myofibers by repeatedly triturating the muscle with a wide-mouth Pasteur pipet. Once 20–30 intact single myofibers have been separated, transfer the muscle bulk to a fresh Petri dish and continue trituration, using a Pasteur pipet with a smaller mouth. Continue the cycles of fiber separation as necessary.
6. Use a Pasteur pipet to transfer separated myofibers into tissue-culture plates coated with 1 mg/mL Matrigel. Plate 5-25 single myofibers per well of a 12-well plate or 10–100 fibers per well of a 6-well plate. Incubate in growth medium at 37°C in 5% CO_2.
7. Confirm the viability of fibers by trypan blue exclusion and infect with adenovirus 24 h after isolation (*see* **Notes 4** and **5**).

3.1.4. Infection of Myoblasts, Myotubes, and Myofibers

1. Thaw adenoviral vectors on ice immediately prior to use.
2. Dilute adenoviral vectors to the desired concentration in Opti-MEM™ serum-free medium (*see* **Note 6**).
3. Aspirate growth medium from myoblasts (2×10^6 per well of a 6-well plate), myotubes, or myofibers in tissue-culture plates and rinse with HBSS.
4. Apply suspension of adenoviral vectors in the smallest volume of Opti-MEM™ necessary to cover the muscle cells or fibers.
5. Incubate for 1 h at 37°C in 5% CO_2 with gentle rocking every 10 min.
6. Aspirate viral infection medium and replace with appropriate growth medium.
7. Continue incubation at 37°C in 5% CO_2 for a further 24–48 h.
8. Perform appropriate assays for transgene expression. In the case of adenoviral vectors expressing β-galactosidase, this will involve histochemical staining using X-gal as the chromogenic substrate. Expression of green fluorescent protein (GFP) can be detected using a fluorescence microscope.

3.2. Adenovirus-Mediated Gene Delivery to Muscle In Vivo

3.2.1. Intramuscular Injection

1. Thaw adenoviral vectors on ice immediately prior to use.
2. Dilute adenoviral vectors to the desired concentration in HBSS (*see* **Note 7**).

3. Anesthetize mice by intraperitoneal injection of ketamine-xylazine (100 and 10 mg/kg body weight, respectively).
4. Inject tibialis anterior or gastrocnemius muscles percutaneously with adenoviral suspension using a syringe and 26 G needle. Inject 25 μL of viral suspension.
5. Sacrifice mice 3–5 d post-infection.
6. Dissect injected and control muscles and snap-freeze.
5. For histochemical analysis, cut 10-μm thick sections using a cryostat.

4. Notes

1. The cells that adhere rapidly to the first collagen-coated flask will mostly be fibroblasts and should be discarded.
2. The small size of immature mice makes the isolation of single myofibers from the EDL muscle technically impossible.
3. Prerinse the Petri dishes with a small volume of horse serum to prevent the fibers from sticking. Flush the Pasteur pipets with DMEM containing 10% horse serum.
4. Those fibers that fail to exclude trypan blue will also appear visibly hypercontracted. Thus, with practice, nonviable fibers can be eliminated by eye.
5. Myofibers should be infected 24 h post-isolation. At this time, those fibers that were badly damaged during isolation will have died, whereas the dedifferentiation of the fiber that occurs in culture will be minimal.
6. It is recommended that initial experiments be performed using the adenoviral vector at a range of concentrations. Multiplicities of infection of 10, 50, and 100 infectious units per cell are suggested for myoblasts, whereas 5×10^5 to 5×10^7 infectious units per well are suggested for myotubes and myofibers. Experiments should be performed with triplicate samples and uninfected cells should serve as a control. It is also recommended that an adenoviral vector expressing a reporter protein such as β-galactosidase or GFP be employed as a control for the efficiency of infection.
7. It is recommended that initial experiments be performed using the adenoviral vector at a range of concentrations. Intramuscular injection of 10^7 to 10^9 infectious units is suggested. It is customary to inject one side of the mouse, so that the contralateral muscle serves as a negative control. It is also recommended that an adenoviral vector expressing a reporter protein such as β-galactosidase or GFP be employed as a control to allow the extent of gene transfer within the muscle to be quantified.

Acknowledgments

Research in the author's laboratory is supported, in part, by a grant from the Muscular Dystrophy Association.

References

1. Douglas, J. T. and Curiel, D. T. (1997) Strategies to accomplish targeted gene delivery to muscle cells employing tropism-modified adenoviral vectors. *Neuromusc.. Disord.* **7,** 284–298.
2. van Deutekom, J. C., Floyd, S. S., Booth, D. K., Oligino, T., Krisky, D., Marconi, P., et al. (1998) Implications of maturation for viral gene delivery to skeletal muscle. *Neuromusc. Disord.* **8,** 135–48.
3. Nalbantoglu, J., Pari, G., Karpati, G., and Holland, P. C. (1999) Expression of the primary coxsackie and adenovirus receptor is downregulated during skeletal muscle maturation and limits the efficacy of adenovirus-mediated gene delivery to muscle cells. *Hum. Gene Ther.* **10,** 1009–1019.
4. Nalbantoglu, J., Larochelle, N., Wolf, E., Karpati, G., Lochmuller, H., and Holland, P. C. (2001) Muscle-specific overexpression of the adenovirus primary receptor CAR overcomes low efficiency of gene transfer to mature skeletal muscle. *J. Virol.* **75,** 4276–4282.
5. Bouri, K., Feero, K., Myerburg, M. M., Wickham, T. J., Kovesdi, I., Hoffman, E. P., and Clemens, P. R. (1999) Polylysine modification of adenoviral fiber protein enhances muscle cell transduction. *Hum. Gene Ther.* **10,** 1633–1640.
6. Feero, W. G., Rosenblatt, J. D., Huard, J., Watkins, S. C., Epperly, M., Clemens, P. R., et al. (1997) Viral gene delivery to skeletal muscle: insights on maturation-dependent loss of fiber infectivity for adenovirus and herpes simplex type 1 viral vectors. *Hum. Gene Ther.* **8,** 371–380.
7. Blau, H. M., Pavlath, G. K., Hardeman, E. C., Chiu, C. P., Silberstein, L., Webster, S. G., et al. (1985) Plasticity of the differentiated state. *Science* **230,** 758–766.
8. Mittereder, N., March, K. L., and Trapnell, B. C. (1996) Evaluation of the concentration and bioactivity of adenovirus vectors for gene therapy. *J. Virol.* **70,** 7498–7509.
9. Rando, T. A. and Blau, H. M. (1994) Primary mouse myoblast purification, characterization, and transplantation for cell-mediated gene therapy. *J. Cell Biol.* **125,** 1275–1287.
10. Qu, Z., Balkir, L., van Deutekom, J. C., Robbins, P. D., Pruchnic, R., and Huard, J. (1998) Development of approaches to improve cell survival in myoblast transfer therapy. *J. Cell Biol.* **142,** 1257–1267.
11. Rosenblatt, J. D., Lunt, A. I., Parry, D. J., and Partridge, T. A. (1995) Culturing satellite cells from living single muscle fiber explants. *In Vitro Dev. Biol.* **31,** 773–779.

4

Delivery of Adenoviral DNA to Mouse Liver

Sheila Connelly and Christine Mech

1. Introduction

The liver represents a major target organ for gene delivery owing to its high biosynthetic capacity and access to the bloodstream. Adenoviral vectors are highly efficient gene-transfer vehicles, making them among the most promising systems for in vivo gene transfer to the liver. Following intravenous administration of adenoviral vectors to a variety of mammalian models, including mice, dogs, and monkeys, hepatocytes are efficiently transduced *(1)*. Several delivery methods to the liver have been described, including portal vein *(2–4)*, hepatic artery *(3,5)*, and peripheral vein infusions *(6)*. This chapter describes the simple, nonsurgical method of intravenous (iv) administration of adenoviral vectors in mice, and an immunohistochemical method to qualitatively evaluate liver transduction efficiency following delivery of an adenoviral vector encoding a β-galactosidase (β-gal) marker gene. Additionally, several alternative methods to verify efficient liver transduction are introduced.

Although systemic vector delivery is facile, a major limitation of this delivery method is the broad tissue distribution. Following a single iv injection in mice, the liver is the most highly transduced organ. However, significant levels of vector DNA have been detected in all organs and tissues evaluated, including lung, spleen, kidney, heart, and gonads *(6)*. Therefore, a variety of parameters must be taken into consideration prior to iv vector delivery to ensure optimal liver transduction and transgene expression efficiency. These include the choice of adenoviral vector backbone, the design of the transgene expres-

From: *Methods in Molecular Biology, vol. 246:*
Gene Delivery to Mammalian Cells: Vol. 2: Viral Gene Transfer Techniques
Edited by: W. C. Heiser © Humana Press Inc., Totowa, NJ

sion cassette, the selection of the animal model, and the optimization of vector dose.

A major limitation to adenoviral vector efficacy is that transgene expression, in general, is transient, and adenoviral vectors are toxic and immunogenic. Therefore, the choice of adenoviral vector backbone is an important consideration. Initial studies have suggested that adenoviral toxicity is mediated in a large part by the cellular immune response to *de novo* synthesis of viral antigens in transduced cells *(7–9)*. Therefore, a strategy to reduce vector-mediated toxicity, and thereby prolong transgene expression, has been the sequential elimination of key early genes from the vector backbone. Because the early region gene products function to upregulate expression of the late region genes, removal of the early genes results in the diminished expression from other essential viral backbone genes. First-generation adenoviral vectors are deficient in the E1 region *(10)*, whereas subsequent generation vectors are deficient in E2a, E2b, and/or E4 in addition to the E1 region *(11–16)*. This gene attenuation strategy had limited success, partly owing to continued low-level expression of viral genes remaining in the attenuated vector backbone, and the unpredictable function of the transgene expression cassettes incorporated in the vectors *(16–19)*. As an alternative approach, vectors devoid of all viral coding regions have been generated *(20–23)*. These vectors, referred to by a variety of names including "gutless," high-capacity, and helper-dependent, are grown in the presence of a helper virus that supplies the proteins required for vector replication and packaging *(24)*. Evaluation of a gutless vector encoding the human factor VIII (FVIII) cDNA in mice demonstrated 10-fold higher FVIII expression levels, longer duration of FVIII expression, and reduced hepatic toxicity compared with an E1/E2a-deficient early generation vector encoding an identical transgene expression cassette *(25)*. However, the initial, dose-dependent hepatic toxicity believed to be associated with the adenoviral capsid protein has not been improved with the gutless vector *(25)*. In baboons, a gutless vector displayed transgene expression sustained for at least 1 yr in two of three treated animals *(20)*. These data suggest that gutless adenoviral vectors represent a significant improvement over early generation adenoviral vectors, owing to the complete elimination of viral genes from the vector backbone. Therefore, to obtain the maximal level and duration of transgene expression with minimal toxicity, the use of a gutless adenoviral vector is recommended. However, gutless vectors are not currently commercially available, and useful data can also be obtained using early generation, E1-deficient vectors.

Another important element in vector design is the choice of the transgene expression cassette. The transgene transcriptional unit consists of the elements required to enable the appropriate expression of the transgene and in most instances, is designed to maximize the expression of the exogenous gene. Transgene expression cassettes consist of the enhancer/promoter, the gene of interest, and a polyadenylation signal. A major determinant of transgene expression persistence is the relative immunogenicity of the transgene protein product in the chosen animal model. For example, reporter proteins, such as β-gal, are highly immunogenic in mice *(26)*, whereas, as expected, endogenous proteins, such as murine erythropoietin, are not immunogenic in most instances *(27)*. The choice of promoter driving expression of the transgene is also an important consideration. A large variety of promoters have been utilized for transgene expression, the choice of which depends on the application and intended target tissue. Strong, constitutively expressed viral promoters such as the cytomegalovirus (CMV) or the Rous sarcoma virus (RSV) promoters have been widely used. In general, such promoters allow maximal expression of the transgene, but in many cases expression is short-term. For example, the CMV promoter was reported to be quickly silenced in mouse liver *(28,29)*. Furthermore, transgene expression from a constitutive promoter allows unrestricted transgene expression in all transduced tissues, potentially increasing vector toxicity and immunogenicity. Indeed, whereas initial transgene expression levels were seven-fold higher, the CMV-promoted vector also displayed increased liver toxicity, decreased persistence of vector DNA in the liver, and a decreased duration of transgene expression, compared with a vector containing the identical transgene driven by a liver-specific promoter *(29)*. The use of liver-specific promoters, such as Apo A1 *(29)*, α_1-anti-trypsin *(20)*, and albumin *(23,30–32)*, has been described. In several cases, the use of a liver-specific promoter reduced vector immunogenicity and toxicity, increased the persistence of transgene expression, and potentially increased vector safety *(29,31,32)*. The mechanism of reduced vector immunogenicity is not known, but it has been speculated that as a result of liver-specific transgene expression, expression in other tissues is significantly reduced and the immune activation of antigen-presenting cells (APCs) may be prevented *(32)*. Therefore, depending on the level of transgene expression and duration of expression required, the use of a liver-specific promoter in place of a ubiquitous promoter is recommended.

An additional approach shown to increase the potency of the transgene in adenoviral vectors is the introduction of genomic elements into the expression cassette. For example, the addition of an intron upstream of a human FVIII cDNA boosted in vivo expression approx 10-fold *(30)*, and the inclusion of an intron upstream of the Apo A1 coding region prolonged the duration of transgene expression *(29)*. Finally, a variety of polyadenylation signals has been incorporated into transgene-expression cassettes. Suggested polyadenylation signals include the bovine growth hormone, Simian virus 40 (SV40), or synthetic polyadenylation signals.

Another important consideration in achieving efficient adenoviral vector-mediated liver transduction and optimal transgene expression is the choice of animal model. Mouse models are probably the most well-studied, and extremely efficient liver transduction following tail-vein injection of adenovirus serotype 5-based vectors has been widely reported. In mice, the liver is the most highly transduced organ, with the lung containing approx 10% the vector level of the liver, and the spleen, kidneys, heart, and other organs containing less than 1% the level of the liver *(6)*. Although rats *(33)*, rabbits *(4)*, pigs *(34)*, dogs *(35–37)*, and monkeys *(38,39)* also display efficient liver transduction following peripheral vein injection, the pattern of vector distribution differs from that of mice. In dogs and monkeys, for example, the spleen and liver are transduced with similar efficiencies *(36,38)*. However, for initial transgene expression studies, mice are the species commonly selected owing to their size, availability, and relative homogeneity as an inbred model. However, different inbred mice display variable immune responses to specific antigens *(40)*, and vary in the efficiency and duration of adenovirus-mediated transgene expression *(40–44)*. Therefore, the choice of mouse strain is also an important consideration. In our experience, C57BL/6 mice represent the optimal mouse strain to achieve long-term liver-mediated expression from an adenoviral vector, because this mouse strain displays an attenuated CTL response to the adenoviral vector backbone genes *(45)*. In contrast, a strong CTL response to the adenoviral vector has been documented in other mouse strains *(7–9)*, which resulted in a rapid loss of vector DNA from the transduced mouse livers, and therefore short-term transgene expression. Furthermore, outbred strains, such as CD-1 mice, which are commonly used for toxicological investigations, may respond differently than inbred strains. Therefore, if a mouse model is chosen, the specific objectives of the study should be carefully considered in the selection of the mouse strain. C57BL/6 mice are recommended for investigations that require sustained transgene expression in liver.

The final component necessary for efficient liver transduction and transgene expression is the choice of vector dose. In general, the higher the vector dose delivered, the higher the initial transgene-expression levels. However, the hepatotoxicity associated with high vector doses resulted in the rapid elimination of transduced cells, and therefore, a rapid decline in transgene expression *(13,46)*. In contrast, when lower, less toxic vector doses were delivered to mice, transgene expression persisted for several months, indicating that dose-dependent vector toxicity limited vector persistence *(13,46)*. Therefore, a dose high enough to achieve efficient transgene expression and low enough to avoid hepatotoxicity is required. To complicate matters further, there does appear to be a lower vector dose limit for effective liver transduction. Several recent reports have documented this "threshold" for efficient liver transduction in mice *(45,47)*. Following tail-vein administration of increasing doses of an adenoviral vector, a nonlinear dose response in transgene-expression levels and liver-transduction efficiency was measured. Transgene expression was not observed at low vector doses *(45)*, whereas a moderate increase in vector dose resulted in extremely high expression levels *(47)*. The current hypothesis to explain this observation is that at low doses, the vector is efficiently removed from the blood prior to reaching the hepatocytes, probably by the reticuloendothelial system, specifically the Kupffer cells in the liver, whereas higher vector doses saturate the Kupffer cells, allowing high level hepatocyte transduction *(45,47)*. Therefore, the use of a vector dose above the threshold for liver transduction, but below the level that causes significant hepatoxicity, is required for efficient liver transduction and expression. The suggested vector dose range that reproducibly results in measurable liver transduction and expression is 10^{12} and 10^{13} vector particles/kg *(25,45)*. Within this vector dose range, the recommended doses are $3–6 \times 10^{12}$ particles/kg. Finally, iv dosing on an individual animal body-weight basis (i.e., vector particles per kilogram, *see* **Table 1**) is recommended to reduce animal to animal variability and to facilitate extrapolation to other species, including humans.

To summarize, the suggested parameters for efficient liver transduction and sustained transgene expression include: (1) A gutless adenoviral vector backbone, although E1-deficient early generation vectors function quite well; (2) a transgene expression cassette consisting of a nonimmunogenic transgene, driven

Table 1
Vector Dosing Relationships[a]

Body weight (g)	Vector dose (particles/kg)	Vector conc. (particles/mL)	Dose volume (mL/kg)	Injection volume (mL)
18	6×10^{11}	6×10^{10}	10	0.18
20	6×10^{11}	6×10^{10}	10	0.20
23	6×10^{11}	6×10^{10}	10	0.23
22	3×10^{11}	3×10^{10}	10	0.22
22	6×10^{11}	6×10^{10}	10	0.22
22	1×10^{12}	1×10^{11}	10	0.22

[a]The relationship between mouse body weight, vector dose, vector concentration, dose volume, and injection volume. The dose volume (10 mL/kg) delivered to the mice is constant; the variables are the body weight (g) and the vector dose (particles/kg). The vector concentration (particles/mL) is adjusted to match the desired vector dose (particles/kg). The upper portion of the table displays the injection volumes for animals of differing body weights each receiving the same vector dose. The lower portion of the table displays the injection volume for animals of the same body weight each receiving a different vector dose. Vector dosing in this manner allows each animal to get an individualized dose, which decreases animal-to-animal variation in liver-transduction efficiency and transgene-expression levels, and allows easier extrapolation to other species, including humans.

by a liver-specific promoter, an intron upstream of the transgene cDNA, and an efficient polyadenylation signal; (3) C57BL/6 mice, between 4 and 8 wk of age; and (4) an optimal vector dose that is above the threshold for efficient liver transduction but below the level that causes toxicity (3–6 × 10^{12} vector particles/kg).

2. Materials

2.1. Sources of Materials

1. Early generation, E1-deficient adenoviral vectors encoding a marker gene, such as green florescent protein (GFP) or β-gal, can be purchased from several sources, such as Strategene (La Jolla, CA) or Clontech (Palo Alto, CA). Also, these sources provide kits with all the materials required to generate adenoviral vectors encoding the transgene expression cassette of choice.
2. Hank's Balanced Salt Solution (HBSS, InVitrogen, Carlsbad, CA) and phosphate-buffered saline (PBS; 10 mM sodium/potassium phosphate, 0.9% sodium chloride, pH 7.4; Zymed laboratories, Inc., San Francisco, CA).
3. Broome rodent restrainer: Harvard Apparatus (Holliston, MA)
4. 10% Neutral-buffered formalin (Surgipath Medical Industries, Richmond, IL).
5. Tissue-Tek VIP Processor, Tissue-Tek Embedding Center, stainless steel base molds, plastic tissue cassettes (Sakura Finetek, Torrance, CA).
6. Blue Ribbon Paraffin (Surgipath Medical Industries, Richmond, IL).
7. Leica RM2155 Microtome (Leica, Wetzlar, Germany).
8. Sequenza slide rack: ThermoShandon (Pittsburgh, PA).
9. Graded alcohols: isopropanol/methanol mixture (Richard Allan Scientific, Kalamazoo, MI).
10. Rabbit-anti-β-galactosidase antibody (ICN Biomedicals, Aurora, OH, Cat. no. 55976).
11. Biotinylated goat-anti-rabbit IgG (Vector Labs, Burlingame, CA, Cat. no. BA-1000).
12. Vector ABC Elite Standard Kit (Vector Labs, Cat. no. PK-6100).
13. DAB Chromogen (Research Genetics, Huntsville, AL, Cat. no. 750118).
14. Hematoxylin, water-based (Research Genetics).

2.2. Solutions

1. Adenovirus lysis buffer: 10 mM Tris HCl, pH 7.4, 1 mM ethylenediaminetetraacetic acid (EDTA), 0.1% sodium dodecyl sulfate (SDS).
2. Adenovirus storage buffer: 200 mM Tris-base, 50 mM HEPES, 10% glycerol, pH 8.0 (with phosphoric acid) (*see* **Note 1**).
3. Methanol-hydrogen peroxide solution: 4 mL 30% hydrogen peroxide plus 200 mL methanol.

4. PBS working solution: 0.1% Tween-20 in PBS.
5. Blocking buffer: 5% bovine serum albumin (BSA), 10% goat serum in PBS.
6. ABC complex: Mix two drops of Solution A and 2 drops of Solution B in 5 mL of PBS (1:80 dilution). Allow this solution to stand for a minimum of 30 min before use.

3. Methods

3.1. Measurement of Vector Concentration

1. Dilute 5 µL of concentrated adenovirus stock into 95 µL of adenovirus lysis buffer (1:20 dilution).
2. Incubate sample at 56°C for 15 min.
3. Measure the OD_{260} in a calibrated spectrophotometer.
4. Multiply OD_{260} × dilution factor. Use the formula 1 OD_{260} = 1.1×10^{12} particles/mL to calculate virus concentration (48).

3.2. Preparation of Vector and Injection of Mice

1. Thaw an aliquot of adenoviral vector on ice. Gently mix when completely thawed. (*see* **Note 2**).
2. Dilute virus to the desired concentration into chilled HBSS. Keep diluted virus on ice until ready to load the injection syringes (*see* **Note 3**).
3. Gently warm the mice under a heat lamp for 5–10 min to facilitate dilation of the lateral tail veins. Alternatively, warm water may be used to dilate the tail veins (*see* **Note 4**).
4. Load 0.5- or 1-mL insulin syringes with fixed 27 G needles with the appropriate volume of the diluted virus (*see* **Note 5**). Carefully remove all air bubbles, and allow virus solution to come to room temperature, approx 10 min.
5. Place heated mouse in a Broome rodent restrainer.
6. Gently rub the selected lateral vein with an alcohol swab.
7. With the bevel up, insert the needle into one of the lateral tail veins at a site close to the tip of the tail. The angle made between the needle and the vein should be as small as possible (*see* **Fig. 1**).
8. Draw back gently on the plunger of the syringe to determine that the needle is in the vein. A small amount of blood in the syringe is a good indication of proper placement (*see* **Fig. 1**).
9. Slowly discharge the contents of the syringe (*see* **Note 6**).
10. As the needle is withdrawn, apply pressure to the injection site until bleeding has abated.
11. Remove animal from the restrainer, and place in cage. Observe animal for 5–10 min to check for an adverse reaction.

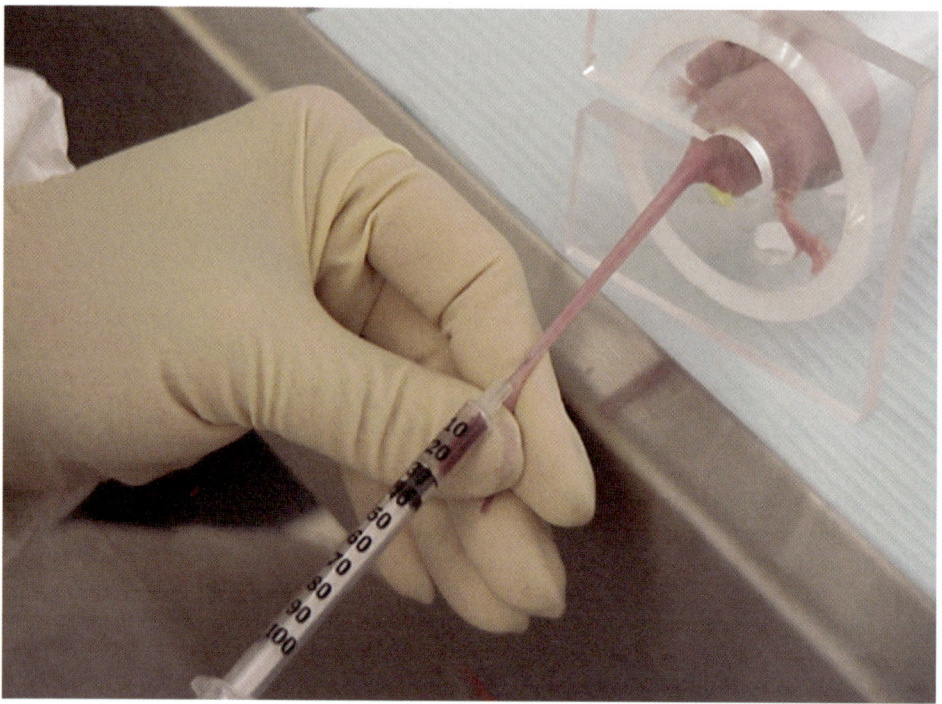

Fig. 1. Demonstration of the proper technique for a tail-vein injection. With the bevel up, insert the needle into the vein at a site close to the tip of the tail. Keep the angle between the needle and the vein as small as possible. Draw back gently on the plunger of the syringe to achieve blood return to ensure that the needle is in the vein. Slowly discharge the contents of the syringe. Stop immediately if pressure is felt upon injection, the formation of a bleb occurs, or another sign of leakage is noted. A second attempt may be required at another site, above the initial site, or in a different vein. Photograph courtesy of Donna Goldsteen.

3.3. Liver Collection, Fixing, and Preparation

1. Euthanize animals via CO_2 asphyxiation or cervical dislocation.
2. Excise entire liver, and immerse into 10% neutral-buffered formalin. Fix livers 24–48 h.
3. Cut livers into 3- to 4-mm sections.
4. Place in a tissue cassette and process to paraffin using a V.I.P. Tissue Processor.
5. Generate a paraffin block by the process of embedding using the Tissue-Tek Embedding Center following standard procedures. Place prepared liver sections carefully within the stainless steel mold, fill with molten paraffin,

and place cassette body on top of the stainless steel mold. Dispense additional paraffin to fill the space created by the cassette body.

6. Transfer the mold to the cold plate and allow the paraffin to solidify. When solid, the paraffin tissue block will separate easily from the stainless steel mold.

3.4. β-Gal Immunohistochemistry Procedure

1. Using paraffin-embedded livers, cut paraffin sections 5-μ thick using a Leica RM 2155 microtome, and place on electrostatically charged slides (*see* **Note 7**).
2. Place slides in a 60°C oven and bake for 1 h to allow sections to adhere to the slides.
3. Deparaffinize slides in three changes of xylene for 5 min each.
4. Rehydrate slides through a series of graded alcohols (100, 95, 70, 0%) prepared in distilled water.
5. Rinse slides in two changes of PBS.
6. Quench endogenous peroxidase activity by placing slides in the methanol-hydrogen peroxide solution for 30 min at room temperature.
7. Wash slides with distilled water three times for 2 min each wash.
8. Load slides into a Sequenza slide staining rack as follows:
   a. Fill a glass-staining dish with PBS.
   b. Immerse slide in PBS and place coverplate on top of slide.
   c. Place the slide/coverplate assembly in the slot in the staining rack.
   d. Fill each slide/coverplate assembly with PBS. If PBS drains too quickly (less than 2 min) repeat **steps b–d**.
9. Allow the PBS to drain from each slide assembly completely, approx 10 min, then apply 200 μL of Blocking solution buffer to each slide and incubate 30 min at room temperature. Do not rinse slide.
10. Apply 200 μL per slide of rabbit anti-β-gal antibody, diluted 1:2600 in blocking solution buffer. Incubate for 1 h at room temperature.
11. Rinse with PBS solution three times, 5 min each wash.
12. Apply 200 μL per slide of biotinylated goat-anti-rabbit IgG, diluted 1:200 in Blocking solution buffer, and incubate 30 min at room temperature.
13. Rinse slides three times with PBS solution, 5 min each wash.
14. Apply 200 μL of prepared ABC complex, and incubate 30 min at room temperature.
15. Rinse with PBS solution three times, 5 min each time.
16. Apply 500 μL of the DAB chromogen to each slide, and incubate for 5 min at room temperature.
17. Rinse slides with PBS solution one time.

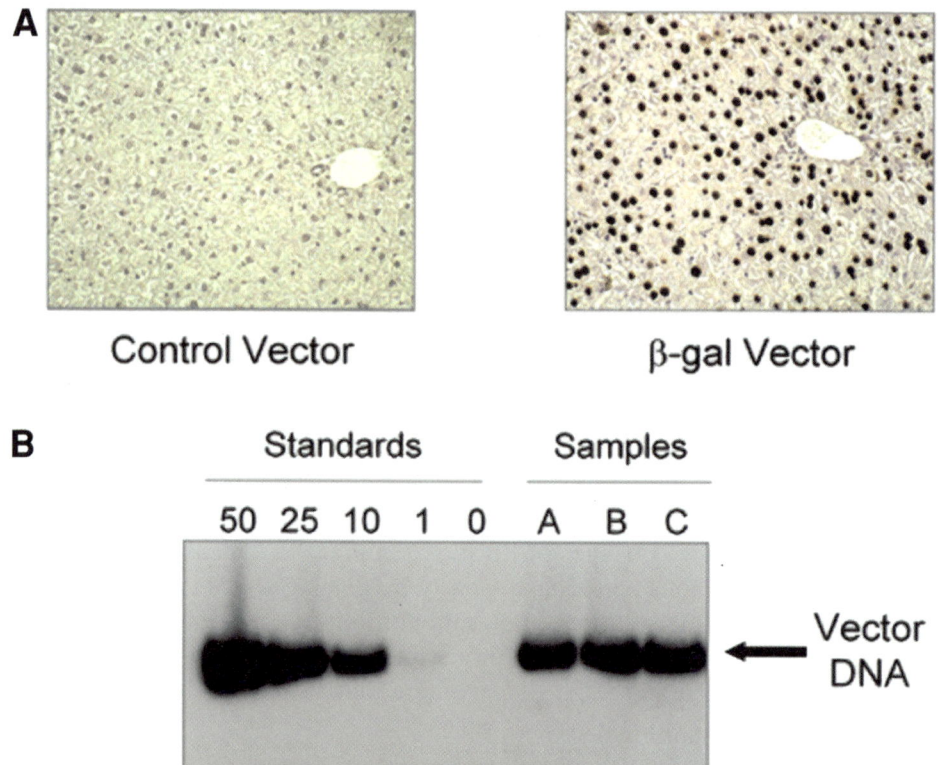

Fig. 2. Analysis of liver-transduction efficiency following tail vein administration of adenoviral vectors to mice. (**A**) β-gal immunohistochemical analysis of mouse liver sections. C57BL/6 mice were treated with 3×10^{11} particles/mouse of an E1-deficient adenoviral vector encoding nuclear localized β-gal (β-gal Vector; *(6)*, or a control vector, labeled Control, lacking the β-gal gene. Seven days after vector administration, livers were collected, and sections were analyzed by immunohistochemistry as described. β-gal positive cells display brown nuclei; negative cells display light purple nuclei. All hepatocytes in the β-gal vector-treated animal display β-gal expression, indicating extremely efficient liver transduction. (**B**) Southern analysis of liver DNA isolated from C57BL/6 mice treated with 3×10^{11} particles/mouse of an early generation vector encoding human factor VIII *(30)*. Each DNA sample (10 µg) was digested with *Bam*HI and subjected to Southern analysis using a ^{32}P-labeled random primed DNA probe encoding a portion of the human FVIII cDNA *(49)*. The lanes labeled Samples contain DNA isolated from three vector-treated mice. The lanes labeled Standards contain viral DNA equivalent to 50, 25, 10, and 1 vector copy per cell and were prepared by adding 3000, 1500, 600, or 60 pg of viral DNA to 10 µg of untreated, control mouse genomic liver DNA and digesting with *Bam*H1 (*see* **Note 9** for a formula useful for the calculation of DNA copy number standards). The data demonstrate 10–20 vector copies per cell in the livers of the treated mice.

18. Remove slides from staining rack and wash in running tap water for 5 min.
19. Lightly counterstain slides in Hematoxylin for 5–10 s.
20. Rinse in running tap water for 2 min.
21. Immerse slides in PBS solution for 1 min.
22. Rinse in running tap water for 2 min.
23. Dehydrate slides through two changes of 95% alcohol and two changes of 100% alcohol for 3 min each time.
24. Clear slides in three changes of xylene 3 min each time.
25. Apply coverslip using xylene-based mounting medium.
26. β-gal positive cells appear dark brown, whereas negative cells appear light purple (**Fig. 2A**; *see* **Note 8** for alternative, quantitative assays to determine liver transduction efficiency).

4. Notes

1. Concentrated E1-deficient adenovirus stocks, between 5×10^{11} to 5×10^{12} particles/mL are stable for at least 2 yr when stored in this buffer at $-80°C$ (M. Bowe and S. Forestell, personal communication). However, storage of virus stocks more concentrated than 1×10^{13} particles/mL is not recommended because viral precipitation and/or inactivation could occur. Freeze/thawing of concentrated virus stocks more than three times is not recommended because loss in infectious viral titer can result (M. Bowe and S. Forestell, personal communication). Therefore, virus should be stored in small aliquots. We find that aliquots of 50 and 200 μL are useful for mouse studies.
2. Concentrated adenovirus stocks should appear milky and translucent. No aggregated particles or precipitate should be visible.
3. The diluted virus, when kept on ice, is stable for at least 4 h. An aliquot of the diluted virus may be tested to confirm vector particle titer and/or in vitro to confirm biological activity.
4. Animals are ready to be injected when they are lying quietly in the cage. Frequent observation during warming is needed to prevent overheating animals, which is recognized by wet hair predominantly on the face and head.
5. The recommended dose volume for iv administration is 10 mL/kg. Dose volumes are adjusted on an individual animal body-weight basis to achieve the desired vector dose (particles/kg). For example, to achieve a dose of 3×10^{12} particles/kg in a 23 g mouse, 0.23 mL of a 3×10^{11} particle/mL vector solution is injected (*see* **Table 1**).
6. Ensure that the needle is in the vein, rather than subcutaneous in the tail, by injecting a small portion of the vector mixture, and carefully looking for liquid accumulation in the tail. Stop immediately in the event pressure is felt upon injection, the formation of a bleb, or other sign of leakage is noted.

A second attempt may be required at another site, above the first site, or in a different vein.

7. Peak transgene expression occurs 4 d to 2 wk following vector administration. A 1-wk time-point is recommended to assess liver-transduction efficiency and transgene-expression levels.

8. The β-gal immunohistochemical procedure described here is a qualitative assessment of liver-transduction and transgene-expression efficiency. To more accurately quantitate β-gal expression in the mouse livers, the Glacto-Light-Plus chemiluminescent assay (Tropix, Inc., Foster City, CA) can be used as described *(44)*. Assays designed to assess more quantitatively liver-transduction efficiency that do not measure transgene expression include direct quantitation of the vector DNA content in the liver, using Southern analysis (**Fig. 2B**; *[49]*), or quantitative polymerase chain reaction (PCR) of genomic DNA directed at viral sequences *(25,44)*.

9. Using Southern analysis, a vector dose of 1.5×10^{13} vector particles/kg results in approx 10–20 vector copies per cell (**Fig. 2B**), whereas a five-fold lower vector dose displays 5–7 vector copies per cell *(46)*. To prepare the vector copy-number control standards, the amount of adenoviral DNA equivalent to one vector copy per cell of mouse liver genomic DNA is calculated as follows. The mouse genome size is approx 6×10^9 bps, whereas the adenoviral genome is 3.6×10^4 bp. Therefore, a ratio of the mouse genome size to the adenovirus genome size indicates that 1.67×10^5 less adenoviral DNA is required in order to be equivalent to that in a mouse cell. If 10 µg of mouse genomic DNA is used per sample, then $10,000$ ng/1.67×10^5 equals 0.0598 ng or 60 pg of DNA. Therefore, one vector copy per cell in 10 µg of genomic DNA is equivalent to 60 pg of adenoviral DNA. A similar calculation can be performed if plasmid DNA is used as the copy-number control, by recalculating the ratio of genome size.

Acknowledgments

The authors thank Donna Goldsteen for the contribution of the mouse tail vein injection protocol and Fig. 1, Drs. Mark Bowe and Sean Forestell for the adenovirus storage stability data, Dr. Russette Lyons for contribution of Table 1, and Drs. Russette Lyons and J. Andrew Bristol for critical review of the manuscript.

References

1. Connelly, S. (1999) Adenoviral vectors for liver-directed gene therapy. *Curr. Opin. Mol. Ther.* **1**, 565–572.

2. Vrancken Peeters, M. J., Perkins, A. L., and Kay, M. A. (1996) Method for multiple portal vein infusions in mice: quantitation of adenovirus-mediated hepatic gene transfer. *Biotechniques* **20**, 278–285.

3. Gerolami, R., Cardoso, J., Bralet, M. P., Cuenod, C. A., Clement, O., Tran, P. L., and Brechot, C. (1998) Enhanced in vivo adenovirus-mediated gene transfer to rat hepatocarcinomas by selective administration into the hepatic artery. *Gene Ther.* **5**, 896–904.

4. Cichon, G., Schmidt, H. H., Benhidjeb, T., Loser, P., Ziemer, S., Haas, R., et al. (1999) Intravenous administration of recombinant adenoviruses causes thrombocytopenia, anemia and erythroblastosis in rabbits. *J. Gene Med.* **1**, 360–371.

5. Maron, D. J., Tada, H., Moscioni, A. D., Tazelaar, J., Fraker, D. L., Wilson, J. M., and Spitz, F.R. (2001) Intra-arterial delivery of a recombinant adenovirus does not increase gene transfer to tumor cells in a rat model of metastatic colorectal carcinoma. *Mol. Ther.* **4**, 29–35.

6. Smith, T. A., Mehaffey, M. G., Kayda, D. B., Saunders, J. M., Yei, S., Trapnell, B. C., et al. (1993) Adenovirus mediated expression of therapeutic plasma levels of human factor IX in mice. *Nat. Genet.* **5**, 397–402.

7. Yang, Y., Nunes, F. A., Berencsi, K., Furth, E. E., Gonczol, E., and Wilson, J. M. (1994) Cellular immunity to viral antigens limits E1-deleted adenoviruses for gene therapy. *Proc. Natl. Acad. Sci. USA* **91**, 4407–4411.

8. Yang, Y., Jooss, K. U., Su, Q., Ertl, H. C., and Wilson, J. M. (1996) Immune responses to viral antigens versus transgene product in the elimination of recombinant adenovirus-infected hepatocytes in vivo. *Gene Ther.* **3**, 137–144.

9. Yang, Y. and Wilson, J. M. (1995) Clearance of adenovirus-infected hepatocytes by MHC class I-restricted CD4+ CTLs in vivo. *J. Immunol.* **155**, 2564–2570.

10. Berkner, K. L. (1988) Development of adenovirus vectors for the expression of heterologous genes. *Biotechniques* **6**, 616–629.

11. Gorziglia, M. I., Kadan, M. J., Yei, S., Lim, J., Lee, G. M., Luthra, R., and Trapnell, B. C. (1996) Elimination of both E1 and E2 from adenovirus vectors further improves prospects for in vivo human gene therapy. *J. Virol.* **70**, 4173–4178.

12. Gorziglia, M. I., Lapcevich, C., Roy, S., Kang, Q., Kadan, M., Wu,V., et al. (1999) Generation of an adenovirus vector lacking E1, E2a, E3, and all of E4 except open reading frame 3. *J. Virol.* **73**, 6048–6055.

13. Morral, N., O'Neal, W., Zhou, H., Langston, C., and Beaudet, A. (1997) Immune responses to reporter proteins and high viral dose limit duration of expression with adenoviral vectors: comparison of E2a wild type and E2a deleted vectors. *Hum. Gene Ther.* **8**, 1275–1286.

14. Amalfitano, A. and Chamberlain, J. S. (1997) Isolation and characterization of packaging cell lines that coexpress the adenovirus E1, DNA polymerase, and preterminal proteins: implications for gene therapy. *Gene Ther.* **4**, 258–263.

15. Armentano, D., Zabner, J., Sacks, C., Sookdeo, C. C., Smith, M. P., St George, J. A., et al. (1997) Effect of the E4 region on the persistence of transgene expression from adenovirus vectors. *J. Virol.* **71**, 2408–2416.

16. Andrews, J. L., Kadan, M. J., Gorziglia, M. I., Kaleko, M., and Connelly, S. (2001) Generation and characterization of E1/E2a/E3/E4-deficient adenoviral vectors encoding human factor VIII. *Mol. Ther.* **3**, 329–336.

17. Brough, D. E., Hsu, C., Kulesa, V. A., Lee, G. M., Cantolupo, L. J., Lizonova, A.,

and Kovesdi, I. (1997) Activation of transgene expression by early region 4 is responsible for a high level of persistent transgene expression from adenovirus vectors in vivo. *J. Virol.* **71**, 9206–9213.

18. Lusky, M., Grave, L., Dieterle, A., Dreyer, D., Christ, M., Ziller, C., et al. (1999) Regulation of adenovirus-mediated transgene expression by the viral E4 gene products: requirement for E4 ORF3. *J. Virol.* **73**, 8308–8319.

19. Christ, M., Louis, B., Stoeckel, F., Dieterle, A., Grave, L., Dreyer, D., et al. (2000) Modulation of the inflammatory properties and hepatotoxicity of recombinant adenovirus vectors by the viral E4 gene products. *Hum. Gene Ther.* **11**, 415–427.

20. Morral, N., O'Neal, W., Rice, K., Leland, M., Kaplan, J., Piedra, P. A., et al. (1999) Administration of helper-dependent adenoviral vectors and sequential delivery of different vector serotype for long-term liver-directed gene transfer in baboons. *Proc. Natl. Acad. Sci. USA* **96**, 12816–12821.

21. Sandig, V., Youil, R., Bett, A. J., Franlin, L. L., Oshima, M., Maione, D., et al. (2000) Optimization of the helper-dependent adenovirus system for production and potency in vivo. *Proc. Natl. Acad. Sci. USA* **97**, 1002–1007.

22. Maione, D., Wiznerowicz, M., Delmastro, P., Cortese, R., Ciliberto, G., La Monica, N., and Savino, R. (2000) Prolonged expression and effective readministration of erythropoietin delivered with a fully deleted adenoviral vector. *Hum. Gene Ther.* **11**, 859–868.

23. Balague, C., Zhou, J., Dai, Y., Alemany, R., Josephs, S. F., Andreason, G., et al. (2000) Sustained high-level expression of full-length human factor VIII and restoration of clotting activity in hemophilic mice using a minimal adenovirus vector. *Blood* **95**, 820–828.

24. Kochanek, S., Schiedner, G., and Volpers, C. (2001) High-capacity 'gutless' adenoviral vectors. *Curr. Opin. Mol. Ther.* **3**, 454–463.

25. Reddy, P. S., Sakhuja, K., Ganesh, S., Yang, L., Kayda, D., Brann, T., et al. (2002) Sustained human factor VIII expression in hemophilia a mice following systemic delivery of a gutless adenoviral vector. *Mol. Ther.* **5**, 63–73.

26. Gao, G. P., Yang, Y., and Wilson, J. M. (1996) Biology of adenovirus vectors with E1 and E4 deletions for liver-directed gene therapy. *J. Virol.* **70**, 8934–8943.

27. Tripathy, S. K., Svensson, E. C., Black, H. B., Goldwasser, E., Margalith, M., Hobart, P. .M., and Leiden, J. M. (1996) Long-term expression of erythropoietin in the systemic circulation of mice after intramuscular injection of a plasmid DNA vector. *Proc. Natl. Acad. Sci. USA* **93**, 10876–10880.

28. Loser, P., Jennings, G. S., Strauss, M., and Sandig, V. (1998) Reactivation of the previously silenced cytomegalovirus major immediate-early promoter in the mouse liver: involvement of NFkappaB. *J. Virol.* **72**, 180–190.

29. De Geest, B., Van Linthout, S., Lox, M., Collen, D., and Holvoet, P. (2000) Sustained expression of human apolipoprotein A-I after adenoviral gene transfer in C57BL/6 mice: role of apolipoprotein A-I promoter, apolipoprotein A-I introns, and human apolipoprotein E enhancer. *Hum. Gene Ther.* **11**, 101–112.

30. Connelly, S., Gardner, J. M., McClelland, A., and Kaleko, M. (1996) High-level tis-

sue-specific expression of functional human factor VIII in mice. *Hum. Gene Ther.* **7**, 183–195.

31. Pastore, L., Morral, N., Zhou, H., Garcia, R., Parks, R. J., Kochanek, S., et al. (1999) Use of a liver-specific promoter reduces immune response to the transgene in adenoviral vectors. *Hum. Gene Ther.* **10**, 1773–1781.

32. Bristol, J. A., Gallo-Penn, A., Andrews, J., Idamakanti, N., Kaleko, M., and Connelly, S. (2001) Adenovirus-mediated factor VIII gene expression results in attenuated anti-factor VIII-specific immunity in hemophilia A mice compared with factor VIII protein infusion. *Hum. Gene Ther.* **12**, 1651–1661.

33. Jaffe, H. A., Danel, C., Longenecker, G., Metzger, M., Setoguchi, Y., Rosenfeld, M. A., et al. (1992) Adenovirus-mediated in vivo gene transfer and expression in normal rat liver. *Nat. Genet.* **1**, 372–378.

34. Hackett, N. R., El Sawy, T., Lee, L. Y., Silva, I., O'Leary, J., Rosengart, T. K., and Crystal, R. G. (2000) Use of quantitative TaqMan real-time PCR to track the time-dependent distribution of gene transfer vectors in vivo. *Mol. Ther.* **2**, 649–656.

35. Kay, M. A., Landen, C. N., Rothenberg, S. R., Taylor, L. A., Leland, F., Wiehle, S., et al. (1994) In vivo hepatic gene therapy: complete albeit transient correction of factor IX deficiency in hemophilia B dogs. *Proc. Natl. Acad. Sci. USA* **91**, 2353–2357.

36. Connelly, S., Mount, J., Mauser, A., Gardner, J. M., Kaleko, M., McClelland, A., and Lothrop, C. D., Jr. (1996) Complete short-term correction of canine hemophilia A by in vivo gene therapy. *Blood* **88**, 3846–3853.

37. Fang, B., Wang, H., Gordon, G., Bellinger, D. A., Read, M. S., Brinkhous, K. M., et al. (1996) Lack of persistence of E1- recombinant adenoviral vectors containing a temperature-sensitive E2A mutation in immunocompetent mice and hemophilia B dogs. *Gene Ther.* **3**, 217–222.

38. Brann, T., Kayda, D., Lyons, R. M., Shirley, P., Roy, S., Kaleko, M., and Smith, T. (1999) Adenoviral vector-mediated expression of physiologic levels of human factor VIII in nonhuman primates. *Hum. Gene Ther.* **10**, 2999–3011.

39. Andrews, J. L., Shirley, P. S., Iverson, W. O., Sherer, A. D., Markovits, J. E., King, L., et al. (2002) Evaluation of the duration of human factor VIII expression in nonhuman primates following systemic delivery of an adenoviral vector. *Hum. Gene Ther.* **13**, *1331–1336.*

40. Michou, A. I., Santoro, L., Christ, M., Julliard, V., Pavirani, A., and Mehtali, M. (1997) Adenovirus-mediated gene transfer: influence of transgene, mouse strain and type of immune response on persistence of transgene expression. *Gene Ther.* **4**, 473–482.

41. Barr, D., Tubb, J., Ferguson, D., Scaria, A., Lieber, A., Wilson, C., et al. (1995) Strain related variations in adenovirally mediated transgene expression from mouse hepatocytes in vivo: comparisons between immunocompetent and immunodeficient inbred strains. *Gene Ther.* **2**, 151–155.

42. Fields, P. A., Armstrong, E., Hagstrom, J. N., Arruda, V. R., Murphy, M. L., Farrell, J. P., et al. (2001) Intravenous administration of an E1/E3-deleted adenoviral vector induces tolerance to factor IX in C57BL/6 mice. *Gene Ther.* **8**, 354–361.

43. Wen, X. Y., Bai, Y., and Stewart, A. K. (2001) Adenovirus-mediated human endostatin gene delivery demonstrates strain-specific antitumor activity and acute dose-dependent toxicity in mice. *Hum. Gene Ther.* **12**, 347–358.

44. Smith, T., Idamaknati, N., Kylefjord, H., Rollence, M., King, L., Kaloss, M., et al. (2002) In vivo hepatic adenoviral gene delivery occurs in a CAR independent fashion. *Mol. Ther.* **5**, 770–779.

45. Bristol, J. A., Shirley, P., Idamakanti, N., Kaleko, M., and Connelly, S. (2000) In vivo dose threshold effect of adenovirus-mediated factor VIII gene therapy in hemophiliac mice. *Mol. Ther.* **2**, 223–232.

46. Connelly, S., Gardner, J. M., Lyons, R. M., McClelland, A., and Kaleko, M. (1996) Sustained expression of therapeutic levels of human factor VIII in mice. *Blood* **87**, 4671–4677.

47. Tao, N., Gao, G. P., Parr, M., Johnston, J., Baradet, T., Wilson, J. M., et al. (2001) Sequestration of adenoviral vector by Kupffer cells leads to a nonlinear dose response of transduction in liver. *Mol. Ther.* **3**, 28–35.

48. Mittereder, N., March, K. L., and Trapnell, B. C. (1996) Evaluation of the concentration and bioactivity of adenovirus vectors for gene therapy. *J. Virol.* **70**, 7498–7509.

49. Connelly, S., Smith, T. A., Dhir, G., Gardner, J. M., Mehaffey, M. G., Zaret, K. S., et al. (1995) In vivo gene delivery and expression of physiological levels of functional human factor VIII in mice. *Hum. Gene Ther.* **6**, 185–193.

5

Delivery of DNA to Lung Airway Epithelium

Daniel J. Weiss

1. Introduction

Delivering exogenous DNA or genes directly to the lung airways offers a unique and appealing opportunity for specifically targeting gene expression to airway and alveolar epithelium. A large body of literature and experience supports the feasibility of this approach. However, airway-directed gene delivery is not as simple as was originally anticipated. The lung has evolved both physical and immunologic barriers that can hinder effective transduction of epithelial cells *(1–3)*. Much current work in lung gene therapy is directed toward overcoming the inflammatory and immune responses provoked by gene-transfer vectors while simultaneously maximizing vector delivery and subsequent gene expression.

Despite these obstacles, successful gene delivery and expression can be accomplished in lung airways. Of currently available viral and nonviral vectors, recombinant adenovirus is the most effective at transducing airway and alveolar epithelium in vivo *(4)*. A variety of recombinant adenovirus vectors have been created with deletions of different endogenous viral-coding regions, including deletion of all viral sequences, i.e., high-capacity vectors. Moreover, adenovirus vectors can be modified by modifying knob and fiber sequences to express different binding determinants as well as by coating the entire adenovirus surface with nonantigenic molecules such as polyethylene glycol (PEG). These surface-modified vectors may interact with different determinants on epithelial cell surfaces and thus be more or less effective for transducing epithelium than unmodified

From: *Methods in Molecular Biology, vol. 246:*
Gene Delivery to Mammalian Cells: Vol. 2: Viral Gene Transfer Techniques
Edited by: W. C. Heiser © Humana Press Inc., Totowa, NJ

vectors. However, for the purpose of this chapter, general delivery techniques and approaches will be considered similar for all varieties of adenovirus vectors.

Successful transduction of airway epithelium is in large part dependent on having a method that provides adequate access of vector to the airways. There are several techniques currently utilized, as summarized in **Table 1.** As noted,

Table 1
General Methods for Adenovirus Vector Delivery to Airway and Alveolar Epithelium

A. Nasal application Deposition of droplets on animal's nares Nasal cannulation and instillation	**Advantages** Technically easy Only brief anesthesia required **Disadvantages** Only works well in obligate nose breathers (i.e., rodents) Poor control of how much vector actually reaches the lower airways
B. Direct intratracheal instillation Transtracheal puncture Tracheal cannulation Bronchoscopy	**Advantages** Direct delivery of vector to trachea and lower airways Bypasses upper airway clearance and defense mechanisms Can target specific regions (lobes or subsegments) in lung, particularly with bronchoscopy **Disadvantages** Heterogenous distribution of vector in lung if instillation is directed to the trachea Limited vector distribution if a specific region is targeted Multiple instillations are required to achieve wide distribution Predominantly targets larger, proximal conducting airways Requires moderate to heavy anesthesia
C. Inhalation of aerosolized vector	**Advantages** More uniform distribution of vector throughout the lung Can target more distal airway and alveolar epithelium depending on aerosol particle size Anesthesia not necessarily required **Disadvantages** Only a small percent of the vector reaches the lower airways Much of the vector is deposited on the surface of the aerosolization device, in the oropharynx, and in the trachea

each method has advantages and disadvantages and use must be tailored towards the specific goal for which gene delivery is being performed. Additionally, there are several adjunct methods that have been described to improve adenovirus vector delivery and effectiveness of epithelial transduction including use of surfactant, perfluorochemical liquid, EGTA, $CaPO_4$, and detergent *(5–13)*. These techniques enhance gene expression and have been easy to utilize in animal models. Use of adjunct approaches is promising but as yet untried for improving attempts at clinical lung gene therapy.

The approaches described below provide detailed methods for the various techniques and should allow investigators to successfully transduce airway and alveolar epithelium.

2. Materials

2.1. Adenovirus Vector

1. Recombinant adenovirus vectors can be obtained from a variety of sources including National Institutes of Health (NIH)-sponsored Vector Core Labs at the Universities of Pennsylvania, Florida, Pittsburgh, and North Carolina, among others. Vectors may also be obtained from several commercial sources including Genzyme Inc. (Framingham MA) and InvivoGen (San Diego, CA).
2. Store vectors at $-70°C$ and thaw just prior to use. It is best to store virus stocks in small aliquots because repeated freeze thaws will inactivate the virus. Prolonged storage at $-70°C$ can also result in loss of activity and many investigators replenish virus stocks after 3–6 mo (*see* **Notes 1** and **2**).
3. Virus storage media includes salt solutions (i.e., normal saline, phosphate-buffered saline [PBS], or Tris-based buffers) or salt solutions containing cryogenic preserving agents such as glycerol, sucrose, or glucose (*see* **Notes 3** and **4**).
4. Virus is generally diluted in normal saline or PBS for instillation into airways. Administration of at least 10^7–10^{10} plaque-forming units (PFUs) (approx 10^8–10^{11} virus particles) is generally required to detect gene expression in lung. The total volume administered depends on the animal being utilized and is discussed further in **Subheading 3**.

2.2. Anesthetic Agents

Animals are anesthetized using agents appropriate for the species and in accordance with institutional IACUC/IRB policies. The goal of anesthesia is adequate sedation and pain control with maintenance of spontaneous breathing and ability to keep the airways patent.

Choice of anesthetic agent depends on the method of accessing the airway. For nasal deposition of vectors in rodents, short-acting inhalational agents such as halothane and isoflurane are commonly used. Insertion of a catheter or bronchoscope into airways generally requires longer-lasting anesthesia utilizing combinations of agents appropriate for each species. Inhalation of aerosolized vector can be done either without anesthetics, for example, rodent total body or nose-only exposure chambers, or with anesthesia when administering aerosolized vector by facemask to nonhuman primates. Representative anesthetic and analgesic agents are listed in **Table 2**.

2.3. Adjunct Agents

Several adjunct agents have been utilized for delivering vectors to lung airways. The rationale and technique for use of each agent is discussed in the methods section.

1. Perfluorochemical liquids: perfluorooctylbromide (perflubron, *LiquiVent*[R], Alliance Pharmaceuticals, San Diego, CA), FC-75 (ACROS Organics, St. Louis, MO), perfluorodecalin (ACROS Organics) *(8,9,17–19)*.
2. Surfactants: Survanta (Ross Laboratories, Columbus, OH), Exosurf[R] (Burroughs Wellcome, Research Triangle Park, NC) *(5–7)*.
3. EGTA (ethylene glycol-*bis*-β-amino-ethyl ether)-*N,N,N′,N′*-tetraacetic acid, Sigma) *(10,11)*.
4. $Ca_3(PO_4)_2$ (Sigma) *(12)*.
5. Detergent: polidocanol (Sigma) *(13)*.

2.4. Other Materials

Several other types of equipment can be utilized for delivery of genes to the lung airways. A brief (nonexhaustive) listing includes:

1. Angiocatheters (Deseret Medical Inc., Sandy, UT).
2. Endotracheal tubes (Mallinckrodt, St. Louis, MO).
3. Bronchoscopes (Olympus, Melville, NY; Pentax, Orangeburg, NY).
4. Face mask devices (Hudson Respiratory Care, Inc., Temecula, CA).
5. Nebulizers (OPTINEB, Air Liquide, Paris, France; MiniHEART, Vortran, Sacramento, CA; PB Raindrop, Puritan–Bennett Corp., Carlsbad, CA).

3. Methods

3.1. Nasal Application

Rodents are obligate nose breathers, thus a substantial proportion of material (i.e., vector solution) deposited on or in the nares will be inhaled and can be deposited in both upper and lower airways. This can be a convenient and noninvasive way of administering vector. If done carefully, reproducibility between animals is generally good.

Table 2
Representative Anesthetic and Analgesic Agents

Agents	Species	Dose[a]	Route[a,b]	Duration of effect[c]
Inhalation agents				
Halothane (Halothane, Halocarbon Labs)	Rodent	To effect	Inhaled	30–60 s
Isoflurane (IsoFLo, Abbott Laboratories)	Rodent	To effect	Inhaled	30–60 s
Injectable agents				
Buprenophine (Buprenex, Reckitt and Colman Products)	Rodent, rabbit, dog, primate	0.01–0.1 mg/kg	im/ip/sc/iv	3–12 h
Ketamine (Ketajet, Phoenix Pharm Inc.)	Rodent, rabbit	20–50 mg/kg	im/iv	30–45 min
	Primate	5–45 mg/kg	im/iv	30–45 min
Tiletimine/Zolazepam (Telazol, Fort Dodge Animal Health)	Mouse	100–160 mg/kg	im/ip	30–45 min
	Rabbit	20–40 mg/kg	im/ip	30–45 min
	Primate	8 mg/kg	im	30–45 min
Tribromoethanol 25% solution (Avertin, Aldrich)	Rodents	1.25–2.5 mg/kg	im/ip	30–60 min
	Dog	400–600 mg/kg	iv	30–60 min
Xylazine (Xyla-ject, Phoenix Pharm Inc.)	Rodent, rabbit	4–8 mg/kg	im/ip	30–60 min
	Primate	1–2 mg/kg	im	30–60 min

[a]Dosing and route of administration should be reviewed with the local veterinarian and IACUC/IRB.
[b]im, intramuscular; ip, intraperitoneal; sc, subcutaneous; iv, intravenous.
[c]Duration of effects are approximate.

1. Place a small bell jar or similar glass chamber in a fume hood or other well-ventilated area.
2. Saturate a small piece of gauze or paper towel with an inhalational anesthetic agent (i.e., halothane or isoflurane) and place it into the jar. Place another layer of gauze or paper over this to prevent direct contact of the animal with the anesthetic (*see* **Note 5**).
3. Place the animal in the jar and observe it until it stops moving. This should take no more than 10–20 s. Quickly remove the animal from the jar. Leaving the animal in the jar for too long can result in excess anesthesia exposure and death from respiratory arrest.
4. Using a pipetting device, deposit small aliquots of the virus solution (5–10 µL) directly on the nares. As the animal inhales each aliquot, immediately place another aliquot on the nares. Alternatively, inject the virus solution directly into the nares. The total volume administered should generally not exceed 50–100 µL for mice and 100–200 µL for rats. If the animal begins to wake during the procedure, briefly place it back in the anesthesia chamber (*see* **Note 6**).

3.2. Transtracheal Puncture

This is a brief and simple operative procedure, used most frequently in rodents, that quickly and easily provides direct access to the trachea *(8)*.

1. Place and secure anesthetized animals in supine position on a suitable operating surface. Provide adequate warmth with use of heated operating surface or overhead lamp. Apply petroleum jelly, boric acid ointment, or other suitable protective agent to the eyes.
2. Pull back the head to extend and expose the ventral surface of the neck. Shave the hair and clean with alcohol or iodine solution (*see* **Note 7**).
3. Using a sterile scalpel blade, make a shallow vertical incision into the skin on the ventral surface of the neck.
4. Using sterile forceps or hemostats, gently separate the subcutaneous fat and soft tissue overlying the trachea. Gently separate the strap muscles over the trachea to expose the trachea.
5. Puncture the trachea just below the larynx with a small gauge syringe (insulin syringe or 25–30 G 1/2-in needle) and instill the vector-containing solution directly into the trachea. Suggested total instillate volumes are 50–100 µL for mice and 100–200 µL for rats. Larger amounts can be tolerated if necessary. Larger animals (rabbits, dogs, primates) can tolerate up to several milliliters of fluid (*see* **Note 8**). The rate of instillation depends on the total volume instilled and on the species. Larger animals (rabbits, dogs, primates) generally tolerate more rapid instillation (i.e., squirting the vector

in). In smaller animals (rats, mice) instill the total volume over 15–60 s (*see* **Note 8**).

6. Once the installation is completed, suture the skin over the trachea and allow the animal to recover from anesthesia. It is not necessary to suture the underlying musculature or soft tissue in smaller animals.
7. Administer analgesic (e.g., buprenorphine) postoperatively according to institutional veterinary and institutional review board policies.

3.3. Direct Tracheal Cannulation

This brief and simple nonoperative procedure provides quick and easy direct access to the trachea. Generally easy to perform in medium sized and large animals, this approach can be technically challenging in smaller animals (i.e., rodents).

1. Place and secure anesthetized animals in supine position or with head slightly elevated on a suitable operating surface. Provide adequate warmth by use of heated operating surface or overhead lamp. Apply petroleum jelly, boric acid ointment, or other suitable protective agent to the eyes.
2. Using a closed curved hemostat, forceps, or retractor, gently pull down the mandible until the vocal cords are visible. It can be helpful to backlight the vocal cords by holding a light source up to the animal's neck.
3. Insert a catheter through the vocal cords into the airway, positioning the inserted tip in the upper or middle portion of the trachea. Several different materials can be used to catheterize the trachea, including plastic angiocatheters, plastic or polypropylene tubing, or, in larger animals (rabbit, dog, primate), an endotracheal tube. Note that several types of catheter materials have been shown to inactivate adenovirus vector infectivity and should be avoided (*see* **Note 9**). These include stainless steel, nitinol, and polycarbonate *(14,15)*. In larger animals (rabbit, dog, primate), the catheter can be directed into one of the mainstem bronchi if desired. This is technically more difficult in rodents.
4. Instill the vector solution directly through the catheter. Suggested total instillate volumes are 50–100 µL for mice and 100–200 µL for rats. Larger amounts can be tolerated if necessary. Larger animals (rabbits, dogs, primates) can tolerate up to several milliliters of fluid (*see* **Note 8**). The rate of installation depends on total volume instilled and on the species. Larger animals generally tolerate more rapid instillation (i.e., squirting the vector in). In smaller animals (rats, mice) instill the total volume over 15–60 s (*see* **Note 8**).
5. After instillation of the virus-containing solution, flush the catheter with a small amount of virus vehicle or buffer to rinse any remaining virus.
6. Remove the catheter and allow the animal to recover from anesthesia.

3.4. Bronchoscopy in Larger Animals

This approach is most commonly utilized in larger animals (dogs, sheep, primates) and has also been utilized in clinical trials of gene delivery to human lung airways. A distinct advantage of bronchoscopy is the ability to visualize the specific lung segment or subsegment into which vector is instilled.

1. Place anesthetized nonhuman primates in seated or semi-recumbent position. Other animals can be in recumbent or supine position.
2. Monitor heart rate and blood pressure throughout the procedure. Monitor oxygen saturation by continuous pulse oximetry. Entering a bronchoscope into the airways can often result in significant desaturation (<90%) with accompanying arrythmia. If so, supplemental oxygen should be administered to maintain saturations greater than 90%.
3. Enter the bronchoscope through the nares or through the mouth and pass the scope through the vocal cords into the airways. The bronchoscope can then be positioned at any desired location in the airways (*see* **Notes 10** and **11**).
4. Administer the vector through the instillation port of the bronchoscope. Vector should be diluted in a minimum volume of 1 mL; smaller volumes can result in a significant portion of the vector solution not being effectively expelled from the tip of the bronchoscope. After instillation, flush the bronchoscope with 1–2 mL of virus vehicle, saline, or air to help expel any remaining vector solution.
5. Remove the bronchoscope and monitor the animal until it recovers from anesthesia. Keep primates upright until they recover fully from anesthesia.

3.5. Adjunct Methods for Direct Intratracheal Instillation

These methods have been demonstrated to enhance lung gene expression in both rodents and rabbits (transtracheal puncture, direct tracheal cannulation) as well as in primates (bronchoscopy). Adjunct methods have not yet been utilized in clinical trials of lung gene delivery.

3.5.1. Perfluorochemical Liquid

Several perfluorochemical (PFC) liquids, including FC-75 and perfluorooctylbromide (perflubron, *LiquiVent*[R]), have been successfully used to enhance gene delivery and expression in airway and alveolar epithelial cells *(8,9,17–19)*. Perfluorodecalin has been used to augment delivery of antibiotics to lung but has not yet been utilized for vector delivery *(20)*. Other PFC liquids could also be conceivably be used for vector delivery.

1. Oxygenate the PFC liquid by bubbling oxygen into it (1–2 L/min) for approx 2 min prior to use *(8)*. The total amount of PFC liquid instilled depends on the experimental goals. In general, 10–15 cc/kg body weight in

rodents and approx 40 cc/kg body weight in primates will fill the entire lung. Approximately 10–20 cc will fill a single lobe in medium-sized primates, i.e., macaques. One advantage of bronchoscopic administration of PFC liquid is that the PFC liquid fluid meniscus can be directly visualized (*see* **Note 12**).

2. Draw the PFC liquid into a separate syringe from that used for the vector. Glass or polypropylene is preferable because PFC liquids may dissolve other materials. PFC liquids are immiscible with aqueous solutions and therefore will not mix with the vector solution.

3. Immediately after instilling the vector (*see* **Subheadings 3.2.**, **3.3.**, and **3.4.**), administer the PFC liquid through the same route (transtracheal puncture site, tracheal cannula, or bronchoscope) utilized for the vector solution.

   a. In smaller animals, administer a small test dose of approx 50 µL at first and observe the animal's breathing pattern. If breathing regularly, administer the rest of the dose as a continuous moderate rate or bolus infusion (250–500 µL over 1–2 min). Too rapid an instillation can result in respiratory arrest (*see* **Notes 13** and **14**).

   b. Administer larger doses initially in the larger animals, for example 20–30 cc in primates. These doses can be administered more rapidly (i.e., over 5–10 s) than in smaller animals. Once the test dose has been evaluated, instill the remainder of the PFC liquid by bolus or continuous instillation.

4. Provide supplemental oxygen during the instillation and anesthesia recovery periods. In rodents, deliver oxygen (5–10 L/min) through a cannula directed towards the animal's nares. In primates, utilize a face mask or face cone (1–2 L/min) (*see* **Note 15**).

3.5.2. Surfactant

Several studies have documented enhanced adenovirus vector delivery and expression in rodent airway and alveolar epithelium when vector was mixed with the synthetic surfactant Survanta (either 10 or 25 mg/mL solutions *[5,6]* or a 50% solution *[7,21]*).

1. Mix the vector with the surfactant solution just prior to instillation.
2. Instill the mixture directly into the lung using any of the above routes (*see* **Subheadings 3.1.–3.4.**). Use a total volume of 2–4 cc/kg body weight of vector-surfactant solution *(5,7,21)* (*see* **Note 16**).

3.5.3. Other Substances: EGTA, CaPO$_4$, Detergent

Several studies have documented enhanced adenovirus vector mediated expression in airway epithelium when vector was mixed with EGTA *(10,11)* or with calcium phosphate *(12)*. EGTA probably enhances delivery by opening tight junctions between epithelial cells, allowing enhanced access of vector to binding sites on basolateral membranes of epithelial cells. Calcium phosphate may increase nonspecific uptake of the vector-$CaPO_4$ precipitate at the apical surface of airway epithelium. Additionally, pretreatment with the detergent polidocanol has been shown to increase adenovirus-mediated gene expression in nasal epithelium *(13)*. As with EGTA, application of polidocanol is thought to increase tight junction permeability between epithelial cells. This approach has not yet been utilized in the lower airways. Use these compounds in conjunction with any of the methods described above (*see* **Subheadings 3.1.–3.4.**).

1. EGTA: Dilute vector into 10–12 m*M* EGTA in hypotonic buffer (10 m*M* HEPES, osmolality 40 mmol/kg) and instill into the lungs using any of the aforementioned methods.
2. Calcium phosphate: Dilute vector to the desired concentration and volume in Eagle's Minimal Essential Media (Sigma), which contains 1.8 m*M* Ca^{+2} and 0.86 m*M* PO_4^{-3}. Add $CaCl_2$ until the final Ca^{+2} concentration is 12 m*M* (*see* **Note 17**). Mix the solution by gentle agitation, incubate at room temperature for 30 min, and then directly administer to the nares or lung using any of the aforementioned methods.
3. Polidocanol: Apply 0.1% polidocanol in PBS to the nasal epithelium for 1 h. Rinse away the solution and apply vector.

3.6. Aerosol Administration

There is no standardized approach to nebulization of vector solutions and administration of the resulting aerosol. A number of different methods and nebulization devices have been described *(11,23–26)*. A reasonable goal of the nebulization equipment is to produce homogenously sized aerosol droplet particles. Droplets with aerodynamic radii of 0.5–5 µm will generally deposit in distal airway and alveolar spaces, whereas droplets with radii of 5–10 µm will deposit in medium-sized and larger airways *(27)*. Larger droplets will tend to deposit in the nares, mouth, or pharynx and are less likely to result in useful gene transfer to lower airways. In addition to particle size, density, and hygroscopicity, other factors, including mode of inhalation and depth and rate of breathing, influence where droplets are deposited *(27)*. Administration of aerosolized vector solutions is inefficient, even with the most sophisticated currently available devices and techniques. It is estimated that at best only 10–30% of aerosolized particles reach the lower airways *(27)*.

For small animals (rodents), whole body or nose-only exposure chambers can be used *(11,24)*. In these devices, the aerosol is pumped into the chamber and the animals breathe it in freely. In general, nose-only exposure chambers are more confining and stressful to the animals and require either anesthesia or a period of acclimation to the chamber prior to aerosol exposure. For larger animals and humans, aerosol can be delivered directly through a tracheal catheter or through a face mask or mouthpiece device *(19,26,27)*.

Two sample protocols for aerosolized vector administration are detailed below.

1. Rodents *(24)*: Place rodents in a plexiglass total body exposure chamber (plexiglass box, 350 cm^3). Place 5 mL of vector suspension (4 × 10^9 PFU/mL in PBS) in the cup of a Puritan Bennett Raindrop nebulizer and generate aerosol by running dry compressed air at 8–10 L/min through the nebulizer. Pump the visible mist directly into the exposure chamber using tubing running directly from the nebulizer to a hole in the chamber top. Continue the exposure until the vector solution runs dry and no further visible mist is observed.

2. Primates *(26)*: Anesthetize primates with ketamine or other suitable agent. Place a snug-fitting veterinary mask with outflow ports to create an air-tight seal around the muzzle. Block the nares with cotton and place a bite block between the teeth. Place 10 mL of vector solution (1.5 × 10^{11} PFU/mL in PBS with 1% sucrose) in the cup of a MiniHeart nebulizer and generate aerosol by running dry compressed air at 2 L/min through the nebulizer. Pump the aerosol directly from the nebulizer to the face mask using 1 inch diameter plastic tubing. Continue the exposure until the vector solution runs dry and no further visible aerosol is observed.

3.7. Vector Delivery to Injured or Diseased Lung

All of the aforementioned techniques have been primarily explored in normal lung. There is less experience with delivery of adenovirus vectors to injured or diseased lungs. In this setting, arguably the more relevant setting in which clinical gene transfer will be attempted, additional barriers in the form of increased mucins and inflammatory secretions, pulmonary edema, and altered airway anatomy can further impede delivery of vector to the epithelial cell surface *(28–30)*. Use of PFC liquid has been shown to enhance adenovirus-mediated gene expression in models of acute and chronic lung injury *(9,18)*. More recently, use of surfactant has also been shown to enhance adenovirus-mediated gene expression in acutely injured lung *(31)*. There is need to further optimize delivery techniques for injured lung.

4. Notes

1. Warming the adenovirus vector to 37°C for 30–40 min prior to instillation has been demonstrated to improve gene expression in brain tissue *(32)*. This approach has not yet been tested for lung.

2. There is a variety of experience with the longevity of viable virus stored at $-70°C$. In general, most virus samples are substantially inactivated after approx 5 freeze-thaw cycles. The investigator is advised to check virus viability and titer by transducing a convenient cell line at periodic intervals.

3. A decrease in the pH of the storage buffer to <8.0 can result in decreased active virus titer, regardless of presence of glycerol or glucose *(33)*. To avoid this, Tris adjusted to pH 8.0 may be a better storage medium than either saline or PBS. Moreover, exposure to dry ice, for example, shipping of frozen virus samples, can result in decreased pH of the virus solution and reduced activity. This can occur even if virus is in sealed cryovials. Placing the cryovial in a heavy-duty plastic freezer bag (Kapak heavy duty sealable bag, VWR, Atlanta, GA) with low permeability to CO_2 can reduce the change in pH *(33)*.

4. There is no particular advantage to either glycerol, sucrose, or glucose as a cryopreservative. Commonly used solutions are 10–50% glycerol, and 10–30% glucose or sucrose. However, if glycerol is used in the storage buffer, it is important to dilute the virus stock such that the final glycerol concentration is less than 5% and preferably less than 1% in the final solution to be administered to the lung. Higher glycerol concentrations can cause significant coating of airway and alveolar surfaces resulting in respiratory distress and death, particularly in smaller animals (rodents).

5. Avoid getting the animals fur or skin wet with the anesthetic agent. This can result in overexposure and respiratory arrest.

6. Rodents can be exposed to anesthesia multiple times in order to complete nasal vector application. However, the fewer the exposures the less chance of inadvertent anesthesia-related toxicity or respiratory arrest.

7. A convenient method of securing a rodent's head in an extended position is to use a rubber band hooked on the animals incisors. Do not overextend the neck as this will cut off airflow through the trachea. Using rubber bands to secure the limbs in an extended position to the operating surface will also help expose the neck.

8. Too rapid instillation or bolus instillation of too large a volume can cause respiratory arrest. This can often be transient and the animal will spontaneously recover after several seconds. Occasionally gentle chest compressions will stimulate resumption of respiratory efforts.

9. Flushing the catheter and syringe tips with albumin (1% bovine serum albumin [BSA] or 5% human serum albumin) prior to vector instillation or, alternatively, adding albumin (0.1% BSA) to the vector solution has been demonstrated to reduce inactivation of virus resulting from exposure to catheter materials *(14)*.

10. Always use a bite-block if the bronchoscope is entered through the mouth.

11. A spray device can be used in conjunction with bronchoscopy *(16)*. Pass the device through the instillation port of the bronchoscope and position it at the main carina or in one of the mainstem bronchi. Instill the vector through the spray device.

12. An advantage to using perfluorooctylbromide (perflubron, *LiquiVent*[R]) is that it is radio-opaque. This allows direct radiographic visualization (chest X-rays) of lung filling with the PFC liquid.

13. Smaller animals are especially prone to transient respiratory arrest with too large an initial PFC volume or with too rapid an initial instillation. The respiratory arrest is usually transient and respiratory rates and efforts recover after several seconds.

14. A visible increase in respiratory excursions and occasionally respiratory rate will be observed after PFC liquid instillation. This reflects alteration in lung mechanics resulting from change of a gas-filled to fluid-filled lung. Breathing pattern will return to normal over several hours as the PFC liquid evaporates and is exhaled.

15. Smaller animals (rodents) are particularly sensitive to hypothermia and hypoxemia during the PFC liquid instillation period and during the anesthesia recovery period. During anesthesia recovery, place animals in a covered cage which is warmed by heat lamps and oxygenated (5–10 L/min) through a cannula directed towards the animals' nares.

16. As change in body position during surfactant administration helps to distribute the surfactant throughout the lungs *(22)*, a more detailed approach has been described in which the vector-surfactant solution is administered with animals (rat) rotated in different positions *(7,21)*. Divide the vector-surfactant solution into four aliquots of 200–250 µL each. Place a small-gauge angiocatheter into the upper part of the rat trachea. Circumferentially compress the thorax to achieve a forced exhalation and instill one aliquot of the vector-surfactant solution. Relax the compression and allow the animal to breathe. Rotate the animal 90° between instillation of each aliquot and repeat at approx 5 min intervals *(7,16)*.

17. Use a concentrated stock of $CaCl_2$, for example, a 1 *M* stock solution, to minimize change in volume of the final vector solution. Always add the $CaCl_2$ last to maximize inclusion of the vector in the $CaPO_4$ precipitate.

References

1. Otake, K., Ennist, D. L., Harrod, K., and Trapnell, B. C. (1998) Nonspecific inflammation inhibits adenovirus-mediated pulmonary gene transfer and expression independent of specific acquired immune responses. *Hum. Gene. Ther.* **9**, 2207–2222.

2. Look, D. C. and S. L. Brody (1999) Engineering viral vectors to subvert the airway defense response. *Am. J. Respir. Cell Mol. Biol.* **20**, 1103–1106.

3. Bromberg, J. S., Debruyne, L. A., and Qin, L. (1998) Interactions between the immune system and gene therapy vectors: Bidirectional regulation of response and expression. *Adv. Immunol.* **69**, 353–409.

4. Factor, P. (2001) Gene therapy for acute diseases. *Mol. Ther.* **4**, 515–524.

5. Jobe, A. H., Ueda, T., Whitsett, J. A., Trapnell, B. C., and Ikegami, M. (1996) Surfactant enhances adenovirus-mediated gene expression in rabbit lungs. *Gene Ther.* **3**, 775–779.

6. Katkin, J. P., Husser, R. C., Langston, C., and Welty, S. E. (1997) Exogenous surfactant enhances the delivery of recombinant adenoviral vectors to the lung. *Hum. Gene. Ther.* **8**, 171–176.

7. Factor, P., Saldias, F., Ridge, K., Dumasius, V., Zabner, J. H., Jaffe, A., et al. (1998) Augmentation of lung liquid clearance via adenovirus-mediated transfer of a Na, K-ATPase betal subunity gene. *J. Clin. Invest.* **102**, 1421–1430.

8. Weiss, D. J., Strandjord, T. P., Jackson, J. C., Clark, J. G., and Liggitt, D. (1999) Perfluorochemical liquid-enhanced adenoviral vector distribution and expression in lungs of spontaneously breathing rodents. *Exp. Lung Res.* **25**, 317–333.

9. Weiss, D. J., Strandjord, T. P., Liggitt, D., and Clark, J. G. (1999) Perflubron enhances adenoviral-mediated gene expression in lungs of transgenic mice with chronic alveolar filling. *Hum. Gene. Ther.* **10**, 2287–2293.

10. Wang, G., Zabner, J., Deering, C., Launspach, J. Shao, J., Bodner, M., Jolly, D. J. et al. (2000) Increasing epithelial junction permeability enhances gene transfer to airway epithelia in vivo. *Am. J. Respir. Cell Mol. Biol.* **22**, 129–138.

11. Chu, Q., St. George, J. A., Lukason, M., Cheng, S. H., Scheule, R. K., and Eastman, S. J. (2001) EGTA enhancement of adenovirus-mediated gene transfer to mouse tracheal epithelium in vivo. *Hum. Gene. Ther.* **12**, 455–467.

12. Fasbender, A., Lee, J. H., Walters, R. W., Moninger, T. O., Zabner, J., and Welsh, M. J. (1998) Incorporation of adenovirus in calcium phosphate precipitates enhances gene transfer to airway epithelia in vitro and in vivo. *J. Clin. Invest.* **102**, 184–193.

13. Parsons, D. W., Grubb, B. R., Johnson, L. G., and Boucher, R. C. (1998) Enhanced in vivo airway gene transfer via transient modification of host barrier properties with a surface-active agent. *Hum. Gene. Ther.* **9**, 2661–2672.

14. Marshall, D. J, Palasis, M., Lepore, J. J., and Leiden, J. M. (2000) Biocompatability of cardiovascular gene delivery catheters with adenovirus vectors: an important determinant of the efficiency of cardiovascular gene transfer. *Mol. Ther.* **1**, 423–429.

15. Tsui, L. V., Zayek, N., Frey, D., Mello, C., Banik, G., Falotico, R., and McArthur, J. G. (2001) Stability of adenoviral vectors following catheter delivery. *Mol. Ther.* **3**,122–125.

16. Harvey, B. G., Leopold, P. L., Hackett, N. R., Grasso, T. M., Williams, P. M., Tucker, A. L., et al. (1999) Airway epithelial CFTR mRNA expression in cystic fibrosis patients after repetitive administration of a recombinant adenovirus. *J. Clin. Invest.* **104**, 1165–1166.

17. Lisby, D. A., Ballard, P. L., Fox, W. W., Wolfson, M. R., Shaffer, T. H., and Gonzales, L. W. (1997) Enhanced distribution of adenovirus-mediated gene transfer to lung parenchyma by perfluorochemical liquid. *Hum. Gene. Ther.* **8**, 919–928.
18. Weiss, D. J., Bonneau, L., and Liggitt, D. (2001) Use of perfluorochemical liquid allows earlier detection and use of less adenovirus vector for gene expression in normal lung and enhances gene expression in acutely injured lung. *Mol. Ther.* **3**, 734–745.
19. Weiss, D. J., Baskin, G. B., Shean, M. K., Blanchard, J. L., and Kolls, J. K. (2002) Use of perflubron to enhance lung gene expression: safety and initial efficacy studies in non-human primates. *Mol. Ther.* **5**, 8–15.
20. Franz, A. R., Rohlke, W., Franke, R. P., Ebsen, M., Pohlandt, F., and Hummler, H. D. (2001) Pulmonary administration of perfluorodecaline-gentamycin and perfluorodecaline-vancomycin emulsions (2001). *Am. J. Resp. Crit. Care Med.* **164**, 1595–1600.
21. Weiss, D. J., Mutlu, G. M., Bonneau, L., Mendez, M., Wang, Y., Dumasius, V., and Factor, P. (2002) Comparison of surfactant and perfluorochemical liquid enhanced adenovirus-mediated gene transfer in normal rat lung. *Mol. Ther.* **6**, 43–49.
22. Spragg, R. G. (2001) Surfactant replacement therapy. *Clin. Chest Med.* **21**, 531–541.
23. Sene, C., Bout, A., Imler, J. L., Schultz, H., Willemot, J. M., Hennebel, V., et al. (1995) Aerosol-mediated delivery of recombinant adenovirus to the airways of non-human primates. *Hum. Gene. Ther.* **6**, 1587–1593.
24. Katkin, J. P., Gilbert, B. E., Langston, C., French, K., and Beaudet, A. L. (1995) Aerosol delivery of β-galactosidase adenoviral vector to the lungs of rodents. *Hum. Gene. Ther.* **6**, 985–995.
25. Bellon, G., Michel-Calemard, L., Thouvenot, D., Jagneaux, V., Poitevin, F., et al. (1997) Aerosol administration of a recombinant adenovirus expressing CFTR to cystic fibrosis patients: a phase I clinical trial. *Hum. Gene. Ther.* **8**, 15–25.
26. McDonald, R. J., Lukason, M. J., Raabe, O. G., Canfield, D. R., Burr, E. A., et al. (1997) Safety of airway gene transfer with Ad2/CFTR2: aerosol administration in the nonhuman primate. *Hum.Gene.Ther.* **8**,411–422.
27. Muir, D. F. C. (1991) Particle deposition, in *The Lung: Scientific Foundations* (Crystal, R.G., ed.), Raven Press Ltd., New York, NY, pp. 1839–1843.
28. van Heeckeren, A., Ferkol, T., and Tosi, M. (1998) Effects of bronchopulmonary inflammation induced by pseudomonas aeruginosa on adenovirus-mediated gene transfer to airway epithelial cells in mice. *Gene Ther.* **5**, 345–351.
29. Stern, M., Caplen, N. J., Browning, J. E., Griesenbach, U., Sorgi, F., Huang, L., et al. (1998) The effect of mucolytic agents on gene transfer across a CF sputum barrier in vitro. *Gene Ther.* **5**, 91–98.
30. Kitson, C., Angel, B., Judd, D., Rothery, S., Severs, N. J., Dewar, A., et al. (1998) The extra-and intracellular barriers to lipid and adenovirus-mediated pulmonary gene transfer in native sheep airway epithelium. *Gene Ther.* **6**, 534–546.
31. Factor, P., Mendez, M., Mutlu, G. M., and Dumasius, V. (2002) Acute hyperoxic lung injury does not impede adenoviral-mediated alveolar gene transfer. *Am. J. Resp. Crit. Care Med.* **165**, 521–526.

32. Kossila M., Jauhianen, S., Laukkanen, M. O., Lehtolainen, P., Jaaskelainen, M., Turunen, P., et al. (2002) Improvement in adenoviral gene transfer efficiency after preincubation at 37°C *in vitro* and *in vivo*. *Mol. Ther.* **5**, 87–93.

33. Nyberg-Hoffman, C. and Aguilar-Cordova, E. (1999) Instability of adenoviral vectors during transport and its implication for clinical studies. *Nature Med.* **5**, 955–957.

6

Delivery of DNA to Pulmonary Endothelium Using Adenoviral Vectors

Paul N. Reynolds

1. Introduction

Delivery of genes to the pulmonary vascular endothelium is a rational approach for the investigation and potential therapy of pulmonary vascular diseases. Furthermore, in view of the exposure of this vascular bed to the entire cardiac output, this technique could be used as an efficient basis to achieve systemic delivery of secreted factors. The attraction of direct gene delivery to endothelium for the therapy of vascular disease has been especially heightened in the last couple of years in view of the new discoveries concerning the genetic basis of primary pulmonary hypertension (PPH) *(1)*. In brief, mutations in the bone morphogenetic protein receptor type 2 (BMPR2, a member of the transforming growth factor-β [TGF-β] family of receptors) gene have been found in many patients with familial PPH. Subsequent in vitro studies have confirmed an association between BMPR2 mutations and abnormal proliferative responses in pulmonary endothelial and smooth-muscle cells *(2)*. Other TGF-β signaling pathways may also be involved in this process, and the mechanisms involved may also have relevance for the more common cases of pulmonary vascular disease secondarily associated with chronic airways obstruction, connective tissue diseases, and perhaps HIV infection. Additionally, new evidence is emerging concerning the role of the vasculature in the pathogenesis of emphysema. Vascular endothelial growth factor (VEGF) and its receptors appear to be key for

From: *Methods in Molecular Biology, vol. 246:*
Gene Delivery to Mammalian Cells: Vol. 2: Viral Gene Transfer Techniques
Edited by: W. C. Heiser © Humana Press Inc., Totowa, NJ

the maintenance of normal alveolar architecture: VEGF receptor inhibition causes emphysema in rats, and both VEGF and its receptor levels are reduced in the lungs of patients with emphysema *(3,4)*. Aside from vascular disease *per se*, the pulmonary bed is frequently the site of metastatic spread of malignancy, and once this occurs conventional therapy options are most often inadequate. Thus, pulmonary vascular targeting could also be envisaged as an approach for metastatic disease, at least in a locoregional context.

From a practical delivery standpoint, the pulmonary vasculature is an attractive target owing to a "first-pass" effect; it is the first major vascular bed seen by substances injected into the peripheral venous system. As discussed in Volume 1, many nonviral vector formulations have a natural propensity to localize to the pulmonary vasculature after systemic injection, and some of this effect may be on a first-pass/aggregation basis. Nevertheless, subsequent studies have shown that there is a degree of specificity to the pulmonary localization of some of these agents, because substantial accumulation in lung vessels is seen even after peripheral artery injection *(5)*. The receptors involved in this phenomenon are yet to be characterized. Despite the pulmonary localization of cationic lipids, however, the actual level of transduction of endothelium remains inefficient, likely owing to endosomal degradation of DNA, which is a frequent limitation with nonviral vectors. Thus, in view of the greater endosomal escape properties of adenoviral vectors, the development of these agents for endothelial gene delivery is rational. Direct comparisons between adenoviral and nonviral systems have been limited, but in one such study, using pulmonary artery catheters in rats, adenovirus was approx 100-fold more efficient than liposomes for the delivery of the β-galactosidase (β-gal) reporter gene *(6)*.

Although adenovirus may be delivered via catheter and brief occlusion to segments of the pulmonary vascular bed, such an approach is not practical for widespread pulmonary vascular gene delivery. Ideally, the vector agent should be injectable into a peripheral vein, and then specifically target the pulmonary endothelium. However, systemic administration of adenovirus in this way results in the vast majority of vector being taken up by the liver—most by Kupffer cells and degraded (a saturable phenomenon) *(7)* and the bulk of the remainder by hepatocytes, with very little transduction occurring in pulmonary vasculature. Thus, the development of specifically targeted Ad vectors is required. We have worked toward the development of such an approach, and have found that both transductional and transcriptional targeting is required for optimal specificity.

1.1. Transductional Targeting

Enhancement of pulmonary endothelial cell infection with a systemically administered Ad vector requires a strategy to improve the binding of Ad to the target cells. In general, Ad vector targeting has been approached using a variety of bi-specific adapter molecules, or by direct genetic modification of the Ad capsid proteins. Molecular adapters may be derived entirely by recombinant techniques, using either anti-Ad single-chain antibodies or soluble Ad receptor (i.e., coxsackievirus and adenovirus receptor [CAR]) to attach the adapter to the virus, linked to a variety of recombinant ligands or single-chain antibodies to bind to the target cell. Despite the apparent technical elegance of recombinant adapters or genetic capsid modification, however, none of these approaches have achieved targeting fidelity in vivo. In fact, the only approach shown to have a degree of targeting fidelity in this setting is the use of a bi-specific, chemically cross-linked, antibody conjugate. This targeting adapter was constructed using the Fab fragment of an anti-Ad antibody to an antibody with established pulmonary vascular targeting capabilities (an antibody, MAb 9B9, against angiotensin-converting enzyme [ACE]) *(8,9)* (**Fig. 1**). Thus, in keeping with the practical nature of the current discussion, this is the approach that is expanded on here.

In the last 5–6 yr there have been many publications describing antibody-based Ad targeting approaches. In addition to ACE targeting, these conjugates have achieved targeting to a range of cellular targets including tumor markers such as epidermal growth factor receptor (EGFR) *(10)*, ErbB2, CC49 *(11)*, and EpCAM *(12)*. Targeting has also been achieved to dendritic cells (DCs) via CD40 *(13)*. In each of these cases, attachment to adenovirus was achieved using a monoclonal antibody (MAb) directed against the knob domain of Ad serotype 5. The rationale for choosing an antibody against the knob domain was based on the knowledge that it is this region that binds to the primary Ad receptor, CAR *(14,15)*. Thus, the concept is to simultaneously ablate the native tropism for CAR while imparting new tropism via the cell-binding antibody. Curiel and co-workers developed such an antibody by immunizing mice with recombinant Ad5 fiber and boosting with recombinant knob. The development of MAbs is beyond the current discussion, but, in essence, a panel of antibodies was generated and screened for neurtralizing ability against adenoviral infection. As a result of this process, an antibody identified as 1D6.14 was characterized and taken forward into targeting studies *(16)*.

1.2. Transcriptional Targeting

Despite the improvements in pulmonary vascular gene delivery seen with the transductional targeting approach, it was clear that the vast majority of vector particles were still being taken up by the liver (as determined by quantitative real-time polymerase chain reaction [PCR] of viral DNA extracted from organs 90 min after injection *(8)*, or more recently by observing the biodistribution of radiolabeled virus, the latter unpublished observations). It is likely that Kupffer

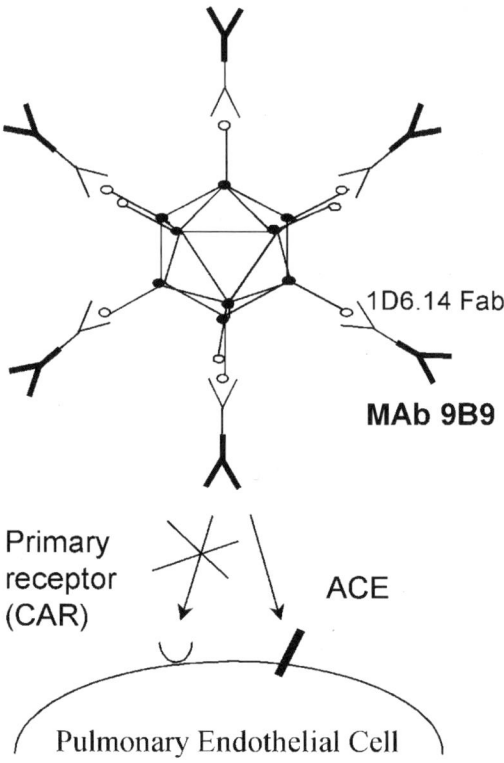

Fig. 1. Retargeting schema. The basic concept of using an adaptor to redirect Ad infection from CAR to ACE is shown.

cells predominate in this, because we were able to show some reduction in hepatocyte transduction. Even so, the residual level of transgene expression in hepatocytes was clearly more than desirable. This finding raised concerns about the stability of the bond between the conjugate and vector in the circulation in vivo. However, it has since been shown that even genetic ablation of the regions of Ad responsible for CAR recognition does not significantly reduce liver uptake *(17,18)*. Mutation of the integrin recognition sequence in the penton base has given variable results *(19)*. More recent data suggests liver uptake may also in part be mediated by interactions between the virus and heparan sulfates *(20)*. Thus, as our conjugate approach only blocks CAR recognition, "background" liver uptake could be occurring by these other means. Clearly, further modifications to the overall targeting approach are needed to fully ablate the natural hepatic tropism of the vector. While these studies are being pursued, we felt it would be worthwhile to investigate the possible synergy that could be obtained by combining transductional targeting with transcriptional targeting using an endothelial-specific promoter. We focused our studies on an evaluation of the promoter for the vascular endothelial growth factor receptor (VEGFR) type 1 (or flt-1) *(21)*. Increasingly, other candidate endothelial-specific promoters are being describe, for instance the promoter for pre-proendothelin-1 *(22)*. Whichever promoter may be selected should of course have minimal activity in hepatocytes.

A detailed description of the cloning steps involved in the construction of a new recombinant Ad vector is beyond the scope of the current discussion. However, in recent years Ad vector construction has been greatly facilitated through the approach of recombination in bacteria to first derive Ad genomes in a plasmid context, which can then be used to generate virus by simple plasmid transfection into appropriate cell lines *(23)*. Several commercial kits and plasmids are now available that even further facilitate this approach. Thus, the cloning of endothelial-specific promoters to drive the transgene of choice in an Ad vector is a relatively straightforward procedure. Problems may arise however, because certain promoters can substantially lose their specificity when placed in an Ad genome, thus the exercise is not entirely trivial. Fortunately, the flt-1 promoter was found to retain good endothelial specificity in this setting. For candidate promoters that do not perform so well, flanking the expression cassette with insulating sequences may help *(24)*.

Once the flt-1 promoter-Ads were constructed, they were combined with Fab-9B9 and injected into rats as before. We found that there was no diminution in the level of transgene expression in pulmonary vasculature compared to the earlier studies where we had used the strong but nonspecific cytomegalovirus (CMV) promoter. Conversely, there was a 10,000-fold drop in hepatocyte transgene expression, thus the overall selectivity ratio of lung vs liver improved by at least 300,000-fold compared to an entirely untargeted system. At high vector doses (likely levels that saturate the phagocytosis of vector by Kupffer cells), we were able to achieve 200-fold more transgene expression in the lung (per mg protein) compared to the liver. We believe that the proof-of principle of the dual-targeting approach not only has relevance for pulmonary vasculature, but the concept could be extended to other cell targets where complementary cell ligands and promoters have been described.

1.3. Other Approaches

Another infection-enhancing strategy that has achieved a degree of utility in vivo is the combination of Ad with nonviral vectors *(25)*. The preparation of nonviral vectors *per se* is described elsewhere. For reasons that are not entirely clear, it has been shown that pre-injection of liposomes via peripheral vein enhance Ad vector transduction of the lungs provided the Ad injection is given at least 5 min after the lipid. This does not seem to be an aggregation issue, because simple mixing of the reagents and co-injection actually reduces transduction. Further work is required to ascertain the mechanisms involved, although this approach could be a useful means to an end in certain settings.

1.4. Conclusion

In summary, the technique discussed here has been able to establish some basic concepts (and limitations) concerning the systemic vascular targeting of Ad vectors, both generally and for the pulmonary endothelium specifically. In the current form, the antibody conjugates are rather crude and further optimizations can be considered. Efforts to generate a single-chain antibody from the 9B9 hybridoma are currently underway, with a view to developing a recombinant targeting adapter. Future efforts with genetically modified vectors may ultimately allow the direct genetic incorporation of a single-chain antibody into the Ad capsid, thus enabling the development of a single component system that may be more amenable to clinical application. For best utility, these approaches could also be combined with strategies to reduce the inflammatory responses caused by Ad vectors (e.g., by using "helper dependent" or "gutless" constructs).

2. Materials

1. Anti-knob MAb 1D6.14 is not commercially available but may be obtained by contacting Dr. David T. Curiel, Director, Division of Human Gene Therapy, University of Alabama, Birmingham, AL.
2. Anti-ACE MAb 9B9 is available from Mono-ACE (River Forest, IL). The antibody we used for ACE targeting is the monoclonal 9B9, developed by Sergei Danilov and colleagues *(26)* (*see* **Note 1**).
3. CHO-ACE cells (clone 2C2), from Mono-ACE. This cell line permits testing ACE-targeting in vitro prior to in vivo assays. In other instances (i.e., for targeting to endothelial markers other than ACE), cultured endothelial cells such as human umbilical vein endothelial cells (HUVECs) might be used.
4. Purified human kidney ACE (Mono-ACE).
5. 0.1 *M* carbonate buffer, pH 9.6.
6. Phosphate-buffered saline (PBS) and PBS-Tween (PBS with 0.5 mL/L Tween-20).
7. AP-conjugated anti-mouse antibody (Jackson ImmunoResearch, West Grove, PA).
8. 4-nitrophenyl phosphate (pNPP) (Sigma, St. Louis, MO).

9. Fab Immunopure preparation kit (or similar) (Pierce, Rockford IL).
10. Borate buffer: 0.015 M Na Borate, 0.15 M NaCl, pH 8.5.
11. Centrifuge concentrator tubes (e.g., Ultrafree-0.5 Centrifugal Filter Units with a molecular weight cut-off of 10,000 and 30,000 from Millipore, Bedford, MA).
12. *N*-succinimidyl 3-(2-pyridyldithio) propionate (SPDP) (either from Sigma or Pierce).
13. Dimethyl sulfoxide (DMSO).
14. Dithiothreitol (DTT) (Bio-Rad, Hercules, CA).
15. Sodium acetate buffer, 1 M, pH 4.5.
16. Cytochrome C (from horse heart, Sigma).
17. Desalting columns (e.g., PD10 columns, from Pharmacia, Uppsala, Sweden).
18. Fast performance liquid chromatography (FPLC) system with a Superose 12 column (Pharmacia).
19. Standard apparatus for Western blotting and transfer (Bio-Rad).
20. Acrylamide, TEMED, ammonium persulphate (APS), and Tris buffers (for making 10% polyacrylamide gels, or ready-made gels can be used).
21. PVDF membrane (Bio-Rad).
22. Nonfat dry milk (NFDM) (Bio-Rad): 1% solution in PBS (to be used as blocker).
23. Bio-Rad detergent compatible (DC) protein assay kit and spectrophometer or enzyme-linked immunosorbent assay (ELISA) plate reader for measuring protein concentration.
24. Recombinant Ad5 knob protein (6-His tagged). Direct inquiries about this reagent to the Division of Human Gene Therapy, University of Alabama at Birmingham.
25. Nickel-Alkaline Phosphatase conjugate (Qiagen Inc., Valencia, CA).
26. Alkaline Phosphatase Immun-Blot Colorimetric Assay Kit (Bio-Rad).
27. Adenoviral vector of choice. This should be purified by cesium chloride centrifugation, then dialysed into storage buffer (10% glycerol/5 mM HEPES in PBS) before being stored in aliquots at −80°C.
28. Luminometer (for luciferase assays if using the luciferase reporter gene).
29. Rats approx 6 wk of age (Harlan Sprague Dawley, Indianapolis, IN).
30. Rodent restrainer (Harvard Instruments, Holliston, MA).
31. Insulin syringes, 0.5 mL (with no dead space and 28 G needle).
32. Heat lamp.
33. CO_2 chamber (for sacrificing rats).
34. Scissors, forceps, etc., for dissection.
35. 20 G Jelco intravenous cannula and giving set (Johnson and Johnson, Arlington, TX).

3. Methods

3.1. Preparation of Fab Fragments from the 1D6.14 MAb

1. Digest at least 10 mg of 1D6.14 MAb into Fab fragments using the Pierce Immunopure Fab preparation kit. Digest the antibody with papain, then purify the Fab fragments from the Fc fragments entirely in accord with the manufacturer's instructions (*see* **Note 2**).
2. Concentrate the purified Fab fragments to 10 mg/mL using an Ultrafree-0.5 Centrifugal Filter Unit with a molecular weight cut-off of 10,000 and exchange the buffer into borate buffer using the concentrator tubes in accord with the manufacturer's instructions.
3. Measure the protein concentration using the Bio-Rad protein assay kit according to the manufacturer's instructions (*see* **Note 3**).

3.2. Preparation of MAb 9B9

Ensure targeting MAb (e.g., 9B9) is concentrated to 10 mg/mL using centrifuge concentrator tube and exchange the buffer into borate buffer.

3.3. Addition of SPDP Linking Moieties to 1D6.14 and MAb 9B9

1. Dissolve the SPDP at a concentration of 2 mg/mL just before use in either 100% ethanol or DMSO (*see* **Note 4**). Prepare a fresh batch for each conjugation.
2. Combine 10 mg of MAb 9B9 (in 1 mL at a concentration of 10 mg/mL) with 62.5 µL of SPDP in a 1.5-mL microfuge tube, and 3.3 mg of Fab 1D6.14 (in 330 µL at a concentration of 10 mg/mL) with 62.5 µL of SPDP in a separate 1.5-mL microfuge tube (*see* **Note 5**).
3. Mix each tube thoroughly then shake at room temperature for 30 min (**Fig. 2**).

3.4. Crosslinking of 1D6.14-SPDP to MAb 9B9-SPDP

1. During the 30-min incubation (**Subheading 3.3., step 3**), equilibrate a desalting PD10 column with borate buffer by loading a total of 10 mL of buffer (in aliquots) and allowing it to flow through by gravity.
2. After the 30-min incubation, add 0.1 vol of 1 *M* sodium acetate buffer, pH 4.5, to the Fab-SPDP to lower the pH (*see* **Note 6**).
3. Add 1 mg of solid DTT to reduce the SPDP and incubate for 5 min at room temperature.
4. Add 5–10 crystals of cytochrome c (*see* **Note 7**) to the Fab-SPDP solution, then load onto the PD10 column.

Antibody Conjugation Schema

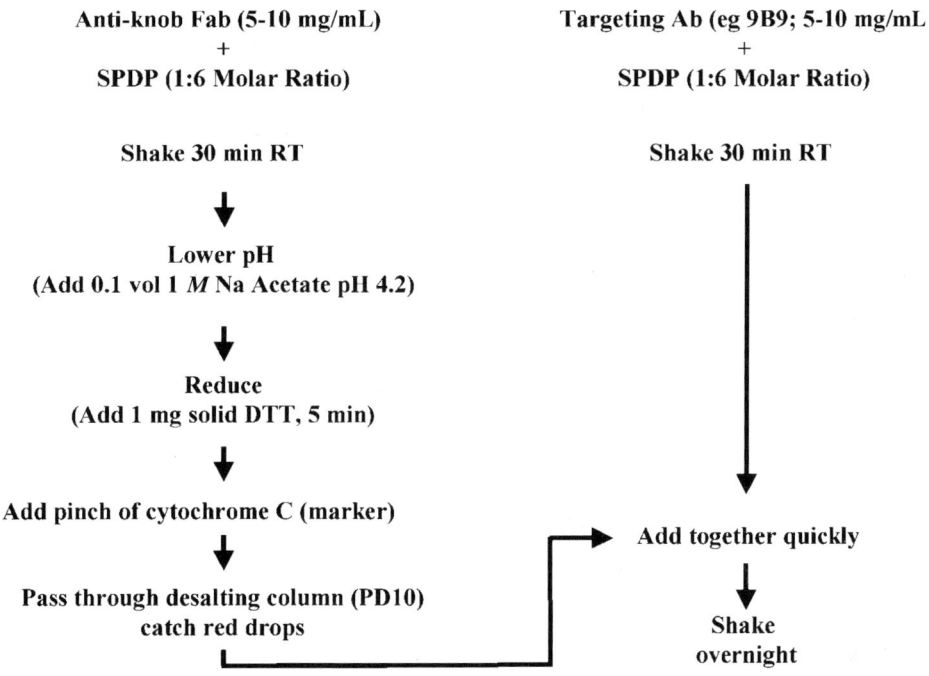

Fig. 2. Schema for construction of antibody conjugate.

5. Once the red solution has entered below the top frit, add 5 mL of borate buffer to wash the solution through the column. Carefully watch the red band move down the column, then collect only the red cytochrome c-labeled drops as they come through (typically less than 0.5 mL volume is collected).
6. Immediately add the Fab-SPDP to the MAb9B9-SPDP and shake overnight at room temperature (*see* **Note 8**).

3.5. Purification of Conjugate

1. Calibrate a Preparation Grade Superose 12 column (250–300 mL column volume) with molecular-weight markers, which may include small aliquots of unconjugated 1D6.14 Fab and MAb 9B9 using a Pharmacia FPLC system and borate buffer (*see* **Fig. 3**).
2. Purify the conjugate mixture from **Subheading 3.4.** using the same calibrated Superose 12 column. Set the flow rate at 0.3 mL/min (*see* **Note 9**).

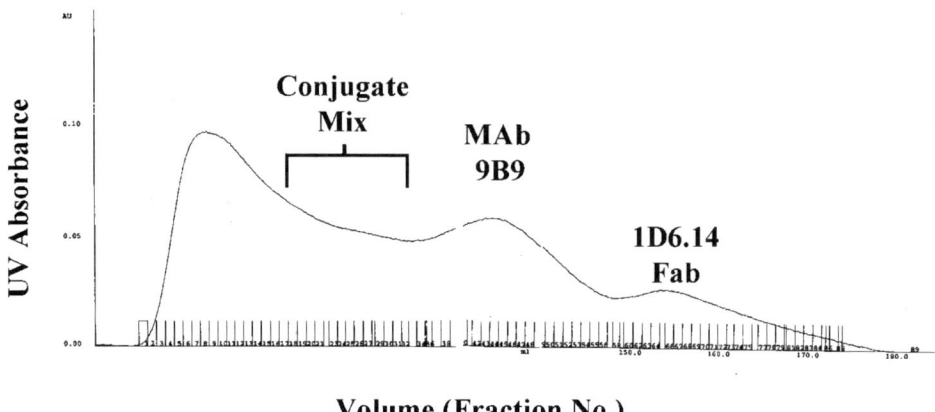

Volume (Fraction No.)

Fig. 3. A typical chromatography profile seen after gel filtration through preparation grade Superose 12 column using a FPLC system. A spectrum of conjugate molecules along with some residual free MAb 9B9 and free Fab is often obtained. The highest molecular-weight elements, corresponding to large aggregates, are avoided then fractions in the region above the free 9B9 are pooled and evaluated.

3.6. Validation of Conjugate Binding to Ad5 Knob

This section briefly summarizes our method to validate that the crosslinking in **Subheading 3.4.** gave rise to a functional conjugate capable of binding Ad5 knob. For additional details, see Reynolds et al. *(8)*.

1. Load 1 μg of conjugate and 1 μg each of 1D6.14 Fab and MAb 9B9 as positive and negative controls, respectively, into separate lanes of a 10% nondenaturing sodium dodecyl sulfate (SDS)/polyacrylamide gel and electrophorese at 160V for 1 h (**Fig. 4A**).
2. Transfer the proteins to PVDF membrane by standard techniques at 300 mA for 1–2 h or 30 mA overnight (**Fig. 4A**).
3. Block the membrane with 1% nonfat dried milk (NFDM), then probe with 6-His tagged recombinant Ad5 knob at 10 μg/mL (in 1% NFDM) for 1 h at room temperature (using just enough volume to cover the membrane, approx 8 mL for a 10 cm × 8 cm membrane). Wash the membrane three times for 5 min each time in 10 mL of PBS-Tween and detect the signal using nickel-alkaline phosphatase conjugate and the Alkaline Phosphatase Immuno-Blot Colorimetric Assay Kit according to the manufacturer's instructions. You should see positive signal for conjugate and 1D6.14 Fab, but not or MAb 9B9 (**Fig. 4B**).

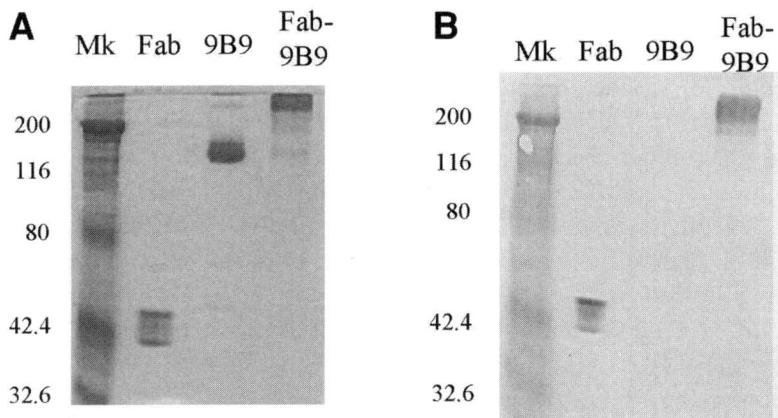

Fig. 4 **(A)** 10% nondenaturing polyacrylamide gel showing 1D6.14 Fab, MAb 9B9 and pooled conjugate. **(B)** Western blot probed with 6-His tagged Ad5 knob showing that the 1D6.14 Fab and the conjugate bind to the knob, but as expected, the parent MAb 9B9 does not. Adapted with permission from **ref. 8**.

3.7. Validation of Conjugate Binding to ACE

1. Coat Nunc maxisorb plates with purified human kidney ACE (50 µL/well, 400 ng/well) overnight at 4°C in 0.1 M carbonate buffer, pH 9.6.
2. The next day, wash the wells three times with PBS-Tween then block with 1% NFDM for 1 h at room temperature.
3. Serially dilute (at least three 10-fold dilutions) conjugate, 9B9 (as positive control) and 1D6.14 Fab as negative control) then add to the wells. Incubate at 4°C overnight.
4. Wash the wells three times with PBS-Tween. Add AP-conjugated anti-mouse antibody and develop color with pNPP according to the manufacturer's instructions. Determine the optical density (OD) in the wells using an ELISA plate reader. You should see binding for MAb 9B9 and conjugate, but not 1D6.14 Fab.

3.8. Validation of ACE-Targeted Gene Delivery In Vitro

1. Plate Chinese hamsterovary (CHO)-ACE cells and CHO cells at 25,000 cells/well into 24-well plates (prepare one plate for each cell line).
2. Twenty-four hours later make seven aliquots of adenoviral vector (containing the luciferase reporter gene) in microfuge tubes. Each aliquot should

contain 2×10^7 plaque forming units (PFU) of adenovirus and the volume should be no more than 20 μL per aliquot.

3. Serially dilute the conjugate. Beginning with 10 μL (approx 3 μg) of conjugate, make five serial 1:10 dilutions (thus having a total of six different conjugation concentrations) in borate buffer. Take 5 μL of each of the six dilutions and add to each of six adenovirus aliquots prepared in **step 2**. To the seventh adenoviral aliquot, add 5 μL of borate buffer to be an untargeted control. Incubate the mixtures for 45 min at room temperature.

4. Bring the volume of each tube to 700 μL with infecting medium (Dulbeco's modified Eagle's medium [DMEM]/F12 50:50 containing 2% fetal calf serum [FCS]).

5. Retrieve the CHO-ACE and CHO cells from tissue-culture incubator. Perform the next steps on triplicate wells, on both plates. To the first three wells on each plate, aspirate the medium and add 100 μL per well of infecting medium without virus to act as uninfected controls. To the next triplicate, aspirate the medium and replace with 100 μL per well of infecting medium containing untargeted virus. Continue this process through the six different virus-conjugate mixtures. Return the plates to the incubator and rock frequently to improve virus-to-cell contact. (A mechanical rocker may be used.)

6. After 1 h, aspirate the infecting medium from each well, then wash with PBS and add complete tissue-culture medium.

7. Twenty-four hours later, assay luciferase activity using the Promega Luciferase Assay System Kit according to the manufacturer's instructions. You should see at least a 10-fold enhancement in gene expression for one of the virus-conjugate mixtures on CHO-ACE cells, but not on CHO cells (*see* **Note 10**).

3.9. In Vivo Gene Delivery to Pulmonary Endothelium

1. Add 10 μg of conjugate (in a volume of 5–10 μL) to 10^{11} viral particles of vector (in a volume of 50–100 μL). Handle the vector using BSL 2 precautions. Adjust amounts up or down as necessary depending on the particular application but keep ratio the same. Incubate 45 min at room temperature (*see* **Note 11**).

2. Place rats in rodent restrainer and warm the tail gently using a heat lamp to achieve vasodilatation.

3. Dilute the adenoviral vector/conjugate mix up to 250 μL with sterile saline and gently draw into an insulin syringe.

4. Inject the solution into the lateral tail vein.

5. Return the rat to its cage. After injection, animals should be cared for using BSL 2 precautions.

3.10. Analysis of Gene Expression

3.10.1. Detection by Immunohistochemistry

1. At the desired time interval, sacrifice rats using CO_2. Peak gene expression with first-generation Ad vectors is typically 3–5 d after vector administration.
2. Make a transverse incision with scissors in the upper abdominal wall, then cut vertically through the diaphragm, taking care not to damage the lung. Continue the vertical cut to the level of the upper sternum (avoid damaging any blood vessels), then open the chest cavity by pulling the two sides of the ribcage forwards and outwards. Hold ribcage sections away from center of chest with surgical clips.
3. Introduce a Jelco 20 G intravenous cannula attached to a 10-mL syringe into the right ventricle by passing the tip directly through the myocardial wall. Aspirate blood to check position.
4. Fill the barrel of a 50-mL syringe and a giving set line with heparinized saline. Remove the 10-mL syringe and connect the cannula to the giving set line. Hold the 50-mL syringe barrel 30 cm above the rat in a retort stand; avoid introducing air into the system.
5. Make a slit in the left ventricle (with scissors) and perfuse the pulmonary vascular bed with 40 mL of heparinized saline by gravity, then 40 mL of 10% buffered formalin.
6. Cannulate the trachea. Identify and free up the trachea from the surrounding muscle and connective tissue by gentle dissection. Pass a ligature behind the trachea and tie it loosely in front. Make a small transverse incision in the anterior trachea, remove the cannula from the right ventricle (still connected to formalin), and place this in the trachea. Hold the cannula in place by tightening the ligature and inflate the lungs with 10% buffered formalin (approx 5 mL).
7. Remove the cannula. Tie off the trachea, then remove the heart and lungs en bloc and continue fixation in formalin in a suitable specimen container for 24 h.
8. The following day, cut lung lobes into vertical slices 2–3 mm thick, then process the tissue into paraffin by standard histopathological techniques. Embed the tissue in paraffin blocks, then section and stain using standard techniques (*see* **Note 12**).

3.10.2. Analysis of Luciferase Activity

1. At the desired time interval, sacrifice rats using CO_2. Peak gene expression with first generation Ad vectors is typically 3–5 d after vector administration.

2. Open the chest cavity, excise the lungs, and place them into 50-mL polypropylene centrifuge tubes.
3. Quick-freeze the tissue in an ethanol-dry ice bath. Frozen tissues may be stored at −80°C.
4. Cool a mortar and pestle in an ethanol/dry ice bath, then grind the tissue to a fine powder.
5. Assay the tissue using a Promega luciferase assay system kit according to the manufacturer's instructions and determine relative light units (RLU) using a luminometer.
6. Assay the protein content of the tissue using the Bio-Rad DC protein assay kit. Express luciferase activity as RLU/mg protein (*see* **Note 13**).

4. Notes

1. The anti-ACE antibody MAb 9B9 has been extensively evaluated for pulmonary targeting over the last 15 yr, including several studies in which it was conjugated to enzymes such as superoxide dismutase (SOD) and catalase *(28)*. Danilov has since developed an expanded panel of anti-ACE antibodies that also show good targeting characteristics, and these too may provide utility for targeted gene delivery *(29)*. Although the available data suggests ACE as the preferred target in many settings because of the pulmonary endothelial selectivity of the available antibodies, other antibodies may also have utility. For example, targeting to PECAM has been successful in improving the pulmonary gene delivery of nonviral vectors (although the actual cell types expressing the transgenes were not accurately determined) *(30)*. Although PECAM is useful for endothelial targeting, it lacks the pulmonary selectivity of the ACE approach. The success in the setting of nonviral systems may relate to a combined effect with the intrinsic pulmonary localizing properties of these agents. However, PECAM targeting has not been fully evaluated in a viral context.
2. When preparing to construct a conjugate using the approach discussed here, it is critical to ensure that enough of the appropriate reagents, 1D6.14 Fab, and the targeting antibody, 9B9, are on hand—preferably at least 3–4 mg of Fab and 10 mg of antibody. In our experience, attempts to make usable amounts of conjugate are futile if you are starting with only microgram quantities of targeting ligand. Simply scaling down the quantities suggested here does not work. Ideally, both Fab and targeting antibody should be concentrated to 10 mg/mL in borate buffer.
3. The rationale for using the Fab fragment rather than intact 1D6.14 is to reduce the theoretical risk of cross-linking adenoviruses. The use of intact

1D6.14 vs Fab in conjugate construction has not systematically been compared, however.

4. Over time, we have come to prefer DMSO because, in certain cases, ethanol seemed to increase protein precipitation during the reaction.

5. In separate reactions, we chemically crosslink SPDP with both MAb 9B9 and Fab 1D6.14 at a molar ratio of 6 SPDP:1 mAb or Fab, then conjugate MAb 9B9 to Fab 1D6.14 at a 1:1 molar ratio. The molecular weight of SPDP is 312; MAb 9B9 is approx 150,000; and Fab 1D6.14 is approx 50,000. Thus, 10 mg of MAb 9B9 would require 3.3 mg of Fab. One mg of MAb is equivalent to 0.00667 μmole (1 mg/150,000 mg/mmole) and the molar concentration of 10 mg/mL MAb is 0.0667 μmole/μL. The molar concentration of 2 mg/mL SPDP is 0.0064 umole/μL (2 mg/mL/312 mg/mmole). Because 6 moles of SPDP are required per mole of either MAb or Fab, combine 0.40 μmole of SPDP (6 × 0.0667 μmole) with 10 mg of MAb or 3.3 mg of Fab. Therefore, mix 10 mg of MAb (in 1 mL) with 62.5 μL of SPDP (0.40 μmole/0.0064 μmole/μL) in a 1.5-mL microfuge tube, and 3.3 mg of Fab with 62.5 μL of SPDP in a separate tube.

6. Lowering the pH protects the internal disulfide bonds of the Fab from the reducing effect of the DTT (so only the SPDP component is reduced).

7. The PD10 column will remove DTT and unbound SPDP. It is critical that the reduced Fab-SPDP is collected off the column in minimal volume (thus ensuring maximal concentration of reagents during the conjugation reaction). We have found the best way to do this is to add cytochrome C to the Fab-SPDP just before passing through the PD10 column. You only need enough to impart a red color to the solution. Then, when this is passed through the column, simply collect just the red drips (the cytochrome C is later removed during gel filtration).

8. Passing the MAb 9B9-SPDP through a PD10 column (without the reduction step) can also be done but this tends to increase reaction volume and does not improve the outcome of the conjugation reaction.

9. A typical chromatographic profile of the conjugate mix is shown in **Fig. 3**, where the fraction collector was set to collect 1-mL fractions. As can be seen, there is a heterogeneous mix of conjugate species. The fractions in the region of an "ideal" molecular weight of 200,000 (i.e., a 1:1 ratio of Fab and MAb 9B9) are pooled, concentrated, and carried forward into further validation studies. A reasonable final concentration of conjugate is 0.3 mg/mL. Minor variations in this basic technique such as using PBS in place of borate buffer may be tried. As stated, in our hands the protocol typically yields an amount of conjugate equivalent to 10–20% of the input weight of Fab and MAb. A balance must be struck between excess con-

jugation (with the increased production of large aggregates and precipitation) and inefficient conjugation (with poor yield and excess free MAb 9B9 and Fab).

10. For in vitro validation we used CHO-ACE cells (clone 2C2), a line developed by Danilov that is naturally low in CAR (owing to their CHO lineage), but stably transfected to express ACE at levels comparable to the pulmonary endothelium in vivo *(31)*. When titration is performed, one typically sees a bell-shaped curve of gene expression (e.g., as RLU if using a vector carrying the luciferase gene). Inadequate amounts of conjugate give low expression (owing to low targeting and lack of CAR on the cells), whereas excess conjugate also reduces gene transfer owing to an excess of free conjugate, which can act as a competitive inhibitor to the Ad-Fab-9B9 complexes. The optimized ratio of virus to conjugate can then be used as a guide for determining the amounts to be used in vivo, although in practice the issue of excess free conjugate is less likely to be an issue in vivo, because it has been shown that approx 1 mg of MAb 9B9 is needed to block ACE binding sites in rat lung in this setting. For other targeting conjugates, there may be cases where the cell lines available for in vitro evaluation may be naturally high in CAR, thus potentially confounding targeting assessment. This problem is simply overcome by pre-incubating cells with an excess of recombinant Ad5 knob to block the "background" CAR-mediated infection. For these experiments, cells are first incubated in unsupplemented DMEM/F12 medium containing 10 µg/mL Ad5 knob for 10 min at room temperature (in a volume of 100 µL per well of a 24-well plate). Vector/conjugate mix is then added, and cells are incubated for another 30 min at room temperature. Then infecting medium is aspirated, cells washed once with PBS, complete medium added, and the cells incubated at 37°C as usual for 24 h before assessment of transgene expression. In this way, any element of "CAR-independent" gene delivery is revealed.

11. In our rat studies, we have used up to 3×10^{11} adenoviral particles (thus at most 30 µg of conjugate). It has previously been shown by Danilov and coworkers that as much as 1 mg of 9B9 is needed to saturate the binding sites in rat lung *(32)*, thus we are well below saturation point. For this reason, we have also not sought to purify Ad-Fab-9B9 complexes from any potentially free Fab-9B9. Efforts to do so using gel filtration by others in our group (using different but similar conjugate approaches; personal communication) have typically reduced transduction efficacy.

12. The precise details of immunohistochemistry detection will vary depending on the transgene delivered, and a full detailed discussion of the various parameters that may need to be adjusted is beyond the scope of the current dis-

cussion. Some gene products may best be detected using fixatives other than formalin, or one may need to use frozen sections. In the latter case the pulmonary vascular bed could still be perfused as above (using either heparinized saline alone or with formalin), then the lungs could be gently inflated with OCT using a syringe to deliver this via the cannula. As an example of staining paraffin sections, we used the following technique to detect carcino-embryonic antigen (CEA) gene delivery. Five micron sections were cut and heat-mounted onto positively charged slides (60°C for 1 h). The sections were de-waxed by passing through three baths of xylene (10 min each), then baths with decreasing ethanol concentrations (100, 95, 90, 70%, each 5 min), and finally Tris wash buffer (0.05 M Tris, 0.15 M NaCl, 0.01% Triton X-100, pH, 7.44). A circle was drawn on the slide around the sections with a wax pencil or PAP-pen. The sections were blocked in 3% goat serum in Tris wash buffer for 1 h at room temperature, then incubated with anti-CEA polyclonal rabbit antibody (Chemicon, Temecula, CA, Cat. no. 46912) diluted 1:2000 with PBE buffer (1% BSA, 1 mM EDTA, 0.15 mM NaN$_3$, in PBS) for 1 h at room temperature. After three washes of 5 min each in Tris wash buffer, sections were incubated with fluorescently labeled secondary antibody (Alexa 488 [green fluorescence] goat anti-rabbit secondary antibody (Molecular Probes, Eugene, OR) diluted 1:1000 in PBE buffer for 1 h at room temperature. Sections were then washed, nuclei were stained with Hoescht 33342 (Molecular Probes) for 10 min at room temperature, then given a further wash and mounted using aqueous mounting medium. Immunofluorescent images were obtained using an Olympus IX 70 inverted microscope or a Leitz Orthoplan with epifluorescence optics and a Photometrics Sensys cooled CCD, high-resolution, monochromatic camera (Roper Scientific, Tucson, AZ) and IPLab Spectrum Image Analysis software (Scanalytics, Fairfax, VA).

13. Using this approach, we have seen significant increases in pulmonary endothelial gene expression compared with unmodified Ads (**Fig. 5**). Although there may be some gains attributable to the first pass effect (i.e., the pulmonary vascular bed is the first capillary bed that is seen by vector injected into the tail vein), we illustrated specificity by comparison with an irrelevant conjugate (made exactly as mentioned previously but combining 1D6.14 Fab with MAb 528, an anti-epidermal growth factor antibody), which did not increase pulmonary gene expression. We also performed blocking in vivo by co-injection of 1 mg of free MAb 9B9 to saturate lung binding sites (which reduced the effect of the anti-ACE conjugate), and also performed left ventricular injection of the vector complexes so that they would reach the pulmonary capillary bed after passing through peripheral capillary beds *(8,9)*.

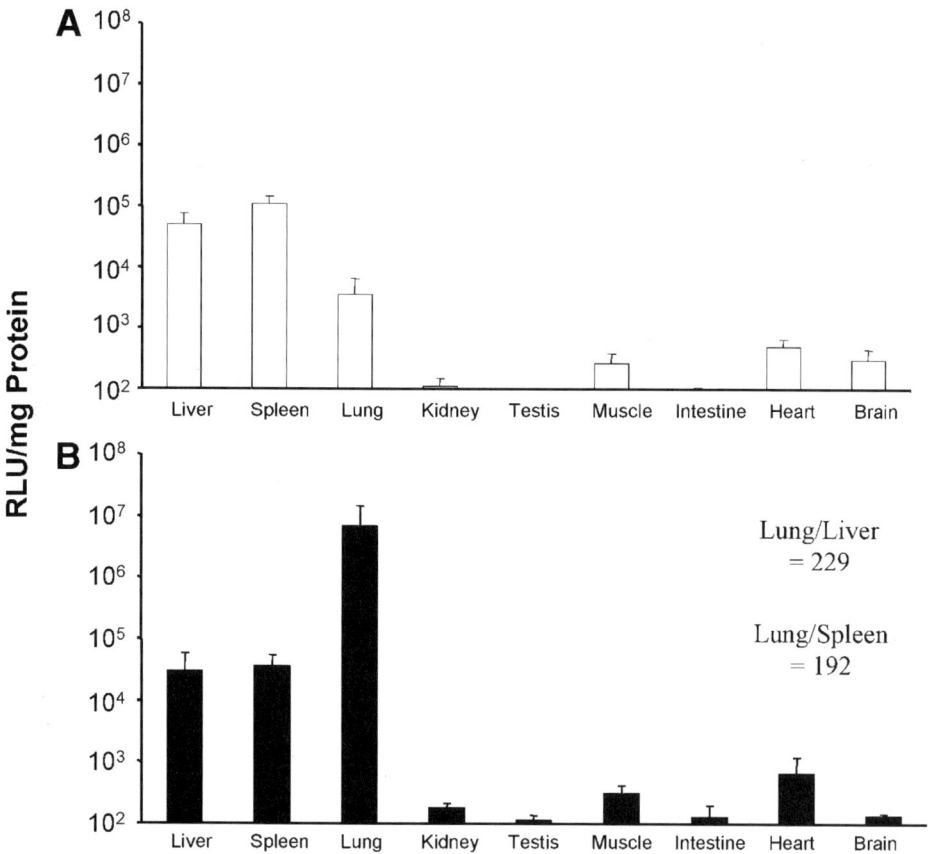

Fig. 5. Combined transductional and transcriptional targeting. Rats were injected (via tail vein) with 3×10^{11} viral particles of AdfltLuc (endothelial-specific promoter driving luciferase expression), either alone **(A)** or in combination with the pulmonary endothelial targeting conjugate Fab-9B9 **(B)**, then sacrificed 3 d later and luciferase activity was determined. Data show the mean ± SD of four rats per group. Adapted with permission from **ref. *9***

Acknowledgments

Supported by NIHRO1 HL67962, AHA and NHMRC.

References

1. Lane, K. B., Machado, R. D., Pauciulo, M. W., Thomson, J. R., Phillips, J. A., 3rd, Loyd, J. E., et al. (2000) Heterozygous germLine mutations in BMPR2, encoding

a TGF-beta receptor, cause familial primary pulmonary hypertension. The International PPH Consortium. *Nat. Genet.* **26**, 81–84.

2. Morrell, N. W., Yang, X., Upton, P.D., Jourdan, K. B., Morgan, N., Sheares, K. K., and Trembath, R. C. (2001) Altered growth responses of pulmonary artery smooth muscle cells from patients with primary pulmonary hypertension to transforming growth factor-beta(1) and bone morphogenetic proteins. *Circulation* **104**, 790–795.

3. Kasahara, Y., Tuder, R. M., Taraseviciene-Stewart, L., Le Cras, T. D., Abman, S., Hirth, P. K., et al. (2000) Inhibition of VEGF receptors causes lung cell apoptosis and emphysema. *J. Clin. Invest.* **106**, 1311–1319.

4. Kasahara, Y., Tuder, R. M., Cool, C. D., Lynch, D. A., Flores, S. C., and Voelkel, N. F. (2001) Endothelial cell death and decreased expression of vascular endothelial growth factor and vascular endothelial growth factor receptor 2 in emphysema. *Am. J. Respir. Crit. Care Med.* **163**, 737–744.

5. McLean, J. W., Fox, E. A., Baluk, P., Bolton, P. B., Haskell, A., Pearlman, R., et al. (1997) Organ-specific endothelial cell uptake of cationic liposome-DNA complexes in mice. *Am. J. Physiol.* **273**, H387–404.

6. Rodman, D. M., San, H., Simari, R., Stephan, D., Tanner, F., Yang, Z., et al. (1997) In vivo gene delivery to the pulmonary circulation in rats: transgene distribution and vascular inflammatory response. *Am. J. Respir. Cell Mol. Biol.* **16**, 640–649.

7. Tao, N., Gao, G. P., Parr, M., Johnston, J., Baradet, T., Wilson, J. M., Barsoum, J., and Fawell, S. E. (2001) Sequestration of adenoviral vector by Kupffer cells leads to a nonlinear dose response of transduction in liver. *Mol. Ther.* **3**, 28–35.

8. Reynolds, P. N., Zinn, K. R., Gavrilyuk, V. D., Balyasnikova, I. V., Rogers, B. E., Buchsbaum, D. J., et al. (2000) A targetable injectable adenoviral vector for selective gene delivery to pulmonary endothelium in vivo. *Mol. Ther.* **2**, 562–578.

9. Reynolds, P. N., Nicklin, S. A., Kaliberova, L., Boatman, B. G., Grizzle, W. E., Balyasnikova, I. V., et al. (2001) Combined transductional and transcriptional targeting improves the specificity of transgene expression in vivo. *Nat. Biotechnol.* **19**, 838–842.

10. Miller, C. R., Buchsbaum, D. J., Reynolds, P. N., Douglas, J. T., Gillespie, G. Y., Mayo, M. S., et al. (1998) Differential susceptibility of primary and established human glioma cells to adenovirus infection: targeting via the epidermal growth factor receptor achieves fiber receptor independent gene transfer. *Cancer Res.* **58**, 5738–5748.

11. Kelly, F. J., Miller, C. R., Buchsbaum, D. J., Gomez-Navarro, J., Barnes, M. N., Alvarez, R. D., and Curiel, D. T. (2000) Selectivity of TAG-72-targeted adenovirus gene transfer to primary ovarian carcinoma cells versus autologous mesothelial cells in vitro. *Clin. Cancer Res.* **6**, 4323–4333.

12. Haisma, H. J., Pinedo, H. M., Rijswijk, A., der Meulen-Muileman, I., Sosnowski, B. A., Ying, W., et al. (1999) Tumor-specific gene transfer via an adenoviral vector targeted to the pan-carcinoma antigen EpCAM. *Gene Ther.* **6**, 1469–1474.

13. Tillman, B. W., Hayes, T. L., DeGruijl, T. D., Douglas, J. T., and Curiel, D. T. (2000) Adenoviral vectors targeted to CD40 enhance the efficacy of dendritic cell-based

vaccination against human papillomavirus 16-induced tumor cells in a murine model. *Cancer Res.* **60,** 5456–5463.

14. Bergelson, J. M., Cunningham, J. A., Droguett, G., Kurt-Jones, E. A., Krithivas, A., Hong, J. S., et al. (1997) Isolation of a common receptor for Coxsackie B viruses and adenoviruses 2 and 5. *Science* **275,** 1320–1323.

15. Tomko, R. P., Xu, R., and Philipson, L. (1997) HCAR and MCAR: the human and mouse cellular receptors for subgroup C adenoviruses and group B coxsackie-viruses. *Proc. Natl. Acad. Sci. USA* **94,** 3352–3356.

16. Douglas, J. T., Rogers, B. E., Rosenfeld, M. E., Michael, S. I., Feng, M., and Curiel, D. T. (1996) Targeted gene delivery by tropism-modified adenoviral vectors. *Nat. Biotechnol.* **14,** 1574–1578.

17. Alemany, R., and Curiel, D. T. (2001) CAR-binding ablation does not change biodistribution and toxicity of adenoviral vectors. *Gene Ther.* **8,** 1347–1353.

18. Smith, T., Idamakanti, N., Kylefjord, H., Rollence, M., King, L., Kaloss, M., et al. (2002) In vivo hepatic adenoviral gene delivery occurs independently of the coxsackievirus-adenovirus receptor. *Mol. Ther.* **5,** 770–779.

19. Einfeld, D. A., Schroeder, R., Roelvink, P. W., Lizonova, A., King, C. R., Kovesdi, I., and Wickham, T. J. (2001) Reducing the native tropism of adenovirus vectors requires removal of both CAR and integrin interactions. *J. Virol.* **75,** 11284–11291.

20. Dechecchi, M. C., Melotti, P., Bonizzato, A., Santacatterina, M., Chilosi, M., and Cabrini, G. (2001) Heparan sulfate glycosaminoglycans are receptors sufficient to mediate the initial binding of adenovirus types 2 and 5. *J. Virol.* **75,** 8772–8780.

21. Nicklin, S. A., Reynolds, P. N., Brosnan, M. J., White, S. J., Curiel, D. T., Dominiczak, A. F., and Baker, A. H. (2001) Analysis of cell-specific promoters for viral gene therapy targeted at the vascular endothelium. *Hypertension* **38,** 65–70.

22. Varda-Bloom, N., Shaish, A., Gonen, A., Levanon, K., Greenbereger, S., Ferber, S., et al. (2001) Tissue-specific gene therapy directed to tumor angiogenesis. *Gene Ther.* **8,** 819–827.

23. He, T. C., Zhou, S., da Costa, L. T., Yu, J., Kinzler, K. W., and Vogelstein, B. (1998) A simplified system for generating recombinant adenoviruses. *Proc. Natl. Acad. Sci. USA* **95,** 2509–2514.

24. Vassaux, G., Hurst, H. C., and Lemoine, N. R. (1999) Insulation of a conditionally expressed transgene in an adenoviral vector. *Gene Ther.* **6,** 1192–1197.

25. Ma, Z., Mi, Z., Wilson, A., Alber, S., Robbins, P. D., Watkins, S., Pitt, B., and Li, S. (2002) Redirecting adenovirus to pulmonary endothelium by cationic liposomes. *Gene Ther.* **9,** 176–182.

26. Danilov, S. M., Muzykantov, V. R., Martynov, A. V., Atochina, E. N., Sakharov, I., Trakht, I. N., and Smirnov, V. N. (1991) Lung is the target organ for a monoclonal antibody to angiotensin- converting enzyme. *Lab. Invest.* **64,** 118–124.

27. Krasnykh, V. N., Mikheeva, G. V., Douglas, J. T., and Curiel, D. T. (1996) Generation of recombinant adenovirus vectors with modified fibers for altering viral tropism. *J. Virol.* **70,** 6839–6846.

28. Muzykantov, V. R., Atochina, E. N., Ischiropoulos, H., Danilov, S. M., and Fisher, A. B. (1996) Immunotargeting of antioxidant enzyme to the pulmonary endothelium. *Proc. Natl. Acad. Sci. USA* **93,** 5213–5218.

29. Balyasnikova, I. V., Yeomans, D. C., McDonald, T. B., and Danilov, S. M. (2002) Antibody-mediated lung endothelium targeting: in vivo model on primates. *Gene Ther.* **9,** 282–290.

30. Li, S., Tan, Y., Viroonchatapan, E., Pitt, B. R., and Huang, L. (2000) Targeted gene delivery to pulmonary endothelium by anti-PECAM antibody. *Am. J. Physiol. Lung Cell. Mol. Physiol.* **278,** L504–511.

31. Balyasnikova, I. V., Gavrilyuk, V. D., McDonald, T., Berkowitz, R., Miletich, D. J., and Danilov, S. M. (1999) Antibody mediated lung endothelium targeting: 1. In vitro model using a cell line expressing angiotensin converting enzyme. *Tumor Target.* **4,** 70–83.

32. Muzykantov, V. R., Martynov, A. V., Puchnina, E. A., and Danilov, S. M. (1989) In vivo administration of glucose oxidase conjugated with monoclonal antibodies to angiotensin-converting enzyme. The tissue distribution, blood clearance, and targeting into rat lungs. *Am. Rev. Respir. Dis.* **139,** 1464–1473.

7

Gene Transfer to Brain and Spinal Cord Using Recombinant Adenoviral Vectors

Joseph M. Alisky and Beverly L. Davidson

1. Introduction

Recombinant adenoviral (Ad) vectors are derived from human adenoviruses: nonenveloped, encapsidated linear, double-stranded DNA viruses that commonly cause respiratory and gastrointestinal infections. Forty-three different human adenovirus serotypes have been characterized *(1)*. Details about production of recombinant Ad vectors are given in Chapter 1. Ad vectors in widespread use are derived from human Ad serotypes 2 and 5 (Ad2 and Ad5), Ad5 being more common for applications in the central nervous system (CNS). Ad5 replication-impaired vectors most often contain deletions in the E1 and E3 regions, with transgenes driven by a variety of promoters including viral promoters, and those that are neuron-specific *(2)*. Recently fiber-modified and "gutless" Ad vectors, and those based on canine adenovirus serotype 2, have been developed for use in brain *(3–9)*.

The main limitation of the original Ad5 vectors is loss of gene expression. Even in the relatively privileged microenvironment of the brain and spinal cord, there is considerable immune response against adenoviral proteins. As a result of immune response, combined with shutdown of cytomegalovirus (CMV) or Rous sarcoma virus (RSV) promoters, Ad5 transgene expression in mouse brain peaks from 4 to 7 d postinfection and declines thereafter *(10–13)*. However, for short-term gene expression, the Ad5 vector has several advantages that make it the vehicle of choice for many applications in the CNS. Ad5 vectors can be gen-

From: *Methods in Molecular Biology, vol. 246:*
Gene Delivery to Mammalian Cells: Vol. 2: Viral Gene Transfer Techniques
Edited by: W. C. Heiser © Humana Press Inc., Totowa, NJ

erated very rapidly in large quantities at high titers. Using shuttle plasmid systems and available adenovirus backbones, research-grade vectors can be produced in as little as 2 wk with titers as high as 1×10^{13} infectious units/mL *(14,15)*. The theoretical packaging limit for Ad5 is up to 34 kb; by contrast adeno-associated virus (AAV) vectors typically can package approx 4.5 kb or less and lentiviral vectors have a maximum capacity of approx 7–8 kb *(16)*. Injections of Ad5 into the brain can transduce a wide variety of targets, including neurons, astrocytes, oligodendrocytes, ependyma, and blood vessels *(13)*. Furthermore, neurons in sites difficult to reach by direct injection can also be transduced via retrograde transport of Ad5 from nerve terminals *(17,18)*. For example, motor neurons in the brainstem and spinal cord can be transduced by Ad5 vectors injected into muscle *(19,20)*. Uptake of Ad5 vectors by neuromuscular terminals can be enhanced through botulism-toxin induced sprouting *(20)*. Neurons in the brainstem can also be transduced from Ad5 vectors injected into the cerebellum or spinal cord *(12,21)*. Also, Draghia and colleagues instilled Ad5 into the nasal cavity of rats and found transduction of neurons in the olfactory bulb, olfactory nucleus, locus coeruleus, and area postrema *(22)*.

Ad5 vectors have tropism for some brain tumors, and can be used in model systems of gene therapy to treat primary and metastatic CNS tumors. In this application, time-limited transgene expression and immune response against transduced cells are advantageous. Clinical trials of Ad vectors to treat gliomas are currently underway *(23,24)*.

In this chapter, we present commonly used methods for gene transfer of Ad5 to targets in the brain and spinal cord of rats and mice, as detailed in **Table 1**. These procedures follow the same basic outline:

1. The first step is injection of vector. Except for intramuscular injection for transduction of motor neurons, vector injection involves surgical exposure of the target region in the brain or spinal cord, followed by delivery of a controlled amount of viral vector using a mechanically controlled pressure injector (see **Note 1**).

2. Attentive postoperative care is important. Immediately following surgery, animals are monitored until awake. Analgesia in the immediate perioperative period is advised. If longer-term gene expression is desired, postoperative immunosuppression is recommended using cyclosporine or FK-506. Lower doses of vector seems to correlate with less immune response; total virus dose below 10^8 transducing units attenuated immune-mediated loss of transgene expression *(25)*. In theory, it is also possible to reduce immune responses to Ad5 by intrathymic injection at birth (*see* **Note 2**).

3. After the desired survival period, animals are deeply anesthetized and sacrificed to obtain brain and spinal cord tissue for histological study. Transcar-

Table 1
Ad5 Transduction in Brain and Spinal Cord of Rats and Mice

Injection site	Cells transduced at site of injection[a]	Cells transduced by retrograde transport	References
Striatum	Mainly glia, small numbers of GABAergic striatal neurons	Corticostriatal, subthalamic, and substantia nigra neurons; callosal, and monoaminergic neurons	(13,17,18,26,27)
Cerebellum	Mainly glia, efficient transduction of ventricular epithelium, almost no neurons at injection site in cerebellar cortex	Large numbers of brainstem cerebellar projecting neurons, small numbers of spinocerebellar neurons, sometimes efficient retrograde transduction of distant Purkinje cells and deep cerebellar nuclei	(21)
Pituitary	Mainly glia	Periventricular and supraoptic hypothalamic nuclei	(52)
Hippocampus	Glia	Pyramidal cells	(53,54)
Spinal cord	Glia, motor, and other ventral horn neurons	Retrograde transduction of brainstem reticulo-, vestibulo-, and rubrospinal neurons, also scattered cerebellar Purkinje cells (from CSF diffusion not retrograde transport)[b]	(12,40)

Table 1
Continued

Injection site	Cells transduced at site of injection[a]	Cells transduced by retrograde transport	References
Muscle	Muscle fibers	Brainstem or spinal cord motor neurons depending on muscle injected	(19,20,42,44)
Cisterna Magna	Cerebral blood vessels and meninges, also spinal cord meninges	None[c]	(27,33,35,40, 47,51)
Ventricle	Ventricular epithelium	No neurons by retrograde transport; however, when vector is injected under conditions of systemic hyperosmolality, scattered cells near ventricles can be transduced	(34)
Brain tumor	Variable, depending on location and type of tumor	Variable, depending on location and type of tumor	(36,37)
Cerebral cortex	Neocortical neurons in multiple layers	None	(55,56)
Hypothalamus	Some supraoptic neurons	Subfornical neurons	(57)

[a]Mainly glia are transduced directly at the injection site, but direct neuronal transduction in the spinal cord appears to be more efficient then other locations in central nervous system.

[b]In rats and mice, corticospinal neurons mainly project to the spinal cord via brainstem interneurons. Hence corticospinal neurons are not efficiently transduced from spinal injections.

[c]Finegold et al. advanced a catheter through the cisterna magna of rats 3 mm into the intrathecal space of the cervical cord and transduced meninges on the dorsal surface of the spinal cord (34).

dial perfusion with 4% paraformaldehyde fixative provides good preservation of histology and many antigens, although it is recommended that each stain and/or immunohistochemical (IHC) approach be refined by testing using other fixatives and fixation times.

4. Tissue sections are prepared from brain and spinal cord tissue using a cryostat. If bacterial β-galactosidase (β-gal) is used as a reporter gene, sections are next processed for β-gal histochemistry to produce vivid blue staining of transduced cells. Whole brain specimens can also be stained for β-gal activity. Another popular reporter gene is enhanced green fluorescent protein (eGFP). No histochemistry is required, and eGFP is ideal for examining live or unfixed cells or tissues. However, because GFP fades over time, IHC stains are recommended for archived sections. In addition, IHC stains or other measures may be employed to assess the effects of gene transfer. As one example, adenoviral-expressed, glial-derived neurotrophic factor in the striatum of rats protected against 6-hydroxydopamine lesions, as measured by counting fluorogold-labeled substantia nigra neurons *(26)*. As another example, adenoviral-mediated expression, of β-glucuronidase in the brains of β-glucuronidase deficient mice resulted in widespread correction of lysosomal storage disease *(27)*.

Material and methods for injection, postoperative care, perfusion/fixation, and preparation of tissue sections are described such that investigators can customize protocols for their own specific aims. All uses and preparations of recombinant viruses, and procedures utilizing live animals, must be approved by the appropriate institutional use committees prior to beginning experiments. Of note, the surgical procedures and tissue-processing methods are applicable to other commonly used vector systems, including AAV *(28)* or lentiviruses *(29,30)*.

2. Materials
2.1. Vector Injection
2.1.1. Injection of Vectors in Striatum, Cerebrum, Hippocampus, Hypothalamus, Ventricles, and Cerebellum

1. Rats or mice obtained from an accredited animal supply company with known birth dates and weight, housed in government-approved facility. We generally use C57BL/6 mice or Sprague-Dawley rats, but other strains can be used. Note, however, that different strains may have different immune response against adenoviral proteins *(10)*. It is advisable to order the animals at least 1 wk before surgery to allow them to adjust from the stress of travel.
2. Ketamine-HCL/xylazine mix for anesthetizing mice: combine 8.9 mL of sterile phosphate-buffered saline (PBS) with 1 mL of 100 mg/mL ketamine

and 0.1 mL of 100 mg/mL xylazine (XYLA-JECT® available from Phoenix Pharmaceuticals, Belmont, CA) for final concentrations of 10 mg/mL ketamine and 1 mg/mL xylazine. PBS is prepared from prepared packets (Sigma, St. Louis, MO) in double-distilled water (ddH$_2$O). The PBS will be pH 7.5, have a concentration of 0.01 *M,* and can be autoclaved for sterilization.

3. Sodium pentobarbital for anesthetizing rats (usually available ready to use; if necessary dilute with PBS for working concentrations).
4. Stereotaxic brain atlas *(31,32)*.
5. A stereotaxic frame (Kopf Instruments, Tujunga, CA) to immobilize the head of the animal. For mice, use small animal stereotaxic frame with mouse adaptor. The same stereotaxic frame can be used for rats, but the rat ear bars must be mounted onto the frame.
6. Underpads such as blue chucks, surgical drapes, or sterilized towels to set up a sterile operating field.
7. Eye ointment containing bacitracin, zinc, and polymixin B sulfate is needed to prevent corneal abrasions while animals are anesthetized for a prolonged period of time with suppressed blink reflex.
8. Electric or disposable razors for removal of hair around the site of injection.
9. Iodine tincture or betadine applicators.
10. Autoclavable surgical instruments and supplies including sterile cotton-tipped applicators, cotton wicks, scalpel and blades, curette (spoon-like instrument), forceps, iridotomy scissors, hemostat, and ronguers can be procured from many different supply companies. The size of ronguers depends on the size of the animal; models are available for mice and rats. A sterile surgical marking pen is helpful for marking stereotaxic coordinates on the skull, available from most surgical supply companies or hospital stores.
11. Sterile insulin syringe with attached 28 1/2G needle.
12. Surgical suture (4.0 sized 30-in silk) with attached needle.
13. Microprocessor-controlled pump with controller (UltraMicroPump, World Precision Instruments, Sarasota, FL) mounted onto the stereotaxic frame to help deliver reproducible volumes of vector at controlled rates (*see* **Note 1**).
14. Drill (Model 395 Type 5, or equivalent, Dremel Moto-Tool, Racine, WI) with 003 bit for all injections requiring access through the skull.
15. Dry sterilizer (Germinator 500, Cellpoint Scientific, Inc., Rockville, MD).
16. Small sterile beaker with 70% ethanol, sterile tubes for making anesthetic solution.
17. Support box taped to the base of the stereotaxic frame, to support the mice or rats.
18. Several 1- or 10-µL glass Hamilton syringes (Reno, NV) with a removable stainless-steel, blunt-ended 33 G needle (Hamilton) for gene transfer in striatum, hypothalamus, ventricles, and cerebral blood vessels.

19. Mannitol prepared as 1 *M* in 0.34 *M* sodium chloride, to be administered at 3 mL/100 gram of body weight of mice 10 min prior to intraventricular administration of Ad5 vector, to improve penetration into the parenchyma (*see* **Note 3**).
20. Several 1-µL glass Hamilton syringes, pulled glass microcapillary tubes, and super glue for cerebellar injections.
21. Dissecting microscope.
22. Sterile PBS prepared as described in **item 2**.
23. Sterile water, autoclaved double-distilled H_2O or sterile, endotoxin-free tissue culture water.
24. Recombinant E1,E3-deleted Ad5 vector at a titer of 1×10^{10} infectious units/mL or greater.

2.1.2. Spinal Cord Injections

Materials are identical to those in **Subheading 2.1.1.** except for a few minor differences:

1. A stereotaxic atlas is not required for injection coordinates but may be helpful for identifying structures in spinal-cord cross-sections when analyzing data.
2. Ronguers, forceps, curettes, and probes are needed to perform a spinal laminectomy to expose a small segment of spinal cord for injection. Dremel drill is not needed.

2.1.3. Cisterna Magna Injections for Transduction of Meninges and Cerebral Blood Vessels

Materials are identical to those in **Subheading 2.1.1.** except for the following differences:

1. A sterile 1-cc syringe, rather than Hamilton glass syringes, with a sterile 27 G needle attached is used to inject vectors.
2. Ad5 vectors have been used for mice *(33)*, whereas either Ad5 *(34)* or Ad2 *(35)* vectors have been used for rats.

2.1.4. Injection of Vector in a Brain and Spinal Tumor Models

Materials are identical to those in **Subheading 2.1.1.** except for the following differences:

1. Our protocols use immunodeficient (nude) rats implanted with LX-1 lung carcinoma cell line in the striatum *(36)* or Fisher rats implanted with 9L rat gliosarcoma cells *(37)*. Details of results obtained with these model systems are described in **Note 4**. Other model systems should be comparable, including nude mice or rats injected with various human tumor cell lines or

immunocompetent rats and mice injected with rat- or mice-derived tumor lines.

2. A 25-μL Hamilton syringe is used for injection of vector in the nude rat lung carcinoma model, whereas 1-μL Hamilton syringes with glass electrode tips identical to cerebellar and spinal injections are used for the rat spinal tumor model.

3. Recombinant E1-deleted Ad5 vector at a titer of 1×10^{10} infectious units/mL or greater, either expressing a reporter gene such as β-gal *(36)* or the tumoricidal herpes-simplex thymidine kinase gene *(37,38)*.

4. For use with the herpes-simplex thymidine kinase in the spinal tumor model, gancylovir (50 mg/kg). Typical results for both model systems are detailed in **Note 4**.

2.1.5. Intramuscular Injection of Vector in Tongue or Limb Muscle

1. Ketamine/xylazine or sodium pentobarbital for mice or rats, respectively, as described in **Subheading 2.1.1.**

2. Disposable sterile 0.5-cc tuberculin syringes to inject larger volumes of vector.

3. A 10- or 25-μL Hamilton syringe with a removable stainless-steel, blunt-ended 33 G needle (Hamilton) to inject small volumes of vector.

4. Autoclaved surgical instruments (scalpel, blades, curette, forceps, iridotomy scissors, hemostat, and probes) and surgical suture (4.0 silk) if limb muscle will be exposed for injection. Injection of tongue requires no suture, and only forceps and curette.

5. Recombinant E1-deleted Ad5 vector at titer of 1×10^{10} infectious units/mL or greater.

6. Botulism toxin (Sigma): 25 pg diluted into 10 μL of sterile PBS for enhanced motor-neuron uptake of Ad5 vectors injected into the tongue of mice as described by Millecamps *(20)*.

2.2. Postoperative Care of Animals

2.2.1. Routine Postoperative Care

1. Sterile 0.9% saline for subcutaneous injection immediately after surgery and on subsequent postoperative days if animals appear dehydrated.

2. A 1-mL tuberculin syringe to administer saline to mice; 10 cc syringes with a 25 G needle for rats.

3. Recovery cage with clean towel bedding with a warming lamp of 50–100 watts placed about 1 meter above the cage.

4. Buprenex® (buprenorphine-HCl) for postoperative analgesia.

5. Unflavored gelatin to facilitate postoperative hydration.
6. Gruel, which is a paste made from standard laboratory chow by grinding it up and adding water, to facilitate postoperative nutrition.

2.2.2. Postoperative Immunosuppression With Cyclosporine or FK-506

1. Cyclosporine (15 mg/kg for mice, 20 mg/kg for rats, available suspended in castor oil from Sandoz Pharmaceuticals, East Hanover, NJ) or FK-506 (1 mg/kg for rats) for daily injections.
2. Disposable, sterile tuberculin syringes.

2.3. Perfusion/Fixation

1. Ketamine/xylazine mix for anesthetizing mice (*see* **Subheading 2.1.1.**).
2. Sodium pentobarbital for anesthetizing rats (*see* **Subheading 2.1.1.**).
3. PBS prepared as described in **Subheading 2.1.1.** Approximately 20 mL of PBS per mouse and 100 mL of PBS per rat are needed.
4. Laboratory grade double-distilled (dd) H_2O.
5. Laboratory-grade paraformaldehyde for paraformaldehyde fixative prepared the day before or day of perfusion/fixation. Weigh paraformaldehyde to give final concentration of 4 g/100 mL of volume (4%). Perfusion of one mouse requires about 25–50 mL of fixative; one rat requires 300–400 mL. For 1 L of fixative, heat 500 mL of ddH_2O in a microwave for a few minutes but not to the point of boiling. Add the dry laboratory-grade paraformaldehyde to the heated water, and then add several drops of 10 *N* sodium hydroxide to quickly dissolve the powdered paraformaldehyde. Then add the remaining volume of ddH_2O, followed by 2.62 g of sodium phosphate monobasic monohydrate and 11.5 g of anhydrous sodium phosphate dibasic per liter of fixative (0.1 *M* phosphate buffer final concentration). Add a few drops of sodium hydroxide to fully dissolve the phosphate buffers, then adjust the pH of the fixative to 7.4. We recommend that the solution be vacuum filtered prior to use. Many laboratories prepare paraformaldehyde for weeks in advance, but we prefer to make it up no more than 24 h in advance of use for optimal preservation of neural antigens and histology.
6. Laboratory-grade sucrose to prepare buffered sucrose solution. Buffered 30% sucrose should be prepared in advance, adding 30 g of sucrose/100 mL of sterile PBS and adjusting the pH to 7.4. It can be stored for several weeks at 4°C.
7. A 0.45-µm bottle top filter (can use filter paper and funnel or various vacuum filters covered with filter paper).
8. Peristaltic perfusion pump.
9. Flexible rubber hosing 2–3 mm in diameter.

10. Butterfly iv catheter, 23 or 25 G.
11. Iridotomy scissors.
12. Blunt-tipped (not rat-toothed) forceps.
13. Styrofoam or wax board.
14. Spray bottle with tap water.
15. Razor blade and scalpel.
16. Ronguers (mouse- or rat-sized).
17. Curette.
18. Disposable tuberculin syringes for anesthetizing mice; disposable 5–10 cc syringes with 23 G needles for rats.
19. Plastic sagittal or coronal slicers (Roboz Surgical Instruments Company, Inc., Rockville MD; mouse coronal, Cat. no. AL1175; mouse sagittal, Cat. no. AL1275).
20. Sagittal slicing molds or coronal slicing molds (Kopf Instruments).

2.4. Preparation and Processing of Tissue Sections

2.4.1. Cutting Frozen Sections

1. Cryostat.
2. Glass slides (Fisher Superfrost glass[+] microscope slides or other appropriately subbed slides to hold tissue sections).
3. Slide boxes.
4. Twenty-four-well culture dishes filled with 0.01 M PBS: needed for collecting thicker (40–50 µm) cryostat sections. Fine needle forceps are necessary for picking up the cryostat sections.
5. PBS with 0.02% sodium azide is recommended for long-term storage of free-floating thick sections.
6. Peel-a-way Tissue Molds (Electron Microscope Sciences, Washington, PA).
7. OCT or comparable tissue histological compound for freezing tissue blocks.

2.4.2. β-Galactosidase Histochemistry: KC Mixer and X-Gal Substrate

1. KC Mixer (items a-f; for 500 mL):
   a. 5.74 g of $K_3 Fe(CN)_6$, 35 mM final.
   b. 7.35 g of $K_4 Fe(CN)_6 \cdot 3H_2O$, 35 mM final.
   c. 1 mL of 1 M $MgCl_2$, 2 mM final.
   d. 0.5 mL of 10% sodium deoxycholate, 0.01% final.
   e. 1.0 mL of 10% NP40, 0.02% final.
   f. 499 mL of PBS, prepared as described in **Subheading 2.1.1.**

   Add the $MgCl_2$ last after all of the aforementioned ingredients are dissolved, filter through a 0.45-µm bottle top filter, and store in a dark or foil-covered bottle at 4°C.

2. 5-bromo-4-chloro-3-indolyl-β-D-galactopyranoside (X-gal) stock, stored with desiccant at $-20°C$, or dissolved at 40 mg/mL in N',N-dimethylformamide and stored at $-20°C$ in small aliquots.
3. If sections are to be counterstained after the β-gal histochemistry, prepare neutral red counterstain the day before or the day of processing. Stock solutions are good for months, but working stain should be fresh. Prepare neutral red stock solution by making a 0.025% solution of dry neutral red in double distilled H_2O (0.25 g of neutral red in 1 L of ddH_2O); filter through a 0.45-μm filter to remove any clumps. Prepare a 0.2 M sodium acetate stock solution. Prepare 0.2 M acetic acid (17 mL of glacial acetic acid and 983 mL of ddH_2O). To make the working neutral red stain solution, combine 2 vol of acetate buffer, pH 5.6 (91 mL 0.2 M sodium acetate, 9.0 mL 0.2 M acetic acid, 100 mL ddH_2O) with 3 vol of 0.025% neutral red stock.

2.4.3. Whole Mount Fixation and β-Galactosidase Histochemistry

Materials are identical to those described in **Subheading 2.4.2.** except that no counterstain reagents (**item 3**) are required.

3. Methods
3.1. Vector Injection
3.1.1. Injection of Vectors in Striatum, Cerebrum, Hippocampus, Hypothalamus, Cerebellum, and Ventricles

3.1.1.1. PREOPERATIVE PREPARATION

1. Autoclave surgical instruments and drapes and prepare appropriate anesthetics as described in **Subheading 2.1.1.**
2. For striatal, cerebral, hippocampal, hypothalamic, and ventricular injections, sterilize 10-μL Hamilton syringes by rinsing several times in 70% alcohol, followed by several rinses in sterile water and finally sterile PBS.
3. For cerebellar injections, attach the pulled microcapillary tubes to the end of blunt 1-μL Hamilton syringes using superglue. Break off the glass electrode about 100 mm above the beginning of the taper and fit the truncated electrode tip snugly in place over the blunt end of the Hamilton syringe; use a cotton-tipped applicator to apply a drop of superglue to the glass to cement the tip in place. After 15 min, sterilize the syringe in the same manner as described in **item 2**. If it is impossible to draw alcohol up into the syringe, gently break off a tiny portion of the tip by pressing it against a sterile cotton-tipped applicator. Sometimes superglue can be drawn into the barrel of the Hamilton syringe if the tip was not snugly in place before applying the glue. If this happens, break off the tip and soak it in acetone until

the superglue is dissolved (minutes to hours). We find it helpful to have several 1-μL syringes on hand so that surgery is not delayed if one syringe is temporarily out of commission.

4. Set the mechanical injector for the appropriate delivery rate and volume for striatal, ventricular, and hippocampal injections at 500 nL/min; cerebellar injections are done under manual control (*see* **Table 2** and **Note 1**).

5. Thaw the frozen vector and keep at room temperature during surgery once syringes, anesthetic, injector, surgical instruments, and drapes are ready. (Ad vectors are stable for up to 6 h at room temperature.)

6. Anesthetize mice with 0.1 mL of ketamine/xylazine cocktail per 10 g of body weight given intraperitoneally (ip); anesthetize rats with 50 mg/kg sodium pentobarbital ip (*see* **Note 1**). Assess adequacy of anesthesia by verifying that the animal does not withdraw its toe when pinched with a forceps. If needed, administer booster doses of 0.1 mL ketamine/xylazine for mice, 10 mg/kg sodium pentobarbital for rats.

7. Mount animals in a rat or mouse-sized stereotaxic apparatus with body resting on a support box so that the head is level with the nose clamp and incisor bar. Mount the animal's front incisor teeth in the incisor bar to provide complete immobilization of the head. The angle of the incisor bar should be such that the skull is level. Some stereotaxic frames also have ear bars that are inserted into the ear canal and tightened in place. Tape the tail to the operating table or stereotaxic apparatus for further stabilization, and reconfirm depth of anesthesia with toe pinch.

8. Once animals are mounted on the stereotaxic apparatus, apply eye ointment to prevent corneal abrasions.

9. Shave the skin with an electric razor to expose the area for incision from the front of the skull to the middle portion of the neck. Thoroughly clean the shaved with the betadine swabs, followed by an alcohol prep pad. As an alternative to an electric razor, use a disposable razor if the betadine is first applied as an astringent.

10. Following shaving, rewash the area with betadine to clear away hair and then swab with an alcohol prep pad. Perform a final anesthesia check and then apply sterile drapes to create a sterile operating field.

11. For intraventricular injections, to achieve transduction of neurons beyond the boundaries of the ventricular epithelium, apply systemic mannitol, ip, 10 min prior to virus injection (*see* **Note 3**).

3.1.1.2. Operative Procedures

Expose the skull over the respective target (i.e., striatum cerebral cortex, ventricles, hypothalamus, pituitary, hippocampus, or cerebellum) and drill a burr hole with a dremel drill.

Table 2
Parameters for Striatal, Cerebral, Ventricular, Hippocampal, Pituitary, Hypothalamic, and Cerebellar Injections.

Target	Stereotaxic coordinates in relation to bregma	Needle depth from dura	Volume and rate[a]
Mouse striatum	0.4 mm anterior and 2.0 mm lateral	3 mm	5 μL at 500 nL/min
Rat striatum	1 mm anterior and 3 mm lateral	4 mm	2–10 μL at 500 nL per min
Mouse lateral ventricle	0.4 mm posterior and 1 mm lateral	2 mm	10 μL at 500 nL/min
Rat lateral ventricle	1–1.8 mm posterior and 1.5 mm lateral	4 mm	100 μL over 30 min
Rat hippocampus[b]	4 mm posterior and 2 mm lateral	4 mm	5 μL at 500 nL/min
Rat neurohypophysis	5.5 mm posterior and 0 mm lateral	10 mm	0.2–0.4 μL at 500 nl/min
Rat periventricular hypothalamus	1.4 mm posterior and 0.6 mm lateral	7.6 mm	0.2–0.4 μL at 500 nL/min
Rat supraoptic hypothalamus	1.3 mm posterior, 5.5 mm lateral[c]	9.6 mm	0.2–0.4 μL at 500 nL/min
Mouse cerebellum	5–7 mm posterior to and 0–1 mm lateral[d]	1 mm	1–6 μL over 3 min
Rat cerebral cortex	3 mm anterior and 5 mm lateral	4 mm	5 μL over 10 min

[a]Except for the cerebellar and ventricular injections, the rate is set preprogrammed on microprocessor-controlled pump.
[b]Nishimura et al. (**54**) and Masumura et al. (**53**) suspended Ad5 vector in 1 *M* mannitol to increase retrograde transport in the hippocampus.
[c]Angle syringe at 20° toward the midline to avoid penetrating ventricle .
[d]Cerebellum can be identified unequivocally from visual landmarks on the skull; posterior to the lambdoidal suture is the occipital bone overlying the cerebellum in both mice and rats.

12. Make a midline incision into the cleaned, shaved skin with a scalpel to expose a wide window so that skin retracts easily without mechanical retractors.

13. Clearly identify the coronal, sagittal, and lambdoidal skull sutures, preferably through a dissecting microscope. The confluence of the sagittal and frontal sutures forms the bregma, which is the landmark for most stereotaxic atlases. More posterior, the confluence of the sagittal and lambdoidal sutures is a landmark for exposure of the cerebellum.

14. Identify stereotaxic coordinates for target of injection from the bregma and lambdoidal landmarks. It may be helpful to mark the spot for drilling using a sterile surgical marker. Coordinates, amount of vector, and injection rate are given in **Table 2**.

15. Prior to drilling, dip the head of the dremel drill in 70% alcohol, followed by sterile water and sterile PBS. Apply the drill to the skull surface and quickly bore down to the dura mater, taking care to stop before the dura is pierced.

16. After drilling, clear away any blood with a sterile cotton-tipped applicator.

17. Draw the vector into the syringe and go slightly above the last marked calibration of the syringe (10 μL or 1 μL, respectively) and then expel down to the 10 or 1 μL mark so that there is no dead space in the syringe. Mount the syringe in the injector and secure in place with clamps.

18. Lower the syringe through the burr hole until the tip (metal or glass) of the needle touches the dura. Read the z-plane coordinates off the stereotaxic device so that the absolute depth can be controlled. Slowly insert the needle into the brain parenchyma to the depth specified in **Table 2**.

19. For striatal, ventricular, hypothalamic, pituitary and hippocampal injections, initiate the microprocessor-controlled pump pre-programmed to deliver the recommended volume at 500 nL/min (**Table 2**). After vector delivery, leave the needle in place for 5 min, then slowly withdraw in several steps, over 5 min.

20. For cerebellar injections, insert through the dura to a depth of 1 mm or less and inject 1 μL at a time over 3–5 min; leave syringe in place for 3–5 min before withdrawing. Up to 6 μL can be injected in as many as three to four different burr holes or up to 2 μL can be injected in two injections in the same burr hole. In our experience, making injections larger than 1 μL per cerebellar injection results in leakage of most of the vector and little if any gene transfer.

21. Remove the mouse or rat from the stereotaxic apparatus and close the scalp incision with 4.0 silk sutures. The dremel hole may first be sealed with sterile bone wax if necessary.

22. Inject mice with1 mL of sterile saline subcutaneously (sc) to replace fluid loss; inject rats with 10 mL of sterile saline sc. Place animals in a cage with absorbent bedding, warmed by a 75-watt lamp about a meter above the cage. Monitor animals periodically until they are free-moving, and then return to animal care facilities.

3.1.2. Injection of Spinal Cord

This protocol is adapted from published papers *(12,39–41)* as well as from procedures used in the Davidson laboratory.

1. Follow preoperative preparation as described in **Subheading 3.1.1.** except the surgical window is going to be above the vertebral column and spinal cord instead of the skull. Prepare a 1-μL Hamilton syringe with a glass electrode tip identical to that for cerebellar injections, and follow anesthesia and stereotaxic immobilization as in **Subheading 3.1.1.** Immobilize the head and tail to provide needed stability for spinal surgeries.
2. Identify the site for spinal injection. Injections are usually made in the lower thoracic-upper lumbar spinal cord, although other sites can be used. Cervical injections can be difficult because of potential for cervical fractures. For the lower thoracic-upper lumbar spinal injections, landmarks are easy to establish in anesthetized mice and rats by palpating the rib cage and following each rib back to its fusion with the spinal cord. Rats and mice have 13 thoracic vertebrae instead of 12 as in humans. The last rib joins with the 13th thoracic vertebra, representing the thoracolumbar junction. The iliac crests represent the end of lumbar vertebra and the beginning of sacral vertebra. For cervical injections, the junction with the skull marks the first cervical vertebra, and the first rib marks the beginning of thoracic vertebra and the end of the cervical cord. Prepare and shave skin as in **Subheading 3.1.1.** to allow for a clean, easily accessible window.
3. Using a scalpel, iridotomy scissors, curettes, and forceps, expose a single vertebral segment with a combination of cutting and blunt dissection. Soft-tissue exposure should be kept to a minimum in order to see a single spinous process and lamina. Be careful not to press down while dissecting, because this can compress heart and vena cava. The most common site for injection in our laboratory is the junction between 13th thoracic and 1st lumbar vertebra or the junction between 1st and 2nd lumbar vertebra. If exposing thoracic vertebra, avoid removing rib or pleura, as this will cause the animal to expire from a pneumothorax.
4. Once the lamina and spinous process are exposed down to the periosteum, break off the spinous process with ronguers. Using ronguers and forceps,

gently wear away the lamina on one side to expose the spinal cord in a small hole that can be widened with the ronguers to expose the entire segment. Work in very small steps and avoid direct downward pressure to prevent damage to the underlying cord. The dorsal spinal artery should be readily visible in the exposed segment.

5. Fill the syringe with vector as described in **Subheading 3.1.1.**
6. Very slowly penetrate the spinal parenchyma with the glass tip 0.5 to 1 mm lateral from the spinal artery.
7. Once the glass tip penetrates the dura as observed through a dissecting scope, for mice slowly descend to a depth of no more than 0.5 to 0.75 mm for cervical and lumbar injections, or 0.25 to 0.5 mm for thoracic and sacral injections. For rats, insert the tip 1.0 to 1.5 mm for cervical and lumbar injections or 0.5 to 0.75 mm for thoracic and sacral cord injections.
8. Inject 1 µL or less over 5 min by manual control, and leave the syringe in place for 3–5 min before withdrawing. The injection can be repeated on the other side of the spinal artery, but make no more than two injections in the exposed segment.
9. Suture the overlying muscle and then suture the exposed skin and inject sc saline as in **Subheading 3.1.1.**
10. Observe animals carefully once the anesthesia has worn off. Postoperatively, animals may have slight foot drop or lower limb weakness that should resolve within 12–24 h after surgery. If an animal is paralyzed after the end of the first full postoperative day, it should be euthanized with an overdose of ketamine/xylazine or sodium pentobarbital.

3.1.3. Injection of Cisterna Magna for Gene Transfer to Cerebral Blood Vessels

This protocol is adapted from Ooboshi et al. *(33)*, and Christenson et al. *(35)* (*see* **Note 5**).

1. Follow preoperative preparation as described in **Subheading 3.1.1.** except that (1) prepare the surgical window from between the ears to the midportion of the neck and (2) mount the animals in the stereotaxic apparatus with the head nose-up, at an angle of 30° along the superior plane of the parietal bone.
2. Using a scalpel, incise the skin from the occipital to the nuchal region and spread with a tissue retractor. Clear away muscular tissue using iridotomy scissors to expose the atlanto-occipital membrane.
3. Attach a 27 G needle to a 1-cc syringe and mount on the manipulating arm of the stereotaxic apparatus.

4. Carefully insert the tip of the needle into the cisterna magna. Stop as soon as penetration occurs to avoid damage to cardio-respiratory centers in the brainstem. Withdraw cerebrospinal fluid CSF (CSF; no more than 100 μL total) via slow aspiration over 1 min.

5. Very carefully withdraw the entire syringe and needle and replace with a fresh 1 cc tuberculin syringe and 27 G needle containing 125 μL freshly thawed adenovirus suspension. Carefully reinsert the syringe into the cisterna magna to the minimum depth necessary for penetration and inject the adenovirus suspension over 5 min.

6. Suture the nuchal skin and muscle and keep the animal in a nose up position in the stereotaxic apparatus for 30 min.

7. Once the animals are removed from the stereotaxic apparatus, inject sterile saline sc in the back (1 mL for mice, 10 mL for rats) to provide hydration.

8. Follow postoperative procedures as outlined in **Subheading 3.2.1**. The usual survival period is 1–7 d but can be longer depending on the nature of the experiment.

9. Investigators may wish to observe β-gal gene transfer in an unsectioned brain to view the entirety of blood vessels and meninges. If so, after perfusion/fixation protocol (**Subheading 3.3.**), process unsectioned brains for whole-mount β-gal histochemistry (**Subheading 3.4.2.**).

3.1.4. Gene Transfer in Brain and Spinal Tumor Models

These protocols require two separate surgeries: the first to implant tumor cells and the second to inject Ad5 vector. Protocols are adapted from Nilaver and coauthors *(36)*, and Colak and colleagues *(37)* (*see* **Note 4**).

3.1.4.1. LUNG CARCINOMA MODEL

1. Prepare LX-1 cells in culture media at a concentration of 9×10^4 cells/μL.

2. Follow preoperative preparation as described in **Subheading 3.1.1.**

3. Inject 10 μL of cell suspension into the striatum of nude rats with a 10 μL blunt-tipped Hamilton syringe using stereotaxic coordinates from **Table 2**. Injection is done over 20 min at 10 μL/20 min.

4. Follow postoperative care as outlined in **Subheading 3.2.1.**

5. Six days after the tumor cell injection, repeat the entire procedure of **Subheading 3.1.1.** for striatal injections using Ad5lacz vector with a total volume of 24 μL injected using a 25-μL syringe. Tumoricidal Ad5 vectors can be substituted if therapeutic protocols are being tested.

6. Following injection of the Ad5 vectors, allow survival period of 2–5 d before doing perfusion/fixation according to the protocol outlined in **Subheading 3.3.**

3.1.4.2. Rat Spinal Tumor Model

1. Prepare 9L gliosarcoma cells in culture media at a concentration of 6.6 × 10^3 cells/µL. Use normal Fisher rats because the gliosarcoma 9L cell line was originally generated in Fisher rats *(37)*.
2. Follow preoperative preparation as described in **Subheading 3.1.3.** for spinal injections. Inject 1–2 µL of tumor cell suspension into the target region, either the T4 or T5 spinal segments identified by counting thoracic vertebra from their junctions with ribs.
3. A week after tumor implantation, re-anesthetize the animals and inject into the same location as the tumor 2 µL of Ad5 expressing herpes thymidine kinase. Some small amount of scar tissue has to be removed prior to the injection.
4. Twelve hours after injection of virus, administer ganciclovir at 50 mg/kg ip twice daily for 6 consecutive days.
5. Allow a survival period of 1 wk or longer, and assess gene-therapy efficacy by looking at function (absence of paralysis as a result of tumor death owing to gene therapy) or by histological study of spinal cord sections (*see* **Subheadings 3.3.** and **3.4.1.**).

3.1.5. Intramuscular Injection

3.1.5.1. Preoperative Preparation

1. Sterilize surgical instruments and sutures. For tongue injections, sterilize forceps and a blunt probe. We use a 10-µL Hamilton syringe sterilized as described in **Subheading 3.1.1.** for small volumes or a presterilized tuberculin syringes for volumes of 100 µL or greater. Anesthetize animals as described in **Subheading 3.1.1.**
2. Inject botulism toxin if performing tongue injections for improved efficiency of retrograde transport in the hypoglossal pathway *(20)*. Inject in the tongue 8 d prior to the vector injection into the identical site. Inject the botulism toxin in the tongue in the same manner as described in **step 5** of this section. Investigators seeking to use botulism toxin in other neuromuscular pathways will have to empirically establish the amount of botulism toxin to inject.

3.1.5.2. Injection Procedure

1. Limb injection in rats is adapted from Finiels et al. *(42)*, where Ad5 vectors were injected in limb muscle with a 4-, 8- or 30-d postinjection survival period. Swab the skin overlying the deltoid, biceps, or gastrocnemius muscle with iodine and then shave to clear the hair. Make an incision over the target muscle using a scalpel. Make three injections of 10 µL each into

separate points along the muscle using a Hamilton syringe under manual control. Make each injection over 2–3 min and then hold the syringe in place for another 2–3 min before removal. Suture the skin and then inject 10 mL of saline either ip or sc, far from the limb injection. Place the animal in warmed cage as described in **Subheading 3.1.1., step 18**, and monitor until the rat is free-moving.

2. Limb injection in mice is adapted from Haase et al. and Warita et al. *(43,44)*, and from the Davidson laboratory. Make an incision in the skin overlying the gastrocnemius, triceps brachii, or thoracic musculature, and inject 25–50 µL of adenovirus into a single site. Suture the incision and inject 1 mL of saline sc away from the limb injection site. Survival period from Haase et al. *(43)* was up to 3 mo. Again volumes of up to 100 µL can be injected using tuberculin syringes (Davidson laboratory, unpublished observations). If injections are done in the neonatal period (before 7 d), expression duration is many months.

3. Tongue injection in mice is adapted from Millecamps and coauthors *(20)*, as well as our own procedures. Gently retract the tongue from the mouth of the fully anesthetized mouse, and inject 10–50 µL of the Ad5 vector with the Hamilton syringe under manual control over 2–3 min, leaving the syringe in place for 2–3 min before withdrawing. If botulism toxin was injected, inject the vector into the identical site. Provide hydration with 1 mL of sterile saline injected ip postoperatively.

3.2. Postoperative Care

3.2.1. Routine Postoperative Care

Check animals on a daily basis for normal spontaneous movement and any sign of respiratory distress or dehydration. Provide gruel paste made from water and lab chow in one corner of cage as well as blocks of unflavored gelatin. The soft gruel and gelatin are especially important after tongue injections. If animals appear dehydrated they can be given ip or sc injections of sterile saline (1 mL for mice, 10 mL for rats). Give postoperative analgesia with buprenorphine twice a day (0.05–0.1 mg/kg for mice and 0.01–0.05 mg/kg for rats) for up to one postoperative week, and promptly euthanize any animal in sustained distress not relieved by analgesia. Finally, occasionally mice and rats will fight to the point of severe injury. If animals are fighting, house them in separate cages.

3.2.2. Postoperative Immunosuppression

1. For rats, give cyclosporine at 20 mg/kg, administered ip daily from postoperative d 2 and as needed *(45)*, or FK-506 ip at 1 mg/kg daily also from postoperative d 2 with continued use as needed.

2. In mice, to prolong Ad5 transgene expression, give cyclosporine at a daily dose of 15 mg/kg ip *(46)*.

3.3. Perfusion/Fixaton

It is recommended that first time histologists work closely with an experienced histologist for all aspects of fixation, freezing, cutting, mounting, and other histological methods.

1. Prepare the perfusion apparatus, which consists of a peristaltic perfusion pump, flexible tubing, and a 22 G-butterfly catheter. Manually compress one end of the flexible tubing and force it inside the connection end of the 22 G-butterfly catheter. The tubing will then expand to form a tight seal. Set up beakers containing PBS and fixative. Place the end without the catheter into the beaker containing PBS; the peristaltic action of the pump will force solution through the tubing and out the cannula. Turn on the pump and verify that the entire circuit is working before anesthetizing any animals. Run PBS through the circuit until all air is removed from the line.

2. Set the perfusion pump next to a large tray containing a perfusion board made of Styrofoam or wax. Or, place the perfusion board over a sink using a plastic support. The basic idea is to have the perfusion done over a surface to collect blood and used fixative. Lay out all instruments where they are easily accessible: anesthetic, pins, scissors for cutting the chest cavity, dissecting equipment (forceps, curette, scalpel, and ronguers), and containers to postfix the brain and spinal cord tissue immediately after removal and dissection.

3. Anesthetize mice and rats with ip injections as outlined in **Subheading 2.1.1.** except that doses are slightly higher to ensure quick onset of deep anesthesia: inject mice with 0.4–0.5 mL of ketamine/xyalzine and rats with 60–75 mg/kg of sodium pentobarbital. If several animals are being perfused on a given session, it is advisable to keep the others removed from where the perfusions are being done. Confirm full depth of anesthesia before proceeding by verifying no withdrawal to toe pinch.

4. Pin fully anesthetized animals to the board through all four limbs. Using dissecting scissors, make an incision just below the rib cage, exposing liver, diaphragm, and rib cage.

5. Cut the rib cage very quickly to expose the heart and lungs. Make one large lateral cut from diaphragm to the upper rib cage. Cut across the top of sternum and then down to retract the rib cage.

6. Cut the left atrium with iridotomy scissors and quickly insert the pointed tip of the butterfly cannula into the apex of the left ventricle.

7. Turn the perfusion pump on and pump 50 mL of saline into mice or 100–200 mL of saline into rats to force out all blood volume. The liver will noticeably blanch.

8. Turn the perfusion pump off and transfer the intake side of the hose to the paraformaldehyde fixative. Pump 50 mL of fixative through mice and 100–200 mL through rats, depending on the size. Perfusion fixation should produce cross-linking of muscle proteins, so that muscles contract and the animal assumes rigor mortis. The aforementioned procedure must be done smoothly so that histology is well-preserved. Common problems include air emboli, perfusing too slowly or too quickly, puncturing the ventricle, or severing the aorta. All of these result in suboptimal perfusion and tissue that is hard to remove and harder to cut, yielding sections of poor histology.

9. Turn off the apparatus and remove the cannula from the heart. If further perfusions are planned, the lines should be rinsed with at least 200–300 mL of PBS. Otherwise flush with 200–300 mL of tap water to clean the lines prior to storage. For processing the brain, continue with **steps 10**, **11**, and **14**; also include **step 12** if only a portion of the brain will be needed. For processing the spinal column, follow **steps 13** and **14**. For staining the brain *en bloc*, follow **steps 10** and **11**, and then proceed to **Subheading 3.4.3.**

10. Decapitate the animal at the atlantoccipital membrane and use ronguers to break way pieces of the skull from the foramen magnum up to the cribiform plate of the ethmoid (essentially from the top back of the skull to the top front) to efficiently remove the brain. Insert a small curette into the front of the brain to gently slide it out, and, using iridotomy scissors, cut away any adherent dura mater. If olfactory bulbs are to be saved, completely remove the cribiform plate of the ethmoid bone with ronguers and use the curette to separate the anterior most poles of the olfactory bulbs from the ethmoid bone.

11. Place the brain immediately into fresh 4% paraformaldehyde fixative for at least 2 h and up to 24 h at 4°C to postfix tissue for optimal cutting.

12. If only particular regions of the brain are needed, place the brains into plastic sagittal or coronal slicers, which contain slits at 1-mm distances for cutting brain blocks. Postfix and cryoprotect these smaller slices the same as whole brains (**step 14**).

13. To remove the spinal cords, decapitate the animals as in **step 10**, using ronguers to widen the opening at the foramen magnum and then remove the lamina and spinous processes progressively down the length of the spinal cord. If only a particular region of the spinal cord is of interest, use a scalpel to transect the vertebral column and spinal cord several segments above the desired region and begin working downward from the new cut point. For instance, transect the vertebral column and canal at the midthoracic region

if the lower thoraci-upper lumbar spinal cord is desired. In this way, the dissection process can be considerably shortened. Once the vertebral canal has been completely exposed, nerve rootlets should be visible. Cut the nerve rootlets with iridotomy scissors to remove the spinal cord. Even if the entire cord is being removed, take it in at least three to four portions rather than all at once to avoid damage. If recovery of the dorsal root ganglia is desired, meticulously dissect all muscle and soft tissue surrounding the vertebral canal. Place the spinal cord into fresh 4% paraformaldehyde fixative for at least 2 h and up to 24 h at 4°C to postfix tissue for optimal cutting.

14. After postfixation, transfer the brain or spinal blocks, or brain slices to buffered 30% sucrose at 4°C for 24–72 h before sectioning on a cryostat (**Subheading 3.4.1.**). When the sucrose has fully infiltrated the tissue, the brain or spinal cord blocks will sink in the 30% sucrose. The sucrose prevents ice crystals from tearing the tissue when it is frozen for sectioning.

3.4. Preparation and Processing of Brain and Spinal Cord Sections

3.4.1. Cutting Frozen Sections

Cut the brains in sagittal, transverse, or coronal planes, depending on the region of study. Note that the sections are made from the bottom face of the block upward. We advise cutting spinal cords in cross section rather than in other planes, in order to visualize all of the lamina.

3.4.1.1. PREPARATION

1. Put blocked brains and spinal cord in Peel-a-way molds and cover with OCT. If the blocks can stand erect without support, the tissue molds can be placed directly into −80°C freezer until cutting.
2. Tissue blocks with only a small flat surface (such as spinal cord or blocks of cerebrum) will not stay upright for the time it takes the OCT to freeze at −80°C. For these, make a bath of dry ice and 95% ethanol, and hold the tissue blocks upright with forceps until they are frozen in place before transferring to the at −80°C freezer.
3. Store OCT-blocked brains indefinitely at −80°C until sectioning. We find it helpful to place paper labels underlying the OCT and extending to the outside of the Peel-a-way molds with identifying data about the specimens.

3.4.1.2. SECTIONING

1. Fully cool the cryostat to equilibrium temperature before sectioning. Transport frozen brain blocks on dry ice from storage to the cryostat. Peel the

plastic mold away from the OCT-blocked brain and use OCT to freeze onto the chuck of the cryostat.

2. Set cryostat thickness mechanically or electronically at 8–50 μm, depending on the application. If sections are 20 μm or thinner they can be transferred directly from the cryostat stage onto glass slides (two to three sections can be placed on one slide). Keep slides in the cryostat while cutting to keep the tissue at a freezing temperature, then transfer the slides to slide boxes and store at −20°C or at least −4°C until processing. For both thick and thin sections, use Fisher Superfrost slides or other coated slides that will tightly adhere to sections, otherwise, sections will float away during histochemical processing.

3. For thicker sections (20–50 μm), use forceps to grab an edge of the section, and quickly place them into a well of a 24-well tissue-culture dish filled with PBS. Keep thick sections in PBS with 0.2% azide at 4°C if they are not going to be mounted right away. To mount thick sections, float them in a Petri dish with PBS; using a thin paintbrush, gently pull them onto the slide. With practice, up to 20–30 brain sections can be mounted onto a single slide; the trick is to hold up the slide to drain off buffer after mounting each section before mounting another. Expensive paintbrushes from microscopy companies are available, but in our experience cheap paintbrushes from general merchandise outlets work just as well.

4. Air-dry thick sections for at least 4–6 h before doing histochemical processing (**Subheading 3.4.2.**).

5. If histochemical processing is to be done within 1–2 wk, keep the sections refrigerated at 4°C; if a longer delay is anticipated, freeze the sections at −20°C, but histology may be suboptimal.

3.4.2. β-Galactosidase Histochemistry

All processing should be done with glass or plastic slide holders and containers, as transition metals in the X-gal mix may react with metal slide holders. For small numbers of slides we prefer 100-mL glass Coplin jars. For large numbers of slides, plastic slide holders that fit into plastic 200 mL rectangular containers work well. Up to 48 slides can be placed back to back in a 24-slide holder that fits into the rectangular containers; 200 mL of solution should be adequate for each step.

1. Rinse slides twice with cold PBS. Make X-gal staining solution by adding X-gal substrate at a dilution of 1/40 (5 mL of X-gal substrate to 200 mL for 48 slides) to the KC mixer.

2. Transfer the slides to the freshly made staining solution (this X-gal/KC mix cannot be made up in advance), and cover the container with aluminum foil

to keep out light. Incubate at 37°C for 4 h. Do not use a tissue culture incubator. A general bacterial incubator works well. Alternatively, a walk-in warm room can be used.

3. Rinse slides three times in PBS, room temperature.
4. Air-dry slides overnight.
5. If counterstaining is desired, rinse in neutral red for 40 s, wash twice with ddH$_2$O, and air dry overnight for identification of landmarks. We find it helpful to counterstain only a portion of the sections, while saving the rest for detection of β-gal positive cells.
6. Overlay the slides with Permount, cover with glass coverslips, and lay slides flat for at least another 24 h before putting into storage. Some investigators dehydrate slides in alcohol and clear with xylenes prior to coverslipping, but this often decreases the intensity of the β-gal stain. If clearing is desired, use Histoclear in place of xylenes, as the X-gal reaction product is fairly soluble in xylene.

3.4.3. Whole Mount β-Galactosidase Histochemistry

This protocol is adapted from Ooboshi et al. and Ooboshi et al. *(35,47)*. Although the fixative concentration in these reports was 2% paraformaldehyde with 0.2% glutaraldehyde (*see* **Note 6**), we have tested the whole mount protocol on mouse brain using 4% paraformaldehyde (Davidson lab, unpublished observations).

1. Prepare KC mixer, X-gal substrate and PBS and 4% paraformaldehyde in advance as described in **Subheading 3.4.2.**
2. After fixation, remove and save the 4% paraformaldehyde fixative, rinse brains twice in room temperature PBS, and then transfer brains to KC mixer with 1/40 X-gal substrate as described in **Subheading 3.4.2.** at 37°C for 4 h. A bright blue stain will appear on the surface of meninges and cerebral blood vessels.
3. Rinse brains again in PBS and save in the original 4% paraformaldehyde fixative at 4°C until photographed.
4. If desired, cryoprotect the wholemount brains in 30% sucrose (**Subheading 3.3., step 14**) and then cut sections on a cryostat. Re-stain the individual tissue sections for β-gal histochemistry (**Subheading 3.4.2.**).

4. Notes

1. In prior work we compared manual injections to microprocessor controlled injections *(48)*. Multiple vector systems were tested. The mechanical injector had significantly greater efficiency of gene transfer and decreased intra-animal variability for all vectors evaluated. Over the past 5 yr, all of

the members of the Davidson laboratory have had similar experience, and for this reason we suggest using mechanical injectors and automated controls except for intramuscular (im) and cerebellar injections. Intramuscular injections are done completely manually with syringes, whereas for cerebellar injections we recommend using the mechanical injection but under manual control, delivering 1 μL over the course of 2–3 min. We have also found certain other steps helpful in making vector injections go more smoothly. We routinely use separate cages for anesthesia with the remaining animals removed from the anesthetized ones. To anesthetize mice and rats, we pick them up by the tail and let the animal grasp the top of the cage with the forepaws. We administer the anesthetic ip and put the animal into a second cage to allow the anesthetic to take effect (usually 10–15 min). It has been our experience that placing newly anesthetized mice and rats back into the original cage causes the other animals to become very agitated. For the same reason, we advise a separate recovery cage once the animals have been taken off the warming lamp.

2. Cyclosporine administered to rats at a dose of 20 mg/kg allowed for the development of significantly reduced immune responses to Ad5 subsequent to injection into the hypothalamus *(45)*. There is no direct example of cyclosporine attenuating the immune response against Ad vectors in mouse brain models, but Petrof and co-workers prolonged expression of Ad5 β-gal in muscle of mice with cyclosporine at a dose of 15 mg/kg *(46)*. Also, Di Polo and colleagues extended expression of Ad5 brain-derived neurotrophic factor in retina of rats using the transplant drug FK-506 at 1 mg/kg *(49)*. In addition to chemical immunosuppression, intrathymic inoculation may prolong Ad transgene expression. DeMatteo et al. *(50)* demonstrated that an injection of 1 μL of Ad5 expressing β-gal to each thymus lobe of 3-d-old C57Bl/6 mice attenuated the cytotoxic T lymphocyte response against subsequent liver administrations of Ad5 β-gal. Hepatocytes expressed β-gal out to 240 d, whereas mice not receiving intrathymic Ad5 lost transgene expression by 30–60 d *(50)*. In theory, this same procedure could work to prolong Ad transgene expression in brain and spinal cord.

3. Ghodsi et al. *(51)* administered 3 mL of mannitol per 100 g of body weight to mice 10 min before injecting 5 μL of Ad5 vector in the lateral ventricle. In the mannitol treated animals, there was scattered transduction of cells in the neuropil surrounding the ventricles; without mannitol, intraventricular injections result in ependyma transduction only.

4. In Nilaver et al., Ad5 expressing β-gal was injected into human lung carcinoma tumors implanted into nude mice. Up to 20% of tumor cells could be transduced by the vector *(36)*. Colak and colleagues tested a tumoricidal model of gene therapy, using Ad5 to express thymidine kinase from

herpes simplex virus within spinal gliosarcoma cells that had been injected into the rat spinal cord. Subsequent treatment with Ganciclovir for 6 d completely eradicated tumors in all treated animals in the short term (18 d after injection). Of the five animals followed out to 6 mo, two were tumor-free *(36)*.

5. Injections of Ad5 expressing β-gal into the cisterna magna of mice and rats resulted in intense β-gal staining of the ventral surface of brain *(33,35)*. Furthermore, cisternal injection can also lead to vector distribution and gene transfer on the pia mater of the spinal cord *(34)*.

6. Published techniques from the literature for β-gal and β-glucuronidase histochemistry following Ad5 gene transfer in the brain use fixative concentrations of 2–4% paraformaldehyde; sometimes 2% paraformaldehyde is combined with 0.2% glutaraldehyde. In the experience of the Davidson laboratory, both 2% and 4% paraformaldehyde work equally well. Even for whole-mount processing, 4% paraformaldehyde with no glutaraldehyde has produced good results. We recommend starting with 4% paraformaldehyde without glutaraldehyde in the protocols listed, with subsequent optimization depending on the histologic stains being used, or the antigens being evaluated.

Acknowledgments

These protocols were developed with support from the NIH (NS34568, HD33531, DK54759), the Roy J. Carver Trust, and the Amyotrophic Lateral Sclerosis Association. We thank Christine McLennan for expert assistance in preparation of this manuscript.

References

1. Horwitz, M. S. (1990) Adenoviridae and their replication, in *Virology*, 2nd ed., (Fields, B. N., and Knipe, D. M., eds.), Raven Press, Ltd, New York, NY, pp. 1679–1721.

2. Millecamps, S., Kiefer, H., Navarro, V., Geoffroy, M. C., Robert, J. J., Finiels, F., et al. (1999) Neuron-restrictive silencer elements mediate neuron specificity of adenoviral gene expression. *Nat. Biotechnol.* **17**, 865–869.

3. Chillon, M., Bosch, A., Zabner, J., Law, L., Armentano, D., Welsh, M. J., and Davidson, B. L. (1999) Group D adenoviruses infect primary central nervous system cells more efficiently than those from Group C. *J. Virol.* **73**, 2537–2540.

4. Xia, H., Anderson, B., Mao, Q., and Davidson, B. L. (2000) Recombinant human adenovirus: targeting to the human transferrin receptor improves gene transfer to brain microcapillary endothelium. *J. Virol.* **74**, 11359–11366.

5. Davidson, B. L. and Bohn, M. C. (1997) Recombinant adenovirus: a gene transfer vector for study and treatment of CNS diseases. *Exp. Neurol.* **144**, 125–130.

6. Heistad, D. D. and Faraci, F. M. (1996) Gene therapy for cerebral vascular disease. *Stroke* **27**, 1688–1693.

7. Soudais, C., Laplace-Builhe, C., Kissa, K., and Kremer, E. J. (2001) Preferential transduction of neurons by canine adenovirus vectors and their efficient retrograde transport in vivo. *FASEB J.* **15**, 2283–2285.

8. Chillon, M. and Kremer, E. J. (2001) Trafficking and propagation of canine adenovirus vectors lacking a known integrin-interacting motif. *Hum. Gene Ther.* **12**, 1815–1823.

9. Umana, P., Gerdes, C. A., Stone, D., Davis, J. R., Ward, D., Castro, M. G., and Lowenstein, P. R. (2001) Efficient FLPe recombinase enables scalable production helper-dependent adenoviral vectors with negligible helper-virus contamination. *Nat. Biotechnol.* **19**, 582–585.

10. Byrnes, A. P., Rusby, J. E., Wood, M. J. A., and Charlton, H. M. (1995) Adenovirus gene transfer causes inflammation in the brain. *Neuroscience* **66**, 1015–1024.

11. Kajiwara, K., Byrnes, A. P., Charlton, H. M., Wood, M. J., and Wood, K. J. (1997) Immune responses to adenoviral vectors during gene transfer in the brain. *Hum. Gene Ther.* **8**, 253–265.

12. Liu, Y., Himes, B. T., Moul, J., Huang, W., Chow, S. Y., Tessler, A., and Fischer, I. (1997) Application of recombinant adenovirus for in vivo gene delivery to spinal cord. *Brain Res.* **768**, 19–29.

13. Davidson, B. L., Allen, E. D., Kozarsky, K. F., Wilson, J. M. and Roessler, B. J. (1993) A model system for *in vivo* gene transfer into the central nervous system using an adenoviral vector. *Nat. Genet.* **3**, 219–223.

14. Anderson, R. D., Haskell, R. E., Xia, H., Roessler, B. J., and Davidson, B. L. (2000) A simple method for the rapid generation of recombinant adenovirus vectors. *Gene Ther.* **7**, 1034–1038.

15. Aoki, K., Barker, C., Danthinne, X., Imperiale, M. J., and Nabel, G. J. (1999) Efficient generation of recombinant adenoviral vectors by Cre-lox recombination in vitro. *Mol. Med.* **5**, 224–231.

16. Alisky, J. M. and Davidson, B. L. (2000) Gene therapy for amyotrophic lateral sclerosis and other motor neuron diseases. *Hum. Gene Ther.* **11**, 2315–2329.

17. Ridoux, V., Robert, J. J., Zhang, X., Perricaudet, M., Mallet, J., and La Salle, G. L. G. (1994) Adenoviral vectors as functional retrograde neuronal tracers. *Brain Res.* **648**, 171–175.

18. Kuo, H., Ingram, D. K., Crystal, R. G., and Mastrangeli, A. (1995) Retrograde transfer of replication deficient recombinant adenovirus vector in the central nervous system for tracing studies. *Brain Res.* **705**, 31–38.

19. Ghadge, G. D., Roos, R. P., Kang, U. J., Wollmann, R., Fishman, P. S., Kalynych, A. M., et al. (1995) CNS gene delivery by retrograde transport of recombinant replication-defective adenoviruses. *Gene Ther.* **2**, 132–137.

20. Millecamps, S., Nicholle, D., Ceballos-Picot, I., Mallet, J. and Barkats, M. (2001) Synaptic sprouting increases the uptake capacities of motoneurons in amyotrophic lateral sclerosis mice. *Proc. Natl. Acad. Sci. USA* **98**, 7582–7587.

21. Terashima, T., Miwa, A., Kanegae, Y., Saito, I., and Okado, H. (1997) Retrograde

and anterograde labeling of cerebellar afferent projection by the injection of recombinant adenoviral vectors into the mouse cerebellar cortex. *Anat. Embryol. (Berl)* **196**, 363–382.

22. Draghia, R., Caillaud, C., Manicom, R., Pavirani, A., Kohn, A., and Poenaru, L. (1995) Gene delivery into the central nervous system by nasal instillation in rats. *Gene Ther.* **2**, 418–423.

23. Sandmair, A. M., Loimas, S., Puranen, P., Immonen, A., Kossila, M., Puranen, M., et al. (2000) Thymidine kinase gene therapy for human malignant glioma, using replication-deficient retroviruses or adenoviruses. *Hum. Gene Ther.* **11**, 2197–2205.

24. Eck, S. L., Alavi, J. B., Judy, K., Phillips, P., Alavi, A., Hackney, D., et al. (2001) Treatment of recurrent or progressive malignant glioma with a recombinant adenovirus expressing human interferon-beta (H5.010CMV*hIFN-b*): a phase I trial. *Hum. Gene Ther.* **12**, 97–113.

25. Thomas, C. E., Birkett, D., Anozie, I., Castro, M. G., and Lowenstein, P. R. (2001) Acute direct adenoviral vector cytotoxicity and chronic, but not acute, inflammatory responses correlate with decreased vector-mediated transgene expression in the brain. *Mol. Ther.* **3**, 36–46.

26. Choi-Lundberg, D. L., Lin, Q., Chang, Y.-N., Chiang, Y. L., Hay, C. M., Mohajeri, H., et al. (1997) Dopaminergic neurons protected from degeneration by GDNF gene therapy. *Science* **275**, 838–841.

27. Ghodsi, A., Stein, C., Derksen, T., Yang, G., Anderson, R. D., and Davidson, B.L. (1998) Extensive b-glucuronidase activity in murine CNS after adenovirus mediated gene transfer to brain. *Hum. Gene Ther.* **9**, 2331–2340.

28. Davidson, B. L., Stein, C. S., Heth, J. A., Martins, I., Kotin, R. M., Derksen, T.A., et al. (2000) Recombinant AAV type 2, 4 and 5 vectors: transduction of variant cell types and regions in the mammalian CNS. *Proc. Natl. Acad. Sci. USA* **97**, 3428-3432.

29. Brooks, A. I., Stein, C. S., Hughes, S. M., Heth, J., McCray, P. B. Jr., Sauter, S.L., et al. (2002) Functional correction of established CNS deficits in an animal model of lysosomal storage disease using feline immunodeficiency virus-based vectors. *Proc. Natl. Acad. Sci. USA* **99,** 6216–6221.

30. DiPasquale, G., Davidson, B.L., Stein, C.S., Martins, I., Scudiero, D., Monks, A., et al. (2003). Identification of PDGFR as a receptor for AAV5 transduction. *Nat. Med.* In press.

31. Franklin, K. B. J. and Paxinos, G. (1997) *The Mouse Brain in Stereotaxic Coordinates*, Academic Press, San Diego, CA, pp. 1–190.

32. Paxinos, G. and Watson, C. (1986) *The Rat brain in Stereotaxic Coordinates*, Academic Press Inc., New York, NY.

33. Christenson, S. D., Lake, K. D., Ooboshi, H., Faraci, F. M., Davidson, B. L., and Heistad, D. D. (1998) Adenovirus-mediated gene transfer in vivo to cerebral blood vessels and perivascular tissue in mice. *Stroke* **29**, 1411–1415.

34. Finegold, A. A., Mannes, A. J., and Iadarola, M. J. (1999) A paracrine paradigm for in vivo gene therapy in the central nervous system: treatment of chronic pain. *Hum. Gene Ther.* **10**, 1251–1257.

35. Ooboshi, H., Welsh, M. J., Rios, C. D., Davidson, B. L., and Heistad, D. D. (1995) Adenovirus-mediated gene transfer in vivo to cerebral blood vessels and perivascular tissue. *Circ. Res.* **77**, 7–13.

36. Nilaver, G., Muldoon, L. L., Kroll, R. A., Pagel, M. A., Breakefield, X. O., Davidson, B. L., and Neuwelt, E. A. (1995) Delivery of herpesvirus and adenovirus to nude rat intracerebral tumors after osmotic blood-brain barrier disruption. *Proc. Natl. Acad. Sci. USA* **92**, 9829–9833.

37. Colak, A., Goodman, J. C., Chen, S. H., Woo, S. L. C., Grossman, R. G., and Shine, H. D. (1995) Adenovirus-mediated gene therapy for experimental spinal cord tumors: tumoricidal efficacy and functional outcome. *Brain Res.* **691**, 76–82.

38. Ross, B. D., Kim, B., and Davidson, B. L. (1995) Assessment of ganciclovir toxicity to experimental intracranial gliomas following recombinant adenoviral-mediated transfer of the herpes simplex virus thymidine kinase gene by magnetic resonance imaging and proton magnetic resonance spectroscopy. *Clin. Cancer Res.* **1**, 651–657.

39. Boulis, N. M., Bhatia, V., Brindle, T. I., Holman, H. T., Krauss, D. J., Blaivas, M., and Hoff, J.T. (1999) Adenoviral nerve growth factor and beta-galactosidase transfer to spinal cord: a behavioral and histological analysis. *J. Neurosurg.* **90(Suppl 1)**, 99–108.

40. Mannes, A. J., Caudle, R. M., O'Connell, B. C., and Iadarola, M. J. (1998) Adenoviral gene transfer to spinal cord neurons: intrathecal vs. intraparenchymal administration. *Brain Res.* **793**, 1–6.

41. Alisky, J. M. and Tolbert, D. L. (1997) Quantitative analysis of converging spinal and cuneate mossy fiber afferent projections to the rat cerebellar anterior lobe. *Neuroscience* **80**, 373–388.

42. Finiels, F., Gimenez, Y., Ribotta, M., Barkats, M., Samolyk, M., Robert, J., et al. (1995) Specific and efficient gene transfer offers new potentialities for the treatment of motor neurone diseases. *NeuroReport* **6**, 2473–2478.

43. Haase, G., Kennel, P., Pettmann, B., Vigne, E., Akli, S., Revah, F., et al. (1997) Gene therapy of murine motor neuron disease using adenoviral vectors for neurotrophic factors. *Nat. Med.* **3**, 429–436.

44. Warita, H., Abe, K., Setoguchi, Y. and Itoyama, Y. (1998) Expression of adenovirus-mediated E. coli lacZ gene in skeletal muscles and spinal motor neurons of transgenic mice with a mutant superoxide dismutase gene. *Neurosci. Lett.* **246**, 153–156.

45. Geddes, B. J., Harding, T. C., Hughes, D. S., Byrnes, A. P., Lightman, S. L., Conde, G. and Uney, J. B. (1996) Persistent transgene expression in the hypothalamus following stereotaxic delivery of a recombinant adenovirus: supresion of the immune response with cyclosporin. *Endocrinology* **137**, 5166–5169.

46. Petrof, B. J., Acsadi, G., Jani, A., Massie, B., Bourdon, J., Matusiewicz, N., et al. (1995) Efficiency and functional consequences of adenovirus-mediated in vivo gene transfer to normal and dystrophic (mdx) mouse diaphragm. *Am. J. Respir. Cell Mol. Biol.* **13**, 508–517.

47. Ooboshi, H., Rios, C. D., and Heistad, D. D. (1997) Novel methods for adenovirus-mediated gene transfer to blood vessels in vivo. *Mol. Cell Biochem.* **172**, 37–46.

48. Brooks, A. I., Halterman, M. W., Chadwick, C. A., Davidson, B. L., Haak-Fredenscho, M., Radel, C., et al. (1997) Reproducible and efficient murine CNS gene delivery using a microprocessor-controlled injector. *J. Neurosci. Methods* **80**, 137–147.

49. DiPolo, A., Lerner, L. E., and Farber, D. B. (1997) Transcriptional activation of the human rod cGMP-phosphodiesterase beta-subunit gene is mediated upstream AP-1 element. *Nucleic Acids Res.* **25**, 3863–3867.

50. DeMatteo, R. P., Chu, G., Ahn, M., Chang, E., Barker, C. F., and Markmann, J. F. (1997) Long-lasting adenovirus transgene expression in mice through neonatal intrathymic tolerance induction without the use of immunosuppression. *J. Virol.* **71**, 5330–5335.

51. Ghodsi, A., Stein, C., Derksen, T., Martins, I., Anderson, R. D., and Davidson, B. L. (1999) Systemic hyperosmolality improves b-glucuronidase distribution and pathology in murine MPS VII brain following intraventricular gene transfer. *Exp. Neurol.* **160**, 109–116.

52. Vasquez, E. C., Beltz, T. G., Haskell, R. E., Johnson, R. F., Meyrelles, S. S., Davidson, B. L., and Johnson, A. K. (2001) Adenovirus-mediated gene delivery to cells of the magnocellular hypothalamo-neurohypophyseal system. *Exp. Neurol.* **167**, 260–271.

53. Masumura, M., Hata, R., Nishimura, I., Uetsuki, T., Sawada, T., and Yoshikawa, K. (2000) Caspase-3 activation and inflammatory responses in rat hippocampus inoculated with a recombinant adenovirus expressing the Alzheimer amyloid precursor protein. *Mol. Brain Res.* **80**, 219–227.

54. Nishimura, I., Uetsuki, T., Dani, S. U., Ohsawa, Y., Saito, I., Okamura, H., et al. (1998) Degeneration in vivo of rat hippocampal neurons by wild-type Alzheimer amyloid precursor protein overexpressed by adenovirus-mediated gene transfer. *J. Neurosci.* **18**, 2387–2398.

55. Abe, K., Setoguchi, Y., Hayashi, T., and Itoyama, Y. (1997) In vivo adenovirus-mediated gene transfer and the expression in ischemic and reperfused rat brain. *Brain Res.* **763**, 191–201.

56. Kitagawa, H., Sasaki, C., Sakai, K., Mori, A., Mitsumoto, Y., Mori, T., et al (1999) Adenovirus-mediated gene transfer of glial cell line-derived neurotrophic factor prevents ischemic brain injury after transient middle cerebral artery occlusion in rats. *J. Cereb. Blood Flow Metab.* **19**, 1336–1344.

57. Vasquez, E. C., Beltz, T. G., Meyrelles, S. S., and Johnson, A. K. (1999) Adenovirus-mediated gene delivery to hypothalamic magnocellular neurons in mice. *Hypertension* **34**, 756–761.

8

Adenovirus-Mediated Gene Transfer to Tumor Cells

Manel Cascalló and Ramon Alemany

1. Introduction

Cell transduction in vitro is only the first step toward proving that a gene-therapy vector can be useful to treat tumors. However, tumor targeting in vivo is now the milestone for gene therapy to succeed against disseminated cancer. Therefore, most valuable information is obtained from studies of vector biodistribution. Owing to the hepatotropism of adenoviral vectors, a particularly important parameter is the tumor/liver ratio. This ratio can be given at the level of gene expression if the amount of transgene expression is measured. To optimize the targeting, however, the levels of viral particles that reach the tumor compared to other organs must be studied. Most of this chapter deals with methods to quantify the virus fate in tumor-bearing animals. We present a radioactive labeling method that can be used to study biodistribution. After a small section dealing with tumor models, we describe methods to quantify different parameters related to adenovirus-mediated tumor targeting.

Multicellular tumor spheroids have been developed as an attempt to design in vitro systems that take into consideration the cellular heterogeneity associated with the three-dimensional arrangement of solid tumors *(1)*. Different methods can be used to generate tumor spheroids. We describe one of the most widely used, the liquid-overlay technique, which is based on the culture of cells or pieces of tumors obtained from surgical operations in dishes covered with a nonadhesive surface. In vivo, subcutaneous (sc) implantation of tumor cells in nude mice constitutes an easy way to evaluate the potency of different antitu-

From: *Methods in Molecular Biology, vol. 246:*
Gene Delivery to Mammalian Cells: Vol. 2: Viral Gene Transfer Techniques
Edited by: W. C. Heiser © Humana Press Inc., Totowa, NJ

moral approaches. The inability of nude mice to develop a complete immune response to xenotransplants provides a method to generate human tumor models of different origin with a rate of acceptance close to 100%. Moreover, the accessibility of the tumor has contributed to the broad use of this technique. The main limitation of the model lays in the poor vascularization of the sc tissue, which impairs tumor spreading. Orthotopic transplantation of solid fragments of human tumors in nude mice allows implantating tumors into the same physiological environment where the tumor cells arise *(2)*. In this site, the cells reproduce their pattern of local growth and distal dissemination, providing a more realistic approach to the behavior of cancers *in vivo*.

Regarding the delivery of the vector, direct delivery into tumors via intratumoral injection is straightforward and is not presented here. It is worth mentioning that these injections should be done in a small volume (20 µL) and that repeated injection often leads to leaks, making it difficult to quantitate the amount of vector actually delivered. We have, however, focused on systemic delivery to tumors. In these protocols, two important parameters need to be determined: vector clearance and biodistribution *(3)*. Besides the detection of a prelabeled vector, the distribution of vector among different organs can be determined at the level of DNA or gene expression. Here we present quantitative polymerase chain reaction (PCR) and luciferase-detection methods to assess these parameters.

In rodents and monkeys, the liver is the main organ to be analyzed for toxicity of adenoviral vectors. In contrast, in other animals, toxicity may be reflected in other organs such as lungs in pigs *(4)*. Hematoxilin-eosin staining of these two organs to check for neutrophil infiltration is a good way to monitor toxicity. Besides the measurement of transaminase levels that we present to check for hepatic toxicity, other parameters may be monitored such as clotting alterations, a drop in blood oxygen levels, or a rise in blood ammonium levels because these may be associated with adenovirus administration. In rodents, an acute toxicity phase has been observed 2 d postinjection and another delayed toxicity phase related to specific immune responses at 1 wk postinjection. As an initial approach, a d 5 measure would give a good estimation of overall toxicity.

Finally several methods dealing with replication-competent vectors are presented. The understanding of viral and cancer biology has progressed in such a way that it has been possible to develop viruses that replicate selectively in tumor cells by genetic engineering *(5)*. These approaches eradicate tumors as a consequence of virus replication. Moreover, replication also allows the amplification of the initial viral load by progeny viruses released after cell lysis. The oncolytic potency of these vectors is usually assessed by quantification of the extent of the cytophatic effect (CPE) in cultures of permissive cells. Staining of attached cells (crystal violet method) or assaying the total protein content (BCA protocol based on the method of O'Carroll et al. *[6]*) are used to determine the extent of the potency of these kinds of vectors. In vivo, the viral load present in a tumor is also an indication of the capacity of replication. Because hexon from the administered virus is degraded within a few days and cannot be detected, hexon staining may also reveal whether viral replication has occurred. Hexon detection can be performed by immunohistochemical methods or by FACS analysis of tumor digests.

2. Materials

1. ^{32}P orthophosphate carrier free, methyl-^{3}H thymidine (Amersham Pharmacia Biotech, Piscataway NJ).
2. Luciferase cell lysis reagent: 25 mM glycylglycine, pH 7.8, 15 mM MgSO$_4$, 4 mM EGTA, 1 mM dithiothreitol (DTT), 1% Triton X-100.
3. Luciferase assay reagent: 25 mM glycylglycine, pH 7.8, 15 mM MgSO$_4$, 4 mM EGTA, 1 mM DTT, 15 mM K$_2$HPO$_4$, pH 7.8, 2 mM ATP, 60 µg/mL luciferin (Promega, Madison, WI).
4. Luminometer (e.g., Monolight 2010 Luminometer, Analytical Luminescence Laboratory, Ann Arbor, MI).
5. 2X PCR Master mix (Applied Biosystems Inc., Foster City, CA, Cat. no. 4304437).
6. Male BALB/c nude mice, weighing 23–27 g. Animals must be housed in a sterile environment; cages and water should be autoclaved, and bedding and food, should be γ-ray sterilized.
7. Prolene 6-0 suture for surgical application.
8. Cell lines and media: the cytopathic effect (CPE) provided by replicative adenoviral vectors can be evaluated in many standard human cell lines depending on the specificity of the replicative vector. 293 cells are commonly used because they are readily infected by adenovirus. Standard growth media, such as Dulbecco's modified Eagle's medium (DMEM), supplemented with 10% heat-inactivated fetal calf serum (FCS) is used for CPE determination by the crystal violet protocol. Use DMEM + 0.5% heat inactivated-FCS when assaying CPE by the BCA method.
9. Fixation solution: Dissolve 4 g of paraformaldehyde in 100 mL of PBS by heating to 65–70°C. Prepare in a fume hood just before use.
10. Crystal violet solution: 0.2 g of crystal violet dissolved in 100 mL of distilled water.
11. Bio-Rad DC Protein Assay Kit (Bio-Rad Laboratories, Hercules, CA, Cat. no. 500-0112).
12. Working solution for BCA™ assay: mix 50 parts of BCA Reagent A (Pierce Chemical, Rockford, IL) with 1 part of BCA Reagent B. Prepare just before use.
13. Real-time thermal cycler (e.g., ABI PRISM 7000 Sequence Detection System, Applied Biosystems).

14. GPT-GOT Transaminase detection kit, containing GPT and GOT substrates and color reagent (Sigma, St. Louis, MO, Cat. no. 505-OP).
15. Blocking solution for hexon detection: 20% goat serum (Sigma, Cat. no. G9023).
16. Primary antibody for hexon detection: PAb α-hexon (Chemicon, Temecula, CA, Cat. no. MAB8052).
17. Secondary antibody for hexon detection: FITC-conjugated goat anti-mouse antibody (Chemicon, Cat. no. AP124F).
18. Pepsin solution: Dissolve 0.1 g pepsin (2000-2400 U/mg protein; ICN Biomedicals Inc., Costa Mesa, CA) in 100 mL of PBS. Adjust pH to 7.5.
19. Two-step sucrose gradient: prepare 10 mL of 60% sucrose in PBS, and 10 mL of 50% sucrose in PBS. Layer 0.8 mL of a 60% solution and 0.8 mL of a 50% solution into 50-mL conical tubes.
20. Vindeløv's propidium iodide (PI) solution: 50 µg/mL PI, 10 mM NaCl, 10 µg/mL RNAse A (DNAse free), 0.1% Igepal CA-630, 10 mM Tris-HCl, pH 8.0.
21. Scintillation cocktail: EcoLume (ICN Biomedicals, Inc.).
22. Tissue-Tek OCT compound (Miles Scientific, Naperville, IL).
23. CsCl solutions for virus purification: 1.5 g/mL CsCl solution in PBS (for 500 mL add 335 g of CsCl to 410 mL of PBS), filter-sterilized. From this solution prepare a 1.35 g/mL solution and a 1.25 g/mL solution by mixing 69 mL or 49 mL with 31 mL or 51 mL of PBS, respectively.
24. Sucrose solutions for virus purification: 20% sucrose and 80% sucrose, each in 10 mM Tris Hcl, pH 8.0, +1mM EDTA.

3. Methods
3.1. Radioactive Vector Labeling

We next describe the use of radioactive vector in vivo to study biodistribution. Vector labeled in this way is also commonly used to study binding or trafficking in vitro *(7)*. Generally, ^{32}P-labeled virus is used for biodistribution assays in vivo and ^{3}H-labeled virus is used for assays in vitro. Alternatively, Cy-3 labeling as described in Leopold et al. *(8)* coupled to fluorescence microscopy detection could be used to avoid the use radioactivity. However, in vivo biodistribution studies are limited by sensitivity and only cells that bind large amounts of virus such as Kupffer cells can be observed. Specific activity should be about 10^{-4} cpm per virion in both methods. The labeled vector is used for biodistribution studies following systemic administration as described in **Subheading 3.3.1.** and detected as described in **Subheading 3.3.3.**

1. Inoculate 293 cells into 150-mm tissue-culture plates so that they will be 80–90% confluent at the time of infection. Usually we inoculate 5 × 10^6 cells in 25 mL of DMEM-10% FCS to each plate to be ready for the next

day. One 90% confluent 150-mm plate contains about 2×10^7 cells. We estimate a yield of 5000 viral particles (vp) per cell. Usually twenty 150 mm plates will yield about 2×10^{12} vp, enough for an in vivo experiment if we consider injecting up to 10^{11} vp/mouse.

2. Without changing the medium, infect cells by adding three plaque-forming units (PFU) of purified vector per cell (multiplicity of infection [MOI] = 3).

3. One day postinfection, add 125 μCi of ^{32}P orthophosphate carrier free or 1 mCi of methyl-^3H thymidine to the 25 mL of medium. In case of ^{32}P labeling, replace the medium with phosphate-free medium before adding the isotope to achieve a higher specific activity medium.

4. When the cytopathic effect is complete (usually at 2 d postinfection), harvest the virus by removing most of the media, then scraping the cells and transferring the remaining medium and cells into 50-mL tubes. Freeze-thaw the tubes three times.

5. Remove the cell debris by centrifugation at 1000g for 5 min. Purify the labeled virus by centrifugation on a CsCl step gradient or a continuous sucrose gradient.

3.1.1. Virus Purification on a CsCl Gradient

The CsCl purification method is the most frequently used.

1. In an ultracentrifuge tube (e.g., Beckman No 344060, 14 × 95 mm, Ultraclear), add 0.5 mL of 1.5 g/mL CsCl solution. Carefully overlay 2.5 mL of the 1.35 g/mL solution, then 2.5 mL of the 1.25 g/mL solution. On top of this step gradient, add 6 mL of the cleared crude extract and centrifuge at 150,000g (35,000 rpm in a SW40Ti rotor) for 4 h at 10°C.

2. A major lower virus band is observed between the 1.25 and 1.35 g/mL layers. Usually another upper band corresponding to immature virus can be observed. Remove the top of the gradient until within a few millimeters of the lower band. Collect this band with a pipet and bring the volume to 11.5 mL with the 1.35 g/mL CsCl solution.

3. Load the mix into a centrifuge tube and centrifuge at 150,000g (35,000 rpm in a SW40Ti rotor) for 16 h at 10°C.

4. Collect the virus band and dialyze twice against PBS and once against PBS/10% Glycerol. Aliquot and store at −80°C.

3.1.2. Sucrose Gradient Centrifugation

The sucrose method is faster and does not need dialysis, therefore better preserving the bioactivity of the virus.

1. Prepare a continuous sucrose gradient in an ultracentrifuge tube using 3.2 mL of 20% sucrose and 3.2 mL of 80% sucrose, each in 10 mM TE, pH 8.0.

2. Carefully layer 6 mL of the cleared, crude extract onto the top of the gradient and centrifuge at 80,000g (26,000 rpm in a SW40 Ti rotor) for 2 h at 4°C.
3. Collect the virus band (2 mL) and load it onto a second continuous gradient prepared from 4 mL of each sucrose solution. Centrifuge for 5 h as before. Collect the virus band. Aliquot and store it at −80°C.

3.2. Three-Dimensional Tumor Models

Diffusion and penetration barriers are greatly responsible for the poor efficacy of cancer gene therapy in a clinical setting *(9)*. Spheroids of tumor cells represent an in vitro model to study adenovirus-mediated gene transfer and oncolysis that takes into account these barriers. As we describe next, spheroids can be derived from cell lines in culture or from pieces of tumors *(10–12)*. Most epithelial-derived tumor cell lines will easily form spheroids. Spheroid formation from fresh tumor pieces will be more dependent on the quality of the specimen.

3.2.1. Tumor Spheroids Derived from Cell Lines

1. Prepare four 75 cm²-flasks bottom-coated with agarose: Weigh 3 g of agarose with a melting temperature above 83°C in a 500 mL screw-cap glass bottle and add 100 mL of sterile water. Autoclave for 15 min at 105°C and keep on a heated plate at 100°C. Add 100 mL of 37°C pre-warmed 2X cell culture medium (2X DMEM + 20% heat-inactivated FCS + L-glutamine + 0.5% gentamycin). Mix and add 4 mL to each 75 cm² flask. Gently, rock the flask during the coating to distribute uniformLy. Let the coated material set for 1 d at room temperature. Store at 4°C.
2. Trypsinize a confluent cell monolayer in a 75 cm²-flask with 3 mL of trypsin/EDTA. Resuspend cells in 10 mL of medium + 10% heat-inactivated FCS + 0.5% gentamycin. Determine the cell density and dilute to 1 × 10⁶ cells/mL
3. Distribute 5 mL of cell suspension (5 × 10⁶ cells) in one agarose-coated flask ("high concentration" flask) and 1 mL (1 × 10⁶ cells) in the other agarose-coated flask ("low concentration" flask).
4. Add medium to bring the liquid volume to 20 mL.
5. Incubate in a humidified incubator at 37°C with 5% CO_2 until the aggregates become well-rounded, compacted and regular in shape (normally 4–5 d for "high concentration" spheroids, 1 wk later for "low concentration").
6. Using a Pasteur pipet transfer the incipient spheroids to a clean bottom-coated agarose flask, avoiding single cells that have not formed aggregates and irregularly shaped aggregates.
7. Keep in culture with the appropriate medium changes until they reach the desired dimensions (usually spheroids with a diameter of 0.4–0.5 mm are

suitable for most gene transduction and oncolysis measurements form in 2–3 wk).

3.2.2. Tumor Spheroids Derived from Fresh Material

1. Prepare 48-well plates bottom-coated with agarose as indicated in **step 1** of **Subheading 3.2.1.** Add 0.1 mL of the agarose-culture medium mix to each well. Store at 4°C.
2. Collect tumor material as fresh as possible from surgical interventions, if possible from non-necrosed well-vascularized areas of the tumor, and transport it on ice in a tube partially filled with DMEM + 10% heat-inactivated FCS + 0.5% gentamycin.
3. In a tissue-culture hood, transfer a small piece of tumor (about 5 mm of diameter) into a plastic culture dish containing 5 mL of DMEM + 10% heat-inactivated FCS + 0.5% gentamycin.
4. Crossing two surgical blades, cut the piece of tumor into smaller pieces (0.4–0.8 mm in diameter).
5. Fill the wells of the agarose-coated 48-well plate with 0.5 mL of medium.
6. Transfer one piece of tumor to each well with a 5-mL pipet.
7. Store the plates in a humidified incubator at 37°C and 5% CO_2 until the spheroids form. This time is quite variable, from 1 to 3 wk depending on tumor type and proliferation rate.

3.2.3. Subcutaneous Tumors in Nude Mice

1. Prepare the tumor cell line as an *in vitro* exponential-phase cell culture at 80–90% confluence in 175 cm^2- flasks.
2. Trypsinize the monolayer with 5 mL of trypsin/EDTA per flask. Neutralize the trypsin with 1 volume of medium-10% FCS and harvest the cells. Remove an aliquot for cell counting. Pellet the cells (500g for 5 min) and resuspend them at 5–20×10^6 cells/150 μL in sterile PBS. Maintain at room temperature.
3. Use 4- to 6-wk-old male BALB/c nude mice; label the animals properly (ear punctures).
4. Subcutaneously inject 150 μL of cell suspension into the flank of each mouse using a 0.5–1 cc syringe with a 25 G needle (*see* **Note 1**). Injecting the needle horizontally into a portion of skin raised between your fingers helps to prevent crossing the connective layers under the skin. A contained bubble of liquid while injecting should be noted between your fingers. No resistance or bleeding should occur.
5. Return the animal to the cage and monitor tumor implantation by palpating animal flanks regularly. The time necessary for tumor implantation is highly dependent on tumor cell type, and it can be adjusted by modifying the number of cells injected.

3.2.4. Orthotopic Implantation of Tumors in Nude Mice

The protocol we present starts from fresh biopsies of human samples or from tumors previously generated in mice (subcutaneously or orthotopically) although it should be noted that serial passages in mice can result in selection of tumor cell types. Owing to the low/moderate rate of tumor xenograft implantation in orthotopic location, it is highly recommended that one piece of the same tumor sample be implanted in a sc position. This sc implant provides a source of tumor with a higher acceptance rate for subsequent orthotopic implantations. This procedure has been successfully used with human pancreatic, hepatic, colorectal, and germinal tumors *(13,14)*.

1. Rinse a piece of tumor from a surgical resection with PBS in a plastic culture dish. Cut the specimen with a surgical blade into small pieces of 10 mg or 2 mm^3. Place them in medium supplemented with penicillin, gentamycin, and fungizone.
2. Puncture one piece with a Prolene 6-0 suture and leave the fragment submerged in medium.
3. Use 4- to 6-wk-old male BALB/c nude mice; label the animals properly (ear punctures).
4. Anesthetize the mice (*see* **Note 2**). Keeping aseptic conditions inside a laminar flow hood, open the abdominal wall to access the desired organ. Very carefully pull out the organ as much as possible.
5. Anchor tumor fragments to the organ through three surgical stitches. The exact locus of tumor implantation is tumor-dependent, but the more irrigated the area of the recipient organ, the better the acceptance.
6. Place the organ with the implant back into the abdominal cavity.
7. Using a syringe without needle, inject 0.5–1 mL of saline solution into the abdominal cavity to help hydration.
8. Suture the abdomen.
9. Monitor tumor implantation by palpation twice a week. The time necessary for tumor growth is highly dependent on tumor aggressiveness, tumor type, and location of the implant. As a general rule, 1–2 mo should be considered for aggressive phenotypes, whereas low to moderate tumors can appear 4 mo or later after surgical intervention.

3.3. In Vivo Delivery of Adenovirus Vectors to Tumors

3.3.1. Systemic Administration of the Vector

1. Use 6- to 8-wk-old male BALB/c nude mice; label the animals properly (ear punctures).
2. Warm the animals with a lamp over the cage for 5 min.
3. Prepare the virus at the desired concentration in PBS at room temperature. For most vectors and mouse strains, the Lethal Dose 50 is around 10^{11}

vp/mouse. Up to 1 mL can be injected, but 0.2 mL is the preferred volume. Taking these parameters into account we usually prepare the virus at 2.5 × 10^{11} vp/mL for a high dose. Load the virus in a 0.5- or 1-mL syringe, insert a 28 G needle, and remove the air from the syringe.

4. Immobilize the animal in a small chamber (*see* **Note 3**). For 25–35 g animals, a 50-mL conical tube with V-shaped cut next to the cap as a tail exit and a hole at the bottom to allow breathing can be used.
5. Localize the tail veins at the sides of the tail and clean the tail with warm water.
6. Inject the needle with the syringe as horizontal as possible. A site at three-quarters the length of the tail towards the tip is a good point to start because, in case of misinjection, it is possible to use the same vein closer to the animal body. The needle should enter 2–3 mm into the vein.
7. Start the injection slowly; the flow should be smooth. If resistance is noticed or the injection site becomes white, it means that the solution is not entering the vein.
8. Upon retraction of the needle, the injection site should bleed if injection is correct. Apply pressure for 30 s to stop bleeding and then return the animal to the cage.

Alternatively, when tail bleeding is necessary after the injection, such as in the pharmacokinetic studies described below, the injection can be done in the cava vein:

9. Anesthetize the mouse and open the abdominal wall. Place a wet and warm cloth on the side of the animal. Grasp the mouse intestine with rinsed and wet gloves and lay it over the cloth to expose the cava vein that runs longitudinally at the dorsal side of the cavity.
10. Prepare the virus as in **step 3** above. Use a 30 or 30.5 G needle bent at a 45° to be able to insert the needle horizontally into the vein.
11. Insert the needle into the vein, then slowly inject the virus (200 µL).
12. Before retracting the needle, to aid clotting, place a small piece of surgicell at the injection site and press it down with a cotton ball.
13. Remove the needle and keep pressure on the injection site for 1 min.
14. Remove the plug and put the intestine into the abdominal cavity.
15. Suture the abdomen.
16. Keep a small amount of virus from the syringe dead volume to determine the titer at time 0.

3.3.2. Pharmacokinetics

1. Prepare heparinized microfuge tubes by placing 5 µL of a heparin solution (200 U/mL) into the inner side of a microfuge tube cap and letting it dry.

2. At different times postinjection (1, 5, 10, 15, 20, 30, 45, and 60 min), bleed the animal from the tail. To allow for repeated bleeding without pain, cut the tail 1 cm from the tip; wash the tip with 200 U/mL heparin solution and dry it with cloth. Control the bleeding by applying light pressure on the tail. If clotting blocks the blood flow, the clot can be washed away with PBS. At each time point, collect one drop of blood into a heparinized cap.
3. Centrifuge for 5 s to bring the blood to the bottom of the microfuge tube; keep the tube on ice until all time points have been collected.
4. Spin down the cells briefly and take 1 µL of plasma to assay the virus concentration.
5. Virus concentration can be measured by PCR as described in **Subheading 3.3.3.2.** Alternatively the virus titer can be measured by plaque assay on 293 cells (see overview chapter). This infection-dependent assay is simplified when the vector expresses GFP or another reporter gene by quantifying transducing units as the number of cells transduced with 1 µL of plasma.
6. Graph virus titer versus time to obtain the blood persistence curve. The slope of the curve represents the clearance rate. The area under the curve represents the effective dose for tumor targeting. About half of a nonmodified vector is cleared from the bloodstream in less than 2 min *(3)*.

3.3.3. Biodistribution

3.3.3.1. DETECTION OF RADIOACTIVE VECTOR

1. Following the steps described for systemic administration of the vector (**Subheading 3.3.1.**), inject up to 10^{11} vp of radioactive vector (**Subheading 3.1.1.**) per mouse.
2. At the desired time point postinjection, anesthetize (*see* **Note 2**) and sacrifice the animals (at least four animals per group with a negative noninjected control). Usually we consider 90 min postinjection as a time point when all particles have reach a stable destination. The assay, however, should also be done at 48 h postinjection in order to compare it with the gene-expression profile.
3. Open the abdominal cavity, remove a small piece each organ (e.g., tumor, muscle, lung, heart, intestine, kidney, fat, uterus, ovary, brain, spleen, and liver) and place the tissue sample in a microfuge tube. Weigh the tissue. To avoid contamination, wash the scissors between organs and leave the spleen and liver as the last organs.
4. For detection of radioactivity, homogenize tissues or tumors in scintillation cocktail. A single use homogenizer can be assembled as described in **Note 8**.

5. Spin down tissue debris by centrifugation at 14,000g for 5 min and measure radioactivity of the cleared lysate in a liquid scintillation counter.
6. Graph radioactivity against amount of tissue. For liver, a nonmodified virus should render about 10^4 dpm/mg.

3.3.3.2. QUANTITATIVE PCR

Inject virus systemically (*see* **Subheading 3.3.1.**), then harvest the desired organs (*see* **Subheading 3.3.3.1.**).

1. Process the tissue using standard SDS-Proteinase K–Phenol/Chloroform extraction: Digest 20 mg of tissue incubating overnight at 55°C with 1% SDS, 100 mM NaCl, 50 mM Tris, pH 8.0, 15 mM EDTA, pH 8.0, and 0.2 mg/mL proteinase K. Add RNase A to 10 µg/mL and incubate 1 h at 37°C. Extract the sample with phenol-chloroform twice and then add 2 volumes of ethanol-2% sodium acetate to precipitate the DNA. Pellet the DNA by centrifugation at 10,000g for 10 min. Resuspend the DNA in 10 mM Tris, pH 8.0, 1 mM EDTA (*see* **Notes 4** and **5**).
2. Quantify purified DNA by spectroscopy. Prepare the PCR reactions with the same amount of sample DNA (50 ng). The differences between tissues are in this way normalized.
3. To obtain a standard curve for quantification, mix 50 ng of DNA from a tissue such as a lung from a noninjected animal to prepare a serial dilution of a known number of viral DNA copies (from 10^8 to 0).
4. Primers to detect E4 are: forward 5′- CACCACCTCCCGGTACCATA, reverse 5′- GGGCTCTCCACTGTCATTGT, and the probe 5′- 6FAM-AAC-CTGCCCGCCGGCTATACACTG-TAMRA (*see* **Note 6**).
5. Prepare amplification reactions in triplicate according to specific TaqMan protocols (*see* **Note 7**): Mix 25 µL of 2X PCR Master mix, 5 µL of 3 µM forward primer, 5 µL of 10 µM reverse primer, 0.5 µL of 10 µM FAM probe, 6 µL of 2.5 mM dNTPs, 3.5 µL of H$_2$O, 5 µL of template (final volume = 50 µL).
6. Load the samples in the thermal cycler programmed as follows: 10 min at 50°C, 10 min at 94°C, followed by 40 cycles of 15 s at 94°C and 1 min at 60°C. Monitor the fluorescence resulting from the annealing of the probe to the PCR products in real time (e.g., using an ABI 7000 thermal cycler). (*See* **Note 7**.)
7. Quantification is based on the cycle number where fluorescence first exceeds baseline (i.e., the threshold cycle, C_t). Plot the C_t of the various vector dilutions to obtain the standard curve. With this relation calculate the vector concentration from the sample C_t. The sample with 10^8 wt viral copies should have a C_t around 14.

3.3.3.3. Luciferase Detection

1. Following systemic injection of the virus (*see* **Subheading 3.3.1.**), anesthetize and sacrifice the animals at the desired time points (*see* **Note 2**; use at least four animals per group along with a negative, noninjected control). For biodistribution studies, transgene expression is normally measured at 48 or 72 h without significant differences within this period.
2. Remove a small piece (about 0.1 g) of each organ as described in **Subheading 3.3.3.1., step 3**.
3. Freeze the tissue by placing the tube in a dry ice-ethanol bath.
4. Grind the tissues using a mortar pestle precooled on dry ice. Place samples over a piece of paper in the mortar to avoid contamination. Transfer the ground powder to a new microfuge tube (*see* **Note 8**).
5. Add 0.5 mL of luciferase cell lysis reagent to the ground tissue and incubate in an orbital shaker at room temperature for 1 h (*see* **Notes 8** and **9**).
6. Spin down the undigested tissue and transfer 0.1 mL of clear supernatant to a new microfuge tube.
7. For each sample, prepare one plastic tube appropriate for the luminometer with 20 µL of clear supernatant.
8. Inject 100 µL of luciferase assay reagent and measure relative light output (RLU) for 10 s in a luminometer (*see* **Note 10**).
9. To normalize the luciferase levels measure the protein content in each cleared lysate using the Bio-Rad DC Protein Assay Kit. For a standard curve prepare a serial dilution of bovine serum albumin (BSA) (1 mg/mL to 0.0625 mg/mL) in luciferase cell lysis reagent.
10. To normalize, divide luciferase activity (RLU) by the amount (mg) of protein in each 20 µL sample of supernatant. For liver, an adenovirus vector with an unmodified capsid containing the luciferase gene expressed from a strong promoter (e.g., CMV) should render about 5×10^6 RLU/mg of protein.

3.3.4. Assessment of Toxicity by Transaminase Levels

1. At the desired time point after vector administration (e.g., 3 d), draw a small amount of blood from the ocular vein or tail vein.
2. Collect blood in an heparinized microfuge tube (*see* **Subheading 3.3.2., step 1**).
3. Pellet the cells (5 s in a microfuge) and transfer 20 µL of serum to a new microfuge tube.
4. Prepare a standard curve for GOT in the range of 0–104 IU/L and for GPT in the range of 0–60 IU/L (*see* **Note 11**).
5. In a microfuge tube, mix 20 µL of GPT substrate with 4 µL of serum or standard GPT activity and incubate for 30 min at 37°C. In another set of

microtubes, mix 20 μL of GOT substrate with 4 μL of serum or standard GOT activity and incubate for 1 h at 37°C.

6. Add 20 μL of color reagent and incubate 20 min at room temperature.
7. Stop the reaction with 0.2 mL of 0.4 *M* NaOH.
8. Read absorbance at 505 nm.
9. Prepare a standard curve is used to calculate the transaminase units of the samples. As a reference point, the normal levels of alanine aminotransferase (GPT) and aspartate aminotransferase (GOT) in mice are around 13 IU/L and 36 IU/L, respectively. An injection of 5×10^{10} vp wild-type adenovirus raises these levels 100-fold by d 3.

3.4. Special Methods for Replicative Vectors

3.4.1. Detection of Cytopathic Effect (CPE)

The replication efficacy of a replication-competent vector is expected to correlate with its oncolytic potency. One way to quantify this parameter is to measure the amount of virions produced from an infected tumor cell (burst size) using a standard plaque assay in 293 cells from the supernatant or cell lysate of infected cells. Alternatively, the replication efficacy can be measured as the progression of cytophatic effects through a population of tumor cells (cultured as monolayers, spheroids, or tumor models as described in **Subheading 3.2.**). As a negative control for tumor-specific replication, normal human epithelial cells should be used (e.g., primary human hepatocytes or bronchial epithelial cells from commercial sources such as Clonetics Corp, San Diego, CA).

3.4.1.1. CPE Detection by Crystal Violet

1. Two to three days prior to use, seed the tumor cells (or normal control cells) to be analyzed for replication sensitivity in 24-well plates (0.7 mL of medium with 10% FCS per well) so that cells are 80–90% confluent when used.
2. Infect cells with six serial dilutions (10-fold) of virus (0.3 mL/well) in duplicate wells. The initial concentration should be around 10,000 vp/mL.
3. Incubate infected cells for 8 d in a humidified incubator at 37°C with 5% CO_2. This time should be empirically determined because different cell types and viruses can differ in their capacity to undergo cytopathic effect.
4. Aspirate the media carefully without disturbing the cell monolayer and wash the wells once carefully by adding 1 mL of PBS. Aspirate the PBS. Carefully add 1 mL of fixation solution and incubate for 20 min at room temperature.
5. Wash twice with distilled H_2O. Stain with crystal violet solution for 5 min at room temperature. Wash twice again with distilled H_2O. The stained cells

(blue color) indicate lack of cytotoxicity and therefore a low replication efficacy. This assay is a semi-quantitative assay for CPE: there should be a gradual decrease from a completely blue well (no CPE) to a transparent, unstained well (complete CPE).

3.4.1.2. CPE Detection by Protein Content (BCA Method)

1. Seed tumor cells (or normal control cells) to be analyzed for replication sensitivity in 96-well plates (30,000 cells in 0.1 mL of medium with 0.5% FCS per well). Incubate overnight in a humidified incubator at 37°C with 5% CO_2.
2. The following day, prepare 11 serial (five-fold) viral dilutions in DMEM + 0.5% FCS from a stock at 1000 PFU/cell.
3. Add 50 µL of each dilution into wells in triplicate without aspirating the medium.
4. Incubate infected cells for 4–5 d at 37°C, 5% CO_2. This time should be empirically determined because different cell types and viruses can differ in their capacity to produce cytopathic effect.
5. Carefully remove the media containing the dead/floating cells by aspiration.
6. Add 200 µL/well of working reagent for BCA assay. Shake the plate gently for 1 min to dissolve the cell membranes.
7. Cover the plate and allow the color to develop for 30 min at 37°C.
8. Measure absorbance at 562 nm on a plate reader (wavelengths from 540–590 nm can be used).
9. Subtract the average reading of the blank standard replicates (wells containing no cells) from all other well replicates.
10. Plot the results as a percentage of protein content with respect to uninfected controls against virus dilution.
11. The titer (CPE U/mL) is the dilution factor where the protein content drops 50% multiplied by 1.5×10^4 (number of cells corresponding to 50% of the total) and divided by 0.05 (volume in mL of the viral solution used). For example, from a viral solution of 3×10^{10} PFU/mL we expect a curve that drops to 50% of protein content at the 10^5 dilution.

3.4.2. Evaluation of Viral Load in Tumors by Hexon Detection Using FACS

Hexon staining by immunohistochemistry has been described as a method to detect virus replication inside tumors (*15,16*). Here we present an alternative method intended to obtain quantitative results.

1. Establish tumor spheroids or tumors in vivo following the protocols described in **Subheading 3.2.** and treat them by intratumoral injection or sys-

temic administration (**Subheading 3.3.**) of a replication-competent adenovirus vector.

2. At the desired experimental time posttreatment, sacrifice the animal in a CO_2 chamber and excise the tumor. An optimal time point to detect replication in a given tumor nodule is when its size decreases.
3. Embed-freeze tumors in OCT: place the sample on a plastic cryomold and cover it with OCT embedding compound; snap-freeze the sample by placing the cryomold on dry ice or on the surface of liquid nitrogen.
4. Using a microtome, cut the whole tumor in 100–200 µm sections and place them in a conical tube with 10 mL of PBS. Leave tumor sections settle for 30 min.
5. Carefully decant PBS and add 1.5 mL of pepsin solution previously warmed to 37°C (*see* **Note 12**). Vortex the sample for 10 s.
6. Digest the tumor slices at 37°C for 30 min, shaking the tubes every 5–10 min. Longer incubations could result in loss of cell integrity.
7. Filter the cell suspension through a 52 µm pore size nylon mesh.
8. Pellet the cells by centrifugation for 25 min at 500g (*see* **Note 13**).
9. Aspirate and discard the supernatant. Resuspend the cellular pellet in 0.5 mL of DMEM + 10% FCS. Count an aliquot in a hemocytometer. Transfer 5×10^5–1×10^6 cells to a microfuge tube.
10. Pellet the cells in a microfuge for 5 min at 500g. Resuspend the pellet in 0.5 mL of goat serum and incubate for 30 min at room temperature with continuous shaking to block nonspecific binding of the primary antibody.
11. Pellet the cells as in **step 10**. Resuspend the pellet in 100 µL of primary antibody against hexon diluted 1:500 in Tris-buffered saline with 0.5% Tween (TBST). Incubate 30 min at RT with continuous shaking.
12. Pellet the cells as in **step 10**. Wash twice with 1 mL of TBST and resuspend in 100 µL of FITC-conjugated goat anti-mouse secondary antibody diluted 1:200 in TBST. Incubate for 30 min at room temperature with continuous shaking.
13. Pellet the cells as in **step 10**. Wash twice with 1 mL of TBST and resuspend in 500 µL of Vindeløv's PI solution.
14. Read fluorescence in a FACS. The positive (infected) cells will be detected as a rightward shifted peak on the fluorescence intensity axis.

4. NOTES

1. Usually two tumors are generated in the flanks of one mouse by subcutaneous injection.
2. Suggested anesthesia: Ketamine 80–100 mg/kg/xylazine 10 mg/kg injected intraperitoneally (ip). Administer 5 min before tumor implantation.

3. Tail vein injection is the common method to administer the vector system-ically into mice. With practice, the procedure is not stressful or painful for the animals and anesthesia should be avoided.

4. Alternatively, use commercial DNA extraction methods such as DNAzol from Invitrogen or DNeasy tissue kits from Qiagen.

5. An alternative way to minimize differences in DNA recovery from differ-ent tissues is to avoid DNA purification. In this case, digest the small piece of tissue at 56°C for 1 h with 20 mM EDTA, 0.1% SDS, 2 mg/mL pro-teinase K, incubate at 90°C for 10 min to inactivate the proteinase. The re-sults are then normalized using the tissue weight instead of the DNA amount.

6. Alternatively, detect IVa2 with forward 5'-GAACCACAGCACAGTGTAT CC and reverse 5'-GGCCCATTGCCATCATTATG primers, and probe 5'-6FAM- TGACCTCCAAGATTTTCCATGCATTCG-TAMRA.

7. Alternatively, if a thermal cycler capable of real time fluorescence detec-tion is not available, label the 5' end of the forward primer with Cy5 and perform a regular PCR in combination with the reverse primer. Run the PCR product on a gel and quantify using a fluorescence imager.

8. Alternatively thaw samples and add 0.2 mL of luciferase cell lysis reagent to the microfuge tube. Homogenize the samples in the microfuge tube with a pestle. A single-use pestle can be assembled by attaching the bottom part of a 0.5-mL microtube to a blunted blue tip. After homogenization, add an additional 0.3 mL of cell lysis reagent.

9. Luciferase cell lysis and assay reagents with a different formula can be pur-chased from Promega (Madison, WI).

10. If readout is high, check that the detection level of the luminometer has not been saturated reading the sample after a 10-fold dilution in lysis reagent. The readout should be 10-fold lower. Otherwise, dilute the sample again and consider the result valid only when a proportional dilution results in a decrease in RLU. Calculate the amount of luciferase units in every 20 µL of supernatant considering these dilutions.

11. This is a scaled-down protocol from the Sigma transaminase detection kit. Note that on page 3 of the Sigma procedure No. 505 to convert SF units to International Units (IU), multiply by 0.48×10^{-3} instead of 0.48 as spec-ified. 1 SF unit $= 4.8 \times 10^{-4}$ IU

12. The length of pepsin incubation during tumor digestion may be dependent on tumor architecture. The concentration reported here is only indicative and should be empirically determined.

13. Cellular debris generated after digestion can be cleared from the sample by centrifugation in a sucrose gradient. Apply the cell suspension from **step 7** to a two-step gradient and spin down for 40 min at 500*g*.

References

1. Santini, M. T. and Rainaldi, G. (1999) Three-dimensional spheroid model in tumor biology. *Pathobiology* **67,** 148–157.
2. Capella, G., Farre, L., Villanueva, A., Reyes, G., Garcia, C., Tarafa, G., and Lluis, F. (1999) Orthotopic models of human pancreatic cancer. *Ann. NY Acad. Sci.* **880,** 103–109.
3. Alemany, R., Suzuki, K., and Curiel, D. T. (2000) Blood clearance rates of adenovirus type 5 in mice. *J. Gen. Virol.* **81,** 2605–2609.
4. Hackett, N. R., El Sawy, T., Lee, L. Y., Silva, I., O'Leary, J., Rosengart, T. K., and Crystal, R. G. (2000) Use of quantitative TaqMan real-time PCR to track the time-dependent distribution of gene transfer vectors in vivo. *Mol. Ther.* **2,** 649–656.
5. Alemany, R., Balague, C., and Curiel, D. T. (2000) Replicative adenoviruses for cancer therapy. *Nat. Biotechnol.* **18,** 723–727.
6. O'Carroll, S. J., Hall, A. R., Myers, C. J., Braithwaite, A. W., and Dix, B. R. (2000) Quantifying adenoviral titers by spectrophotometry. *Biotechniques* **28,** 408–412
7. Wickham, T. J., Tzeng, E., Shears L. L., Roelvink, P. W., Li, Y., Lee, G. M., et al. (1997) Increased in vitro and in vivo gene transfer by adenovirus vectors containing chimeric fiber proteins. *J. Virol.* **71,** 8221–8229
8. Leopold, P. L., Ferris, B., Grinberg, I., Worgall, S., Hackett, N. R., and Crystal, R. G. (1998) Fluorescent virions: dynamic tracking of the pathway of adenoviral gene transfer vectors in living cells. *Hum. Gene Ther.* **9,** 367–378.
9. Sterman, D. H., Treat, J., Litzky, L. A., Amin, K. M., Coonrod, L., Molnar-Kimber, K., et al. (1998) Adenovirus-mediated herpes simplex virus thymidine kinase/ganciclovir gene therapy in patients with localized malignancy: results of a phase I clinical trial in malignant mesothelioma. *Hum. Gene Ther.* **9,** 1083–1092
10. Fujiwara, T., Grimm, E. A., Mukhopadhyay, T., Zhang, W. W., Owen-Schaub, L. B., and Roth, J. A. (1994) Induction of chemosensitivity in human lung cancer cells in vivo by adenovirus-mediated transfer of the wild-type p53 gene. *Cancer Res.* **54,** 2287–2291.
11. van Beusechem, V. W., Grill, J., Mastenbroek, D. C., Wickham, T. J., Roelvink, P. W., Haisma, H. J., et al. (2002) Efficient and selective gene transfer into primary human brain tumors by using single-chain antibody-targeted adenoviral vectors with native tropism abolished. *J. Virol.* **76,** 2753–2762.
12. Grill, J., Van Beusechem, V. W., Van Der Valk, P., Dirven, C. M., Leonhart, A., Pherai, D. S., et al. (2001) Combined targeting of adenoviruses to integrins and epidermal growth factor receptors increases gene transfer into primary glioma cells and spheroids. *Clin. Cancer Res.* **7,** 641–650.
13. Reyes, G., Villanueva, A., Garcia, C., Sancho, F. J., Piulats, J., Lluis, F., and Capella, G. (1996) Orthotopic xenografts of human pancreatic carcinomas acquire genetic aberrations during dissemination in nude mice. *Cancer Res.* **56,** 5713–5719.
14. Tarafa, G., Villanueva, A., Farre, L., Rodriguez, J., Musulen, E., Reyes, G., et al. (2000) DCC and SMAD4 alterations in human colorectal and pancreatic tumor dissemination. *Oncogene* **19,** 546–555.

15. Alemany, R. and Zhang, W. (1999) in Methods in Molecular Medicine, Gene Therapy: Methods and Protocols, Humana Press, Totowa, NJ.
16. Suzuki, K., Fueyo, J., Krasnykh, V., Reynolds, P. N., Curiel, D. T., and Alemany, R. (2001) A conditionally replicative adenovirus with enhanced infectivity shows improved oncolytic potency. *Clin. Cancer Res.* **7,** 120–126.

9

Adenovirus-Mediated Gene Delivery to Dendritic Cells

Laura Timares, Joanne T. Douglas, Bryan W. Tillman, Victor Krasnykh, and David T. Curiel

1. Introduction

Dendritic cells (DCs) are "professional" antigen-presenting cells (APCs) that are uniquely capable of activating and instructing a naive immune system to mount a specific cellular and humoral response. Recognition of this crucial function makes the development of technologies for DC-based immuno-therapies a priority for the treatment of a wide variety of diseases. The most immediate impact of this emerging technology will be in the treatment of cancer and the development of third generation vaccines to protect against viral and intracellular pathogens. In addition to elicitation of immune responses, DCs also function to maintain tolerance to "self." Once the biological basis for this important function is understood, future applications of DC-based immuotherapies may be developed to ameliorate autoimmune diseases or enhance acceptance of transplanted organs. The feasibility of "engineering" the function of DCs has been realized by recent advances in ex vivo methodologies that allow selective DC propagation, antigen loading, and genetic modification in vitro for subsequent therapeutic transfer into the host. Ultimately, the ability to genetically modify these cells will allow us to design DC-mediated interventions that will direct predictable control of either immune activation or tolerance in vivo.

DCs are nondividing cells found mainly in two stages, immature and mature. In peripheral tissues, they are immature and function as macropinocytotic cells that monitor the environment and pick up foreign antigens; they do not stimu-

From: *Methods in Molecular Biology, vol. 246:*
Gene Delivery to Mammalian Cells: Vol. 2: Viral Gene Transfer Techniques
Edited by: W. C. Heiser © Humana Press Inc., Totowa, NJ

late good T-cell responses at this stage. However, once foreign antigens are detected and engulfed, DCs leave the periphery and migrate to draining lymph nodes, where they finally mature into professional APCs, able to activate T cells. In the mature stage, DCs can no longer take up or process exogenous foreign antigens.

Numerous reports have documented that infection with recombinant, replication-defective adenovirus (Ad) vectors encoding a transgene is an efficient method for gene transfer into murine or human DCs. Antigen-specific T-cell proliferation, interleukin (IL)-12 secretion, and activation of cytotoxic T lymphocyte (CTL) responses have been observed in vitro and in vivo when Ad-modified DCs are cultured with responder cells or injected into animals. DCs derived from either $CD34^+$ or $CD14^+$ precursor cells have been successfully transfected with Ad with efficiencies ranging from 30% to greater than 90%. The level of transgene expression is dependent on virus dose and stage of DC maturation. Mature DCs are resistant to Ad transfection and the level of gene expression per transfected cell is also reduced *(1)*. Therefore, for efficient Ad-mediated gene transfer, DCs should be infected when they are in the immature stage.

1.1. Advantages and Disadvantages of Ad-Mediated Gene Transfer into DC

1.1.1. Ad-Infected DCs Are Resistant to Neutralizing Antibodies

Pre-existing neutralizing antibody to Ad is ubiquitous in the general population owing to widespread exposure to Ad. However, the epitopes recognized by pre-existing antibody depend on the intact secondary structure of the Ad virions. When Ad virions infect DCs, they are digested into peptides by antigen-processing pathways. A number of reports, using different animal model systems, demonstrate that repeated immunizations with adenoviral-based DC therapy have improved vaccine efficacy, despite the presence of pre-existing anti-Ad neutralizing antibody *(2–5)*. The current body of literature, therefore, lends support for the use of Ad-infected DCs as potent immunotherapy vehicles.

1.1.2. CAR-Deficient DCs Are Poor Cellular Targets for Ad

The rate-limiting step for entry of adenovirus into cells is binding to the primary receptor, the coxsackie adenovirus receptor (CAR), followed by interaction with $\alpha_V\beta$ class integrins, which results in endosome formation and internalization. Recent evidence suggests that DCs and their precursors are CAR deficient *(5–7)*. Given that one of the main functions of immature DCs is to sample the environment through macropinocytosis, it is likely that Ad is taken up largely by this process. Consequently, efficient gene transfer into DCs necessitates high concentrations of infectious Ad virus particles. The requirement for such high levels of virus can lead to reduced DC function, either through induction of suppressor functions, apoptosis, or reduced presentation of transgene products by favoring Ad peptide processing. The future of Ad-mediated gene transfer into DCs will rely on technologies that reduce the number of viral particles needed for efficient gene delivery by targeting Ad to DC-specific receptors. Receptor-mediated uptake of Ad in DCs can result in increased gene expression and, depending on the receptor targeted, induce DC activation and increased antigen presentation. Some examples of this will be discussed below. However, when using the high concentrations of unmodified recombinant Ad vectors required for gene transfer, it is prudent to monitor the effect of Ad infection on DC viability and function in addition to transgene expression.

1.2. Dendritic Cell Sources

Methods for the isolation and/or generation of human or murine DCs are outside the scope of this methodology paper. However, the Ad-mediated gene-transfer protocol described here can be applied to any source of immature DCs. For example, murine or human subsets of lymphoid (derived from CD34$^+$ precursors grown from cord blood or bone marrow for 10–14 d in the presence of GM-CSF) or myeloid (derived from CD14$^+$ precursor cells and cultured for 7 d in the presence of GM-CSF and IL-4) origin *(8–10)*. In standard cultures, immature and mature DCs are nonadherent cells. Recently, Munn et al. described a new adherent subset in human monocyte-derived cultures that may have enhanced potency *(10a)*. Cognizant that DCs are not fully characterized, and new cell subsets are still emerging, this protocol will focus on infecting nonadherent DCs. Additionally, a number of growth factor-dependent murine DC lines have been increasingly utilized and serve as relevant models of primary and in vitro cultured DCs *(11–14)*. In this chapter, we show examples of one such murine DC line, XS106 *(15)*.

Although controversial, an important consideration when infecting human DCs is whether or not serum is present, and if so, what the source of serum is. The ability of Ad to directly activate DCs may be dramatically influenced by the serum source. Some investigators have reported that autologous, heterologous, or xenobiotic (e.g., fetal bovine) serum can inhibit Ad infection and subsequent human DC activation *(16)*, whereas others have demonstrated pronounced Ad-mediated DC activation in the presence of serum *(17)*. In our experience at the University of Alabama at Birmingham, we occasionally observed modest activation by Ad-5 infection of monkey DCs (J.M. Thomas, personal communication), but unimpressive activation of murine or human DCs (L. Timares and P. Triozzi, personal communication) in the absence or presence of bovine or human serum.

Depending on the nature of the study, DCs may be activated after infection with Ad by addition of inflammatory stimuli (such as IL-1, tumor necrosis factor-α [TNFα], or LPS) *(18)* to obtain mature DCs. However, it has been observed that immature DCs can be easily activated simply by in vitro handling during isolation and processing, and upon introduction into the host *(19)*. This observation emphasizes the importance of including appropriate controls to account for "mechanical/processing"-induced activation of DCs.

2. Materials

2.1. Solutions

1. Phosphate-buffered saline (PBS): without calcium, without magnesium (reduces cell aggregation).
2. Infection medium: RPMI with 2% serum (or any appropriate medium for DC culture with 2% serum appropriate for the DC population under study).
3. Culture medium: complete RPMI-1640 supplemented with relevant growth factors and 10% serum appropriate for the DC population under study.
4. Staining buffer: PBS containing 0.1% BSA and 0.01% sodium azide.
5. Annexin V binding buffer (BD Biosciences PharMingen, San Diego, CA); 1X working solution is 10 mM HEPES/NaOH, pH 7.4, 140 mM NaCl, 2.5 mM CaCl$_2$.

2.2. Reagents

2.2.1. Cells

Immature DCs are nonadherent and therefore grow as suspension cultures. If contaminant-adherent cells (e.g., monocytes) are present, they will compete for Ad particles. See references in **Subheading 1.2.** for DC generation and isolation methods.

2.2.2. Adenovirus Stocks (see **Note 1**)

The reporter Ad vectors utilized here (*see* **Note 2**) were generated in our lab and are termed AdEasyCMV-GFP (Ad-GFP) and AdEasyCMV-YFP (Ad-YFP). The AdEasy cloning kit can be obtained from Stratagene (La Jolla, CA). Always thaw Ad stocks (typically stored in small aliquots at −80°C) immediately before use and keep on ice prior to dilution and addition to cells. Stock concentrations should have a titer of ≥10^{10} multiplicity of infection (MOI)/mL and ≥10^{11} particles/mL.

2.2.3. Reagents for Apoptosis Assays

1. Staurosporine (Sigma, St. Louis, MO), at a final concentration of 30–50 nM in culture medium to induce apoptosis and provide a positive control for apoptotic cells.

2. Phycoerythrin-conjugated (PE)-Annexin V (orange emission spectra) with Annexin V Binding Buffer (BD Biosciences PharMingen): used to stain cell populations expressing the GFP reporter gene (green emission spectra) undergoing "early stages" of apoptosis.
3. Annexin V Kit from BD Biosciences PharMingen (Via-Probe™) containing the "far-red" emission spectrum dye, 7-aminoactinomycin D (7-AAD), or 7-AAD from Sigma (100X stock 7-AAD solution 500 µg/mL): use to stain "late stage" apoptotic and necrotic cells; this may be used in combination with PE-Annexin V.

2.2.4. Staining Reagents for DC Activation

1. PE-anti-CD40, PE-anti-CD80, PE-anti-CD86, and PE-anti-class II major histocompatibililty complex (MHC) molecules (for the murine system, this locus is designated I-A/I-E; for the human system, it is termed HLA-DR). PE-conjugated isotype controls (e.g., the murine PE-anti-CD40 clone 3/23 (Cat. no. 553791) is a Rat IgG2a,κ antibody; therefore, as an isotype control use PE-Rat IgG2a,κ (Cat. no. 553930)) should be used at the same concentration as the PE-Ab to determine background fluorescence levels. All antibodies (Abs) can be obtained from BD Biosciences PharMingen.
2. Fc-Block™ (CD16/CD32, murine Fcγ II/II Receptor,) is available from BD Biosciences PharMingen).
3. Mouse gamma globulin (~10 mg/mL) is available from Jackson Immuo-Research Labs (West Grove, PA).
4. Normal human AB serum is available from United States Biological (Swampscott, MA).

2.2.5. Flow Cytometry

1. FacScan™ or FacsCalibur™ flow cytometers are available from Becton-Dickinson BioSciences (Palo Alto, CA).
2. Flow cytometry analysis tubes are available from Falcon/Becton-Dickinson (Franklin Lakes, NJ; Cat. no. 35-2058).
3. Polypropylene mesh (~150 µm) is available from Small Parts Inc. (Miami Lakes, FL).

3. Methods

The following protocols are provided for different applications. **Subheading 3.1.** is a basic protocol for Ad infection and simple monitoring of Ad-GFP gene-transfer efficiency by determining the frequency of cells expressing GFP using flow cytometry. **Subheading 3.2.** provides a protocol for simultaneously assessing both GFP reporter gene expression and DC activation status. **Subheading 3.3.** provides a protocol for simultaneously assessing GFP gene expression and apoptosis induction (Ad toxicity). When first developing an infection protocol, attention should be paid to the level of DC activation and to the toxic effects on the cell population: follow the infection protocol in **Subheading 3.1.** and the staining protocols in **Subheadings 3.2.** and **3.3.** After the protocol has been optimized and the cell population has been well-characterized (with a high purity of DCs), then the basic application of simply monitoring GFP expression levels in **Subheading 3.1.** can be applied routinely.

3.1. Adenovirus Infection for Flow Cytometric Analysis of Gene Transfer Efficiency (see Note 3)

1. Wash nonadherent immature DCs in Infection Medium at room temperature. Resuspend the cells in Infection Medium at a concentration of 6×10^6 cells/mL. (Scale up the cell concentration three-fold to 18×10^6 cells/mL if DC activation analysis is planned; *see* **Notes 4 and 4A.**)
2. For each experimental and control condition to be tested, transfer 25 μL (1.5×10^5) of cells (or three-fold more if cytometric analysis of DC activation is planned) per sterile microfuge tube, prepunctured using a sterile 18 G needle in a few places on top to provide air exchange during culture. When appropriate, set up triplicate samples per test condition.
3. Add 25 μL of the Ad-GFP vector at 100–1000 MOI (or 1000–10,000 PPC; *see* **Note 1**) diluted in Infection Medium. Add 25 μL of Infection Medium only to mock-infected samples.
4. Incubate 1 h at 37°C, 5% CO_2 with agitation every 10–15 min. Longer incubation can modestly improve transfection efficiency but can also lead to toxicity (*see* **Note 4b**).
5. Remove unbound Ad by adding 1 mL of cold PBS per sample. Pellet the cells in a microfuge at $200g$ for 6 min. Carefully aspirate from the side opposite the angled cell pellet. Keep the cell pellet undisturbed by leaving approx 50 μL of supernatant. Repeat the wash (*see* **Note 4b**).
6. Resuspend the cells in 0.5 mL of pre-warmed (37°C) growth medium (with the appropriate serum and the necessary cytokines for DC growth).
7. Place the microfuge tubes in a rack and put the rack in a 37°C humidified incubator with 5% CO_2 at an oblique angle to increase the surface area for the cells to settle on the side of the tube; this increases gas exchange and viability. Incubate 18–48 h. (Peak reporter gene expression usually is found about 24 h postinfection.) For multicolor analysis of both reporter gene green fluorescent protein (GFP) expression and DC activation proceed to **Subheading 3.2.**
8. To analyze reporter gene (GFP) expression by single color flow cytometric analysis, pellet the cells at $200g$ for 6 min and carefully aspirate the super-

natant leaving the cell pellet intact. For multicolor analysis of both reporter gene (GFP) expression and apoptosis, proceed to **Subheading 3.3.**

9. Resuspend the cells in 300–400 µL of Staining Buffer and transfer to labeled flow cytometry analysis tubes. During transfer, pass the cells through 150-µm polypropylene mesh placed at the top of the tube to filter out aggregates.

10. Analyze the cells by single color cytometric analysis (*see* **Note 5**).

3.2. Dual Analysis of Ad-Transfection Efficiency and DC Activation by Multiparameter Flow Cytometric Analysis of GFP and Activation of Antigen Expression (see Notes 4a and 5)

1. After Ad infection and overnight incubation of 4.5×10^5 cells per sample (**Subheading 3.1., step 7**; also, *see* **Note 4a**), split each cell culture into three microfuge tubes labeled appropriately (e.g., PE-Ab stained, PE-isotype control, Unstained), add 500 µL of cold Staining Buffer to each sample, briefly vortex, and pellet the cells at 200g for 6 min.

2a. For murine cultures: to Block Fc-receptors on DCs and any contaminant cells, add 5 µL of Fc-Block™ plus approx 10 µg of mouse gamma globulin in a small volume (e.g., < 5 µL) directly to the cell pellet of each sample. This amount of mouse gamma globulin should be greater than 10-fold the concentration of staining Ab used. Flick the tubes to loosen the cell pellets. Incubate 15 min on ice.

2b. For human cultures: to Block Fc-receptors on human DC cells, add 10% human AB serum in 50 µL of PBS. Flick the tubes to loosen the cell pellets. Incubate 15 min on ice.

3a. For mouse samples: add 50 µL of PE-conjugated Ab (e.g., specific for DC activation markers such as PE-anti-CD40 clone 3/23) in Staining Buffer, their respective PE-isotype control (e.g., PE-Rat IgG2a,κ isotype), or Staining Buffer to the cells treated with Fc-block™. Vortex the tubes gently. Incubate for 30–60 min on ice.

3b. For human samples: wash out AB serum by adding 1 mL of Staining Buffer. Pellet the cells and discard the supernatant. Stain the cells by adding 50 µL of PE-conjugated Ab (e.g., PE-anti-CD40 clone 5C3), isotype control (e.g., PE-mouse IgG1,κ isotype), or Staining Buffer; incubate the cells following the manufacturer's instructions.

4. Pellet the cells at 200g for 6 min, carefully aspirate the supernatant, add 500 µL–1 mL of Staining Buffer, then pellet the cells again. Repeat the wash steps.

5. Resuspend the cells in 300–400 µL of Staining Buffer and transfer the cells to labeled flow cytometry analysis tubes. During the transfer, pass the cells

through 150-µm polypropylene mesh placed at the top of the tube to filter out aggregates (*see* **Note 4a**).

6. Analyze the cells by multicolor cytometric analysis (*see* **Notes 6c, 7, and 8b**).

3.3. Dual Assessment of Ad-Transfection Efficiency and DC Viability by Flow Cytometric Analysis of GFP Expresssion and Apoptotic Cells (see Note 6)

1. Wash pelleted cells (**Subheading 3.1.**, **step 8**) by resuspending cell pellet in 1 mL of cold PBS; repeat the centrifugation and aspiration steps 200*g* for 6 min.

2. Add 45 µL of Annexin V Binding Buffer and 5 µL of PE-Annexin V per sample and gently flick the tubes to resuspend the cells and mix.

3. Incubate 15 min at room temperature (~25°C) protected from light.

4. Wash the cells by adding 500 µL of Annexin V Binding Buffer and pellet the cells 200*g* for 6 min.

5. Resuspend the cell pellet in 300–400 µL of Annexin V Binding Buffer containing 7-AAD (5 µg/mL) and transfer to labeled flow cytometry analysis tubes.

6. During transfer, pass the cells through 150-µm polypropylene mesh to filter out aggregates.

7. Analyze the cells by multicolor cytometric analysis (*see* **Notes 6 and 8a**).

4. Notes

1. Units: MOI versus PPC. The term "multiplicity of infection" (MOI) relates to the number of infectious virus particles in a virus preparation that a single target cell is exposed to. The term "particles per cell" (PPC) refers to the number of viral particles, regardless of virus integrity, to which a single target cell is exposed. Classically, the titer of infectious particles is determined on the susceptible packaging cell line HEK 293. In a good preparation, the infectious unit concentration is usually 1–2 logs lower than the concentration of total virus particles. Poor-quality virus preparations will have considerably lower infectious units to particle-number ratios and this can have a detrimental impact on DC gene transfer and function. Unfortunately, the definition of infectious particles in a virus preparation can vary depending on the particular assay studied in a given report. This variation in the definition of MOI has caused considerable confusion in the literature. Therefore, it is important to carefully define MOI and to judiciously report the particle to infectious units number ratio of the virus preparation used.

2. Reporter genes. The reporter proteins most frequently utilized to monitor Ad gene delivery are bacterial β-galactosidase (β-gal), firefly (*Photinus pu-*

ralis), or sea pansy (*Renella renifomis*) luciferase, and GFP derived from the bioluminescent jelly fish *Aequorea victoria*. The bacterial β-gal gene is less favored because of the poor sensitivity of the colorimetric assay and the significant background from endogenous activity in mammalian cells and tissues. (Endogenous activity can be reduced by treatment of the cells or cell extracts with heat or high pH *[20]*.) Luciferase activity is measured in cell lysates, and, owing to the short half-lives of its mRNA and protein (30 min and 3–5 h, respectively), is directly proportional to the level of gene expression attained, but no information on the frequency of cells infected can be gleaned. The luciferase gene is an ideal reporter for examining the strength of a test promoter. GFP reporter-gene expression has emerged as one of the most useful reporters for indicating the number of cells successfully transfected by Ad infection. By determining the level of luciferase activity and the frequency of GFP positive cells, one can determine the relative gene expression level per transfected cell *(15)*. GFP has a great advantage because its expression allows tracking of Ad-transduced cells after injection into animal models. Furthermore, the excitation and emission spectra are compatible with flow cytometric analysis, providing an objective measure of percent GFP positive cells in a given population. A number of "enhanced" GFP genes (EGFP) are commercially available that have improved characteristics for mammalian cell expression through optimizing the codon usage for mammalian cells, reducing hydrophobicity to reduce toxicity, and, with a series of point mutations, changing the excitation and emission spectra resulting in multiple color indicators. Two such genes are EGFP and enhanced yellow fluorescent protein (EYFP) from Clontech (Palo Alto, CA).

3. Monitoring Ad-GFP transduction effeciency by flow cytometry. Flow cytometric analysis of viable cells reproducibly provides a better GFP signal when compared to cells fixed in 1% paraformaldehyde-PBS. However, depending on the level of expression, the cell preparation, and the GFP gene being used, samples may be fixed in paraformaldehyde for analysis at a later time. (This can be applied to samples stained with Abs as in **Section 3.2.**) Store fresh or fixed samples at 4°C and protect from light prior to analysis.

4. Bulk infection. This protocol provides a method to determine the optimal Ad vector infection conditions by measuring gene transfer of either a GFP gene or a luciferase gene. GFP expression is measured by assessing the fluorescence of intact cells by flow cytometry; luciferase activity is measured by determining the light output of cell extracts using a luminometer. These assays should be done in replicate sets for each condition. Hence, infection and culture in microfuge tubes is convenient for subsequent analysis. To infect bulk cultures of DCs for other experimental purposes (e.g., adoptive

transfer experiments in mice), this protocol may be easily scaled up by infecting DCs in loosely capped 5-mL (Falcon, cat. no. 35-2063) or 14-mL (Falcon, cat. no. 35-2006) polypropylene tubes. Polypropylene reduces cell adherence during infection and overnight culture and enhances recovery. Infect the cells in as small a volume as possible (e.g., 100 µL per 5×10^6 cells in a 5-mL tube) with frequent agitation.

    a. If simultaneous assessment of GFP transfection efficiency and DC activation for each Ad treatment is planned, scale up the cell density at least three-fold to 18×10^6 cells per mL (4.5×10^5 cells per microfuge tube in approx 25 µL) per infection condition to provide enough cells for subsequent analysis. Increase the Ad vector concentration accordingly. After 24 h, split the sample into three tubes, each containing approx 1.5×10^5 cells after infection, to serve as: (i) unstained (which may contain GFP positive transfected cells), (ii) background staining (PE-isotype control), and (iii) specific Ab stained (e.g., PE-CD40) samples for two-color flow cytometric analysis.

    b. Because DCs principally rely on macropinocytotic mechanisms to internalize Ad, most of the relevant exposure occurs when Ad is added at high concentrations for 1–2 h. Longer incubation in this small volume can induce toxicity. An option is to skip the wash step (**Subheading 3.1.**, **step 5**) after Ad incubation and directly add culture medium for the overnight culture. This will reduce the Ad concentration 10-fold, resulting in a modest increase in gene-transfer efficiency with reduced toxic effects.

5. Monitoring DC activation by examining phenotype. The extent of DC activation by Ad treatment can be easily monitored by comparing the expression level of DC-specific markers (normally expressed at negative or low levels on immature DCs) on mock-treated control and Ad-treated DCs using flow cytometry (frequently termed FACS [fluorescence activated cell sorter] analysis). The following markers are upregulated 10–1000-fold upon DC activation: class II MHC (murine I-A/I-E molecules or human DR molecules), co-receptor molecules such as CD80 (murine B7.1) or CD86 (murine B7.2), or the activation molecule CD40. Upregulation of activation markers indicates that DCs have undergone functional maturation and can now effectively present antigen to T cells and secrete cytokines such as IL-12. (Note that it is not yet known if highly mature DCs migrate as effectively as immature DCs to lymph nodes upon adoptive transfer to the host.)

6. Multiparameter cytometric analysis of GFP expression and apoptosis.

    a. For analysis of apoptosis, a positive control sample with some dead cells must be stained with the same concentrations of reagents used in the test samples in order to adjust the flow cytometer fluorescence gain and compensation settings. Treating a sample of cells ($\geq 6 \times 10^5$ in 1 mL) overnight with 30–50 n*M* staurosporine provides such a sample. In addition, a GFP-positive control culture of cells (containing $\geq 6 \times 10^5$ cells) provides additional set-up parameters for compensation settings for cells expressing GFP. The staurosporine treated cells and the positive GFP control culture should each be divided into two sets of four tubes labeled: unstained, Annexin V only, 7-AAD only, and Annexin and 7-AAD to set up the cytometer. With these set-up controls available, all of the Ad-transfected test samples can be stained directly with both Annexin V and 7-AAD without scaling up, allowing simultaneous analysis of GFP expression and apoptosis.

b. Apoptosis dyes require viable cell flow cytometric analysis within an hour of staining. Keep samples at 4°C and protected from light prior to analysis.

c. 7-AAD can be used in conjunction with PE stained cells (e.g., PE-Annexin V or PE-conjugated Abs). Propidium iodide (PI) used at 10 µg/mL final concentration (Sigma) also stains cells "red" but cannot be used in conjunction with PE staining because the emission spectrum overlaps with PE. PI is typically used only in combination with GFP stained cells and can be used as an alternative stain to 7-AAD to reveal dead cells.

7. Examples of analysis of transgene expression.

a. GFP-frequency of transduced cells by flow cytometry. **Figure 1** shows the level of reporter gene expression in murine DCs 24 h following infection with increasing numbers of Ad5-GFP. Significant gene expression is noted when relatively high doses of 10,000 PPC were used. XS106 is a relatively immature murine DC line exhibiting low but sig-

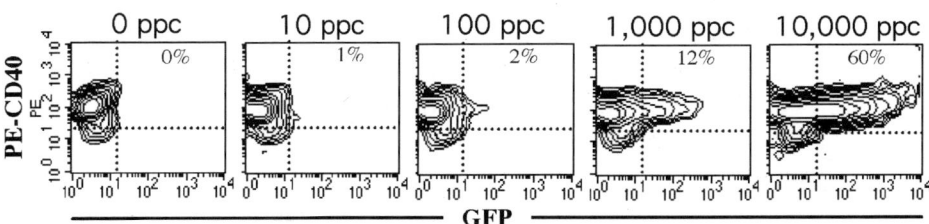

Fig. 1. Wild-type Ad 5-mediated transfection of the GFP gene in murine DCs does not induce DC activation and requires high levels of virion particles. The murine DC line, XS106 *(15)*, was treated with the indicated PPC of Ad-GFP for 1 h at 37°C, then washed and further incubated for a total of 24 h. Cells were subsequently stained with DC-specific PE-CD40 antibody to determine cell activation, then immediately analyzed on a FacScan Flow Cytometer. Ten thousand events are shown. PPC to MOI ratio was 12.

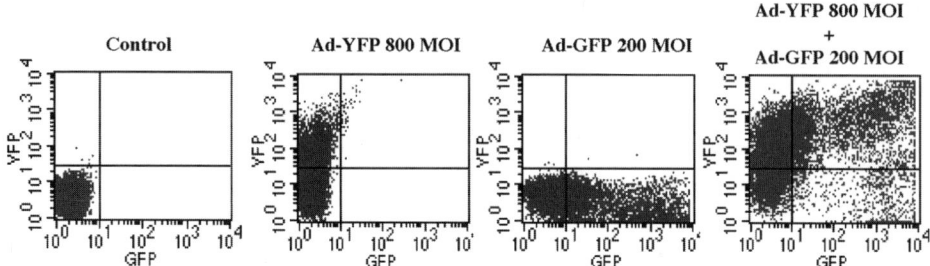

Fig. 2. Identification of DC subsets expressing genes of two Ad constructs. The murine DC line, XS106, was treated with either Ad-YFP at a MOI of 800, or Ad-GFP at a MOI of 200, or both at once for 1 h. Cells were washed and further incubated for 24 h total, then analyzed by flow cytometry. The majority of the GFP-positive population (58% in panel 3) also expressed YFP (86% YFP positive), whereas only 60% of the YFP population expressed GFP. Ten thousand events are shown. PPC to MOI ratio was 12 for Ad-GFP and 20 for Ad-YFP.

nificant levels of CD40 expression in the "resting" state, but note that CD40 upregulation was not observed at any dose of Ad, indicating that this Ad stock did not induce activation. (The PPC to MOI ratio was 12 for this stock.)

b. Identification of cells expressing two Ad-delivered genes. Because only a fraction of DCs are routinely transfected, it is impossible to identify the cells expressing a nonreporter gene. This can only be achieved when a single Ad construct permits dual expression of both a reporter gene and the nonreporter gene of interest (e.g., utilizing an internal ribosome entry site [IRES sequence]). An alternative strategy allows dual infection of two separate Ad vector stocks, one containing a reporter gene and the other Ad vector used at higher MOI (at least three-fold higher) encoding the gene of interest. **Fig. 2** illustrates that a DC population is heterogeneous in its susceptibility to Ad infection. A subset of cells is highly susceptible to Ad transfection and this population is more likely to be doubly infected with a second Ad construct when present at a higher MOI. As seen in Figure 2, gating on the minority of GFP positive cells selects those that also express the gene of interest (83%), in this case YFP. (The PPC to MOI ratio in this experiment is 12 for Ad-GFP and 20 for Ad-YFP.)

8. Examples of assessment of viability and function.

a. Apoptosis assay. Ad induced toxicity assessed by flow cytometry. To achieve high levels of gene transfer in primary DCs with Ad requires a

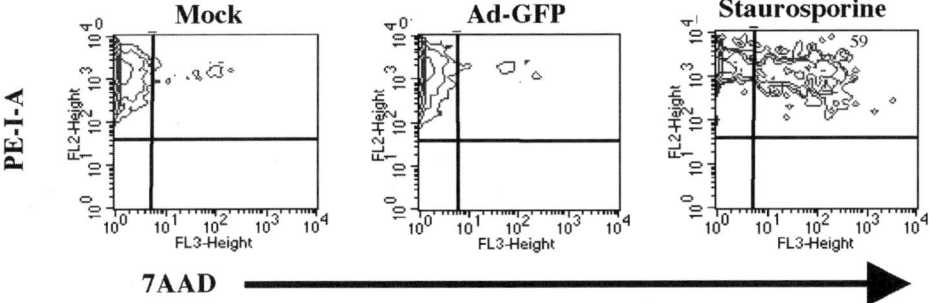

Fig. 3. Checking apoptosis of Ad treated DC. The murine DC line, XS106, was infected overnight with 1,000 PPC of Ad5-GFP, then stained with 7-AAD (5 Bg/mL) and analyzed by flow cytometry. Staurosporine (30 n*M*) was added overnight in some samples as a positive control for apoptosis. Anti I-A (MHC class II) Ab was used to verify DCs. Note that levels of 7-AAD staining in Ad treated cells equaled background levels seen in mock-infected cells. Panels depicting Ad-GFP and staurosporine treatment show only GFP gated cells. PPC to MOI ratio was 12. Ten thousand events are shown.

high MOI over a prolonged culture period *(21)* and is frequently associated with DC apoptosis *(22)*. Depending on the purity of the virus stock and the health of the DC culture, it is important to establish that the Ad-infection protocol itself does not induce DC apoptosis. One can determine the health of the whole DC population, or the viability of those DCs that specifically express Ad-encoded reporter genes, such as GFP. **Figure 3** shows that the Ad infected GFP-positive population (GFP-gated cells are shown in panels 2 and 3) of murine DCs shown to be I-A positive (MHC class II) remains viable after treatment with 1000 PPC.

   b. Activation assay. Ad targeted to DC receptors enhances gene transfer. Novel targeting technologies have been developed with the aim of enhancing Ad-transduction efficiency in DCs through targeted Ad interaction with receptors on DCs. Two such modified Ad systems are shown here. The CD40 antibody "coated" Ad uses a bispecific antibody with affinities for both the Ad capsid fiber knob and CD40, generated by chemical conjugation of a Fab fragment from a neutralizing anti-fiber knob MAb to an anti-CD40 MAb *(5)*. The Fab-anti-CD40 conjugate is used to retarget the natural binding of any adenoviral vector to CD40-expressing DCs. Not only does this CD40-Ad enhance transduction efficiency in primary human and murine DCs, but significantly, CD40 ligation activates them to become mature *(5,7)*. Although less specific in its targeting, another Ad vector has been genetically altered to contain an Arg-Gly-Asp

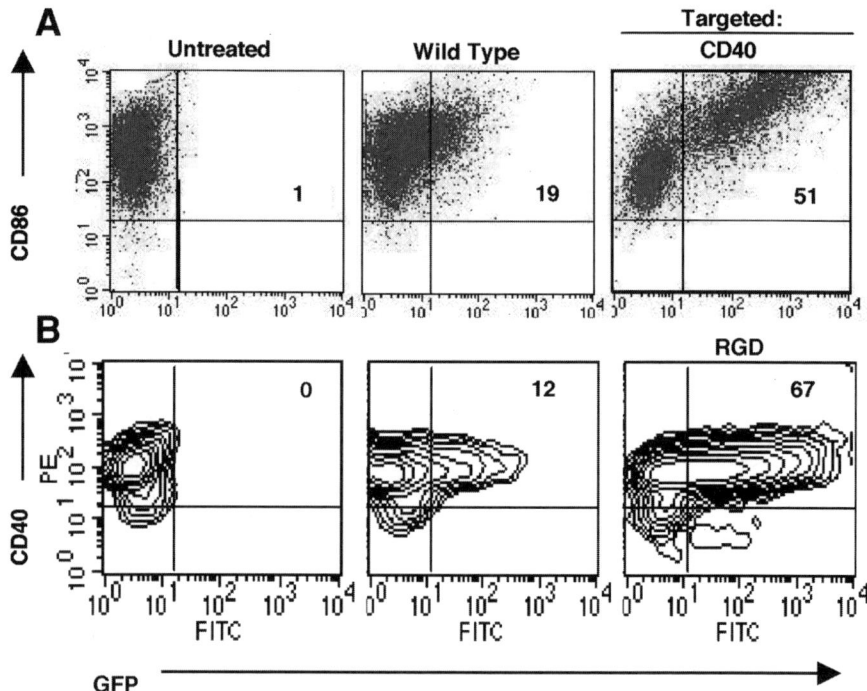

Fig. 4. Targeted Ad enhances gene transfer in DCs and, depending on the molecule targeted, can induce DC activation. XS106 cells were treated with GFP encoded untargeted wt-Ad, targeted CD40-Ad, or RGD-Ad at 1000 PPC for 1 h, washed and cultured for 24 h. Cells were harvested and stained with **(A)** PE-B7.2 (CD86) or **(B)** PE anti-CD40, then analyzed by two-color flow cytometry. Triplicate cultures were analyzed with identical results. PPC to MOI ratio was 12 for CD40-Ad and 15 for RGD-Ad. Ten thousand events are shown.

(RGD) sequence motif in the H1 loop of the fiber knob *(23)* and therefore binds to RGD ligands, such as integrins expressed on DCs *(24)*. **Figure 4** shows that after 1 h infection with wt-Ad only 12–19% GFP positive cells were detected, and that these cells expressed control levels of the co-stimulatory molecules CD86 and CD40. However, redirected CD40-targeted Ad efficiently infected these murine DC within 1 h, resulting in greater than 50% GFP positive cells; additionally, these cells showed more than a 10-fold increase in CD86 expression levels (**Fig. 4A**, right panel) indicating that DC activation was achieved. In contrast, the RGD-Ad vector transferred the GFP gene to DCs quite effectively (67% of the cells expressed GFP) without activating the DCs because no increase in CD40 expression levels was observed (**Fig. 4B**, right panel).

References

1. Zhong, L., Granelli-Piperno, A., Choi, Y., and Steinman, R. M. (1999) Recombinant adenovirus is an efficient and non-perturbing genetic vector for human dendritic cells. *Eur. J. Immunol.* **29,** 964–972.
2. Brossart, P., Goldrath, A. W., Butz, E. A., Martin, S., and Bevan, M. J. (1997) Virus-mediated delivery of antigenic epitopes into dendritic cells as a means to induce CTL. *J. Immunol.* **158,** 3270–3276.
3. Wan, Y., Emtage, P., Foley, R., Carter, R., and Gauldie, J. (1999) Murine dendritic cells transduced with an adenoviral vector expressing a defined tumor antigen can over come anti-adenovirus neutralizing immunity and induce effective tumor regression *Int. J. Oncol.* **14,** 771–776.
4. Kaplan, J. M., Yu, Q., Piraino, S. T., Pennington, S. E., Shankara, S., Woodworth, L. A., and Roberts, B. L. (1999) Induction of antitumor immunity with dendritic cells transduced with adenovirus vector-encoding endogenous tumor-associated antigens. *J. Immunol.* **163,** 699–707.
5. Tillman, B. W., de Gruijl, T. D., Luykx-deBakker, S. A., Scheper, R. J., Pinedo, H. M., Curiel, T. J., et al. (1999) Maturation of dendritic cells accompanies high-efficiency gene transfer by a CD40-targeted adenoviral vector. *J. Immunol.* **162,** 6378–6383.
6. Rea, D., Schagen, F. H. E., Hoeben, R. C., Mehtali, M., Havenga, M. J. E., Toes, R. E. M., et al. (1999) Adnoviruses activate human dendritic cells without polarizaiton toward a T-helper type 1-inducing subset. *J. Virol.* **73,** 10245–10253.
7. Tillman, B. W., Hayes, T. L., DeGruijl, T. D., Douglas, J. T., and Curiel, D. T. (2000) Adenoviral vectors targeted to CD40 enhance the efficacy of dendritic cell-based vaccination against human papillomavirus 16-induced tumor cells in a murine model. *Cancer Res.* **60,** 5456–5463.
8. Caux, C., Vanbervliet, B., Massacrier, C., Dezutter-Dambuyant, C., de Saint-Vis, B., Jacquet, C., et al. (1996) CD34+ hematopoietic progenitors from human cord blood differentiate along two independent dendritic cell pathways in response to GM-CSF+TNF alpha. *J. Exp. Med.* **184,** 695–706.
9. Schuler, G., Lutz, M., Bender, A., Thurner, B., Roder, C., Young, J. W., and Romani N. (1999) A guide to the isolation and propogation of dendritic cells, in *Dendritic Cells* (Lotz, M. T., and Thomson, A. W., eds.), Academic Press, New York, NY, pp. 515–553.
10. Inaba, K., Swiggard, W. J., Steinman, R. M., Romani, N., and Schuler, G. (1998) Generation of dendritic cells from proliferating mouse bone marrow progenitors, Unit 3.7, in *Current Protocols of Immunology* (Coico, R., ed.), Wiley, New York pp. 7–15.
10a. Munn, D.H., Sharma, M. D., Lee, J. R., et al. (2002) Potential regulatory function of human dendritic cells expressing indoleamine 2,3-dioxygenase. *Science* **297,** 1867–1870.
11. Ebe, A., Schleidegger, S., Strunk, D., and Stingl, G. (1994). Fetal skin-derived MHC class I+, MHC class II-dendritic cells stimulate MHC class I-restricted responses of unprimed CD8+ T cells. *J. Immunol.* **153,** 2878–2889.

12. Xu, S., Ariizumi, K., Edelbaum, D., Bergstresser, P., and Takashima, R. (1995) Successive generation of antigen-presenting, dendritic cell lines from murine epidermis. *J. Immunol.* **154,** 2697–2705.
13. Jacob T., Saitoh, A., and Udey, M. C. (1997) E-caderhin-mediated adhesion involving Langerhans cell-like dendritic cells expanded from murine fetal skin. *J. Immunol.* **159,** 2693–2701.
14. Winzler, C., Roveere, P., Rescigno, M., Granucci, F., Penna, G., Adorini, L., et al. (1997) Maturation stages of mouse dendritic cells in growth factor dependent long-term cultures. *J. Exp. Med.* **185,** 317–328.
15. Timares, L., Takashima A., and Johnston S. A. (1998) Quantitative analysis of the immunopotency of genetically transfected dendritic cells. *Proc. Natl. Acad. Sci. USA* **95,** 13147–13152.
16. Lyakh, L. A., Koski, G. K., Young, H. A., Spence, S. E., Cohen, P. A., and Rice, N. R. (2002) Adenovirus type 5 vectors induce dendritic cell differentiation in human CD14(+) monocytes cultured under serum-free conditions. *Blood* **99,** 600–608.
17. Morelli, A. E., Larregina, A. T., Ganster, R. W., Zahorchak, A. F., Plowey, J. M., Takayama, T., et al. (2000) Recombinant adenovirus induces maturation of dendritic cells via an NF-kappaB-dependent pathway. *J. Virol.* **74,** 9617–9628.
18. Romani, N., Reider, D., Heuer, M., Ebner, S., Kampgen, E., Eibl, B., et al. (1996) Generation of mature dendritic cells from human blood-an improved method with special regard to clinical applicability. *J. Immunol. Methods* **196,** 137–151.
19. Gallucci, S., Lolkema, M., and Matzinger, P. (1999) Natural adjuvants: endogenous activators of dendritic cells. *Nat. Medicine* **5,** 1249–1255.
20. Alam, J. and Cook, J. (1990) Reporter genes: Application to the study of mammalian gene transcription. *Anal. Biochem.* **188,** 245–354.
21. Ranieri, E., Herr, W., Gambotto, A., Olson, W., Rowe, D., Robbins, P. D., et al. (1999) Dendritic cells transduced with an adenovirus vector encoding Epstein-Barr virus latent membrane protein 2B: a new modality for vaccination. *J. Virol.* **73,** 10416–10425.
22. Brand, K., Klocke, R., Possling, A., Paul, D., and Strauss, M. (1999) Induction of apoptosis and G2/M arrest by infection with replication-deficient adenovirus at high multiplicity of infection. *Gene Ther.* **6,** 1054–1063.
23. Krasnykh, V., Dmitriev, I., Mikheeva, G., Miller, C. R., Belousova, N., and Curiel, D. T. (1998) Characterization of an adenovirus vector containing a heterologous peptide epitope in the HI loop of the fiber knob. *J. Virol.* **72,** 1844–1852.
24. Asada-Mikami, R., Heike, Y., Kanai, S., Azuma, M., Shirakawa, K., Takaue, Y., et al. (2001) Efficient gene transduction by RGD-fiber modified recombinant adenovirus into dendritic cells. *Japan. J. Cancer Res.* **92,** 321–327.

II

Delivery Using Adeno-Associated Viruses

10

Overview of Adeno-Associated Viral Vectors

Thomas M. Daly

1. Introduction

The use of adeno-associated virus (AAV) as a gene transfer vector has been steadily increasing over the past several years. AAV vectors have been particularly useful for applications where sustained gene expression is required. Prolonged in vivo expression following AAV treatment has been seen in the liver *(1,2)*, brain *(3,4)*, skeletal muscle *(5,6)*, lung *(7,8)*, and hematopoietic stem cells *(9,10)* of animal models. Therapeutic benefit from AAV treatment has been shown in a number of preclinical models of disease, including animal models of coagulopathies *(11,12)*, lysosomal storage diseases *(13,14)*, vision defects *(15,16)*, and amino acid disorders *(17)*. Clinical trials using AAV for the treatment of hemophilia B have begun, and early reports from these trials have been promising *(18)*. In this introductory chapter to AAV, we will provide a brief overview of the molecular biology of this virus, an overview of methods of vector production, and a brief summary of the use of alternate AAV serotypes. The following chapters will then focus on specific methods and techniques for AAV transduction of the organs listed previously.

2. Background

AAV is a human parvovirus that is replication-defective in the absence of helper functions *(19)*. Viral particles are small (20–30 nm) and nonenveloped, containing single-stranded DNA molecules with plus and minus strands packaged with equal efficiency. Exposure to AAV appears to be common in humans,

From: *Methods in Molecular Biology, vol. 246:*
Gene Delivery to Mammalian Cells: Vol. 2: Viral Gene Transfer Techniques
Edited by: W. C. Heiser © Humana Press Inc., Totowa, NJ

with seropositivity rates for antibodies directed against AAV2 capsid proteins range from 50 to 80% in adult populations *(20)*. However, no human disease has been associated with AAV infection, suggesting that the wild-type virus is nonpathogenic in humans. Six serotypes of AAV have been described, with AAV2 being the most widely used for gene-transfer studies *(21)*. (For the remainder of this chapter, AAV will refer to AAV2 unless otherwise specified. The remaining serotypes will be discussed in more detail at the end of the chapter.)

The viral genome of wild-type AAV is approx 4.7 kb long, and codes for two families of viral proteins produced by alternate splicing from transcripts driven by 3 promoters (**Fig. 1**). The 5′ and 3′ ends of the viral genome are demarcated by inverted terminal repeats (ITR). These ITRs contain palindromic sequences, which can form an internal T-shaped structure to self-prime for second strand synthesis during viral replication *(19)*. The 5′ and 3′ ITRs are the only portion of the viral genome required *in cis* for packaging *(22,23)*. The remainder of the AAV genome codes for two families of proteins, *rep* and *cap*. The three *cap* proteins (VP1, VP2, and VP3) are derived by alternative splicing of transcripts from a single promoter (p40), with VP3 being the dominant form. The 5′ end of the viral genome consists of the *rep* family of proteins. There are four species of rep proteins (Rep 78/68 and Rep 52/40), which are derived from alternative splicing of transcripts generated from the p5 and p19 promoters, respectively.

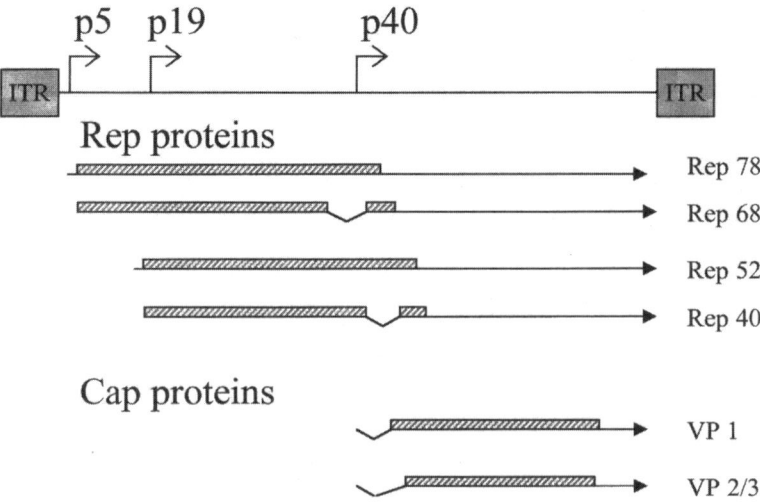

Fig. 1. Genomic structure of wild-type AAV. Seven transcripts are generated from three distinct viral promoters. The 5′ and 3′ inverted terminal repeats (ITR) are the only portion of the wild-type genome required for viral packaging.

These molecules play a role in viral replication as described later. Both rep and cap functions can be provided in *trans*.

The life cycle of wt AAV is covered briefly here with emphasis on aspects that affect its use as a gene-transfer vector. Viral entry by AAV2 is mediated by binding to heparin sulfate proteoglycans *(24)*. The $\alpha_v\beta_5$ integrin and fibroblast growth factor receptor-1 (FGFR-1) may also be involved, although this has been disputed by some groups *(25,26)*. The distribution of these molecules on many different cell types helps to explain the wide target range of AAV2 vectors. Following cell entry and transfer to the nuclear compartment, the single-stranded AAV genomes are slowly converted to double-stranded molecules. This process is dependent on cellular DNA polymerases, and appears to be the rate-limiting step in expression from AAV vectors in vitro *(27)*. In the presence of helper functions such as those provided by adenovirus, transcription from the p5, p19, and p40 promoters is upregulated, and rep and cap proteins are produced. Rep78/68 can then bind to the rep-binding sequence (rbs) present in the 5′ ITR, and direct cleavage of the growing double-stranded DNA molecule at the adjoining terminal resolution site (trs). This leads to the production of single-stranded AAV molecules, which can then be packaged to produce infectious progeny.

In the absence of helper functions, AAV viral promoters are inactive and rep and cap proteins are not actively synthesized, although a small amount of rep proteins may be present. However, double-stranded DNA molecules are still produced from the input viral genomes, as this process does not depend on virally coded proteins. These dsDNA molecules can be degraded, integrate into the host genome in either a random or targeted manner *(28,29)*, or persist as an episomal form *(30)*. The relative occurrence of these forms is still under investigation, and appears to vary between tissues. However, it is this persistence of dsDNA in the absence of a productive viral infection that allows AAV to be used as a successful gene-transfer vector.

3. Production of Recombinant AAV Vectors

A number of protocols are available for the production and purification of recombinant AAV vector stocks. Construction of recombinant AAV vector plasmids is a straightforward process. As mentioned earlier, only the 5′ and 3′ ITR sequences are required for viral packaging. Therefore, the entire wild-type coding sequence can be excised, and replaced with a therapeutic construct. A commonly used plasmid for this purpose is pSub201, which has been engineered to contain flanking XbaI sites internal to the ITRs to allow easy removal of the wild-type sequence *(31)*. In addition, AAV plasmids containing common promoter cassettes such as the CMV promoter are commercially available, simplifying the vector development process. Perhaps the biggest limitation in the design of AAV vectors is the size constraints of recombinant AAV plasmids. Constructs that are much larger than the wild-type 4.7 kB sequence do not package well, and vector titers decrease sharply. This size limitation prevents the effective gene transfer of very large cDNAs such as dystrophin or factor VIII in a single AAV construct. However, recent studies have suggested that transduction with two AAV vectors that each contain a portion of a coding sequence can lead to production of an intact protein following heterodimerization *(32,33)*. Although the mechanism of this finding is still under study, such techniques may allow the size limitations of AAV vectors to be overcome.

Production of recombinant AAV is accomplished by the introduction of the AAV vector plasmid into cells where rep, cap, and adenoviral helper functions are present. This is most commonly done by cotransfection of the AAV vector plasmid along with helper plasmids containing the wt AAV rep/cap sequences. Adenoviral helper functions can be supplied either by superinfection with wild-type adenovirus, or by cotransfection of a third plasmid containing the minimal adenoviral proteins required for AAV production (E2A, E4ORF6, VA). Because separation of wild-type Ad from recombinant AAV is time-consuming and adenoviral contamination of viral solutions can be problematic, most laboratories now use a plasmid-based approach. Two commonly used systems are the pXX2/pXX6 system, which supply the Ad helper functions and AAV rep/cap on separate plasmids *(34)*, and pDG or pSH3/5, which provide both Ad and AAV functions on single plasmids *(35,36)*. As an alternative to cotransfection techniques, packaging cell lines have been developed that contain integrated copies of rep/cap. Although such lines are more difficult to establish because of the cytostatic effects of rep production, conditionally producing cell lines have been established by a number of groups *(37,38)*.

Isolation of recombinant AAV from cellular lysates can be performed by many different techniques. Early methods took advantage of the relatively low density of AAV particles relative to contaminating adenovirus, and separated AAV vectors by centrifugation on a series of isopycnic cesium chloride gradients. Such methods were labor-intensive, and the long periods of centrifugation required for purification led to decreased titers. The use of adenoviral-free production methods has simplified the purification process somewhat, allowing use of more rapid separation techniques. One example uses an initial separation of AAV from cellular proteins on a stepwise iodixanol gradient, followed by further purification of AAV via binding to a heparin sulfate column *(39)*. This method can be done rapidly, with relatively little equipment, and produces good-quality virus for the average user. Large-scale purification of clinical-grade virus can also be done using high-performance liquid chromatography (HPLC) methods over heparin sulfate columns *(39)*. These methods allow virus production to be scaled up significantly, but do require a more significant investment in equipment.

4. AAV Serotypes Other Than AAV2

Although the majority of studies to date have been done using AAV2, an emerging body of work is being published using vectors based on other AAV serotypes. These non-AAV2 vectors have advantages in overcoming some of the limitations inherent in AAV2, and may also expand the target range and function of AAV vectors as a gene-transfer tool. A prime motivation for the development of non-AAV2 vectors has been the relatively high incidence of anti-AAV2 antibodies in the general population *(40)*. It has been shown in animal models that administration of AAV2 vectors stimulates a humoral response against capsid proteins, and that expression following readministration of the same vector is blocked *(41–43)*. Thus, the high level of individuals seropositive for AAV2 in the general population could potentially limit the efficacy of AAV2 for widespread human use. Development of alternate serotypes could potentially provide a method for readministration in diseases where multiple treatments would be required *(41)*.

Sequence data for AAV serotypes 1 through 6 has been published, and much of the variation between serotypes lies in the 3′ end of *rep* and in four discrete regions of the *cap* proteins *(44)*. Regions important for viral replication such as the rbs, trs, orf start sites, and intronic splice junctions are relatively conserved between serotypes. In addition, the ITR sequences are highly conserved. This has the benefit from a vector-production standpoint that AAV2 vector plasmids can be efficiently pseudotyped with serotypes 1–4 merely by providing an alternate rep/cap helper plasmid in *trans* during viral production *(45)*. Such methods have been used to produce AAV1, AAV3, AAV4, and AAV5 serotyped vectors *(46)*. This simplifies comparison of vector performance between serotypes, by allowing identical vector sequences to be used in multiple serotypes.

The variations in the capsid proteins between serotypes can impart different binding characteristics, altering the biodistribution and transduction efficiency of non-AAV2 vectors. For example, AAV4 and AAV5 do not use heparin sulfate as a primary receptor, but instead require α2,3 sialic acid for effective binding *(47,48)*. In addition, in vitro and in vivo efficiency can vary between serotypes. For example, although AAV2 shows a higher rate of transduction in 293 cells in vitro, other serotypes have been shown to be highly potent in vivo in intramuscular (im) injections, both in levels of expression and percentage of cells transduced *(46)*. As these vectors become more widely available, they may offer more options for investigators targeting specific organs.

In summary, AAV vectors are useful tools for in vivo gene transfer, particularly in situations where long-term expression of gene is required. The following chapters will provide a more detailed look at methods for the use of AAV in a variety of contexts.

References

1. Xiao, W., Berta, S. C., Lu, M. M., Moscioni, A. D., Tazelaar, J., and Wilson, J. M. (1998) Adeno-associated virus as a vector for liver-directed gene therapy. *J. Virol.* **72,** 10222–10226.
2. Snyder, R. O., Miao, C., Meuse, L., Tubb, J., Donahue, B. A., Lin, H. F., et al. (1999) Correction of hemophilia B in canine and murine models using recombinant adeno-associated viral vectors [see comments]. *Nat. Med.* **5,** 64–70.
3. During, M. J., Samulski, R. J., Elsworth, J. D., Kaplitt, M. G., Leone, P., Xiao, X., et al. (1998) In vivo expression of therapeutic human genes for dopamine production in the caudates of MPTP-treated monkeys using an AAV vector. *Gene Ther.* **5,** 820–827.
4. Chen, H., McCarty, D. M., Bruce, A. T., and Suzuki, K. (1998) Gene transfer and expression in oligodendrocytes under the control of myelin basic protein transcriptional control region mediated by adeno-associated virus. *Gene Ther.* **5,** 50–58.
5. Bohl, D., Naffakh, N., and Heard, J. M. (1997) Long-term control of erythropoietin secretion by doxycycline in mice transplanted with engineered primary myoblasts [see comments]. *Nat. Med.* **3,** 299–305.
6. Rivera, V. M., Ye, X., Courage, N. L., Sachar, J., Cerasoli, F., Jr., Wilson, J. M., and Gilman, M., et al. (1999) Long-term regulated expression of growth hormone in mice after intramuscular gene transfer. *Proc. Natl. Acad. Sci. USA* **96,** 8657–8662.
7. Conrad, C. K., Allen, S. S., Afione, S. A., Reynolds, T. C., Beck, S. E., Fee-Maki, M., et al. (1996) Safety of single-dose administration of an adeno-associated virus (AAV)-CFTR vector in the primate lung. *Gene Ther.* **3,** 658–668.
8. Beck, S. E., Jones, L. A., Chesnut, K., Walsh, S. M., Reynolds, T. C., Carter, B. J., et al. (1999) Repeated delivery of adeno-associated virus vectors to the rabbit airway. *J. Virol.* **73,** 9446–9455.
9. Ponnazhagan, S., Mukherjee, P., Wang, X. S., Qing, K., Kube, D. M., Mah, C., et al. (1997) Adeno-associated virus type 2-mediated transduction in primary human bone marrow-derived CD34+ hematopoietic progenitor cells: donor variation and correlation of transgene expression with cellular differentiation. *J. Virol.* **71,** 8262–8267.
10. Tan, M., Qing, K., Zhou, S., Yoder, M. C., and Srivastava, A. (2001) Adeno-associated virus 2-mediated transduction and erythroid lineage-restricted long-term expression of the human beta-globin gene in hematopoietic cells from homozygous beta-thalassemic mice. *Mol. Ther.* **3,** 940–946.
11. Xu, L., Daly, T., Gao, C., Flotte, T. R., Song, S., Byrne, B. J., et al. (2001) CMV-beta-actin promoter directs higher expression from an adeno-associated viral vector in the liver than the cytomegalovirus or elongation factor 1 alpha promoter and results in therapeutic levels of human factor X in mice. *Hum. Gene Ther.* **12,** 563–573.

12. Chao, H., Monahan, P. E., Liu, Y., Samulski, R. J., and Walsh, C. E. (2001) Sustained and complete phenotype correction of Hemophilia B mice following intramuscular injection of AAV1 serotype vectors. *Mol. Ther.* **4,** 217–222.

13. Jung, S. C., Han, I. P., Limaye, A., Xu, R., Gelderman, M. P., Zerfas, P., et al. (2001) Adeno-associated viral vector-mediated gene transfer results in long-term enzymatic and functional correction in multiple organs of Fabry mice. *Proc. Natl. Acad. Sci. USA* **98,** 2676–2681.

14. Daly, T. M., Ohlemiller, K. K., Roberts, M. S., Vogler, C. A., and Sands, M. S. (2001) Prevention of systemic clinical disease in MPS VII mice following AAV-mediated neonatal gene transfer. *Gene Ther.* **8,** 1291–1298.

15. Liang, F. Q., Dejneka, N. S., Cohen, D. R., Krasnoperova, N. V., Lem, J., Maguire, A. M., et al. (2001) AAV-mediated delivery of ciliary neurotrophic factor prolongs photoreceptor survival in the rhodopsin knockout mouse. *Mol. Ther.* **3,** 241–248.

16. Acland, G. M., Aguirre, G. D., Ray, J., Zhang, Q., Aleman, T. S., Cideciyan, A. V., et al. (2001) Gene therapy restores vision in a canine model of childhood blindness. *Nat. Genet.* **28,** 92–95.

17. Chen, S. J., Tazelaar, J., Moscioni, A. D., and Wilson, J. M. (2000) In vivo selection of hepatocytes transduced with adeno-associated viral vectors. *Mol. Ther.* **1,** 414–422.

18. Kay, M. A., Miao, C. H., Ohashi, K., Arruda, V., McClelland, A., Couto, L. B., et al. (2001) A proposed rAAV-Liver-directed clinical trial for hemophilia B. *Mol Ther.* **3,** S33.

19. Berns, K. I. and Linden, R. M. (1995) The cryptic life style of adeno-associated virus. *Bioessays.* 17, 237–245.

20. Erles, K., Sebokova, P., and Schlehofer, J. R. (1999) Update on the prevalence of serum antibodies (IgG and IgM) to adeno-associated virus (AAV). *J. Med. Virol.* 59, 406–411.

21. Hermonat, P. L. and Muzyczka, N. (1984) Use of adeno-associated virus as a mammalian DNA cloning vector: transduction of neomycin resistance into mammalian tissue culture cells. *Proc. Natl. Acad. Sci. USA* **81,** 6466–6470.

22. Xiao, X., Xiao, W., Li, J., and Samulski, R. J. (1997) A novel 165-base-pair terminal repeat sequence is the sole cis requirement for the adeno-associated virus life cycle. *J. Virol.* **71,** 941–948.

23. Samulski, R. J., Berns, K. I., Tan, M., and Muzyczka, N. (1982) Cloning of adeno-associated virus into pBR322: rescue of intact virus from the recombinant plasmid in human cells. *Proc. Natl. Acad. Sci USA* **79,** 2077–2081.

24. Summerford, C., and Samulski, R. J. (1998) Membrane-associated heparan sulfate proteoglycan is a receptor for adeno-associated virus type 2 virions. *J. Virol.* **72,** 1438–1445.

25. Summerford, C., Bartlett, J. S., and Samulski, R. J. (1999) AlphaVbeta5 integrin: a co-receptor for adeno-associated virus type 2 infection [see comments]. *Nat. Med.* **5,** 78–82.

26. Qiu, J. and Brown, K. E. (1999) Integrin $\alpha_V\beta_5$ is not involved in adeno-associated virus type 2 (AAV2) infection. *Virology* **264,** 436–440.

27. Ferrari, F. K., Samulski, T., Shenk, T., and Samulski, R. J. (1996) Second-strand synthesis is a rate-limiting step for efficient transduction by recombinant adeno-associated virus vectors. *J. Virol.* **70,** 3227–3234.

28. Russell, D. W. and Hirata, R. K. (1998) Human gene targeting by viral vectors. *Nat. Genet.* **18,** 325–330.

29. Nakai, H., Iwaki, Y., Kay, M. A., and Couto, L. B. (1999) Isolation of recombinant adeno-associated virus vector-cellular DNA junctions from mouse liver. *J. Virol.* **73,** 5438–5447.

30. Miao, C. H., Nakai, H., Thompson, A. R., Storm, T. A., Chiu, W., Snyder, R. O., and Kay, M. A. (2000) Nonrandom transduction of recombinant adeno-associated virus vectors in mouse hepatocytes in vivo: cell cycling does not influence hepatocyte transduction. *J. Virol.* **74,** 3793–3803.

31. Samulski, R. J., Chang, L. S., and Shenk, T. (1987) A recombinant plasmid from which an infectious adeno-associated virus genome can be excised in vitro and its use to study viral replication. *J. Virol.* **61,** 3096–3101.

32. Yan, Z., Zhang, Y., Duan, D., and Engelhardt, J. F. (2000) Trans-splicing vectors expand the utility of adeno-associated virus for gene therapy. *Proc. Natl. Acad. Sci. USA* **97,** 6716–6721.

33. Duan, D., Yue, Y., and Engelhardt, J. F. (2001) Expanding AAV packaging capacity with trans-splicing or overlapping vectors: a quantitative comparison. *Mol. Ther.* **4,** 383–391.

34. Xiao, X., Li, J., and Samulski, R. J. (1998) Production of high-titer recombinant adeno-associated virus vectors in the absence of helper adenovirus. *J. Virol.* **72,** 2224–2232.

35. Grimm, D., Kern, A., Rittner, K., and Kleinschmidt, J. A. (1998) Novel tools for production and purification of recombinant adeno-associated viral vectors. *Hum. Gene Ther.* **9,** 2745–2760.

36. Collaco, R. F., Cao, X., and Trempe, J. P. (1999) A helper virus-free packaging system for recombinant adeno-associated virus vectors. *Gene* **238,** 397–405.

37. Liu, X., Voulgaropoulou, F., Chen, R., Johnson, P. R., and Clark, K. R. (2000) Selective Rep-Cap gene amplification as a mechanism for high-titer recombinant AAV production from stable cell lines. *Mol. Ther.* **2,** 394–403.

38. Chadeuf, G., Favre, D., Tessier, J., Provost, N., Nony, P., Kleinschmidt, J., et al. (2000) Efficient recombinant adeno-associated virus production by a stable rep-cap HeLa cell line correlates with adenovirus-induced amplification of the integrated rep-cap genome. *J. Gene Med.* **2,** 260–268.

39. Zolotukhin, S., Byrne, B. J., Mason, E., Zolotukhin, I., Potter, M., Chesnut, K., et al. (1999) Recombinant adeno-associated virus purification using novel methods improves infectious titer and yield. *Gene Ther.* **6,** 973–985.

40. Moskalenko, M., Chen, L., van Roey, M., Donahue, B. A., Snyder, R. O., McArthur, J. G., and Patel, S. D. (2000) Epitope mapping of human anti-adeno-associated virus type 2 neutralizing antibodies: implications for gene therapy and virus structure. *J. Virol.* **74,** 1761–1766.

41. Halbert, C. L., Rutledge, E. A., Allen, J. M., Russell, D. W., and Miller, A. D. (2000) Repeat transduction in the mouse lung by using adeno-associated virus vectors with different serotypes. *J. Virol.* **74,** 1524–1532.

42. Manning, W. C., Zhou, S., Bland, M. P., Escobedo, J. A., and Dwarki, V. (1998) Transient immunosuppression allows transgene expression following readministration of adeno-associated viral vectors. *Hum. Gene Ther.* **9,** 477–485.

43. Halbert, C. L., Standaert, T. A., Wilson, C. B., and Miller, A. D. (1998) Successful readministration of adeno-associated virus vectors to the mouse lung requires transient immunosuppression during the initial exposure. *J. Virol.* **72,** 9795–9805.

44. Rutledge, E. A., Halbert, C. L., and Russell, D. W. (1998) Infectious clones and vectors derived from adeno-associated virus (AAV) serotypes other than AAV type 2. *J. Virol.* **72,** 309–319.

45. Hildinger, M., Auricchio, A., Gao, G., Wang, L., Chirmule, N., and Wilson, J. M. (2001) Hybrid vectors based on adeno-associated virus serotypes 2 and 5 for muscle-directed gene transfer. *J. Virol.* **75,** 6199–6203.

46. Chao, H., Liu, Y., Rabinowitz, J., Li, C., Samulski, R. J., and Walsh, C. E. (2000). Several log increase in therapeutic transgene delivery by distinct adeno-associated viral serotype vectors. *Mol. Ther.* **2,** 619–623.

47. Kaludov, N., Brown, K. E., Walters, R. W., Zabner, J., and Chiorini, J. A. (2001) Adeno-associated virus serotype 4 (AAV4) and AAV5 both require sialic acid binding for hemagglutination and efficient transduction but differ in sialic acid linkage specificity. *J. Virol.* **75,** 6884–6893.

48. Walters, R. W., Yi, S. M., Keshavjee, S., Brown, K. E., Welsh, M. J., Chiorini, J. A., and Zabner, J. (2001) Binding of adeno-associated virus type 5 to 2,3-linked sialic acid is required for gene transfer. *J. Biol. Chem.* **276,** 20610–20616.

11

AAV Vector Delivery to Cells in Culture

Andrew Smith, Roy Collaco, and James P. Trempe

1. Introduction

Adeno-associated virus (AAV) gene delivery vectors are being investigated as vehicles for gene therapy for a wide variety of hereditary and acquired human diseases. AAV's inability to self-propagate, ability to be maintained as an episome in the transduced cell, and relatively innocuous effects on the immune system make it the vector of choice for prolonged in vivo gene expression. AAV type 2 is the most commonly used serotype for gene delivery. AAV2 vectors will deliver DNA to a wide variety of cell types. The development of vectors derived from the other five serotypes has expanded the tissue tropism of the AAV vector system (*1–6*). Tropism depends on the presence of cell-surface receptor elements on the target cell. For AAV2, heparin sulfate proteoglycan (*7*), $\alpha_2\beta_V$ integrin (*8*) and the fibroblast growth factor receptor-1 (FGFR-1) (*9*) are believed to mediate the initial internalization steps in infection. The ubiquity of these cell-surface components confers a wide tropism on AAV2 vectors.

Despite the wide tropism, barriers exist in some cell and tissue types that limit transduction efficiency. Some cells lack necessary receptor components, rendering them refractory to AAV2 transduction (*10*). Polarized airway epithelial cells, when grown in culture at an air-liquid interface, lack virus receptors on the apical side but retain them on the basal side of the cell (*11,12*). This explains the limited in vivo transduction efficiency of AAV2 vectors in the lung (*13*). One of the advantages of AAV vectors is their ability to transduce dividing and nondividing cells in vivo. In cell culture, quiescent cells will internalize and express the vector, but the level of expression is dramatically improved

From: *Methods in Molecular Biology, vol. 246:*
Gene Delivery to Mammalian Cells: Vol. 2: Viral Gene Transfer Techniques
Edited by: W. C. Heiser © Humana Press Inc., Totowa, NJ

when DNA synthesis is stimulated *(14,15)*. The single-strand vector DNA must be converted to a double-stranded conformation to attain transcriptional competence. Co-expression of the adenovirus (Ad) *E4* orf 6 protein or induction of a cellular DNA repair response stimulates this conversion *(13,14,16)*. Another means of enhancing vector expression is via treatment with inhibitors of the epidermal growth factor receptor tyrosine kinase (EGFRTK) *(17)*. The EGFRTK phosphorylates a single-strand D-sequence binding protein (ssD-BP) that interacts with the AAV terminal repeat DNA element. When de-phosphorylated, ssD-BP no longer interacts with the D-sequence, thus enabling doublestrand conversion. Tyrphostins and genistein inhibit EGFRTK, resulting in significant increases in transgene expression *(18)*. Other exogenous agents such as inhibitors of endosomal and proteolytic pathways also stimulate transgene expression *(19,20)*.

Analyses of AAV vector functionality and dissection of the biochemical mechanisms of transduction requires vector infection of cells in culture. In this chapter, we will review the methods for AAV vector transduction and the assessment of efficacy in standard tissue culture. We will also describe the use of Ad co-infection and Tyrphostin-1 treatment to enhance vector transduction efficiency.

2. Materials

2.1. Sources of Chemicals

1. Luciferase from *Photinus pyralis* (activity: $10–25 \times 10^6$ light units per mg protein) and Tyrphostin-1 are available from Sigma (St. Louis, MO). Store luciferase at $-70°C$ and Tyrphostin-1 at $-20°C$.
2. Luciferase Assay System with Reporter Lysis Buffer (1 vial lyophilized Luciferase Assay Substrate, 10 mL Luciferase Assay Buffer, 30 mL Reporter Lysis 5X Buffer) is available from Promega (Madison, WI). Store these reagents at $-70°C$.
3. DC Protein Assay Kit II (250 mL Reagent A, 2000 mL Reagent B, 5 mL Reagent S, bovine serum albumin [BSA] standard) is available from Bio-Rad (Hercules, CA).
4. 5-bromo-4-chloro-3-indoyl-β-D-galactopyranoside (X-gal) is available from Gold Biotechnology (St. Louis, MO).
5. Heparin sulfate agarose, Type 1 (cat. no. H6508) for vector purification is available from Sigma (St. Louis, MO).

2.2. Plasmids, Cells, Adenovirus, and AAV Vectors

1. Plasmids.
   a. pSH3 is used for AAV vector packaging *(21)*. This plasmid contains the Ad *E4*, *E2a*, and *VA* genes and the AAV2 genome lacking the terminal repeat elements.

b. pAV*luc* and pAV*gal* are AAV vector plasmids containing the firefly luciferase gene and the *Escherichia coli* β-galactosidase (β-gal) gene, respectively, driven by the cytomegavirus (CMV) promoter flanked by the AAV terminal repeats. The pAV*luc* and pAV*gal* plasmids were used to prepare the vAV*luc* and vAV*gal* vectors *(21)*.

2. Cells.

a. Hela (ATCC-CCL2) and 293 (ATCC-CRL1573) cells, used for vector preparation and virus titration, are available from ATCC (Manassas, VA).

b. Undifferentiated SV40 T-Ag-transformed cystic fibrosis airway epithelial IB3 cells *(22)* are used for some transduction experiments.

3. Virus.

a. Ad5 is used for stimulation of AAV vector transduction.

b. Recombinant adeno-associated virus (rAAV), vAV*luc* and vAV*gal*, are packaged using the helper virus-free methodology *(21)*. For preliminary experiments, the use of a crude vector preparation (cell lysate from a packaging experiment) may be employed. If vectors are packaged on a larger scale (i.e., using more than ten 10-cm culture dishes), they are purified by CsCl ultracentrifugations or by heparin agarose affinity chromatography *(23)*. Determine the virus titer as described in **Subheading 3.1**.

2.3. Solutions and Culture Medium

1. Cell-culture Media.

a. Hela and 293 cells are grown in MEM supplemented with 10% fetal bovine serum (FBS) and 100 µg/mL gentamycin.

b. IB-3 cells are grown in LHC-8 medium available from Invitrogen (Carlsbad, CA). The media is supplemented with 100 µg/mL streptomycin, 100 U/mL penicillin, 10 µg/mL gentamycin, and 2.5 µg/mL amphotericin.

2. Tyrphostin-1 stock solution (100X): 50 m*M* Tyrphostin-1 dissolved in dimethyl sulfoxide (DMSO). Store at −20°C.

3. Tyrphostin-1 media solution (1X): For 1 mL of Tyrphostin-1 media solution (1X), sequentially add 10 µL Tyrphostin-1 stock solution (100X), 50 µL ethanol, and 940 µL LHC-8 culture medium to a sterile tube. Prepare immediately before use.

4. Phosphate-buffered saline (PBS): 137 m*M* NaCl, 2.7 m*M* KCl, 4.3 m*M* Na_2HPO_4, 1.4 m*M* KH_2PO_4 at a final pH of 7.3. PBS-Mg also contains 5 m*M* $MgCl_2$.

5. Reporter lysis 5X buffer: 125 m*M* Tris-phosphate, pH 7.8, 5% Triton® X-100. Store at −20°C away from direct sunlight. Dilute with four volumes of water prior to use.

6. Luciferase assay reagent: A mixture prepared by reconstituting Luciferase Assay Substrate with 10 mL of luciferase assay buffer. Aliquot and store at −70°C.
7. Fixing solution: 2% formaldehyde, 0.2% glutaraldehyde in PBS.
8. β-Gal staining solution: 5 mM potassium ferrocyanide, 5 mM potassium ferricyanide, 1 mM MgCl$_2$, and 1 mg/mL X-Gal in dimethyl sulfoxide
9. Virus preservation medium: 10 mM Tris-HCl, pH 8.0, 100 mM NaCl, 1 mg/mL BSA, 50% glycerol.

3. Methods

3.1. Recombinant Adeno-Associated Virus (rAAV) Transduction

1. Plate cells (e.g., 293, IB-3 or Hela cells) in 24-well dishes at approx 4 × 10^4 cells per well using the appropriate medium. Incubate the cultures overnight in a standard humidified, eukaryotic tissue-culture incubator.
2. Twenty-four hours after inoculation, remove the culture medium and add serial dilutions of the packaged rAAV vector (*see* **Note 1**) in PBS-Mg or culture medium in a volume of 0.1 ml. If required, Ad infection may be performed simultaneously at this point (*see* **Subheading 3.2.**, and **Note 2**). Add Ad at a concentration of five plaque forming units (PFU)/cell.
3. Incubate plates at 37°C for at least 1 h in a humidified incubator (*see* **Note 3**).
4. Remove the vector and replace with cell-culture medium.
5. Incubate for 24–48 h in a tissue-culture incubator.
6. Examine cells transduced with vectors expressing *gfp* by fluorescent microscopy at any time after vector transduction.
7. To assay cells transduced with X-gal- or luciferase-expressing vectors, remove medium from wells and rinse wells with PBS-Mg. Fix cells for staining with X-gal (*see* **Subheading 3.6.**) or harvest cells for luciferase assays (*see* **Subheading 3.4.**).
8. For vector titration, each green fluorescent- or blue-stained cell is defined as 1 IU.

3.2. Adenovirus Infection

1. Plate and incubate cells as described in **Subheading 3.1.** Remove culture medium from the wells and replace with an appropriate amount of serum-free medium. Prepare serial dilutions of Ad5 and add to the cells at a multiplicity of infection (MOI) of 1–100. Resuspend Ad5 in PBS, or serum-free medium.
2. Incubate plates at 37°C for at least 1 h in a humidified incubator (*see* **Notes 2** and **3**).
3. Remove the inoculum and replace with complete cell-culture medium.

4. Incubate for 24–48 h in a tissue-culture incubator (*see* **Note 4**).
5. Assay cells as described in **Subheading 3.1.**

3.3. Tyrphostin-1 Treatment of IB-3 Cells

1. Inoculate a 48-well plastic tissue-culture plate with 5×10^4 IB-3 cells per well in LHC-8 culture medium containing 10% FBS and grow to confluence (2 d) (*see* **Note 5**).
2. Aspirate the culture medium and carefully add 250 µL of serum-free LHC-8 culture medium containing 1–10 I.U. per cell of vAV*luc* to the wells.
3. Incubate for 1 h at 37°C.
4. Prepare the Tyrphostin-1 media solution (1X) (*see* **Note 6**).
5. Aspirate the culture medium and replace with the Tryphostin-1 media solution (1X).
6. Incubate for 1 h at 37°C.
7. Aspirate the culture medium and add LHC-8 culture medium to the wells. Incubate the cells for 24–48 h at 37°C in a humidified tissue-culture incubator.

3.4. Luciferase Assays

1. Determine the linear range of light detection on the luminometer prior to first use as described by the manufacturer (*see* **Note 7**).
2. Aspirate the culture medium from the cells and rinse with PBS-Mg.
3. Remove the PBS-Mg and add enough 1X reporter lysis buffer to cover the monolayer (250 µL per well of a 48-well plate). Incubate 5 min and freeze the entire plate at −70°C for complete lysis of the cells and storage of the samples.
4. Thaw the cell lysates on ice.
5. Dispense 20 µL of the lysates into separate luminometer tubes. Dispense diluted recombinant luciferase into another luminometer tube as a positive control for the assay.
6. Program the luminometer to perform a 2-s delay in measurement followed by a 10-s measurement for enzyme activity.
7. Prime the injector of your luminometer at least three times with the luciferase assay reagent.
8. Place the luminometer tube in the luminometer and begin the reading by injecting 100 µL of luciferase assay reagent into the tube. Record the reading from each tube. Standardize luciferase activity (relative light units [RLU]) in the sample to the total amount of protein in the sample (RLU/µg protein).

3.5. DC Protein Assay

1. Add 20 µL of Reagent S to each mL of Reagent A (both from the Bio-Rad DC Protein Assay kit) that will be needed for the assay (Working Reagent A′). Assay each sample and standard in triplicate, and use 25 µL of Working Reagent A′ for each measurement (*see* **Note 8**).
2. Reconstitute the BSA protein standard in water and store at 4°C. Generate a standard curve each time the assay is performed by preparing at least five dilutions of the BSA containing between 0.2 mg/mL and 5.0 mg/mL.
3. Dilute samples 10-fold with 1X reporter lysis buffer, and pipet 5 µL of standards and samples into wells of a 96-well microtiter plate.
4. Add 25 µL of Working Reagent A′ to each well.
5. Add 200 µL of Reagent B to each well and gently agitate the plate to mix the reagents. Be careful to avoid bubbles, and remove them if they form.
6. Incubate for 15 min at room temperature, and read the absorbance of the samples and standards at 750 nm.

3.6. β-Gal Staining of rAAVβGal Transduced Cells

1. Rinse the transduced cell monolayer with PBS-Mg.
2. Add 0.5 mL of fixing solution and incubate for 5 min at room temperature to fix the cells.
3. Remove the fixing solution and overlay the cells with 0.5 mL of β-gal staining solution.
4. Wrap the multiwell plate in aluminum foil and incubate at 37°C in an incubator for 12–48 h.
5. Observe the cells under a microscope. Each blue cell corresponds to a transduced cell. This enables a determination of the number of infectious units (IU) of that rAAV preparation.

4. Notes

1. In our laboratory, we prepare most of our recombinant vectors using a two-plasmid transfection procedure *(21)*. Briefly, we co-transfect the vector plasmid containing the gene to be packaged (e.g., pAV*gal* or pAV*luc*) with a packaging plasmid, pSH3 or pSH5, onto 293 cells. Two to three d later, the cells are harvested and the virus is purified away from cell debris by successive CsCl gradients or by heparin affinity chromatography *(23)*. The vector produced by any of these different methods is analyzed to determine the infectious titer, which is often reported as IU. For titration of vectors that express visually detectable reporter genes (*lacZ*, GFP, *gfp*), determination of IU and experimental transductions are performed as described in Methods. For titration of vectors expressing other genes (e.g., *luc*) the infectious center assay is used *(23)*.

2. Virus infections of tissue-culture cells are frequently initiated by incubating the virus with cells in serum-free medium. After allowing virus adsorption for an appropriate length of time, the inoculum is removed and replaced with normal medium. One of the reasons for using serum-free medium is that components in the serum (complement, immunoglobulins, etc.) may interfere with virus adsorption. To determine if serum affected AAV vector transduction, Hela cells were transduced with 100 MOI of vAV*luc* vector in the presence of medium containing 10% normal or heat-inactivated FBS, or no FBS at all. A similar set of transductions was performed on Ad-infected Hela cells. One h after adding the virus, the inoculum was removed and replaced with normal medium. Forty-eight h later, the cultures were harvested and assayed for luciferase activity. There was no significant difference in the level of enzyme activity between the three FBS conditions (**Fig. 1**). The presence of Ad in the transduction resulted in more than a 100-fold increase in luciferase activity. However there was no significant difference between the FBS conditions in the presence of Ad (**Fig. 1**). These results indicate that AAV2 vectors are able to adsorb to, and transduce, Hela cells in the presence or absence of FBS with comparable efficiency.

3. For vector titration, triplicate wells are used for each dilution. The culture plates are returned to the incubator for at least 1 h. An appropriate volume of fluid should be used to allow for complete coverage of the cell monolayer. Placing the plates on a rocking platform improves vector and cell contact. Alternatively, gentle, manual rocking of the plates every 10–15 min is sufficient. The vector can be left on the culture until the end of the experiment or the inoculum can be removed after 1 h.

4. Ad infection dramatically increases vector transduction (**Fig. 1**) *(13,16)*. Although AAV vectors are not toxic to cells in culture, concurrent Ad infection results in cytopathic effects between 24–48 h after infection. The cultures should be observed daily for evidence of toxicity such as an increase in cellular birefringence, rounding, and detaching from the dish.

5. The use of Tyrphostin-1 to enhance AAV transgene expression has been pioneered by others *(18)*. We have found that Tyrphostin-1 stimulates transduction in all cells that contain the EGFR-1 gene. We use IB-3 cells here because they are highly responsive to the stimulatory effects of Tyrphostin-1. An example of the dramatic increase in transduction is shown in **Fig. 2** where 500 μ*M* Tyrphostin-1 induces transduction nearly 800-fold in IB-3 cells.

6. Tyrphostin-1 and its vehicle (1% DMSO, 5% ethanol) are toxic to some cell types (e.g., 293 and Hela cells). Some cell lines may require lower concentrations of Tyrphostin-1 and/or its vehicle. The suggested 500 μ*M* concentration of Tyrphostin-1 and its vehicle can be proportionally reduced in

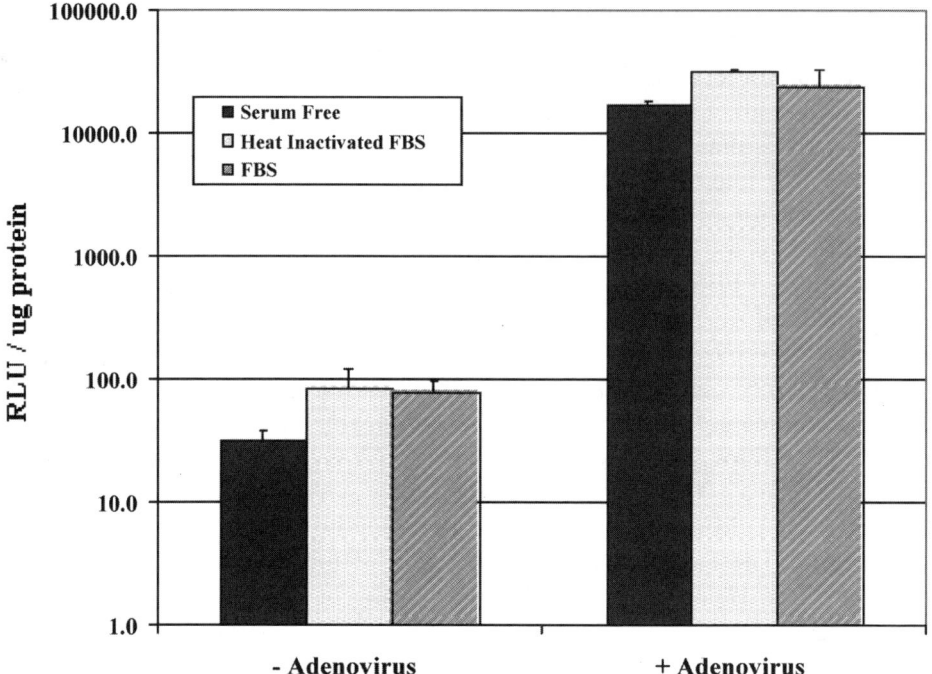

Fig. 1. Analysis of recombinant AAV2 vector transduction under different FBS conditions. Hela cells were incubated with 100 MOI of vAV*luc* for 1 h in the presence of 10% FBS, heat-inactivated FBS or no serum. One h after adsorption, the inoculum was replaced with normal medium containing 10% FBS. The cultures were incubated for 48 h, cell lysates were prepared and luciferase activity assessed. Alternatively, Ad2 was added to the inoculum at a MOI of 5. The transductions were performed in triplicate and the results reported as relative light units (RLU) per microgram of total protein in the lysates. Protein assays were performed using the DC protein assay.

these cases to avoid excessive toxicity in sensitive cell lines while maintaining solubility of Tyrphostin-1. Keep in mind that the enhancement in rAAV transgene expression by Tyrphostin-1 is proportional to the concentration of Tyrphostin-1.

7. The luciferase assay system may be used with manual luminometers or luminometers with injectors as per the manufacturer's protocols. Determining the linear range of light detection for the luminometer is essential because luminometers can be saturated at high light intensities. Make serial dilutions of luciferase (either purified recombinant luciferase or cell-culture lysate) in 1X reporter lysis buffer supplemented with 1 mg/mL BSA to avoid adsorption of the enzyme. A positive control for the luciferase assay

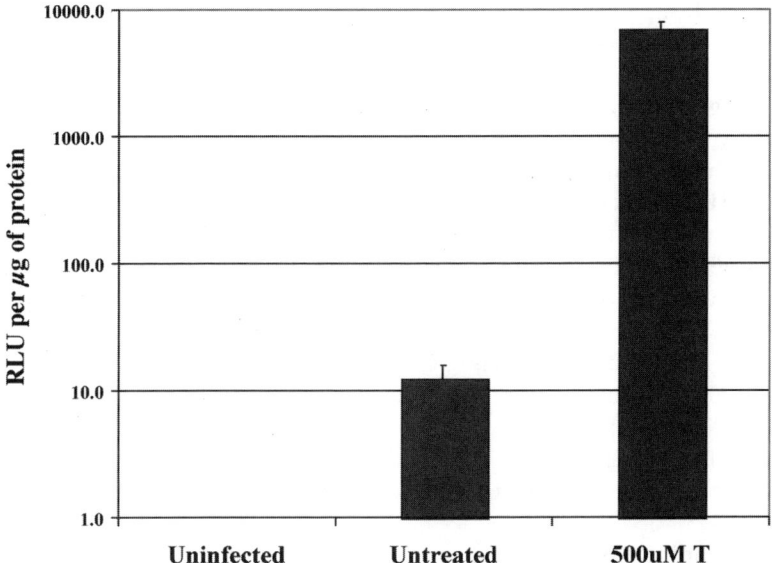

Fig. 2. Analysis of recombinant AAV2 vector transduction after Tyrphostin-1 treatment. IB-3 cells were incubated with 100 MOI of vAV*luc* for 1 h. The medium was replaced and one set of cultures was incubated for 2 h in the presence of 500 μ*M* Tyrphostin-1. The Untreated sample was incubated in vehicle alone. The medium was then replaced with normal medium and the cultures were incubated for 48 h. Cell lysates were prepared and luciferase activity assessed. The transductions were performed in triplicate and the results reported as relative light units (RLU) per microgram of total protein in the lysates. Protein assays were performed using the DC protein assay

can be obtained by transfecting cells with a luciferase reporter plasmid and assaying the lysate from the transfected cells.
8. We have found that the detergent-compatible protein assay from Bio-Rad is far more reliable when analyzing cellular extracts prepared by detergent lysis. Experimental variation between identical samples is negligible compared with results obtained from the Bradford assay. This protein assay system is strongly recommended for standardizing protein content in the performance of luciferase assays that commonly use Triton X-100 for cell lysis.

References

1. Chiorini, J. A., Yang, L., Liu, Y., Safer, B., and Kotin, R. M. (1997) Cloning of adeno-associated virus type 4 (AAV4) and generation of recombinant AAV4 particles. *J. Virol.* **71**, 6823–6833.

2. Chiorini, J. A., Kim, F., Yang, L., and Kotin, R. M. (1999) Cloning and characterization of adeno-associated virus type 5. *J. Virol.* **73,** 1309–1319.

3. Duan, D., Yan, Z., Yue, Y., Ding, W., and Engelhardt, J. F. (2001) Enhancement of muscle gene delivery with pseudotyped adeno-associated virus type 5 correlates with myoblast differentiation. *J. Virol.* **75,** 7662–7671.

4. Hildinger, M., Auricchio, A., Gao, G., Wang, L., Chirmule, N., and Wilson, J. M. (2001) Hybrid vectors based on adeno-associated virus serotypes 2 and 5 for muscle-directed gene transfer. *J. Virol.* **75,** 6199–6203.

5. Rutledge, E. A., Halbert, C. L., and Russell, D. W. (1998) Infectious clones and vectors derived from adeno-associated virus (AAV) serotypes other than AAV type 2. *J. Virol.* **72,** 309–319.

6. Xiao, W., Chirmule, N., Berta, S. C., McCullough, B., Gao, G., and Wilson, J. M. (1999) Gene therapy vectors based on adeno-associated virus type 1. *J. Virol.* **73,** 3994–4003.

7. Summerford, C. and Samulski, R. J. (1998) Membrane-associated heparin sulfate proteoglycan is a receptor for adeno-associated virus type 2 virions. *J. Virol.* **72,** 1438–1445.

8. Summerford, C., Bartlett, J. S., and Samulski, R. J. (1999) AlphaV Beta5 integrin: a co-receptor for adeno-associated virus type 2 infection. *Nat. Med.* **5,** 78–82.

9. Qing, K., Mah, C., Hansen, J., Zhou, S., Dwarki, V., and Srivastava, A. (1999) Human fibroblast growth factor receptor 1 is a co-receptor for infection by adeno-associated virus 2. *Nat. Med.* **5,** 71–77.

10. Ponnazhagan, S., Mukherjee, P., Wang, X. S., Qing, K., Kube, D. M., Mah, C., et al. (1997) Adeno-associated virus type 2-mediated transduction in primary human bone marrow-derived CD34+ hematopoietic progenitor cells: donor variation and correlation of transgene expression with cellular differentiation. *J. Virol.* **71,** 8262–8267.

11. Duan, D., Yue, Y., Yan, Z., McCray, P. B., Jr., and Engelhardt, J. F. (1998) Polarity influences the efficiency of recombinant adenoassociated virus infection in differentiated airway epithelia. *Hum. Gene Ther.* **9,** 2761–2776.

12. Zabner, J., Seiler, M., Walters, R., Kotin, R. M., Fulgeras, W., Davidson, B. L., and Chiorini, J. A. (2000) Adeno-associated virus type 5 (AAV5) but not AAV2 binds to the apical surfaces of airway epithelia and facilitates gene transfer. *J. Virol.* **74,** 3852–3858.

13. Fisher, K. J., Gao, G.-P., Weitzman, M. D., DeMatteo, R., Burda, J. F., and Wilson, J. M. (1996) Transduction with recombinant adeno-associated virus for gene therapy is limited by leading -strand synthesis. *J. Virol.* **70,** 520–532.

14. Alexander, I. E., Russell, D. W., and Miller, A. D. (1994) DNA-damaging agents greatly increase the transduction of nondividing cells by adeno-associated virus vectors. *J. Virol.* **68,** 8282–8287.

15. Russell, D. W., Alexander, I. E., and Miller, A. D. (1995) DNA synthesis and topoisomerase inhibitors increase transduction by adeno-associated virus vectors. *Proc. Natl. Acad. Sci. USA* **92,** 5719–5723.

16. Ferrari, F. K., Samulski, T., Shenk, T., and Samulski, R. J. (1996) Second-strand

synthesis is a rate-limiting step for efficient transduction by recombinant adeno-associated virus vectors. *J. Virol.* **70,** 3227–3234.

17. Qing, K. Y., Wang, X.-S., Kube, D. M., Ponnazhagan, S., Bajpai, A., and Srivastava, A. (1997) Role of tyrosine phosphorylation of a cellular protein in adeno-associated virus 2-mediated transgene expression. *Proc. Natl. Acad. Sci. USA* **94,** 10879–10884.

18. Mah, C., Qing, K., Khuntirat, B., Ponnazhagan, P., Wang, X.-S., Kube, D. M., et al. (1998) Adeno-associated virus type 2-mediated gene transfer: Role of epidermal growth factor receptor protein tyrosine kinase in transgene expression. *J. Virol.* **72,** 9835–9843.

19. Douar, A.-M., Poulard, K., Stockholm, D., and Danos, O. (2001) Intracellular trafficking of adeno-assoicated virus vectors: routing to the late endosomal compartment and proteosome degradation. *J. Virol.* **75,** 1824–1833.

20. Duan, D., Yue, Y., Yan, Z., and Engelhardt, J. F. (2000) Endosomal processing limits gene transfer to polarized airway epithelia by adeno-associated virus. *J. Clin. Invest.* **105,** 1573–1587.

21. Collaco, R. F., Cao, X., and Trempe, J. P. (1999) A helper virus-free packaging system for recombinant adeno-associated virus vectors. *Gene* **238,** 397–405.

22. Zeitlin, P. L., Lu, L., Rhim, J., Cutting, G., Stetten, G., Keiffer, K. A., et al. (1991) A cystic fibrosis bronchial epithelial cell line: immortalization by adeno-12-SV40 infection. *Am. J. Resp. Cell Mol. Biol.* **4,** 313–319.

23. Zolotukhin, S., Byrne, B. J., Mason, E., Zolotukhin, I., Potter, M., Chesnut, K., et al. (1999) Recombinant adeno-associated virus purification using novel methods improves infectious titer and yield. *Gene Ther.* **6,** 973–985.

12

AAV-Mediated Gene Transfer to Skeletal Muscle

Roland W. Herzog

1. Introduction

Adeno-associated viral (AAV) vectors are derived from a nonpathogenic, replication-deficient virus with a small (~4.7-kb) single-stranded DNA genome. AAV vectors are devoid of viral-coding sequences and may efficiently transfer genes to nondividing cells such as muscle fibers or hepatocytes following in vivo transduction (1–7). Recombinant AAV can be administered to skeletal muscle of experimental animals and, as recently documented in a Phase I clinical trial, to humans at high vector doses without local or systemic toxicity (8,9). The potential of the vector to activate cytotoxic T lymphocytes is greatly reduced compared with some other viral vectors, thereby reducing the risk of inflammation at the site of gene transfer (7,10,11). Sustained expression of therapeutic transgenes such as coagulation factor IX (F.IX), erythropoietin, leptin, insulin-like growth factor (IGF), sarcoglycans, mini-dystrophin genes, α_1-antitrypsin, and others have been demonstrated (2,12–18). Efficient gene transfer to myofibers by intramuscular (im) injection has been shown in several species including mice, hamsters, dogs, and nonhuman primates (6–8,13,19). These studies resulted in various levels of correction of the disease phenoypes in small and large animal models of hemophilia B (F.IX deficiency), muscular dystrophy, obesity, age-related atrophy, and β-thalassemia (8,12,13,15,17,18,20–25).

AAV vectors can be produced in HEK-293 cells in the absence of a helper virus by a triple transfection procedure that relies on the use of two helper plasmids (26,27). The vector plasmid is constructed by inserting an expression cas-

From: *Methods in Molecular Biology, vol. 246:*
Gene Delivery to Mammalian Cells: Vol. 2: Viral Gene Transfer Techniques
Edited by: W. C. Heiser © Humana Press Inc., Totowa, NJ

sette for the therapeutic gene between the two 145-bp AAV inverted terminal repeats (ITRs). The expression cassette typically contains an enhancer/promoter combination, a short intron (e.g., globin, cytomegalovirus [CMV], or F.IX splice donor/acceptor sequences), a cDNA encoding the therapeutic gene, and a polyadenylation signal. The total size of the vector (including ITRs) should be 5 kb or less in order not to exceed the packaging limit of AAV. The most extensively studied AAV vectors to date are based on serotype-2 (there are eight published AAV serotypes). Whereas AAV-2 vectors have been successfully used in animal studies to transfer therapeutic genes, further increase of transduction efficiency in murine skeletal muscle by 1–2 logs has been reported for vectors based on AAV-1 or AAV-5 serotypes *(28–30)*.

The main interest of our laboratory is to develop gene-therapy strategies for treatment of the X-linked disorder hemophilia B. Although F.IX is normally made in the liver, muscle fibers are capable of producing biologically active F.IX after gene transfer *(31)*. Intramuscular administration is a technically simple and noninvasive procedure and thus attractive for in vivo gene transfer. Transduced muscle fibers secrete the F.IX transgene product, resulting in an increase of systemic F.IX levels that can be monitored and quantitated using plasma samples drawn from the animal. Therefore, this system is also useful to address basic questions about efficiency and stability of gene transfer and transgene expression, which can be followed over time without having to sacrifice the animal for quantitation of transgene expression.

2. Materials

2.1. Vector Production

1. Three plasmids are required for production of recombinant AAV containing the therapeutic gene of interest (*see* **Note 1**):
   a. pAAV-F.IX encoding the F.IX expression cassette flanked by AAV-2 ITRs.
   b. pAd-Help including adenoviral genes E2A, E4, and VA genes (*see* **Note 2**).
   c. pAAV-Help encoding wild-type AAV Rep and Cap genes (*see* **Note 3**).
2. Transfection reagents:
   a. 2X HEPES buffer: weigh 4.77 g of HEPES (free acid) and 8.77 g of NaCl; dissolve in 400 mL of dd H_2O, adjust pH 7.05 using NaOH, then adjust the final volume to 500 mL. Autoclave and store at room temperature.
   b. 70 mM Na_2HPO_4. Autoclave and store at room temperature.
   c. 1.0 M $CaCl_2$. Filter-sterilize and store at $-20°C$.

3. Human embryonic kidney 293 cells free of wild-type AAV (HEK-293, ATCC, Manassas, VA).
4. Dulbecco's modified Eagle's medium (DMEM), fetal bovine serum (FBS), penicillin, and streptomycin are available from GIBCO/Invitrogen (Carlsbad, CA).
5. Branson sonifier or dry ice/ethanol bath for cell lysis.
6. 10 mM Tris-HCl, pH 8.0, and HEPES-buffered saline (HBS), pH 7.8. Filter-sterilize.
7. RNase A and DNase from Roche Biochemicals (Indianapolis, IN) or Benzonase®; 10% (w/v) sodium deoxycholate (filter-sterilized).
8. CsCl.
9. Ultracentrifuge with SW28 and Ty70.1 (Beckman Coulter, Fullerton, CA) or equivalent rotors for CsCl gradient centrifugation. Ultraclear centrifuge tubes (Beckman Coulter). Beckman fraction collector with piercing needle (clean and autoclave needle prior to use).
10. Milton Roy Abbe-3L refractometer for measurement of refractive index of collected fractions.
11. Slide-A-Lyzer dialysis cassettes (Pierce, Rockford, IL).
12. Apparatus and reagents for slot-blot hybridization.

2.2. Animal Experiments

1. Mouse strains: Rag-1 mice deficient in functional B and T cells and CD4-deficient mice are available from Jackson Laboratory (our experience with both strains is with animals on a C57BL/6 genetic background). Hemophilia B mice (F.IX knockout mice) with a large gene deletion including the promoter and the first three exons of the F.IX gene may be obtained from Dr. Darryl Stafford, UNC-Chapel Hill *(32)*.
2. Materials for im injections: Hamilton syringe and needle (both autoclaved), sterile instruments (scissors, forceps), ketamine/xylazine and/or metofane anesthesia, 4-0 vicryl suture, 70% alcohol, betadine.

2.3. Analysis of Gene Expression

1. Heparinized microcapillary tubes for retro-orbital bleeds. For alternative bleeding protocols, *see* Note 4.
2. Enzyme-linked immunosorbent assay (ELISA) plate reader.
3. Antibodies: monoclonal mouse anti-human F.IX (clone HIX-1, Sigma, St. Louis, MO), goat anti-human F.IX (affinity purified with or without horseradish peroxidase conjugate, Affinity Biologicals, Hamilton, Ontario, Canada), rabbit anti-goat IgG-FITC (Dako Corp., Carpinteria, CA).
4. ELISA buffers:

a. Coating buffer, pH 9.2.
b. Wash buffer (WB): 1X PBS, pH 7.4, with 0.05% Tween-20.
c. Blocking and sample buffer: dissolve nonfat dry milk in WB to a final concentration of 5% (w/v). Prepare fresh.
d. Buffer for detecting antibody: dissolve bovine serum albumin (BSA, Sigma) in WB to a final concentration of 5% (w/v) and store at $-20°C$.

5. Substrate solution: 0.01 M sodium citrate, pH 4.5, store at 4°C. Just prior to use, dissolve o-phenylene diamine substrate (Sigma) at a final concentration of 1 mg/mL, add 2.5 μL H_2O_2 (30% stock) per 12 mL of solution.
6. F.IX standard: Pooled normal human plasma (Verify 1, Organon Teknika, Durham, NC).
7. Preparation of muscle necropsy: liquid nitrogen, methyl-butane, Tissue Tek OCT compound (Sakura Finetek, Torrance, CA).
8. Immunohistochemistry of muscle sections: staining trays, freshly prepared 3% (w/v) paraformaldehyde in 1X phosphate-buffered saline (PBS), pH 7.4, methanol, 1X PBS, pH7.4; 1% and 3% BSA in 1X PBS, pH 7.4, fluoromount G (Fisher Scientific, Pittsburgh, PA) or similar mounting media.

3. Methods

3.1. Vector Production

1. Inoculate HEK-293 cells into fifty 150-mm tissue-culture plates. Incubate at 37°C in a humidified 5%-CO_2 incubator until the cells are 80–90% confluent. Change media; add 20 mL of fresh media to each plate.
2. Prepare calcium phosphate precipitate of plasmid DNA. Allow all reagents to warm to room temperature. Use the following amounts of reagents for transfection of fifty 150-mm plates. (Additional plates can be transfected by proportionally increasing the volumes of this transfection cocktail.) Add 1.25 mL of 70 mM Na_2HPO_4 to 62.5 mL of 2X HEPES buffer. Transfer 62.5 mL of this mix to a 200-mL conical tube. In a separate (100-mL) tube, add 2.5 mg of total plasmid DNA (including all three plasmids at an equimolar ratio; if the three plasmids are of similar size, simply use 833 μg of each plasmid). Add water to bring the volume to 46.9 mL; add 15.6 mL of 1 M Ca_2Cl. Draw the entire DNA/Ca_2Cl solution into a 50-mL pipet and add dropwise to the HEPES solution while vortexing the HEPES solution continuously. Once mixed, incubate at room temperature for 25 min to allow DNA precipitation to occur.
3. Briefly vortex the DNA precipitate and add dropwise onto conditioned media (2.5 mL/plate) using a 10-mL pipet.
4. Change media 8–12 h after transfection.

5. Harvest cells on d 4 posttransfection and spin them down at 1000*g* for 20–30 min at 4°C. Resuspend the cells in 10 m*M* Tris-HCl, pH 8.0, and subsequently store in a sterile conical tube at −80°C (use 5–10 mL per 20 plates).

6. Thaw cell suspension, lyse cells by repeated (three times) freeze-thaw cycles using an ethanol/dry ice bath and a 37°C bath, or by three rounds of ultra-sonication using a Branson sonifier *(33)*. Keep the cell suspension on ice and sonicate for 1.5 min with a 1/4-in probe (at an output of 2–2.5).

7. Subsequently, add 5 mg of RNase and 5 mg of DNase. Incubate at 37°C for 30 min. Add sodium deoxycholate to the lysate at a final concentration of 0.5% (w/v), and incubate at 37°C for 10 min before returning the lysate to the ice bath for an additional 10 min. Add and dissolve CsCl (0.454 g/mL of lysate). *See* **Note 5** for alternative purification protocols.

8. Depending on the volume of cell lysate, prepare one or two ultraclear tubes for a Beckman SW28 ultracentrifugation rotor as follows (one tube can hold up to 20 mL of cell lysate): add 9 mL of 1.45 g/mL CsCl, underlay with 9 mL of 1.6 g/mL CsCl, then carefully load the cell lysate onto the top of the gradient. Centrifuge in a pre-cooled ultracentrifuge using pre-cooled rotor buckets for 24 h spin at 131,000*g* (27,000 rpm in a Beckman SW28 rotor) at 4°C.

9. Collect 1-mL fractions from the gradient. Pool those fractions potentially containing AAV vector (i.e., fractions with a density of 1.37–1.42 g/mL as determined with a refractometer).

10. Depending on the volume of the lysate, prepare one or two 5/8-in × 3-in ultracentrifuge tubes as follows (one tube can hold approx 10 mL of vector): add 1–2 mL of 1.6 g/mL CsCl solution, then fill up the rest of the tube with the pooled fractions from the previous centrifugation. If there is additional room in the tube, fill with 1.41 g/mL CsCl. Centrifuge in a pre-cooled ultracentrifuge for 12–16 h at 331,000*g* (60,000 rpm in a Beckman 70.1Ti rotor) at 4°C.

11. Collect 0.5-mL fractions from the gradient; pool those fractions with a density of 1.37–1.42 g/mL; repeat the centrifugation as in **step 10**.

12. Following the third centrifugation, combine those fractions with a density of 1.37–1.42 g/mL; transfer into a Slide-a-lyzer cassette for dialysis. Place the cassette in 1.5–2 l of cold HBS, pH 7.8, followed by three rounds of dialysis at 4°C (2 h each).

13. Remove the vector solution from the cassette, add glycerol or sorbitol to a final concentration of 5% to osmotically stabilize it, filter-sterilize using a 0.2-μm syringe filter, aliquot, and store at −80°C in sterile cryovials.

14. Determine the vector yield as vector genomes ([vg]/mL) by quantitative dot-blot hybridization on plasmid standards and DNA extracted from vec-

tor by proteinase K digest *(2)* (*see* **Note 6**). Alternatively, quantitate the yield of vector by TaqMan PCR or infectious vector particles can be determined by infectious center assay *(34,35)*.

3.2. IM Injections

The method described here for im injection of mice can also be carried out in hemophilia B mice. These animals generally tolerate this procedure well and do not show bleeding that would otherwise require plasma infusion in order to temporarily restore hemostasis. For a large animal, such as hemophilia B dogs, im injections are usually done percutaneously, and thus do not require a skin incision (*see* **Note 7**). A different method of delivering an AAV vector to skeletal muscle (intravascular instead of im) is summarized in **Note 8**.

1. Thaw the AAV vector at room temperature and dilute in sterile HBS or PBS to 5×10^{10}–5×10^{12} vg/mL (150 µL/mouse) as required and keep on ice.
2. Anesthetize mice by intra-peritoneal injection of ketamine/xylazine (80/16 mg/kg, respectively) or by inhalation of metofane gas. If using metofane, keep mice under anesthesia by inhalation of metofane gas (soak cloth with metofane solution and place in a 50-mL conical tube, allow mouse to breathe anesthesia by placing head at the mouth of the open tube).
3. Shave both hind limbs and rub with 70% alcohol. Place the mouse on its back on sterile drape.
4. Make an approx 1-cm incision in the skin of one hind limb (just above the knee) *(7,14)*. The lower part of the quadriceps muscle (above the knee) and the upper part of the tibialis anterior muscle (below the knee) should be visible.
5. Draw vector into a Hamilton syringe and inject into the quadriceps muscle (50 µL) and the tibialis anterior muscle (25 µL). For these injections, insert the needle parallel to the muscle, pull back to about the middle of the muscle, and inject vector slowly. Injection of the tibialis anterior muscle is obvious from the slow "kicking" motion resulting in extension of the leg during injection. For a larger muscle such as the quadriceps, the needle may be pulled back slowly while injecting to distribute the vector over a wider area.
6. Following the injection, close the skin by suture. Repeat the procedure in the second hind limb.
7. Apply Betadine at the sites closed with suture. Allow injected mice to wake up and return them to the cage.

3.3. Analysis of Systemic F.IX Expression

1. Bleed mice biweekly from the retro-orbital plexus (~200 µL/sample) using heparinized microcapillary tubes. Collect blood from the capillaries in microfuge tubes, place on ice, and spin in a microcentrifuge at 7700*g* (9000

rpm) at 4°C for 10 min to remove blood cells. Transfer plasma to fresh tubes and store at -20°C or -80°C.

2. Determine systemic levels of human F.IX by ELISA specific for the human F.IX antigen. Coat microtiter plates overnight with monoclonal anti-human-F.IX (1:850 dilution, 50 µL/well).

3. The following day, wash the plate with wash buffer, block with blocking buffer (200 µL/well) for several hrs at 4°C, and wash again. Dilute plasma samples and F.IX standards) in blocking buffer. Use dilute normal human plasma for generation of a standard curve. The normal F.IX level in human plasma is about 5000 ng/mL. Prepare a standard curve comprised of two-fold serial dilutions from 200 ng/mL to 3.125 ng/mL. Load standards and samples onto the plate (in duplicate, 50 µL/well) and incubate overnight at 4°C.

4. Wash the plate. Detect human F.IX antigen using a horseradish peroxidase-conjugated goat anti-F.IX (1:2,300 dilution in WB with 5% BSA buffer, 100 µL/well). Incubate 2 h at 37°C, wash the plate, and apply substrate solution (100 µL/well). After 15–20 min, measure F.IX concentration at OD_{450}.

5. Use plate reader software to calculate F.IX plasma levels based on standard curve (F.IX concentration vs OD_{450} on a semi-log scale) and dilution factors. To study muscle-directed gene transfer of an expression cassette for human F.IX, we have been using Rag-1 or CD4-deficient mice to avoid an antibody response against the human transgene product that could block systemic expression *(14,22)*. Both strains (on a C57BL/6 background) show similar levels of F.IX expression following muscle-directed gene transfer (*see* **Note 9** for use of immunocompetent mice). F.IX levels are vector dose-dependent. After a gradual rise of expression over a 6- to 8-wk period, a stable plateau is reached (*see* **Fig. 1**). Using the CMV enhancer/promoter for expression of F.IX and an AAV-2 vector, a vector dose of approx 1×10^{11} vg/animal distributed over four sites of injection as described in **Subheading 3.2.** typically resulted in expression of approx 150 ng human F.IX/mL mouse plasma *(36,37)*.

3.4. Analysis of Local Gene Transfer and F.IX Expression

1. Sacrifice the experimental mouse (e.g., by CO_2 inhalation), open the skin of the injected limb with scissors, and excise the injected muscles using a scalpel and a pair of forceps.

2. In order to preserve muscle for subsequent cryosectioning, pour methyl butane into a beaker and cool by holding the beaker in liquid nitrogen until the methyl butane forms white crystal-like structures at the bottom of the beaker. Place muscle onto a small drop of OCT compound on a labeled piece of foil (the drop is simply to hold the tissue on the foil; excess OCT may introduce freeze artifact). Using forceps, submerge the foil containing muscle tissue into the cold methyl butane bath for 7–10 s, and then immediately transfer into liquid nitrogen (*see also* **Note 10**). The identical pro-

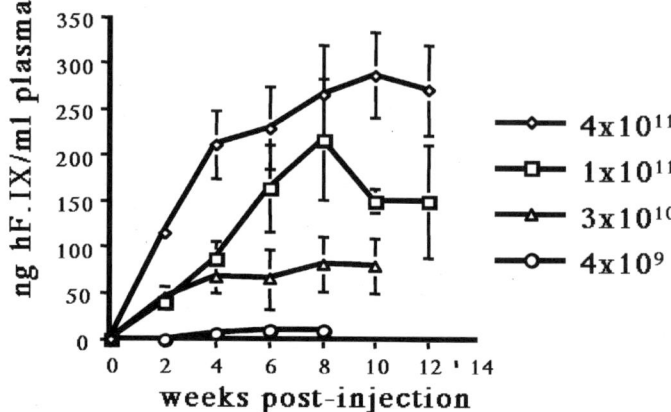

Fig. 1. Dose response of systemic human factor IX (hF.IX) expression from AAV-CMV-hF.IX-3 vector following intramuscular injection into Rag-1 mice (at four sites of hindlimbs). Expression of the hF.IX cDNA is under transcriptional control of the CMV IE enhancer/promoter. The vector also contains a 1.4-kb portion of intron I of the hF.IX gene and the SV40 polyadenylation signal. Each line represents average values from at least four animals. Vertical bars are standard deviation. Doses are vector genomes as measured by quantitative dot-blot hybridization.

cedure can be used to freeze pieces of muscle from a biopsy obtained from a larger animal (*see* **Note 11**).

3. Prepare 6-μm cryosections for immunohistochemistry or hematoxylin and eosin staining.
4. Fix sections for 15 min in 3% paraformaldehyde in 1X PBS, pH 7.4; rinse in PBS for 5 min; incubate in methanol for 10 min, and wash three times in PBS. (Perform these and all subsequent incubations at room temperature.)
5. Block with PBS/3% BSA for 1 h.
6. Incubate sections for 90 min with affinity purified goat anti-human F.IX (dilution of 1:400–1:1000 in PBS/1% BSA). After three washes (10 min each) in PBS/1% BSA, apply the secondary antibody (FITC-conjugated rabbit anti-goat IgG, diluted 1:50 in PBS/1% BSA) for 90 min.
7. After three additional washes in PBS/1% BSA, rinse the sections in distilled water, air-dry, and mount with mounting media. Typical F.IX fluorescence stains of muscle cross sections are shown in **Fig. 2A, B** (*see* also **Note 12**).

4. Notes

1. We routinely use Qiagen (Valencia, CA) Mega and Giga kits for production of endotoxin-free plasmid DNA. This is of particular importance for ex-

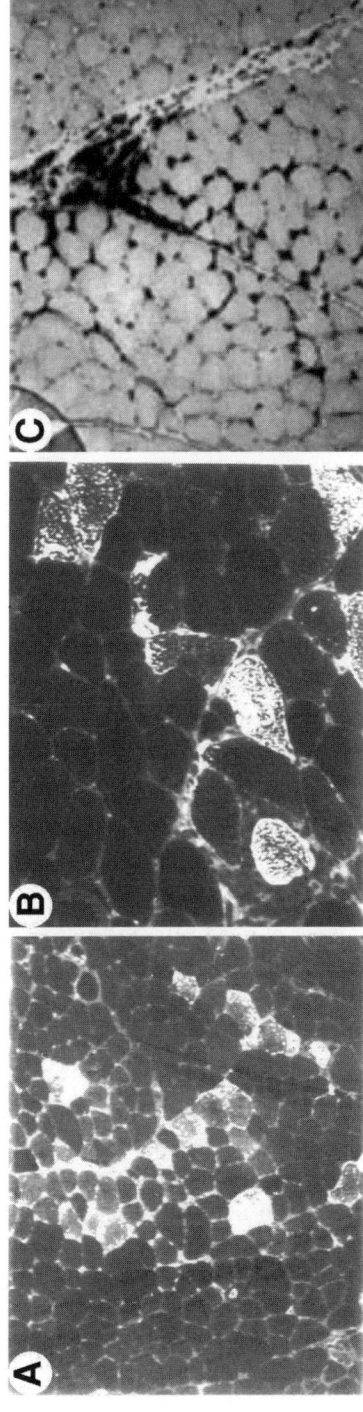

Fig. 2. Cross-sections of skeletal muscle transduced with AAV-F.IX vector. (**A,B**). Immunofluorescent antibody stain for human factor IX expressed by mouse muscle fibers 3 mo after im injection of AAV-CMV-hF.IX vector. Extracellular signal likely represents binding of F.IX to collagen IV in the extracellular matrix after secretion by transduced fibers. (**C**). Hematoxylin and eosin stain of canine muscle illustrating carbon particle marking at the site of AAV vector injection. Original magnification: (A, C) 100×, (B) 200×.

periments in dogs. Vector preparations may be tested with the Limulus amoebocyte lysate assay for contamination with endotoxin (Sigma).

2. Examples for plasmids encoding adenoviral helper functions are given by Matsuhita et al. and Xiao et al. *(26,27)*.

3. The AAV helper plasmid should preferably be a plasmid with reduced potential for generation of pseudo wild-type AAV that may occur through recombination between the rep/cap and the vector plasmid. Examples of such plasmids were constructed by Matsuhita et al. and Cao et al. *(38,39)*. The most commonly applied strategy to produce AAV vectors of different serotypes is to maintain AAV-2 ITRs in the vector plasmids and the AAV-2 rep functions in the rep/cap helper plasmid. The choice of capsid sequences in the rep/cap plasmid determines the serotype, as vectors flanked by AAV-2 ITRs can generally be packaged quite efficiently into capsids of other serotypes.

4. Hemophilia B mice are typically bled from the tail vein in order to determine the biological activity of F.IX (blood cannot be collected in heparin for this purpose). The mouse is anesthetized, and the tail is clipped approx 1.5 cm from the tip with a single clean cut using sterile scalpel. Blood is allowed to drip into an Eppendorf tube containing 20 µL of citrate buffer (3.8% [w/v] sodium citrate in H_2O) *(40)*. Once 200 µL of blood are collected, the tail is sutured and bleeding stopped by chemical cauterization using silver nitrate sticks. Blood cells are removed as described in **Subheading 3.3.**, and F.IX activity is measured by correction of F.IX deficient plasma in an activated partial thromboplastin time (aPTT) assay using a fibrometer *(22)*.

5. Alternatives for vector purification: CsCl gradients can be applied to AAV vector of any serotype; iodixanol gradient centrifugation is used by several laboratories *(41)*; AAV-2 vectors can be purified using heparin affinity or other types of column chromatography *(35,42,43)*.

6. Prepare the vector and assay by slot blot as follows. Digest the vector with proteinase K at 37°C for 1 h in a total volume of 80 µL: 2 µL of vector, 38 µL of water, and 40 µL of 2X proteinase K buffer (proteinase K at 1 mg/mL in 1 M Tris, pH 8.0, 2 mM EDTA, and 1% SDS). Following digestion, extract the DNA once with an equal volume of phenol-chloroform (1:1) and once with chloroform. Transfer the aqueous phase to a new microfuge tube. Add 20 µg of glycogen and sodium acetate to a final concentration of 0.3 M. Add 2.2 volumes of ethanol, incubate on ice for 1 h, then spin 10 min in a microfuge at full speed to pellet the DNA. Discard the supernatant and dry the pellet briefly. Resuspend the pellet in 10 µL of 10 mM Tris, pH 8.0, 1 mM EDTA. Distribute 5, 2.5, 1, and 0.5 µL into microfuge tubes containing 150 µL of spotting solution (0.4 N NaOH, 25 mM EDTA, 0.00008%

Bromothymol Blue). For standards, use the plasmid pAAV-F.IX (*see* **Subheading 2.1.1.**) that contains the entire vector genome. Prepare twofold dilutions of this plasmid from 40 ng to 1.25 ng in 5 µL of water and add 150 µL of spotting solution to each sample. Prepare blots using Biodyne B nylon membranes and a slot-blot apparatus (Bio-Rad Laboratories, Hercules, CA) following the manufacturer's instructions. Prepare the probe using 100 ng of pAAV-F.IX and 50 µCi of ^{32}P-dCTP using a Prime-it II labeling kit (Stratagene, LaJolla, CA) following the manufacturer's instructions; purify the plasmid using a MicroSpin G-50 column (Amersham-Pharmacia, Piscataway, NJ). Denature the probe immediately before use by heating for 10 min in a boiling water bath then immediately cooling it on ice. Pre-hybridize the membrane for 20 min in 100 mL of hybridization solution (5X SSPE, 0.5% w/v SDS; 20X SSPE = $3M$ NaCl, 0.2 M NaH$_2$PO$_4$, 0.02 M EDTA, pH 7.4) at 65°C in a shaking water bath. Discard the hybridization solution, then hybridize for 20 min in 30 mL of hybridization solution containing 100 ng of denatured probe at 65°C. Rinse the membrane twice in 100 mL of wash solution (1.5X SSPE, 0.5% w/v SDS). Wash the membrane for 15 min twice in 100 mL of wash solution at 65°C. Place the membrane in a seal-a-meal bag, seal and expose to XAR film for 2–16 h at −80°C. Scan the film on a densitometer to estimate the amount of vector in each of the samples. To determine the amount of vector in the sample, first use the formula 1 µg of double-stranded 5 kb DNA is equivalent to 0.3 pmol of DNA. For pAAV-F.IX, a 7 kb plasmid, 1 µg is equivalent to $(5/7) \times 0.3$ pmol = 0.21 pmol. If, for example, 1 µL of vector contained 30 ng (= 0.03 µg) of plasmid standard, this is equivalent to 0.0063 pmol (0.03×0.21 pmol). To convert to vector genomes, use Avagadro's number to convert from moles to molecules: 0.0063 pmol is equivalent to 3.8×10^9 molecules ($[0.0063 \times 10^{-12}] \times [6.02 \times 10^{23}]$) of double-stranded DNA. Because AAV is single-stranded, multiply this number by 2; the concentration therefore 7.6×10^9 molecules (vector genomes)/µL. For 1 mL of vector, the concentration is 7.6×10^{12} vg/mL.

Using a probe specific to the backbone of pAAV-F.IX, the slot-blot procedure can also be used to assess and quantitate the possibility of contamination of vector preparations with input plasmid DNA from the initial transfection step in HEK-293 cells (**Subheading 3.1.3.**), that might have escaped DNase digestion (**Subheading 3.1.7.**). An additional DNase digest prior to proteinase K treatment of the vector may avoid overestimation of viral titer by significant contamination with plasmid DNA.

7. Intramuscular injections in large animals: To date, we have performed im injections for AAV-mediated gene transfer in more than 10 Hemophilia B dogs of two different colonies. Dogs from both colonies have severe dis-

ease with no circulating F.IX antigen or activity (44,45). All animals tolerated the procedure well, and there was generally no need for plasma infusion. The veterinary staff prepared the animals by shaving and sterilizing
the sites of injection, and sedated the dogs using general anesthesia. Vector
was administered by percutaneous im injections at multiple sites using sterile 25 G needles and a volume of 250–500 µL/site. In order to avoid injection into a major blood vessel, the procedure may be carried out under ul-
trasound guidance (8).

8. Alternative method for vector delivery to skeletal muscle: for gene transfer
   in certain disease models such as muscular dystrophy, it may be necessary
   to deliver the therapeutic gene to a large number of muscle fibers. This may
   be achieved by an isolated limb-perfusion technique. This method is described in detail for a hamster model by Greelish et al. (13). In brief, the
   vasculature of one limb is isolated temporarily by applying tourniquets and
   microvasular clamps. Permeability of the vasculature is increased by infusion of papaverine in 10 mM histamine/PBS, pH 7.4, into a cannulated vessel. Vector is diluted in 10 mM histamine/PBS and infused 5 min later followed by perfusion with PBS. Vessels are electrocoagulated as the cannulae
   are withdrawn, and the incision closed with suture.

9. Formation of anti-F.IX: If immunocompetent animals are used, these may
   form antibodies against the F.IX transgene product. Methods to assay for
   such immune responses (Bethesda assay, immunocapture assay, Western
   blot) are described in detail by Fields et al. (10,22) and Herzog et al. (8,42).
   In order to avoid an antibody response against F.IX (these mice also lack tolerance to murine F.IX, likely owing to the gene deletion), one can generate
   immune deficient hemophilia B mice by crossing with Rag-1 or CD4 KO
   mice. Alternatively, biweekly intraperitoneal (ip) injections of the immune
   suppressive drug cyclophosphamide at a dose of 50 mg/kg starting from the
   day of vector administration up to wk 6 may block anti-F.IX formation (22).

10. For extraction of nucleic acids, muscle tissue is simply snap-frozen in liquid
    nitrogen. Isolation of DNA or RNA for Southern-blot analysis, PCR, or RT-
    PCR can be performed using commercial kits and following the manufac-
    turer's instructions (e.g., Pure Gene genomic DNA isolation kit from Gentra
    Sytems, Minneapolis, MN; Trizol reagent [Invitrogen] for RNA isolation).

11. Locating the site of gene transfer in muscle of a large animal: because it
    may be difficult to locate the site of gene transfer in a large muscle, we
    often include carbon particles in the vector solution for injection of large
    animals (generally only in some of the injection sites) (8). Black ink from
    a pencil or fountain pen manufacturer is spun in a centrifuge to recover the
    carbon particles. These are washed with 100% ethanol for sterilization, air-
    dried under a sterile hood, and re-suspended in sterile PBS. The carbon par-

ticle suspension (1/10 of final volume) is mixed with the vector solution just prior to injection. We did not find any inflammation at the site of carbon deposit in injected muscle tissue (**Fig. 2C**). Transgene expression and presence of carbon particles is generally co-localized. Expression is usually limited to an area of 1–1.5 cm in diameter from the center of the injection.

12. Transduction of different fiber types: Tibialis anterior and quadriceps muscles in a mouse are exclusively fast-twitch fibers. Immunohistochemistry typically reveals a "patchy" pattern of gene transfer with strongly transduced fibers next to untransduced or weakly transduced fibers following im injection of an AAV-2 vector. These data suggest a limiting step for transduction of myofibers with AAV-2. Pruchnic et al. have shown that AAV-2 preferentially transduces slow-twitch fibers and is comparatively less efficient in transduction of fast-twitch fibers *(46)*. This can be demonstrated by injection of the vector into the soleus (>90% slow-twitch fibers) and gastrocnemius (a mix of both fiber types) muscles. These fiber types can be differentiated histochemically by staining with monoclonal antibodies specific for the slow or fast isoform of myosin. Slow-twitch fibers appear to overexpress heparin sulphate proteoglycan, the primary receptor for AAV-2.

References

1. Clark, K. R., Sferra, T. J., and Johnson, P. R. (1997) Recombinant adeno-associated viral vectors mediate long-term transgene expression in muscle. *Hum. Gene Ther.* **8**, 659–669.
2. Kessler, P. D., Podsakoff, G. M., Chen, X., McQuiston, S. A., Colosi, P. C., Matelis, L. A., et al. (1996) Gene delivery to skeletal muscle results in sustained expression and systemic delivery of a therapeutic protein. *Proc. Natl. Acad. Sci. USA* **93**, 14082–14087.
3. Nakai, H., Herzog, R., Hagstrom, J. N., Kung, J., Walter, J., Tai, S., et al. (1998) AAV-mediated gene transfer of human blood coagulation factor IX into mouse liver. *Blood* **91**, 4600–4607.
4. Snyder, R. O., Miao, C. H., Patijn, G. A., Spratt, S. K., Danos, O., Nagy, D., et al. (1997) Persistent and therapeutic concentrations of human factor IX in mice after hepatic gene transfer of recombinant AAV vectors. *Nat. Genet.* **16**, 270–276.
5. Snyder, R. O., Spratt, S. K., Lagarde, C., Bohl, D., Kaspar, B., Sloan, B., et al. (1997) Efficient and stable adeno-associated virus-mediated transduction in the skeletal muscle of adult immunocompetent mice. *Hum. Gene Ther.* **8**, 1891–1900.
6. Xiao, X., Li, J., and Samulski, R. J. (1996) Efficient long-term gene transfer into muscle tissue of immunocompetent mice by adeno-associated virus vector. *J. Virol.* **70**, 8098–8108.
7. Fisher, K. J., Jooss, K., Alston, J., Yang, Y., Ehlen-Haecker, S., High, K., et al. (1997) Recombinant adeno-associated virus for muscle directed gene therapy. *Nat. Med.* **3**, 306–312.

8. Herzog, R. W., Yang, E. Y., Couto, L. B., Hagstrom, J. N., Elwell, D., Fields, P. A., et al. (1999) Long-term correction of canine hemophilia B by gene transfer of blood coagulation factor IX mediated by adeno-associated viral vector. *Nat. Med.* **5**, 56–63.
9. Kay, M. A., Manno, C. S., Ragni, M. V., Larson, P. J., Couto, L. B., McClelland, A., et al. (2000) Evidence for gene transfer and expression of factor IX in haemophilia B patients treated with an AAV vector. *Nat. Genet.* **24**, 257–261.
10. Fields, P. A., Kowalczyk, D. W., Arruda, V. R., McCleland, M. L., Hagstrom, J. N., K.J., P., Ertl, H. C. J., et al. (2000) Choice of vector determines T cell subsets involved in immune responses against the secreted transgene product factor IX. *Mol. Ther.* **1**, 225–235.
11. Jooss, K., Yang, Y., Fisher, K. J., and Wilson, J. M. (1998) Transduction of dendritic cells by DNA viral vectors directs the immune response to transgene products in muscle fibers. *J. Virol.* **72**, 4212–4223.
12. Cordier, L., Hack, A. A., Scott, M. O., Barton-Davis, E. R., Gao, G., Wilson, J. M., et al. (2000) Rescue of skeletal muscles of gamma-sarcoglycan-deficient mice with adeno-associated virus-mediated gene transfer. *Mol. Ther.* **1**, 119–129.
13. Greelish, J. P., Su, L. T., Lankford, E. B., Burkman, J. M., Chen, H., Konig, S. K., et al. (1999) Stable restoration of the sarcoglycan complex in dystrophic muscle perfused with histamine and a recombinant adeno-associated viral vector. *Nat. Med.* **5**, 439–443.
14. Herzog, R. W., Hagstrom, J. N., Kung, Z.-H., Tai, S. J., Wilson, J. M., Fisher, K. J., and High, K. A. (1997) Stable gene transfer and expression of human blood coagulation factor IX after intramuscular injection of recombinant adeno-associated virus. *Proc. Natl. Acad. Sci. USA* **94**, 5804–5809.
15. Murphy, J. E., Zhou, S., Giese, K., Williams, L. T., Escobedo, J. A., and Dwarki, V. J. (1997) Long-term correction of obesity and diabetes in genetically obese mice by a single intramuscular injection of recombinant adeno-associated virus encoding mouse leptin. *Proc. Natl. Acad. Sci. USA* **94**, 13921–13926.
16. Song, S., Morgan, M., Ellis, T., Poirier, A., Chesnut, K., Wang, J., et al. (1998) Sustained secretion of human alpha-1-antitrypsin from murine muscle transduced with adeno-associated virus vectors. *Proc. Natl. Acad. Sci. USA* **95**, 14384–14388.
17. Wang, B., Li, J., and Xiao, X. (2000) Adeno-associated virus vector carrying human minidystrophin genes effectively ameliorates muscular dystrophy in mdx mouse model. *Proc. Natl. Acad. Sci. USA* **97**, 13714–13719.
18. Barton-Davis, E. R., Shoturma, D. I., Musaro, A., Rosenthal, N., and Sweeney, H. L. (1998) Viral mediated expression of insulin-like growth factor I blocks the aging-related loss of skeletal muscle function. *Proc. Natl. Acad. Sci. USA* **95**, 15603–15607.
19. Ye, X., Rivera, V. M., Zoltick, P., Cerasoli, F., Jr., Schnell, M. A., Gao, G., et al. (1999) Regulated delivery of therapeutic proteins after in vivo somatic cell gene transfer. *Science* **283**, 88–91.
20. Bohl, D., Bosch, A., Cardona, A., Salvetti, A., and Heard, J. M. (2000) Improvement of erythropoiesis in beta-thalassemic mice by continuous erythropoietin delivery from muscle. *Blood* **95**, 2793–2798.

21. Chao, H., Samulski, R., Bellinger, D., Monahan, P., Nichols, T., and Walsh, C. (1999) Persistent expression of canine factor IX in hemophilia B canines. *Gene Ther.* **6**, 1695–1704.

22. Fields, P. A., Arruda, V. R., Armstrong, E., Chu, K., Mingozzi, F., Hagstrom, J. N., et al. (2001) Risk and prevention of anti-factor IX formation in AAV-mediated gene transfer in context of a large factor IX gene deletion. *Mol. Ther.* **4**, 201–210.

23. Herzog, R. W., Mount, J. D., Arruda, V. R., High, K. A., and Lothrop, C. D. J. (2001) Muscle-directed gene transfer and transient immune suppression result in sustained partial correction of canine hemophilia B caused by a null mutation. *Mol. Ther.* **4**, 192–200.

24. Li, J., Dressman, D., Tsao, Y. P., Sakamoto, A., Hoffman, E. P., and Xiao, X. (1999) rAAV vector-mediated sarcogylcan gene transfer in a hamster model for limb girdle muscular dystrophy. *Gene Ther.* **6**, 74–82.

25. Xiao, X., Li, J., Tsao, Y. P., Dressman, D., Hoffman, E. P., and Watchko, J. F. (2000) Full functional rescue of a complete muscle (TA) in dystrophic hamsters by adeno-associated virus vector-directed gene therapy. *J. Virol.* **74**, 1436–1442.

26. Matsushita, T., Elliger, S., Elliger, C., Podsakoff, G., Villarreal, L., Kurtzman, G. J., et al. (1998) Adeno-associated virus vectors can be efficiently produced without helper virus. *Gene Ther.* **5**, 938–945.

27. Xiao, X., Li, J., and Samulski, R. J. (1998) Production of high-titer recombinant adeno-associated virus vectors in the absence of helper adenovirus. *J. Virol.* **72**, 2224–2232.

28. Chao, H., Liu, Y., Rabinowitz, J., Li, C., Samulski, R. J., and Walsh, C. E. (2000) Several log increase in therapeutic transgene delivery by distinct adeno-associated viral serotype vectors. *Mol. Ther.* **2**, 619–623.

29. Duan, D., Yan, Z., Yue, Y., Ding, W., and Engelhardt, J. F. (2001) Enhancement of muscle gene delivery with pseudotyped adeno-associated virus type 5 correlates with myoblast differentiation. *J. Virol.* **75**, 7662–7671.

30. Xiao, W., Chirmule, N., Berta, S. C., McCullough, B., Gao, G., and Wilson, J. M. (1999) Gene therapy vectors based on adeno-associated virus type 1. *J. Virol.* **73**, 3994–4003.

31. Arruda, V. R., Hagstrom, J. N., Deitch, J., Heiman-Patterson, T., Camire, R. M., Chu, K., et al. (2001) Post-translational modifications or recombinant myotube-synthesized human factor IX. *Blood* **97**, 130–138.

32. Lin, H. F., Maeda, N., Smithies, O., Straight, D. L., and Stafford, D. W. (1997) A coagulation factor IX-deficient mouse model for human hemophilia B. *Blood* **90**, 3962–3966.

33. Fisher, K. J., Gao, G. P., Weitzman, M. D., DeMatteo, R., Burda, J. F., and Wilson, J. M. (1996) Transduction with recombinant adeno-associated virus for gene therapy is limited by leading-strand synthesis. *J. Virol.* **70**, 520–532.

34. Bartlett, J. S. and Samulski, R. J. (1996) Methods for the construction and propagation of recombinant adeno-associated virus vectors. *Methods Mol. Med.* **7**, 25–41.

35. Clark, K. R., Liu, X., McGrath, J. P., and Johnson, P. R. (1999) Highly purified re-

combinant adeno-associated virus vectors are biologically active and free of detectable helper and wild-type viruses. *Hum. Gene Ther.* **10**, 1031–1039.

36. Hagstrom, J. N., Couto, L. B., Scallan, C., Burton, M., Arruda, V. R., Fields, P. A., et al. (2000) Enhanced muscle-derived expression of coagulation factor IX from a skeletal actin/CMV hybrid promoter. *Blood* **95**, 2536–2542.

37. Herzog, R. W., and High, K. A. (1999) Adeno-associated virus-mediated gene transfer of factor IX for treatment of hemophilia B by gene therapy. *Thromb. Haemost.* **82**, 540–546.

38. Cao, L., Liu, Y., During, M. J., and Xiao, W. (2000) High-titer, wild-type free recombinant adeno-associated virus vector production using intron-containing helper plasmids. *J. Virol.* **74**, 11456–11463.

39. Matsushita, T., Godwin, S., Surosky, R., and Colosi, P. (1999) *Proceedings of the 2nd Annual Meeting of the American Society of Gene Therapy.* Washington, DC.

40. Kung, J., Hagstrom, J., Cass, D., Tai, S., Lin, H. F., Stafford, D. W., and High, K. A. (1998) Human F.IX corrects the bleeding diathesis of mice with hemophilia B. *Blood* **91**, 784–790.

41. Zolotukhin, S., Byrne, B. J., Mason, E., Zolotukhin, I., Potter, M., Chesnut, K., et al. (1999) Recombinant adeno-associated virus purification using novel methods improves infectious titer and yield. *Gene Ther.* **6**, 973–985.

42. Auricchio, A., Hildinger, M., O'Connor, E., Gao, G. P., and Wilson, J. M. (2001) Isolation of highly infectious and pure adeno-associated virus type 2 vectors with a single-step gravity-flow column. *Hum. Gene Ther.* **12**, 71–76.

43. Gao, G., Qu, G., Burnham, M. S., Huang, J., Chirmule, N., Joshi, B., et al. (2000) Purification of recombinant adeno-associated virus vectors by column chromatography and its performance in vivo. *Hum. Gene Ther.* **11**, 2079–2091.

44. Evans, J. P., Brinkhous, K. M., Brayer, G. D., Reisner, H. W., and High, K. A. (1989) Canine hemophilia B resulting from a point mutation with unusual consequences. *Proc Natl Acad Sci USA* **86**, 10095–10099.

45. Mauser, A. E., Whitlark, J., Whitney, K. M., and Lothrop, C. D., Jr. (1996) A deletion mutation causes hemophilia B in Lhasa Apso dogs. *Blood* **88**, 3451–3455.

46. Pruchnic, R., Cao, B., Peterson, Z. Q., Xiao, X., Li, J., Samulski, R. J., et al. (2000) The use of adeno-associated virus to circumvent the maturation-dependent viral transduction of muscle fibers. *Hum. Gene Ther.* **11**, 521–536.

13

AAV-Mediated Gene Transfer to the Liver

Thomas M. Daly

1. Introduction

The liver is a frequent target of gene-transfer experiments, because of its central role in many metabolic and synthetic pathways. For applications where prolonged expression of genes in the liver is required, adeno-associated virus (AAV) has proven to be an effective tool for in vivo gene transfer. High-level, persistent hepatic expression has been achieved in a number of experimental systems following a single treatment with AAV in murine and larger animal models *(1,2)*. This prolonged expression is particularly useful for the treatment of genetic diseases such as the inborn errors of metabolism, where lifelong expression of the deficient enzyme may be required. Therapeutic benefits using AAV vectors have been demonstrated in animal models of amino acid disorders, lysosomal storage diseases, and coagulopathies *(3–5)*, and Phase I clinical trials are proposed for the treatment of hemophilia B *(6)*.

Gene transfer to the murine liver using AAV is achieved by intravenous (iv) injection of recombinant virus, either via a peripheral or portal vein. The liver is the primary organ transduced following intravenous injection of AAV, although other tissues such as heart and lung may also take up virus to a lesser extent when peripheral injection sites are used *(7)*. Portal-vein injection can reduce the amount of extra-hepatic transduction, and allows a larger dose of virus to be delivered to the liver. However, this technique requires surgical expertise, and can only be performed on adult mice.

Our laboratory is interested in inborn errors of metabolism, diseases that

From: *Methods in Molecular Biology, vol. 246:*
Gene Delivery to Mammalian Cells: Vol. 2: Viral Gene Transfer Techniques
Edited by: W. C. Heiser © Humana Press Inc., Totowa, NJ

often present early in life. The following procedure describes a peripheral injection method used to achieve hepatic transduction with AAV in neonatal mice. Injections are performed via the superficial temporal vein (*see* **Note 1**) *(8)*. The use of newborn animals allows a very high dose/kg to be given, resulting in high levels of liver transduction. This process requires no special equipment, and results in rapid, high-level expression of transduced genes within a week of injection, and persistence of activity for at least 1 yr *(4)*.

2. Material

1. AAV vector. The virus may be prepared by any method (*see* **Note 2**), but should be suspended in a sterile biologically compatible diluent such as Opti-MEM without phenol red (Gibco BRL).
2. 1 cc Hamilton syringe with 30 G needle.
3. Magnifier lamp with flourescent light source (2×–6×). Although the injection can be performed without magnification, the use of a magnifying lamp greatly simplifies the procedure.

3. Methods

The injection process requires two people, with one person to hold the newborn mouse and a second person to inject. The procedure should be performed in a hood, and is most easily done with both people seated.

1. Wipe down hood with 70% ethanol, and position injection equipment (magnification lamp, chairs, needles, etc) prior to removing mice from cage (*see* **Note 3**).
2. Prime a 1-cc syringe with a 30 G needle using 100 µL of sterile OPTI-MEM, and eject fluid. Draw 100 µL of AAV solution into syringe, avoiding the introduction of air bubbles into the syringe.
3. Remove the entire litter containing mice to be injected from cage, and place on a bed of paper towels in hood. Close the original cage containing parents and place aside.
4. The first researcher should select the mouse to be injected and immobilize as described in **Note 4** and **Fig. 1**. Be extremely careful not to restrict the animal's breathing during the hold.
5. While the first researcher immobilizes the animal, the second researcher should insert the needle, bevel side up, under the skin approx 1–2 mm to one side of the superficial temporal vein, parallel to the vessel. Advance the needle so the bevel is just under the skin.
6. Visualizing the needle through the transparent skin, slide the needle above the superficial temporal vein, bevel upward, and advance the tip of the needle into the vein until you see the tip of the needle obscured by blood flow.

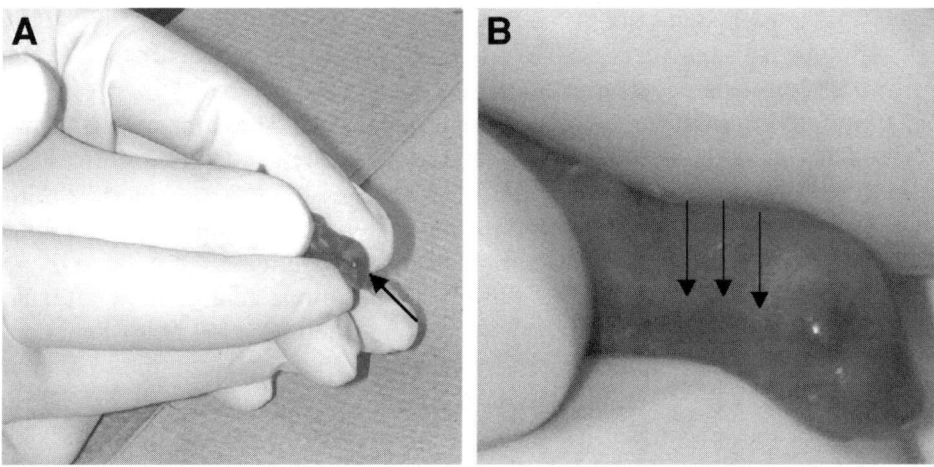

Fig. 1. Neonatal injection technique. (**A**) The mouse should be immobilized between the thumb and index finger of the left hand, while the right index finger is used to extend the neck. The direction of approach for the injection is indicated by the arrow. (**B**) The superficial temporal vein is highlighted.

7. Upon entry into the vessel, slowly (3–5 s) inject 100 µL of AAV solution. (*see* **Note 5**).
8. Remove the needle, and apply direct pressure with a Kim-Wipe for 1–2 min until bleeding has stopped. Do not press too hard on animal's head.
9. Continue with next animal to be injected until all animals in the litter have been injected. The entire litter should be returned to the dam at the same time (*see* **Note 6**).

4. Notes

1. The superficial temporal vein is a prominent vessel located just below and behind the eye, running toward the neck (**Fig. 1**). It is visible through the skin from birth until d 3–4 of life, when the accumulation of pigment in the newborn skin makes visualization difficult. Injections are best performed on d 2 of life. Although it is possible to inject the vessel earlier (d 1), we have noticed a higher incidence of litters being killed by the parents when mice are injected on the first day of life.
2. If AAV has been prepared using cesium chloride gradients, the virus must be purified by dialysis prior to injection. Even very small amounts of cesium are toxic to newborn mice when given in an injection of this volume, and will result in the rapid death of the animal. Virus should be thawed shortly before use and kept on ice until time of injection. We have seen de-

tectable liver transduction using titers as low as 5×10^7 infectious units (IU)/mL *(9)*, but titers of 0.5–1.0×10^9 IU/mL give much better results (>50% of hepatocytes transduced 1 week after injection) *(4)*.

3. Mortality from the injection process is uncommon. However, excessive manipulation of the newborn mice can cause the parents to kill the entire litter. Therefore, the time that the litter is away from the parents should be minimized as much as possible, and further manipulation of the mice (i.e., cage changes) following injection should be avoided until at least the next day.

4. To immobilize the newborn mouse, hold the animal gently in the left hand, with the dorsal surface of the mouse resting against the left index finger and the head near the tip of the left index finger. Use the left thumb to restrain the animal's right forelimb. With the right index finger, gently extend the neck of the mouse, so that the superficial temporal vein is clearly visible and runs in a straight line into the neck (*see* **Fig. 1**). A gentle tension should be maintained on the skin of the head, in order to allow the needle to enter smoothly.

5. In some cases, one can see the vein blanch as the solution is injected, but this is not always apparent. The main indicator of a successful injection is a lack of swelling at the injection site. If the needle has slipped out of the vein, the injection site will swell as fluid is injected into the subcutaneous tissue. If this happens, the injection should be aborted, as it is very difficult to reenter the vein once you have lost the initial injection.

6. An excellent way to practice this technique is by injecting a solution of nontoxic green food coloring (diluted 1:4 with sterile PBS), which will allow you to track the injection. Following a successful injection, the entire mouse will rapidly turn green as the dye is distributed throughout the body. An unsuccessful injection will produce a localized green area around the injection site. The dye is rapidly excreted, and injected mice will appear normal by the following day.

References

1. Koeberl, D. D., Alexander, I. E., Halbert, C. L., Russell, D. W., and Miller, A. D. (1997) Persistent expression of human clotting factor IX from mouse liver after intravenous injection of adeno-associated virus vectors. *Proc. Natl. Acad. Sci. USA* **94**, 1426–1431.
2. Snyder, R. O., Miao, C. H., Patijn, G. A., Spratt, S. K., Danos, O., Nagy, D., et al. (1997) Persistent and therapeutic concentrations of human factor IX in mice after hepatic gene transfer of recombinant AAV vectors. *Nat. Genet.* **16**, 270–276.
3. Chen, S. J., Tazelaar, J., Moscioni, A. D., and Wilson, J. M. (2000) In vivo selection of hepatocytes transduced with adeno-associated viral vectors. *Mol. Ther.* **1**, 414–422.

4. Daly, T. M., Ohlemiller, K. K., Roberts, M. S., Vogler, C. A., and Sands, M. S. (2001) Prevention of systemic clinical disease in MPS VII mice following AAV-mediated neonatal gene transfer. *Gene Ther.* **8**, 1291–1298.

5. Jung, S. C., Han, I. P., Limaye, A., Xu, R., Gelderman, M. P., Zerfas, P., et al. (2001) Adeno-associated viral vector-mediated gene transfer results in long-term enzymatic and functional correction in multiple organs of Fabry mice. *Proc. Natl. Acad. Sci. USA* **98**, 2676–2681.

6. Kay, M. A., Miao, C. H., Ohashi, K., Arruda, V., McClelland, A., Couto, L. B., et al. (2001) A proposed rAAV-Liver-directed clinical trial for hemophilia B. *Mol. Ther.* **3**, S33.

7. Ponnazhagan, S., Mukherjee, P., Yoder, M. C., Wang, X. S., Zhou, S. Z., Kaplan, J., et al. (1997) Adeno-associated virus 2-mediated gene transfer in vivo: organ-tropism and expression of transduced sequences in mice. *Gene* **190**, 203–210.

8. Sands, M. S. and Barker, J. E. (1999) Percutaneous intravenous injection in neonatal mice. *Lab. Animal Sci.* **49**, 328–330.

9. Daly, T. M., Vogler, C., Levy, B., Haskins, M. E., and Sands, M. S. (1999) Neonatal gene transfer leads to widespread correction of pathology in a murine model of lysosomal storage disease. *Proc. Natl. Acad. Sci. USA* **96**, 2296–2300.

14

AAV-Mediated Gene Transfer to Mouse Lungs

Christine L. Halbert and A. Dusty Miller

1. Introduction

The ability of adeno-associated viral (AAV) vectors to promote persistent gene expression in nondividing cells in multiple somatic tissues of animals *(1–4)* makes them excellent tools for gene transfer. One tissue of interest for gene transfer is the lung epithelium, which is afflicted in cystic fibrosis (CF). However, although initial animal studies done with vectors based on AAV type 2 have demonstrated transduction in multiple cells types in the lung, the rates were modest in alveolar cells and much lower rates in airway epitheila and required high particle numbers *(5–7)*. In contrast, an AAV6 encapsidated vector showed preferential transduction of epithelial cells in large and small airways *(8)* at rates that exceeded the 5% efficiency rate predicted to have a therapeutic value for CF gene therapy *(9)*. In fact, recent studies using vectors based on other AAV types showed that types 1–6 have different tissue tropisms *(10–15)*, and that types 5 and 6 are more efficient than type 2 in lung epithelium *(8,14)*. In mouse lung, an AAV2 vector gave modest transduction rates.

For animal gene-transfer studies, a panel of different AAV vectors can be generated by utilizing different capsids. The production of AAV vectors having different capsids is relatively simple because AAV types 1, 2, 3, and 4 and 6 have many regions of homology that are important for vector replication and packaging *(15–18)*. Owing to these similarities, vectors can be generated from different combinations of *rep*, *cap*, and vector genome from the various AAV types. However, for simplicity, an AAV2 vector can be packaged in all capsid

From: *Methods in Molecular Biology, vol. 246:*
Gene Delivery to Mammalian Cells: Vol. 2: Viral Gene Transfer Techniques
Edited by: W. C. Heiser © Humana Press Inc., Totowa, NJ

types. The region with the greatest variation lies in the *cap* gene, and the transduction efficiency achieved by these vectors in vivo is primarily determined by the capsid or the vector pseudotype. Thus in approaching in vivo studies, pseudotyping of one vector, such as an AAV2 vector, into the different capsid types is a logical first step. After determining which capsid type performs best, attempts to further increase the efficiency can be made, for example, by modifying the internal packaged vector to have the inverted terminal repeat (ITR) sequence derived from the same AAV type as the capsid.

The procedures described in this chapter are suitable for production of AAV2 and AAV6 vectors or pseudotype vectors in sufficient quantity and purity for lung experiments in small rodents. Focus has been placed on these two capsid types because AAV6 performs well in the airway epithelium *(8)*, whereas the AAV2 capsid performs better in alveolar cells *(8,19)*. Production of AAV 5 is not described because it does not bind to heparin, which is utilized in our procedure. However, from our limited experience, we have found AAV5 to be stable in cesium chloride, as reported previously *(14,17)*. AAV2 and AAV6 vectors expressing human placental alkaline phosphatase (AP) are generated by cotransfection of several plasmids into 293 cells. Concentration and purification of AAV vectors are done by centrifugation and affinity columns *(20)*. Inoculation of mice is noninvasive and is performed under mild gas anesthesia, and the *in situ* histochemical staining allows efficient and easy assay of transduction in various cell populations of the lung.

2. Materials

2.1. Sources of Enzymes, Reagents, and Supplies

1. NaCl, CaCl$_2$, NaH$_2$PO$_4$, Na$_2$HPO$_4$, HEPES, sucrose, sodium dodecyl sulfate (SDS; also called sodium lauryl sulfate), and Tris-Cl are available from Sigma (St. Louis, MO).
2. Deoxyribosenuclease (DNase), tRNA, and proteinase K are available from Invitrogen (Carlsbad, CA).
3. Random-primed DNA labeling kits are available from Boehringer Mannheim (Indianapolis, IN).
4. Nitro blue tetrazolium (NBT) and 5-bromo-4-chloro-3-indolyl phosphate disodium salt (x-phos) are available from Sigma.
5. HiTrap heparin 1 mL columns are available from Amersham (Cat. no. 17-0406-01) (Arlington Heights, IL).
6. Spectra/Por (dialysis tubing MW co 50,000) is available from Spectrum Laboratories (Rancho Dominguez, CA).
7. 25 × 89 mm polyallomer centrifuge tubes (Cat. no. 326823) are available from Beckman (Palo Alto, CA).
8. Isoflurane is available from Abbott Laboratories (North Chicago, IL).

9. Laboratory Animal Anesthesia system may be obtained from VetEquip (Pleasanton, CA).
10. Mice. We routinely use C57BL/6 mice at 8–10 wk of age for experiments, although other strains such as BALB/c, C3H, RagII, and 129/SV may also be used. All strains are available from Taconic (Germantown, NY).

2.2. Solutions and Culture Medium

1. Tissue-culture solutions: Dubecco's modified Eagle's medium (DMEM), 0.05% trypsin with ethylenediaminetetraacetic acid (EDTA), Dulbecco's phosphate-buffered saline (PBS), and Hanks' Balanced Salt Solution (HBSS) are all available from Invitrogen.
2. Tissue culture medium: DMEM supplemented with 10% fetal bovine serum (FBS).
3. Transfection solutions: 500 mM HEPES-NaOH, pH 7.1; 2.0 M NaCl; 2 M CaCl$_2$; 150 mM Na$_2$HPO$_4$-NaH$_2$PO$_4$, pH 7.0.
4. Purification solutions: 40% sucrose in PBS.
5. Heparin column buffers: HBSS and HBSS containing 300 mM, 500 mM, or 1 M NaCl.
6. Fixatives: 3.7% formaldehyde in PBS for fixation of tissue-culture cells; 2% paraformaldehyde in PBS for fixation of mouse lungs.
7. AP staining reagents: 50 mg/mL NBT in 30% absolute ethanol; 10 mg/mL x-phosphate in water.
8. AP buffer: 100 mM Tris-HCl, pH 8.5, 100 mM NaCl, 50 mM MgCl$_2$.
9. AP staining solution: AP buffer containing 1 mg/mL NBT and 0.1 mg/mL x-phosphate.

2.3. Plasmids

1. Amplify plasmids in recombination-deficient bacteria *(21)* according to standard methods *(22)* and purify the supercoiled form by ion-exchange column (Qiagen) or CsCl gradient centrifugation.
2. ARAP4-2, an AAV2-based vector, contains the human placental alkaline phosphatase (AP) cDNA expressed from a Rous sarcoma virus (RSV) promoter and enhancer sequences, and contains the SV40 polyadenylation sequences. ARAP4-2, pCMVE4orf6, and the AAV2 packaging plasmids, MTrep2 and CMVcap2, and the AAV6 plasmids pARAP4-6, pMTrep6, and pCMVcap6 have been previously described *(8,23)*. Other plasmids for AAV production can be substituted.

2.4. Cells

1. Human embryonic kidney 293 cells *(24)* and human HT-1080 fibrosarcoma cells (ATCC CCL 121) are available from American Type Culture Collec-

tion (Manassas, VA) and are maintained in DMEM supplemented with 10% FBS, penicillin, and streptomycin at 37°C in an atmosphere of 5% CO_2/air.

3. Methods

3.1. Vector Production and Purification

1. Day 1. Trypsinize subconfluent cells, count cells, and seed 4×10^6 cells per dish into one hundred 10-cm tissue-culture dishes in 10 mL of cell-culture medium so that the cell monolayer will be approx 75 % confluent the next day.
2. Day 2. Prepare the following DNA-$CaCl_2$ mixture for 100 transfections in a disposable 50-mL tube.

    a. 500 μg of pMTrep packaging plasmid.
    b. 500 μg of pCMVcap packaging plasmid.
    c. 500 μg of pARAP4 vector plasmid.
    d. 500 μg of pCMVE4orf6 helper plasmid.
    e. 6.25 mL of 2.0 M $CaCl_2$.
    Bring the final volume to 50 mL with distilled water.

    In a separate tube, prepare the Precipitation Mixture:

    a. 5 mL of 500 mM HEPES-NaOH, pH 7.1.
    b. 6.25 mL of 2.0 M NaCl.
    c. 0.5 mL of 150 mM Na_2HPO_4-NaH_2PO_4, pH 7.0.
    Bring the final volume to 50 mL with sterile distilled water.

    Place 10 mL of the aforementioned mixture in each of five 50-mL tubes.

3. Add 10 mL of the DNA-$CaCl_2$ Mixture drop-wise to an equal volume of precipitation mixture while vortexing. Incubate for 30 min at room temperature to allow a fine precipitate to form. Add 1 mL of DNA precipitate to each 10 cm dish of cells. Inoculate 20 dishes at a time (i.e., from one tube of precipitated DNA), then return the dishes to the incubator. Repeat until all dishes have been treated.
4. Day 5. Cells will exhibit cytopathic effect of viral production manifested as a swollen and beadlike appearance. Aspirate the medium without disturbing the cells from 20 dishes. Harvest the cells from 20 dishes in 20–25 mL of DMEM. To do this, add 10 mL of DMEM to the first of 20 dishes, harvest the cells by pipetting medium gently three to four times over the cells to dislodge them, then add the cell-containing medium to the next dish. Before harvesting the cells in the second dish, take another 10 mL of DMEM and add it to the first dish to rinse and collect the remaining cells. Then harvest the cells from the second dish by gently pipetting 10 mL of cells and culture medium (from the first dish) onto the cells in the second dish. Re-

peat for 20 plates, using 10 mL to harvest and 10 mL to obtain residual cells. After 20 plates have been washed, approx 25 mL of cells in DMEM is collected (the volume is more than 20 mL owing to residual medium left in the plates after aspiration). Repeat for all 100 dishes. Prepare five tubes, each containing approx 20–25 mL of cells in DMEM. Transfer the tubes to a dry ice/ethanol bath (*see* **Note 1**).

5. Freeze and thaw the cell suspensions three times by transferring the tubes between a dry ice/ethanol bath and a 37°C water bath, vortexing after each thaw. Centrifuge the crude lysates at 6900g for 30 min at 4°C (*see* **Note 2**). Remove the medium from the cell-debris pellet and keep the medium on ice. Carefully overlay the medium (20–22 mL) on 15 mL of a 40% sucrose solution in six 25 × 89 mm polyallomer centrifuge tubes. Centrifuge the preparation at 131,000g for 18 h at 4°C rotors (*see* **Note 3**).

6. Day 6. Remove and discard the supernatant. Add 0.5 mL of HBSS to each tube to cover the pellets. Let the viral pellets resuspend for 20 min at room temperature. Pool the vector pellets from the six tubes in a total of 3 mL of HBSS (*see* **Note 4**).

7. Purification of AAV2 or AAV6 vectors over a heparin column. Elute the storage buffer from a heparin column by applying 5 mL of HBSS under gentle pressure using a 5-cc syringe. Equilibrate the column with another 3 mL of HBSS; elute the HBSS at approximately one drop per second. Leave the column with the last approx 1 mL of HBSS for 20 min at room temperature. Apply the vector solution using a 5-cc syringe in 1 mL aliquots; allow 10 min for each 1 mL aliquot to bind. Wash the column with 3 volumes (3 mL) each of HBSS, HBSS containing 0.3 M NaCl, 0.5 M NaCl, and 1.0 M NaCl. Catch the flow-through in 1-mL fractions (*see* **Note 5**). For AAV6 vectors, load the 0.3 M fractions, and for AAV2, load the 0.5 M fractions in dialysis tubing and dialyze in 1 L of HBSS for 2 h at 4°C. Freeze aliquots of the vector at −80°C.

3.2. Vector Characterization

3.2.1. Analysis of Transducing Titer

Determine the infectious titer of ARAP4 vector stocks by using HT-1080 cells as targets for transduction.

1. For each vector to be assayed from **Subheading 3.1.**, plate 5 × 10⁴ HT-1080 cells in 2 mL of medium into each well of 6-well dishes. Incubate overnight at 37°C in an incubator with 5% CO_2.

2. Make 10-fold serial dilutions of the vector ranging from 10^{-2} to 10^{-9} by diluting 10 µL of vector the from the heparin fractions into 1 mL of DMEM, then making 10-fold serial dilutions. Inoculate 100 µL of each di-

lution to the cells in duplicate wells; include two wells as a control without virus. Incubate the cells for 3 d at 37°C in an incubator with 5% CO_2.

3. Remove the medium from the cells. Fix the cells in 2 mL of PBS containing 3.7% formaldehyde for 5–10 min followed by washing three times in 3 mL of PBS (5 min each). Inactivate endogenous AP activity by heating the plates at 68°C for 1 h in the final PBS wash. Aspirate the final wash. Add 2 mL of AP staining solution and incubate the plates overnight at room temperature. AP^+ cells will stain dark purple in the cytoplasm and membrane. Count foci of AP^+ cells to obtain AP^+ Focus-Forming Units (FFU) to determine the transducing titer of the vector stock (expressed as FFU/mL).

3.2.2. Analysis of Genome Equivalents

Perform Southern analysis on vector stocks to determine the number of genome-containing particles in the vector preparation (22).

1. Incubate 2 µL of vector preparation (10^8 genome-containing particles) of an AAV2- or AAV6-AP vector (10^7 or 10^9 AP^+ FFU, respectively) with 100 U of DNase in a total volume of 200 µL at 37°C for 15 min.
2. Add 1.5 µL of 0.5 M EDTA and 10 µg of tRNA to the vector, then inactivate the DNase at 68°C for 10 min.
3. Add 2 µL of 10% SDS, and 2 µL of proteinase K (20 mg/mL), then incubate the solution at 56°C for 3 h.
4. Extract the virion DNA twice, each time with an equal volume phenol/chloroform (1:1). Transfer the aqueous phase to a clean microfuge tube.
5. Add 1/10 volume of 3 M sodium acetate and two volumes of ethanol to precipitate the virion DNA. Incubate at -20°C for 16 h. Centrifuge the virion DNA in a microfuge (20 min at full speed and 4°C), aspirate the supernatant, then resuspend the pellet in 20 µL of water and incubate at 65°C for 15 min. Vortex the sample briefly and recentrifuge for a few seconds to pellet the DNA suspension.
6. Load 5 µL of virion DNA on a 1% neutral agarose gel to visualize its size as well as to quantify the amount of DNA. Include on the gel vector plasmid-control DNA, digested with a restriction enzyme that releases the vector from the plasmid backbone, at 50 pg, 500 pg, and 5 ng (representing approx 10^7, 10^8, and 10^9 single-stranded genomes, respectively, for a 6 Kb plasmid). Transfer the DNA from the gel to a nylon membrane, denature the DNA, and hybridize to probe containing the AP cDNA purified by gel isolation and labeled with ^{32}P using a random-primed DNA labeling kit. Vector DNA bands are visible after a 4 h exposure. A typical AAV vector stock contains approx 10^{11}–10^{12} particles per mL. Particle-to-infectivity, as determined by FFU in HT1080 cells, is approx 1×10^2 and 2×10^4 for AAV2 and AAV6 pseudo-

type AAV-AP (such as ARAP4), respectively. It is better to normalize differ-ent vector pseudotypes according to genome equivalents rather than trans-ducing titer for animal administration because the titer is a function of the target cell and varies for each cell line with each vector and vector pseudo-type.

3.3. Vector Administration

1. Based on the analysis of the virus stock in **Subheading 3.2.2.**, prepare an aliquot of the vector at 10^{10}–10^{11} genome equivalents in 50–75 µL in HBSS; place the aliquot at room temperature.
2. Set up a laboratory animal anesthesia system according to standard proce-dures provided by the manufacturer. Sedate the mice in the anesthesia chamber containing isofluorane (2.5%) in oxygen at a flow rate of 1 L/min.
3. After sedation, remove the animal from the chamber and place it in a supine position with nares pointing up. If the animal exhibits signs of waking by twitching its whiskers, continue to deliver the anesthesia mixture by use of a nose cone. When the animal does not exhibit a jerking response to a light pinch on the footpad, remove the nose cone and deliver the vector in a 50–75 µL aliquot into the nares of the mouse (*see* **Note 6**).
4. After administration of the vector, watch the mouse until it becomes alert; steady breathing should resume in less than 1 min. Place the animal gently into its cage with its head tilted upright during recovery; do not attempt to invert the animal immediately after vector inoculation.

3.4. Fixation and Histochemical Staining of Lungs

1. Four weeks after vector administration euthanize the mouse using CO_2.
2. Exsanguinate the animal via cardiac puncture using a 1-mL syringe and a 0.5 in 25 G needle.
3. Open the chest cavity and expose the lungs and trachea. Puncture the tra-chea with a 20 G needle. Insert a blunted 22 G needle into one end of a catheter approx 7 cm long with a diameter of 0.8–1 mm (slightly less than the diameter of the mouse trachea). Insert the catheter into the trachea and attach the trachea to the catheter by tying a string around the trachea just below the inserted catheter.
4. Extirpate the lung and trachea (with attached catheter and needle).
5. Connect the needle on the catheter to a 30-mL syringe filled with 2% paraformaldehyde and fitted with a three-way stopcock. Attach a 1-mL sy-ringe to the side opening of the three-way stopcock to drain the trapped air in the lung. Extraction of air using the 1-mL syringe creates negative pres-sure in the lung, which promotes the flow of fixative into the lung to ade-quately fix and preserve the normal cell architecture.

6. When the lobes of the lung have been adequately inflated with fixative, close the stopcock and immerse the lung in 15 mL of fixative for 2–3 h at room temperature. Remove the catheter and syringes.

7. Dissect the lung so that each lobe is cut along its main bronchial airway. Rinse the lung tissue three times in dPBS (30 min each rinse), then place it in a 15-mL conical tube with 7 mL of dPBS. Incubate the tube in a 68°C water bath for 70 min to heat-inactivate endogenous AP. Afterwards, rinse the tissue in AP buffer for 15 min and stain overnight in AP staining solution at room temperature. Staining is done in the dark because the staining reagents are light-sensitive.

8. Remove the AP staining solution by rinsing the tissue for 1–2 h in PBS.

9. Photograph AP staining in the gross tissue using a dissecting microscope.

10. To quantify the transduction efficiency, cut the stained tissue slices from the entire lung into 3 mm blocks, then embed the blocks in paraffin. Using a microtome, cut 5-µm thick sections such that a slice from the block contains a representative sample from the entire lung. Counterstain some of the sections with nuclear-fast red (*see* **Fig. 1**, **Note 7**) and others with hemotoxylin and eosin (H&E). Quantitate transduction efficiency by counting the number of AP+ cells per section on the slides stained with nuclear-fast red. The AP+ cells are more difficult to see in the slides counterstained with H&E, however, H&E staining can give a better visualization of cell morphology and histopathology. To determine the number of AP+ alveolar cells, count AP+-stained cells in 10 random 1 mm² areas in several of the 5-µm sections. For all other cell types, determine the number of AP+ cells by counting the AP+-stained cells for each particular cell type in all of the 5 µm sections from a 3-mm block of tissue; the total area may be most easily quantified by scanning stained slides (e.g., using Adobe Photoshop) and is approx 0.5–1 cm²/slide. Express transduction efficiency as AP+ cells/cm² (*see* **Fig. 2** and **Note 7**).

4. Notes

1. Recently, we used adenovirus helper plasmids that contain adenovirus E2a and VA in addition to the E4orf6 gene. Generally, cells exhibit a higher cytopathic effect with the helper plasmid containing additional adenovirus genes that a significant percentage of the cells are floating and/or lysed by day 3. In such a case, both cells and medium can be harvested. The collection volume is greater than if only the cell monolayer is harvested, which makes purification much more tedious. However, the vector production is higher than with adenovirus E4orf6 (two- to fivefold). Therefore, one can decrease the number of total; plates transfected.

2. This is equivalent to 6000 rpm in an HS-4 rotor.

3. This is equivalent to 27,000 rpm in an SW28 rotor and requires two rotors.

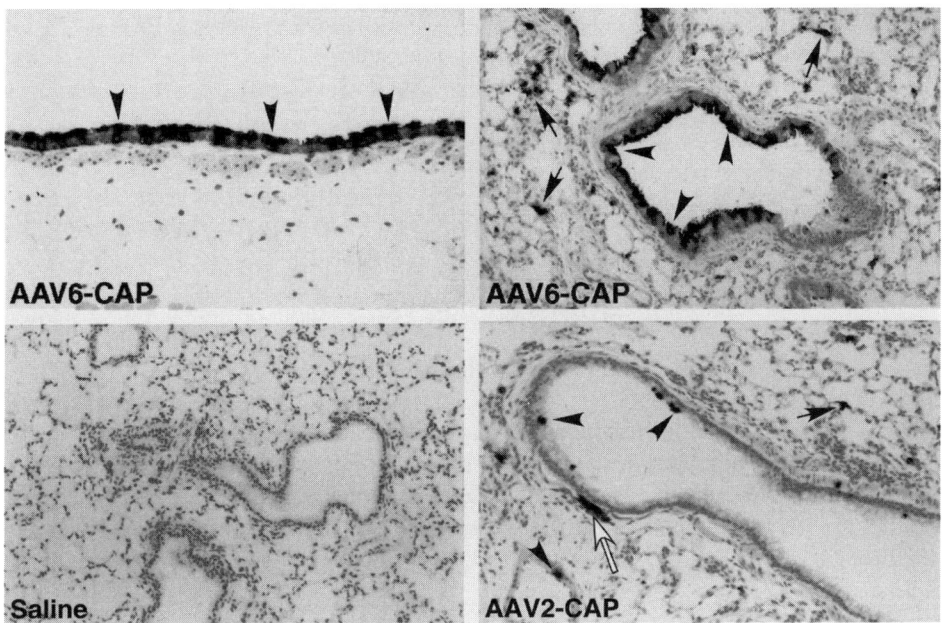

Fig. 1. Histochemical detection of AP expression in mouse lungs 1 mo after vector exposure. Saline-treated lungs did not exhibit AP⁺ cells, whereas the lungs of mice given AAV vector (7×10^{11} genome-containing particles) exhibited AP⁺ cells that stained dark purple: AP⁺ alveolar cells (arrows), airway epithelial cells (arrowheads), and smooth-muscle cells (clear arrows). Tissue was stained *in situ* for AP and counter-stained in slide sections with nuclear-fast red. Original magnifications of photographs are 200×.

4. If aggregates are seen in the vector solution, gently pass the viral solution through a 22 G needle and centrifuge at maximum speed in a tabletop centrifuge for 5 min. Transfer the clarified virus preparation to a new tube.

5. We usually collect 1-mL samples from each wash and keep them separate for titering (**Subheading 3.2.1.**). For AAV6 vectors, the first two 1-mL samples from the 0.3 *M* NaCl wash, and, for AAV2 vectors, the first two 1-mL samples from the 0.5 *M* NaCl wash, usually have the highest titers. Alternatively, the entire 3-mL fraction from each wash may be pooled and assayed in bulk. We occasionally collect and assay the higher concentration NaCl washes to confirm that all of the vector has been eluted.

6. To inoculate the vector, lay the mouse down with its nose pointing up and with the tip of the pipet close to the nares of the mouse. The vector will form as droplets on the pipet tip, with each droplet being spontaneously aspirated by the mouse. The entire delivery should not exceed 5 s.

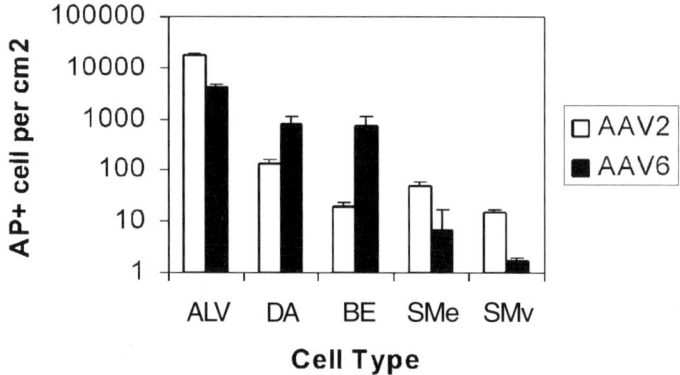

Fig. 2. Quantitation of transduction rates in mouse lungs 1 mo after vector exposure. Vector capsid pseudotypes are indicated. Abbreviations for cell types are indicated at the bottom: ALV, alveolar cells; DA, distal airway epithelial cells; BE, bronchial airway epithelial cells (having underlying cartilage); SMe, smooth muscle cells underlying epithelium; and SMv, smooth muscle in vascular walls. Mean and SEM are given. $n = 3$ animals in each group.

7. This protocol has been used in studies comparing transduction by AAV2 and AAV6 pseudotype vectors in mouse lung *(8)*. In one experiment, a vector having an AAV2 capsid was compared to one having an AAV6 capsid. Mouse lungs treated with saline did not exhibit AP$^+$-staining cells (**Fig. 1**, lower left panel), whereas those treated with an AAV6-encapsidated vector exhibited abundant AP$^+$-staining cells in the airway epithelium and in alveolar cells (**Fig. 1**, top panels). The transduction efficiency in airways was high, with up to 80% of the cells in some airways expressing AP. In contrast, mouse lungs treated with the AAV2 vector exhibited much lower AP staining in the airway epithelium (**Fig. 1**, lower right panel).

   Quantitation of the number of AP$^+$ cells shows that the AAV2 encapsidated vector transduced alveolar cells at 100-fold and 1000-fold higher rates than it transduced cells of the distal airway and bronchial epithelium, respectively (**Fig. 2**).

   Transduction of smooth-muscle cells by both AAV2 and AAV6 was also lower than transduction of alveolar cells. The AAV6-encapsidated vector showed a lower transduction efficiency of alveolar cells and smooth-muscle cells compared with the AAV2-encapsidated vector. In contrast, the transduction of epithelial cells in distal airways and larger bronchial airways was 10-fold and 100-fold higher, respectively, for AAV6-encapsidated vector than for AAV2 encapsidated vector. Administration of a lower dose of vector (10^{11} genome-containing particles) gave similar relative transduction efficiency for both the AAV2 and AAV6 vectors (data not shown). The experiments in mouse lungs showed that AAV6 capsids consistently performed more efficiently than did AAV2 capsids at mediating transduction of airway epithelial cells.

References

1. Flotte, T. R., Afione, S. A., Conrad, C., McGrath, S. A., Solow, R., Oka, H., et al. (1993) Stable in vivo expression of the cystic fibrosis transmembrane conductance regulator with an adeno-associated virus vector. *Proc. Natl. Acad. Sci. USA* USA **90,** 10613–10617.
2. Herzog, R. W., Yang, E. Y., Couto, L. B., Hagstrom, J. N., Elwell, D., Fields, P. A., et al. (1999) Long-term correction of canine hemophilia B by gene transfer of blood coagulation factor IX mediated by adeno-associated viral vector. *Nat. Med.* **5,** 56–63.
3. Muzyczka, N. (1992) Use of adeno-associated virus as a general transduction vector for mammalian cells. *Curr. Topics Microbiol. Immunol.* **158,** 97–129.
4. Snyder, R. O., Miao, C. H., Patijn, G. A, Spratt, S. K., Danos, O., Nagy, D., et al. (1997) Persistent and therapeutic concentrations of human factor IX in mice after hepatic gene transfer of recombinant AAV vectors. *Nat. Genet.* **16,** 270–276.
5. Fisher, K. J., Gao, G., Weitzman, M. D., DeMatteo, R., Burda, J. F., and Wilson, J. M. (1996) Transduction with recombinant adeno-associated virus for gene therapy is limited by leading-strand synthesis. *J. Virol.* **70,** 520–532
6. Halbert, C. L. et al. (1997) Transduction by adeno-associated virus vectors in the rabbit airway: efficiency, persistence, and readministration. *J. Virol.* **71,** 5932–5941.
7. Halbert, C.L., Standaert, T.A., Wilson, C.B., and Miller, A.D. (1998). Successful readministration of AAV vectors to the mouse lung requires transient immunosuppression during the initial exposure. *J. Virol.* **72,** 9795–9805.
8. Halbert, C. L., Allen, J. M., and Miller, A. D. (2001) Adeno-associated virus type 6 (AAV6) vectors mediate efficient transduction of airway epithelial cells in mouse lungs compared to that of AAV2 vectors. *J. Virol.* **75,** 6615–6624.
9. Johnson, L. G., Olsen, J. C., Sarkadi, B., Moore, K. L., Swanstrom, R., and Boucher, R.C. (1992) Efficiency of gene transfer for restoration of normal airway epithelial function in cystic fibrosis. *Nature Genet.* **2,** 21–25.
10. Chao, H., Liu, Y., Rabinowitz, J., Li, C., Samuloki, R. S., and Walsh, C. E. (2000) Several log increase in therapeutic transgene delivery by distinct adeno-associated viral serotype vectors. *Mol. Ther.* **2,** 619–623.
11. Davidson, B. L., Stein, C. S., Heth, J. A., et al. (2000) Recombinant adeno-associated virus type 2, 4, and 5 vectors: transduction of variant cell types and regions in the mammalian central nervous system. *Proc. Natl. Acad. Sci. USA* **97,** 3428–3432.
12. Halbert, C.L., Rutledge, E. A., Allen, J. M., Russell, D. W., and Miller, A. D. (2000) Repeat transduction in the mouse lung by using adeno-associated virus vectors with different serotypes. *J. Virol.* **74,** 1524–1532.

13. Hildinger, M., Auricchio, A., Gao, G., Wang, L., Chirmule, N., and Wilson, S.M. (2001) Hybrid vectors based on adeno-associated virus serotypes 2 and 5 for muscle-directed Gene Transfer. *J. Virol.* **75,** 6199–6203.

14. Zabner, J., Seiler, M., Walters, R., et al. (2000) Adeno-associated virus type 5 (AAV5) but not AAV2 binds to the apical surfaces of airway epithelia and facilitates gene transfer. *J. Virol.* **74,** 3852–3858.

15. Xiao, W., Chirmule, N., Berta, S. C., McCullough, B., Gao, G., and Wilson, J. M. (1999) Gene therapy vectors based on adeno-associated virus type 1. *J. Virol.* **73,** 3994–4003.

16. Chiorini, J. A., Yang, L., Liu, Y., Safer, B., and Kotin, R. M. (1997) Cloning of adeno-associated virus type 4 (AAV4) and generation of recombinant AAV4 particles. *J. Virol.* **86,** 6823–6833.

17. Chiorini, J. A., F. Kim, L. Yang, and Kotin, R. M. (1999) Cloning and characterization of adeno-associated virus type 5. *J. Virol.* **73,** 1309–1319.

18. Rutledge, E. A., Halbert, C. L., and Russell, D. W. (1998) Infectious clones and vectors derived from adeno-associated virus (AAV) serotypes other than AAV type 2. *J. Virol.* **72**:309–319.

19. Zeitlin, P. L., Chu, S., Conrad, C., McVeigh, U., Ferguson, K., Flotte, T. R., and Guggino, W. B. (1995) Alveolar stem cell transduction by an adeno-associated viral vector. *Gene Ther.* **2,** 623–663

20. Zolotukhin, S., Byrne, J., Mason, E., Zolotukhin, I., Potter, M., Chesnut, K. C., et al. (1999) Recombinant adeno-associated virus purification using novel methods improves infectious titer and yield. *Gene Ther.* **6,** 973–985.

21. Boissy, R. and Astell, C. R. (1985) An Escherichia coli recBCsbcBrecF host permits the deletion-resistant propagation of plasmid clones containing the 5'-terminal palindrome of minute virus of mice. *Gene* **35**:179–185.

22. Sambrook, J., Fritsch, E. F., and Maniatis T. (1989) *Molecular Cloning: A Laboratory Manual*, 2nd ed. Cold Spring Harbor Laboratory Press, Cold Spring Harbor, NY.

23. Allen, J. M., Halbert, C. L., and Miller, A. D. (2000) Improved adeno-associated virus vector production with transfection of a single helper adenovirus gene, E4orf6. *Mol. Ther.* **1,** 88–95.

24. Graham, F. L. and Smiley, J. (1977) Characteristics of a human cell line transformed by DNA from human adenovirus type 5. *J. Gen. Virol.* **36,** 59–72.

15

Gene Delivery to the Mammalian Heart Using AAV Vectors

Danny Chu, Patricia A. Thistlethwaite, Christopher C. Sullivan, Mirta S. Grifman, and Matthew D. Weitzman

1. Introduction

There are a large number of cardiovascular diseases that could be treated by myocardial gene transfer *(1,2)*. These include congestive heart failure, ischemic heart disease, and cardiomyopathy. In addition to its potential for treatment of disease, myocardial gene transfer is useful for the analysis of gene expression and promoter function and for generating animal models of human disease such as pulmonary hypertension. The ideal vector for myocardial gene therapy should give efficient and stable transduction of cardiomyocytes in vivo. Recombinant adenovirus vectors have been used to transduce cardiomyocytes in rodents, rabbits, pigs, and humans by both intramyocardial injection and intracoronary infusion *(3–5)*. Although efficient transduction can be obtained with adenovirus vectors, immune responses and elimination of transduced cells results in only transient expression in immunocompetent hosts. Vectors based on recombinant adeno-associated virus (rAAV) offer a number of attractive features and are emerging as promising gene transfer vehicles for many in vivo applications.

Experiments with reporter genes have shown that rAAV vectors can efficiently transduce murine and rat cardiac myocytes in vitro *(6–8)*. There are two major methods of in vivo gene delivery to mammalian hearts using a viral vector system. The first method involves direct intramyocardial injection of the

From: *Methods in Molecular Biology, vol. 246:*
Gene Delivery to Mammalian Cells: Vol. 2: Viral Gene Transfer Techniques
Edited by: W. C. Heiser © Humana Press Inc., Totowa, NJ

virus into the cardiac muscle via a thoracotomy incision. This method allows direct organ targeting of gene delivery as well as location specificity within that organ. The gene of interest delivered by this particular method is not expressed anywhere in the animal other than the site of injection in the heart, as viral particles do not enter the systemic circulation. The second method for myocardial gene delivery is perfusion of the virus into the coronary circulation. This method of gene delivery has proven not to be organ-specific and often results in hepatic accumulation of the virus in preference to the heart. Direct myocardial injection with rAAV vectors results in efficient transduction (40% around the needle injection site) of cardiomyocytes in vivo *(8,9)*. In addition, direct injection of rAAV expressing vascular endothelial growth factor (VEGF) into mouse myocardium has been used to induce angiogenesis in a model of ischemic myocardium *(8)*. Efficient transduction by rAAV in vivo has also been observed after intracoronary perfusion in mice *(6)* and pigs *(10)*. Unlike adenovirus vectors, no inflammatory cell infiltration or myocyte necrosis is observed with rAAV vectors and transgene expression is stable for at least 6 mo *(6,8–10)*.

In this chapter we describe protocols for gene delivery to the heart using rAAV vectors. This chapter will enable the reader to become proficient in delivering AAV carrying the gene of interest into rodent hearts via direct intramyocardial injection of the virus. We produce rAAV vectors in the absence of contaminating helper virus by using the triple transfection method *(11)*. Cells (human 293 cell line) are transfected with a vector plasmid containing the transgene cassette flanked by viral inverted terminal repeats (ITRs), a packaging plasmid (pXX2) that expresses the AAV Rep and Cap proteins, and a helper plasmid (pXX6) that supplies adenoviral gene products to enable replication and packaging. Vectors are harvested in cell lysates and purified by iodixanol gradient ultracentrifugation. We titer the virus by quantitating genomic particles using Southern blotting or real-time polymerase chain reaction (PCR) and assess transgene expression by limiting dilution transduction. Protocols to purify rAAV vectors have been previously published *(12–14)* and are therefore not reproduced here. Instead, specific steps of rodent anesthesia and techniques of survival surgery will be discussed in depth.

2. Materials

1. AAV carrying gene of interest; at least 1×10^{11} particles/mL titer.
2. Albino Fischer rats 10–12 wk old (Harlan, San Diego, CA).
3. Ketamine 100 mg/mL (Fort Dodge Animal Health, Fort Dodge, IA).
4. Xylazine 20 mg/mL (VEDCO, Inc., St. Joseph, MO).
5. Antisedan 5 mg/mL (Animal Health, Exton, PA).
6. Oto-opthalmoscope with nasal speculum attachment (Welch-Allyn, San Diego, CA).

7. 14 G angiocatheter (1-in length).
8. 7-in long small-caliber, soft, flexible guide-wire.
9. Puralube Vet Artificial tear ointment (Allpets.com, Hazelton, PA).
10. Animal ear hole puncher (Fisher Scientific, Pittsburgh, PA).
11. Insulin syringes.
12. Harvard Rodent Ventilator (Harvard Apparatus Inc., Holliston, MA).
13. Standard small animal surgical instruments (including small self-retaining retractor).
14. Electric hair shaver (Fisher Scientific).
15. 70% EtOH.
16. Battery operated electro-cautery (optional) (Solan, Accu-Temp, San Diego, CA).
17. 3-cc syringe with 26 G diameter rubber flexible tubing attached.
18. 4-0 Prolene sutures noncutting needle.
19. 2 × 2 gauze.
20. 3-0 Vicryl cutting needle.
21. Animal scale (Fisher Scientific).
22. X-gal solution: 1 mg/mL of X-gal, 5 mM potassium ferro-cyanide, 5 mM potassium ferric-cyanide, 2 mM magnesium chloride (Sigma-Aldrich, St. Louis, MO).
23. Kinematics PT2000 Polytron Tissue Homogenizer (Fisher Scientific).
24. Buffer A: 50 mM Tris-HCl, pH 7.5, 1 mM EDTA, 1 mM EGTA, 0.5 mM Na$_3$VO$_4$, 1% Triton X-100, 50 mM sodium chloride, 5 mM sodium pyrophosphate, 10 mM sodium glycerophosphate. (Add following at time of use: 0.1 mM PMSF, 0.1% 2-mercaptoethanol, 1 μg/mL of aprotinin, pepstatin, leupeptin) (Sigma-Aldrich, St. Louis, MO).
25. RNAzol™ B (TelTest, Inc., Friendswood, TX).
26. Turboblotter Transfer System (Schleicher & Scheull, Keene, NH).

3. Methods

1. The first step for successful in vivo gene delivery is anesthetizing the animal *(15)*. Determine the appropriate amount of anesthetics based on the weight of the animal (*see* **Table 1**). Using intraperitoneal (ip) injection, deliver the appropriate dose of a mixture of Ketamine (100 mg/mL) and Xylazine (20 mg/mL). The amount of ip anesthesia should sedate the animal within approx 15–20 min (*see* **Note 1**).
2. Apply artificial eye lubricant to the animal's eyes in order to prevent them from drying during the procedure.
3. As soon as the animal is appropriately sedated, perform an endotracheal intubation with the help of a nasal speculum device (*see* **Fig. 1B**) and a thin, flexible guide-wire (*see* **Note 2**). Once the animal's vocal cords are visual-

Table 1
Rodent Anesthesia Table

Body weight (g)	Anesthesia mix (mL)*	Body weight (g)	Anesthesia mix (mL)*
150	0.150	330	0.330
155	0.155	335	0.335
160	0.160	340	0.340
165	0.165	345	0.345
170	0.170	350	0.350
175	0.175	355	0.355
180	0.180	360	0.360
185	0.185	365	0.365
190	0.190	370	0.370
195	0.195	375	0.375
200	0.200	380	0.380
205	0.205	385	0.385
210	0.210	390	0.390
215	0.215	395	0.395
220	0.220	400	0.400
225	0.225	405	0.405
230	0.230	410	0.410
235	0.235	415	0.415
240	0.240	420	0.420
245	0.245	425	0.425
250	0.250	430	0.430
255	0.255	435	0.435
260	0.260	440	0.440
265	0.265	445	0.445
270	0.270	450	0.450
275	0.275	455	0.455
280	0.280	460	0.460
285	0.285	465	0.465
290	0.290	470	0.470
295	0.295	475	0.475
300	0.300	480	0.480
305	0.305	485	0.485
310	0.310	490	0.490
315	0.315	495	0.495
320	0.320	500	0.500
325	0.325		

* Anesthesia mix: 1.0 mL Ketamine (100 mg/mL) + 1.0 mL Xylazine (20 mg/mL). Deliver by injection 50 mg/kg Ketamine + 10 mg/kg Xylazine.

ized, insert the guide-wire through the vocal cords. With the guide-wire in the vocal cords and the speculum removed, carefully thread a 15 G angio-catheter over the guide-wire through the vocal cords into the animal's trachea. Promptly remove the guide-wire and connect the angiocatheter to the Harvard Animal Ventilator for a test breath (*see* **Note 3**).

4. Set the ventilator at a minute-ventilation of approx 120 mL/min. For example, a typical setting would be a tidal volume of 2.0 mL and a respiratory rate of 60.

5. Position the animal in a semi-decubitus position with left side up (*see* **Fig. 1C**). Use an electric shaver to remove the excess fur from the planned incision site. Briefly clean the skin and its surrounding area with 70% EtOH.

6. Perform a left postero-lateral thoracotomy incision to access the heart (*see* **Fig. 1D**). Once the skin is incised, the next layer of muscle is the latissmus dorsi muscle. Incise this layer to reveal the intercostal muscles and the serratus anterior muscle (*see* **Fig. 1E**).

7. Use the serratus anterior muscle as a landmark to enter the appropriate intercostal space. Once the intercostal space is entered, the heart and lung are easily identifiable.

8. Insert a small self-retaining rib spreader and gently spread the ribs apart for improved operative visualization. Open the thin layer of pericardial membrane with either scissors or a blade (*see* **Note 4**).

9. Once the pericardial sac is incised, prepare the virus carrying the gene of interest for injection. Use 200 µL of virus for each animal from a stock that is at least 1×10^{11} particles/mL (*see* **Note 5**). Directly inject the virus into the left ventricular apex of the heart at a shallow angle (*see* **Note 6**). Divide the vector into three separate injections in order to deliver the gene to a confluent uniform area along the left ventricular apex (*see* **Fig. 1F**).

10. Once hemostasis is satisfactorily achieved (*see* **Note 7**), begin closure of the intercostal muscles and ribs with 4-0 Prolene or nylon (nonabsorbable) suture. Prior to tying these intercostal sutures, insert flexible small-caliber suction tubing into the thoracic cavity.

11. After tying the intercostal sutures, use a syringe to evacuate the air/fluid in the thoracic cavity in order to regain the normal negative physiologic intrathoracic pressure.

12. Close the skin incision with running 3-0 Vicryl (absorbable) suture. Antisedan can be given ip at this point to reverse the anesthesia (2.5 mg/kg for each rat).

13. When the animal is making spontaneous breathing efforts as evident by diaphragmatic movements, disconnect the ventilator from the endotracheal tube. Do not remove the endotracheal tube if there is no evidence of spontaneous movement of the animal's neck (an indication that the airway muscles

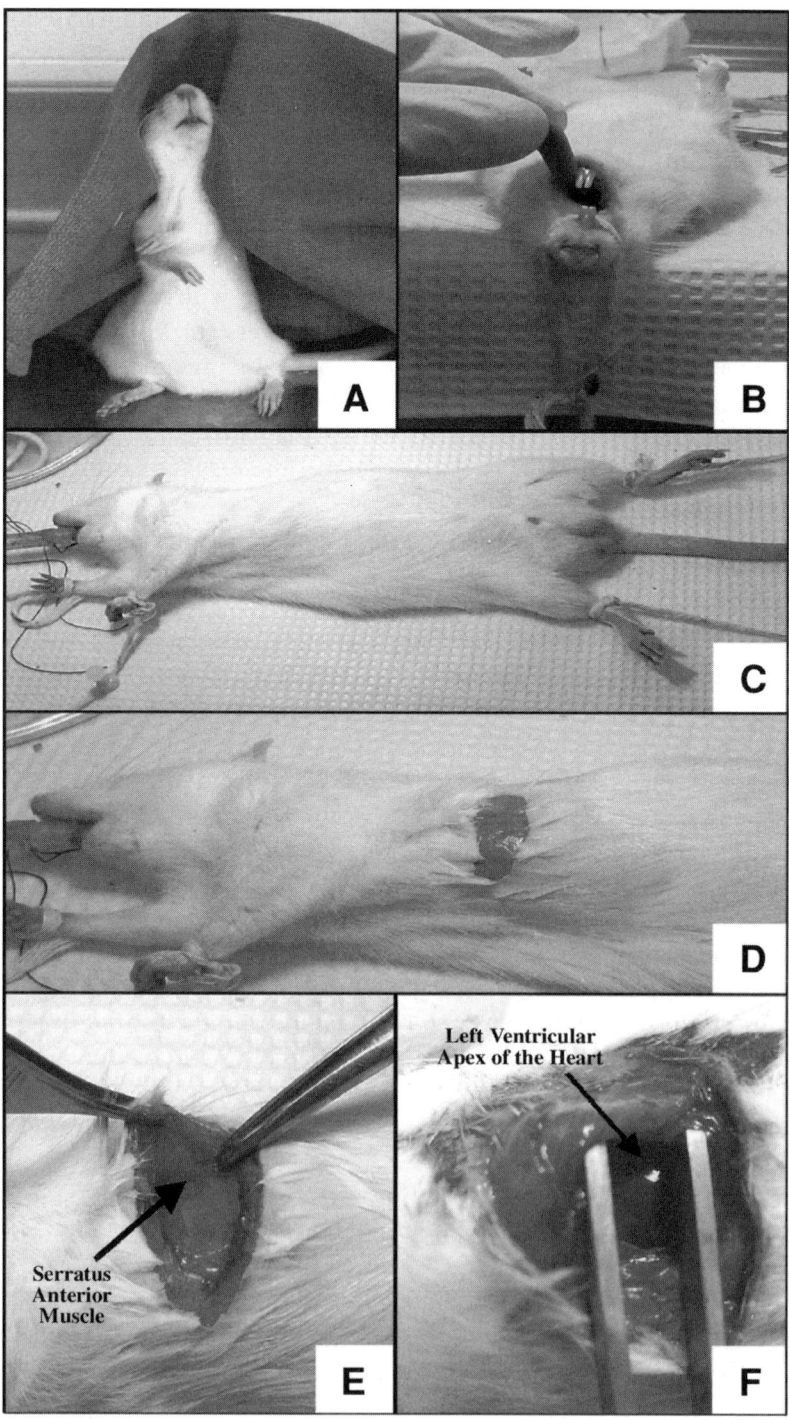

are functioning spontaneously to protect the animal from aspirating gastrointestinal contents) (*see* **Note 8**). An alternative method of delivery to the mammalian heart involves intracoronary perfusion of the virus *(16,17)*. The steps are similar to direct intramyocardial injection (*see* **Note 9**).

14. Sacrifice animals at various time postinjection and surgically remove heart (*see* **Note 10**). The heart and other organs can be processed for analysis of gene expression at the whole organ level (*see* **step 15**), at the cellular level by histochemistry of fixed sections (*see* **step 16**), at the protein level by Western blot analysis (*see* **step 17**), or at the RNA level by Northern (*see* **step 18**).

15. The explanted hearts can be analyzed for transgene expression at the gross level. For animals injected with rAAV.LacZ expression of β-galactosidase (β-gal) can be assessed by X-gal staining the whole organ (*see* **Note 10**). Wash heart three times briefly with PBS and fixed overnight at 4°C in 4% paraldehyde solution. Rinse three times briefly in PBS and stained with X-gal solution overnight at 37°C (*see* **Fig. 2**).

16. For analysis at the cellular level by histochemistry, wash heart three times with PBS solution and fix overnight at 4°C in 4% paraldehyde solution. Serially dehydrate heart with ethanol and embed in paraffin blocks for sectioning. Cut 5-μm thick sections of the hearts for staining and analysis. Expression of β-gal can be assessed by X-gal staining. Stain sections with hematoxylin and eosin and observe for inflammatory infiltrates by microscopy (*see* **Note 11**).

17. For protein analysis wash heart three times with cold PBS after explanting and store at −80°C until further use. Weigh out approx 100 mg of tissue from the apex of the heart. Homogenize tissue with Polytron Tissue Homogenizer in 2 mL of cold Buffer A and then centrifuged at 15,800g at 4°C for 30 min. Collect the supernatant and store at −80°C. Determine protein concentration using the Bradford assay and run approximately 40 μg of total protein from each sample on a 7.5% Tris-Glycine Sodium dodecyl sulfate polyacrylamide gel electrophoresis (SDS-PAGE) gel. Transfer to a nitrocellulose membrane and perform Western blot analysis to the transgene. Include an antibody to a cellular protein (e.g., actin) as a loading control.

Fig. 1. Surgical procedures for intramyocardial injection. (**A**) Technique for immobilization of rats for anesthetic injection. (**B**) Position of animal for intubation. (**C**) Position of animal for left posterolateral thoracotomy. (**D**) Skin incision of animal approx 1 fingerbreadth above the xiphoid cartilage. (**E**) Serratus anterior muscle as a landmark for entering the appropriate intercostal space. (**F**) Left ventricular apex of the animal's heart shown after entering the left thoracic cavity.

7 days   **30 days**

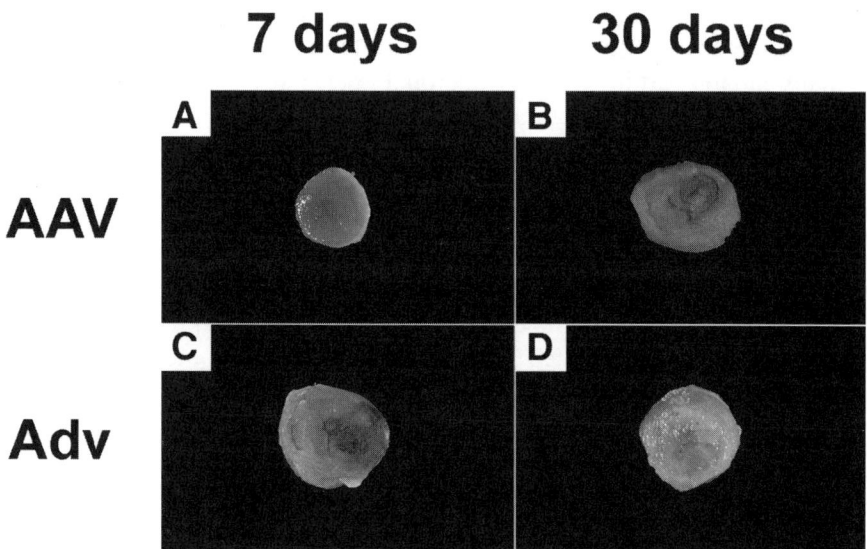

Fig. 2. Transduction of rat hearts by recombinant AAV and Ad vectors. Hearts were harvested from rats after intramyocardial injection of rAAV.LacZ or Ad.LacZ at 7 and 30 d postinjection. Gross cross-section photographs are shown of hearts stained for β-galactosidase with X-gal.

18. For RNA analysis, wash heart three times with cold PBS after explanting and store at −80°C until further use. Weigh out approx 300 mg of tissue from the apex of the heart. Homogenize tissue with a Polytron Tissue Homogenizer in 6 mL of cold RNAzol™ B and continue RNA isolation according to the manufacturer's protocol. Dissolve the RNA in 50 µL of DEPC-treated water and determine the concentration by spectrophotometic quantitation. Run approx 20 µg of total RNA from each sample on a 1.2% agarose-formaldehyde gel. Transfer RNA to a nylon membrane according to the Turboblotter Transfer System protocol for 6 h then immobilize the RNA by UV cross-linking. Prehybridization/hybridization should be carried out using standard Northern protocols. Perform Northerns for the transgene as well as an internal control (e.g., actin).

4. Notes

1. There are multiple ways of securing the rat for ip injection of anesthetics. The best and least traumatic way is to use a towel to cover the animal's entire body while holding its tail gently. Hold the rat along with the towel cov-

ering its eyes in order to calm the animal (**Fig. 1A**). The towel should also protect the researcher from being bitten. Hold the animal's head between the index and the middle fingers securely. Use the thumb and the other fingers to securely hold the animal's front legs. Release the tail and hold the animal standing upright on its hind legs on the table with its abdomen showing and its eyes covered with the towel. At this point, the animal should not be struggling at all. The rat's abdomen should be easily accessible in this forced upright standing position. The nondominant hand should be holding the animal in such an upright position while the dominant hand is used for ip injection directly in the middle of the abdomen. Take care to avoid injury to the liver and/or the bladder by aspirating the syringe prior to injection of its contents.

2. After the animal is appropriately sedated with anesthesia, the intubation should proceed as soon as possible because the anesthetics can cause respiratory depression quite rapidly. During the insertion of the guide-wire beyond the vocal cords, multiple failed attempts may cause increasing laryngeal secretions and bleeding, making visualization of the vocal cords more difficult. A small cotton swab may be used to absorb these fluids and provide better visual identification of the vocal cords for further attempts.

3. When the intubation tube is in the trachea, the chest cavity should rise and fall with each breath given by the ventilator. One finger should be placed on the animal's upper abdomen to feel for possible gastrointestinal intubation. If the esophagus is intubated instead of the trachea, one should be able to feel the stomach being inflated rapidly and progressively. If this happens, the ventilator should be disconnected IMMEDIATELY, and another attempt to intubate the trachea should be made promptly.

4. Prior to entering the thoracic cavity, it may be useful to decrease the tidal volume and increase the respiratory rate on the ventilator in order to obtain better visual exposure of the heart.

5. We have found that the absolute minimum concentration of AAV to achieve successful in vivo delivery to the rodent heart is 1×10^{11} particles/mL. Concentrations of AAV lower than 1×10^{11} particles/mL will not yield adequate gene expression.

6. When injecting the virus into the left ventricular apex of the myocardium, the needle should be approx 0.5–1.0 mm beneath the myocardium at a very shallow angle in order to avoid injecting the virus into the left ventricular cavity. A way to ensure proper intra-myocardial injection is to visualize the brief blanching (whitening) of the myocardium during the injections. Blanching of the myocardium simply implies that the myocardial muscle wall is being compressed briefly by the gene delivery. If a viral injection occurs without blanching, the virus is simply being introduced into the left ventricular cav-

ity and directly into the systemic circulation with each stroke of the heart. Therefore, it is very important to visualize the blanching of the myocardium during viral injections. Hemostasis is achieved by packing the thoracic cavity gently with some 2 × 2 gauze.

7. Battery operated electro-cautery can be used to achieve hemostasis if necessary. However, it is to be used with extreme caution because the 70% EtOH used to cleanse the animals skin is extremely flammable.

8. Ten percent of the animals usually die within the first 48 h postoperatively from anesthesia complication, aspiration pneumonia, and cardiac arrhythmia. These complications are usually hard to avoid. Survival rate for the animals following this type of procedure is usually 90% even in the most experienced hands.

9. Both intramyocardial injection and intracoronary perfusion methods of gene delivery require a thoracotomy incision to ensure adequate gene transfer. The steps are similar to direct intramyocardial injection method with a few exceptions. Rats are anesthetized in a similar fashion as described in **steps 1–4, Subheading 3**. Intraoperatively, the aortic arch is circumscribed with a suture but not tied. A small 27 G angiocatheter is inserted into the left ventricle at the apex of the heart. The angiocatheter is gently threaded rostrally just past the aortic valve, so that the tip of the angiocatheter is caudal to the suture circumscribing the aortic arch. The aortic suture is briefly tied down when the virus is infused into the angiocatheter to ensure the virus is directed into the coronary circulation and not the systemic circulation. When the infusion is completed, the aortic suture and angiocatheter are removed. Hemostasis is achieved by applying a 2 × 2 gauze on the left ventricular apex, the site where the angiocatheter entered the left ventricular space. The chest is closed with sutures after evacuating the pleural air as described in **steps 10–12, Subheading 3**. This method has been used to transduce murine and porcine myocardium using rAAV vectors *(6,10,16,17)*.

10. For rAAV the earliest gene expression will be detectable at 3–4 wk postdelivery. The duration of gene expression varies from 3 to 12 mo according to the literature *(6–12)*. In comparison, injection with Ad.LacZ results in high levels of gene expression within the first week, but falls off to undetectable levels by 1 mo postinjection (*see* **Fig. 2**).

11. There is little to no inflammatory effect on the myocardium after rAAV injection.

References

1. Nabel, E. G. (1995) Gene therapy for cardiovascular disease. *Circulation* **91(2)**, 541–548.

2. Alexander, M. Y., Webster, K. A., McDonald, P. H., and Prentice, H. M. (1999). Gene transfer and models of gene therapy for the myocardium. *Clin. Exp. Pharmacol. Physiol.* **26(9)**, 661–668.

3. Guzman, R. J., Lemarchand, P., Crystal, R. G., Epstein, S. E., and Finkel, T. (1993) Efficient gene transfer into myocardium by direct injection of adenovirus vectors. *Circ. Res.* **73(6)**, 1202–1207.

4. French, B. A., Mazur, W., Geske, R. S., and Bolli, R. (1994) Direct in vivo gene transfer into porcine myocardium using replication-deficient adenoviral vectors. *Circulation* **90(5)**, 2414–2424.

5. Barr, E., Carroll, J., Kalynych, A. M., Tripathy, S. K., Kozarsky, K., Wilson, J. M. and Leiden, J. M. (1994) Efficient catheter-mediated gene transfer into the heart using replication-defective adenovirus. *Gene. Ther.* **1(1)**, 51–58.

6. Svensson, E. C., Marshall, D. J., Woodard, K., Lin, H., Jiang, F., Chu, L., and Leiden, J. M. (1999) Efficient and stable transduction of cardiomyocytes after intramyocardial injection or intracoronary perfusion with recombinant adeno-associated virus vectors. *Circulation* **99(2)**, 201–205.

7. Maeda, Y., Ikeda, U., Shimpo, M., Ueno, S., Ogasawara, Y., Urabe, M., et al. (1998) Efficient gene transfer into cardiac myocytes using adeno-associated virus (AAV) vectors. *J. Mol. Cell Cardiol.* **30(7)**, 1341–1348.

8. Su, H., Lu, R. and Kan, Y. W. (2000) Adeno-associated viral vector-mediated vascular endothelial growth factor gene transfer induces neovascular formation in ischemic heart. *Proc. Natl. Acad. Sci USA* **97(25)**, 13801–13806.

9. Wright, M. J., Wightman, L. M., Lilley, C., de Alwis, M., Hart, S. L., Miller, A., et al. (2001) In vivo myocardial gene transfer: optimization, evaluation and direct comparison of gene transfer vectors. *Basic Res. Cardiol.* **96(3)**, 227–236.

10. Kaplitt, M. G., Xiao, X., Samulski, R. J., Li, J., Ojamaa, K., Klein, I. L., et al. (1996) Long-term gene transfer in porcine myocardium after coronary infusion of an adeno-associated virus vector. *Ann. Thorac. Surg.* **62(6)**, 1669–1676.

11. Xiao, X., Li, J., and Samulski, R. J. (1998) Production of high-titer recombinant adeno-associated virus vectors in the absence of helper adenovirus. *J. Virol.* **72(3)**, 2224–2232.

12. Zolotukhin, S., Byrne, B. J., Mason, E., Zolotukhin, I., Potter, M., Chesnut, K., et al. (1999) Recombinant adeno-associated virus purification using novel methods improves infectious titer and yield. *Gene Ther.* **6(6)**, 973–985.

13. Clark, K. R., Liu, X., McGrath, J. P., and Johnson, P. R. (1999) Highly purified recombinant adeno-associated virus vectors are biologically active and free of detectable helper and wild-type viruses. *Hum. Gene Ther.* **10(6)**, 1031–1039.

14. Hauswirth, W. W., Lewin, A. S., Zolotukhin, S., and Muzyczka, N. (2000) Production and purification of recombinant adeno-associated virus. *Methods Enzymol.* **316**, 743–761.

15. Wixson, S. K. and Smiler, K. L. (1997) Anesthesia and analgesia in rodents, in *Anesthesia and Analgesia in Laboratory Animals* (Kohn, D. F., Wixson, S. K., White, W. J., and Benson, G. J., eds.), Academic Press, New York, NY, pp. 174–175.

16. Lai, N. C., Roth, D. M., Gao, M. H., Fine, S., Head, B. P., Zhu, J., et al. (2000) Intracoronary delivery of adenovirus encoding adenylyl cyclase VI increases left ventricular function and cAMP-generating capacity. *Circulation* **102(19)**, 2396–2401.
17. Ito, B. R., Roth, D. M., Chenoweth, D. E., Lefer, A. M., and Engler, R. L. (1989) Thromboxane is produced in response to intracoronary infusions of complement C5a in pigs. Cyclooxygenase blockade does not reduce the myocardial ischemia and leukocyte accumulation. *Circ. Res.* **65(5)**, 1220–1232.

16

Gene Delivery to the Mouse Brain
with Adeno-Associated Virus

Marco A. Passini, Deborah J. Watson, and John H. Wolfe

1. Introduction

The efficient transduction of postmitotic cells by adeno-associated virus (AAV) makes it an excellent vector to deliver marker, functional, or therapeutic genes to the mammalian brain. An attractive feature of AAV is that all the viral-coding sequences are removed when engineering the recombinant genome, thereby limiting the extent of cell toxicity and immune response that are often associated with viral gene transcription *(1)*.

Of the seven described AAV serotypes, AAV serotype-2 (AAV2) is the most studied gene-transfer vehicle for in the mammalian brain. A feature of AAV2 transduction in the brain is that the vector remains confined to the injection site and predominately infects neurons rather than glia *(2–8)*. The limited diffusion of AAV2 vectors is beneficial for controlled gene delivery. For instance, targeting therapeutic genes only to brain structures showing pathology would eliminate complications associated with vector diffusion and subsequent expression in healthy structures, and is an important consideration when designing treatment strategies for localized neurodegenerative diseases. The same is true for other experimental paradigms, such as investigating the function of genes in specific brain structures or using marker genes in tract-tracing experiments. Although AAV2 vectors were shown to remain predominately at the injection site, one study demonstrated that the vector itself may undergo axonal transport in inter-regional systems *(9)*.

From: *Methods in Molecular Biology, vol. 246:*
Gene Delivery to Mammalian Cells: Vol. 2: Viral Gene Transfer Techniques
Edited by: W. C. Heiser © Humana Press Inc., Totowa, NJ

The seven AAV serotypes differ from one another by their capsid proteins *(10–14)*. The fundamental basis for viral infection involves the interaction between the capsid proteins with molecules on the surface of the host cell. Thus, the different AAV serotypes target different cell-surface receptors for binding, which may lead to complementary and unique transduction patterns in the central nervous system (CNS). This was illustrated by the finding that AAV2 and AAV5 use different attachment receptors on cell surfaces *(15–17)*, which explains the different transduction patterns observed for both serotypes in the brain *(7,18)* and retina *(19)*, as well as lung *(20)* and muscle *(21)*. Furthermore, AAV5 was shown to diffuse further through the mouse brain parenchyma compared to AAV2 *(7)*. Although the attachment receptors for the other AAV serotypes have not been identified, AAV4 shows a strong preference for transducing ependymal cells of the brain rather than the underlying parenchyma, and also transduces the retina very efficiently following subretinal injection into adult rats *(19)*. AAV1 and has yet to be examined in the brain, but has been shown to transduce the retina with moderate efficiency *(19)*.

Successful gene transfer to the CNS must also rely on activity from the promoter supporting expression. Previous reports demonstrated that the duration of AAV vector expression declined in the adult rodent brain when using viral promoters *(3,22–26)*. This could be owing to the inability of certain brain regions to support transcription following the methylation of viral regulatory sequences, as observed with the cytomegalovirus (CMV) promoter *(27)*. The downregulation of CMV was attenuated when the chicken β-actin enhancer was attached to CMV to produce a hybrid promoter *(5)*. When a pure mammalian regulatory promoter was engineered into an AAV genomic vector, such as the neuronal-specific enolase and platelet-derived growth factor promoters, long-term expression was observed in the CNS *(23–26,28)*. However, the relatively large size of both the pure or hybrid promoters is a disadvantage when engineering recombinant genomes owing to size limitations in AAV packaging reactions *(29)*. Our laboratory has used a small, eukaryotic housekeeping promoter (human β-glucuronidase) *(30,31)* capable of long-term expression in multiple brain structures *(8,32)*. The 378 base-pair human β-glucuronidase promoter would therefore provide a distinct advantage in somatic gene-transfer experiments involving large cDNAs.

Gene transfer with AAV2 to the developing brain has also been performed in newborn mice *(32–34)*. The smaller size of the neonatal brain is better suited for achieving global delivery of the transferred gene, because intraparenchymal injection of a volume of AAV2 would cover a larger area of the developing brain compared with a similar injected volume into an adult brain. Furthermore, intraventricular injection of AAV2 results in widespread brain transduction when the vector is administered into the cerebrospinal fluid (CSF) during neonatal development *(32)*, whereas similar intraventricular injection into the adult brain results in a very limited transduction pattern *(7)*. In addition to achieving widespread gene delivery, administration of AAV2 vectors during neonatal development may provide a better clinical outcome for a wide variety of inherited CNS diseases by blocking the onset of pathology or by reducing the severity of the disease.

In this chapter, we describe how to deliver genes to the adult brain. Any AAV serotype can be used with this procedure. The protocol for neonatal gene delivery is found elsewhere and will not be covered in this chapter *(32–35)*. The information obtained in this chapter should aid in AAV-mediated somatic gene-transfer experiments to the adult mammalian brain.

2. Materials

2.1. AAV Vectors for Brain Injections

1. AAV virions generated by triple transfection *(36,37)* work well in gene-transfer experiments to the mammalian brain (*see* **Note 1**). Titers ranging from 10^{12}–10^{13} genomic particles/mL should be used to ensure high in vivo transduction efficiency.
2. Sterile, autoclaved glycerol should be added to the viral stock solution to a final concentration of 5%. Aliquots ranging from 20–50 µL should be made and frozen at $-80°C$ to avoid multiple freeze-thaw cycles.
3. On the day of injection, remove tube(s) from $-80°C$ and thaw on ice. Load the viral solution into the Hamilton syringe when ready to begin surgery.

2.2. Brain Surgery

1. Anesthesia. Ketamine (100 mg/mL, Ketaset, Fort Dodge Animal Health, Fort Dodge, Iowa) and Xylazine (20 mg/mL, Phoenix Pharmaceutical Inc., St. Joseph, MO), 0.9% sterile saline, disposable sterile syringes with 27 G needles (Fisher, Pittsburgh, PA).
2. Surgical Preparation. Small electric shaver and disposable scalpel (Webster Veterinary Supply, Sterling, MA), ethanol swabs (Fisher), stereotaxic frame (Kopf Small Animal Stereotaxic Frame Model 900, David Kopf Instruments, Tujunga, CA).
3. Drilling and Injection. Drill and drill bits (Foredom FM3545 control with MH145 hand-piece, Kopf), Hamilton syringe with 26–30 G needle for vector injections (Fisher), and pump (Stoeltling 310; Stoelting Co, Wood Dale, IL).
4. Manual. Manuals and review articles for stereotaxic surgery are helpful in illustrating the steps of surgery, and are available *(38,39)*.
5. Atlas. A mouse brain atlas will be helpful in determining the stereotaxic coordinates for the brain structures of interest *(40)*.

6. Postoperative care. Absorbable suture thread (Vicryl with C3 needle, Ethicon, Inc. Somerville, NJ) and heating pad (Gaymar Industries, Orchard Park, NY).

2.3. Transcardial Perfusion and Dissection of Brain from Skull

1. Perfusion. Dissecting tray, sterile 1X PBS, fixative (4% paraformaldehyde/1X PBS, pH, 7.4), large scissors for cutting the skin, fine scissors for cutting the abdominal wall and diaphragm, forceps, butterfly needle (23 × 3/4 in infusion set, Abbott Laboratories no. 4565, Chicago, IL), three-way valve, infusion pump, and tubing.
2. Dissection. Bone crusher (RS8282, Roboz Surgical Instrument Company, Inc., Rockville, MD).

2.4. Preparation of Brain Tissue for Cyrosectioning

1. 30% Sucrose, 100% optimal cutting temperature (OCT) solution (Sakura Finetek USA, Torrance, CA), disposable plastic molds (Polysciences, Inc., Warrington, PA), slides (Fisher).

2.5. Reagents for Detection of β-Glucuronidase on Frozen Tissue Sections

1. Solution 1 (Chloral hydrate formalin fixative): 1% chloral hydrate in 20% (v/v) neutral-buffered formalin in distilled water: Mix 0.1 g chloral hydrate (Sigma C8383, St. Louis, MO), 20 mL neutral-buffered formalin, 80 mL deionized water. Chloral hydrate is extremely hygroscopic. A DEA license is required for chloral hydrate purchase and use. Store at 4°C.
2. Solution 2: Chloral-formal-acetone fixative). Mix seven parts acetone to three parts solution 1.The solution will be cloudy when first mixed. Put at 37°C until clear, then filter through 0.45 µm filter. Store at 4°C.
3. Solution 3: 0.2 M sodium acetate buffer. Mix 20 mL of 1 N HCl, 40 mL of 1 M sodium acetate and 140 mL deionized water. Adjust pH to 4.5. Store at 4°C.
   a. Solution 3a: 0.05 M sodium acetate, pH 4.5. Dilute solution 3 at 1:4 with deionized water and adjust the pH to 4.5. Store at 4°C.
   b. Solution 3b: 0.05 M sodium acetate, pH 5.2. Dilute solution 3 at 1:4 with deionized water and adjust the pH to 5.2. Store at 4°C.
4. Solution 4 (0.25 mM napthol AS-BI β-D-glucuronide in 0.05 M sodium acetate buffer, pH 4.5): Add 13.7 mg napthol AS-BI β-D-glucuronide (Sigma N1875) to 100 mL of solution 3a. Dissolve with stirring at 37°C but do not raise the temperature above 50°C. Store at 4°C.
5. Solution 5: 0.25 mM napthol AS-BI β-D-glucuronide in 0.05 M sodium acetate buffer, pH 5.2. Add 13.7 mg napthol AS-BI β-D-glucuronide (Sigma

N1875) to 100 mL of solution 3b. Dissolve with stirring at 37°C but do not raise the temperature above 50°C. Store at 4°C.

6. Components of the substrate solution.

    a. Solution 6a (4% pararosaniline chloride). Dissolve 0.04 g pararosaniline chloride (Sigma P3750) in 1.0 mL 2 *N* HCl. Make 1 d before use and vortex well to mix. Store at 4°C for up to 3 wk.

    b. Solution 6b (4% sodium nitrite). Dissolve 0.04 g sodium nitrite (Fisher S347-250) in 1 mL deionized water. Make 1 d before use and vortex well to mix. Store at 4°C for up to 1 wk.

7. To make the filtered substrate solution (prepare 10–30 min before staining): Mix equal volumes of solution 6a and 6b (add the pararosaniline to the sodium nitrite to minimize precipitation). Add 1 volume of the mixture to 500 volumes of solution 5 (e.g., 20 µL of the 6a/6b mixture to 10 mL of solution 5). Filter through a 0.45-µm filter before use.

8. 1% aqueous methyl green extracted three to four times with $CHCl_3$.

9. Aqua Polymount (Polysciences, Inc.).

3. Methods

3.1. Brain Surgery

1. Weigh the adult mouse to determine the amount of anesthesia to inject into the intraperitoneal cavity (*see* **Note 2**). Use a dose of 100 mg/kg of ketamine and 5 mg/kg of Xylazine. When the mouse does not respond to pinching of the footpad, it is anesthetized.

2. Shave the head with an electric razor and swab the scalp with 70% ethanol. Make a straight incision along the midline of the head to expose the underlying skull.

3. Mount the mouse on a stereotaxic frame, and make sure that the head is level. Ear bars should also be used to fix the head in place to maintain proper 3-D alignment for the stereotaxic coordinates.

4. Align the drill directly over bregma. This morphological landmark is the intersection between the coronal and sagital suture lines. Using the coordinates of bregma as a reference point, move the drill using the anterior–posterior and medial–lateral knobs to the desired coordinates of the brain structure of interest.

5. Carefully move the skin to one side with a sterile cotton swab. Activate the drill and carefully lower it using the dorsal–ventral knob. As the drill touches the outside of the skull, carefully continue turning the knob so that the drill presses onto the skull with more pressure. Raise the drill on a consistent basis to determine if complete penetration of the skull occurred. Once the drill-hole is made, raise the drill and remove it from the stereo-

taxic frame and replace it with a Hamilton syringe that contains the AAV2 vector solution.

6. Lower the Hamilton syringe into the drill-hole so that the tip of the needle is approx 0.5 mm below the top base of the skull. Consider this location as the top of the pial surface of brain.

7. Lower the needle into the parenchyma to the desired depth (*see* **Note 3**). After reaching the target site, inject 1–2 µL of the AAV solution at a rate of 0.2 µL/min.

8. Once the injection is complete, leave the needle in place for 1–2 min to ensure complete absorption of the virus to the surrounding area. Then, raise the needle slowly to prevent back-flow of the virus out of the injection hole.

9. Remove the mouse from the stereotaxic frame and suture the scalp. Place the cage on a heating pad and cover the cage floor with a paper towel. Place the mouse on the paper towel and wet a small handful of food and place near the mouse. The mouse will lick the water off the food when it begins to recover from the surgery to prevent becoming dehydrated. The mouse should wake up from surgery after 30 min and be groggy in the cage between 1–4 h postoperation. Once the mouse is moving, place the cage back in the ventilated rack.

3.2. Transcardial Perfusion and Dissection of Brain from Skull

Transcardial perfusion results in fixation of molecules in the brain and overall maintenance of morphology.

1. Deeply anesthetize the mice with a dose of 0.1 mL of 5.0 mg of ketamine and 1.0 mg of xylazine.

2. Carefully make an incision along the belly of the mouse that extends to the top of the chest. Hold the abdominal wall with forceps and cut the wall with fine scissors. Do not cut the organs below.

3. When the ribcage is reached, cut the diaphragm on the left side to enter the chest cavity. Do not initially cut the right side of the diaphragm because you may puncture the heart.

4. The heart should be beating. Locate and cut the right atrium to allow for the blood to drain during the perfusion.

5. Holding the heart with forceps, locate the left ventricle and carefully insert a butterfly needle and perfuse with 1X PBS. Do not insert the needle too deep. The liver should turn light in color. Perfuse 10–15 mL of 1X PBS over a 5-min period.

6. After perfusing with 1X PBS, perfuse with approx 30–35 mL of fixative solution over a 20-min span (*see* **Note 4**). The perfusion should be done in the hood to avoid inhaling the fixative. Remove the needle when completed.

7. Cut the skin over the head and expose the skull. Using a bone crusher, make a small break at the back of the skull. This should expose the medulla. Carefully, work your away from the back of the brain to the front with the bone crushers. Forceps may be used to help remove the bone as you are moving around the skull. Remove the brain and place it in fresh fixative solution and incubate overnight at 4°C.

3.3. Preparation of Brain for Cyrosectioning

1. Transfer the brain to a fresh tube containing 30% sucrose at 4°C (*see* **Note 5**). The brain will initially float. Incubate the brain until it sinks, which takes approx 24–36 h.
2. Remove the sunken brain from the 30% sucrose solution and transfer it to a plastic mold containing 100% OCT. Let the brain infiltrate in the 100% OCT at room temperature for 1 h.
3. Make a flat bed of crushed dry-ice (−80°C) (*see* **Note 6**). In the plastic mold, stand the brain on the olfactory bulb for coronal, on its side for sagital, or flat for horizontal sections. The 100% OCT is very viscous and the brain will maintain its correct suspended position for a small time period before tipping over. Use forceps to help the brain maintain the desired orientation if it begins to tip.
4. Place the plastic mold on top of the dry-ice and keep monitoring the brain so that it doesn't tip before freezing in the block. Use forceps to guide it back into the proper orientation. After the block is completely frozen, transfer the block to −80°C for storage.
5. On the night before cyrosectioning, transfer the block to −20°C to equilibrate to the cutting temperature. Using a cryotome, cut sections ranging from 5–20 µm in thickness.
6. Slides designated for β-galactosidase or other enzyme staining reactions, or immuno-histochemistry should be stored at −20°C. Slides designated for *in situ* hybridization should be stored at −80°C. Protocols pertaining to the analysis of gene-transfer experiments to the brain for β-galactosidase *(41)* and β-glucuronidase *(6,42)* enzyme staining reactions, immunohistochemistry *(3,22,23)*, and *in situ* hybridization *(32,43,44)* are available.

3.4. Analysis of β-Glucuronidase Expression on Frozen Tissue Sections

1. Bring the slides to room temperature (20 min) and draw a ring around the sections with a hydrophobic slide marker (such as PAP pen, Kiyota Intl. Inc).
2. Immerse the sections in solution 2 for 30 min at 4°C.
3. Wash the slides in three changes of solution 3a. It is essential to rehydrate the slides well. If the slides are in a rack, 3X 10 min incubations with agitation at 4°C will suffice.

4. Incubate the slides in solution 4 for 4 h to overnight at 4°C.

5. Remove the slides, aspirate the remaining fluid, and remark the hydrophobic well if necessary.

6. Place the slides horizontally in a flat tray and add the filtered substrate solution to cover the sections.

7. Incubate the trays horizontally overnight in a humidified 37°C incubator.

8. Stop the reaction by rinsing the slides well in distilled water. Counterstain with 1% methyl green if desired.

9. Cells positive for β-glucuronidase activity will be visualized by a red precipitate. For long-term storage, mount a coverslip with Aqua Polymount and store at room temperature.

4. Notes

1. DNA harvested by Qiagen (Valencia, CA) maxi-prep is suitable to be used in triple transfection protocols when generating recombinant AAV virions.

2. Inhalant anesthesia can be substituted for injectable anesthesia in fragile mice strains. In this case, a precision isoflurane vaporizor should be used at a range of 1.5–3.0% (SurgiVet Inc., Waukesha, WI).

3. If injections are to be done in multiple structures along the dorsoventral axis, inject the ventralmost structure first and then raise the needle dorsally to the subsequent site(s).

4. Improper perfusion will result in the brain being malleable to touch when dissecting the tissue from the skull. Improper perfusion may also contribute to a lacy brain, which can interfere with the analysis.

5. Another major contributor to a lacy brain is insufficient cryoprotection. If one were to freeze the brain before it has completely sunken in the 30% sucrose solution, the nondisplaced water molecules would form ice crystals during the freezing process. The formation of these ice crystals would produce holes in the brain.

6. Isopentane-liquid nitrogen can be substituted for dry ice during the freezing process *(40)*.

Acknowledgments

This work was supported by NIH grants DK46637, NS38690, and DK63973 (JHW), institutional NRSA training grants in gene therapy (DK 7748) (MAP) and neurovirology (NS07180) (DJW), and an individual NRSA postdoctoral Fellowship (NS 11024) (to DJW).

References

1. Xiao, X., Li, J., McCown, T. J., and Samulski, R. J. (1997) Gene transfer by adenoassociated virus vectors into the central nervous system. *Exp. Neurol.* **144,** 113–124.

2. Kaplitt, M. G., Leone, P., Samulski, R. J., Xiao, X., Pfaff, D. W., O'Malley, K. L., and During, M. J. (1994) Long-term gene expression and phenotypic correction using adeno-associated virus vectors in the mammalian brain. *Nat. Genet.* **8,** 148–154.

3. Chamberlin, N. L., Du, B., de Lacalle, S., and Saper, C. B. (1998) Recombinant adeno-associated virus vector: use for transgene expression and anterograde tract tracing in the CNS. *Brain Res.* **793,** 169–175.

4. Bartlett, J. S., Samulski, R. J., and McCown, T. J. (1998) Selective and rapid uptake of adeno-associated virus type 2 in brain. *Hum. Gene Ther.* **9,** 1181–1186.

5. Mandel, R. J., Rendahl, K. G., Spratt, S. K., Snyder, R. O., Cohen, L. K., and Leff, S. E. (1998) Characterization of intrastriatal recombinant adeno-associated virus-mediated gene transfer of human tyrosine hydroxylase and human GTP-cyclohydrolase I in a rat model of parkinson's disease. *J. Neurosci.* **18,** 4271–4284.

6. Skorupa, A. F., Fisher, K. J., Wilson, J. M., Parente, M. K., and Wolfe, J. H. (1999) Sustained production of β-glucuronidase from localized sites after AAV vector gene transfer results in widespread distribution of enzyme and reversal of lysosomal storage lesions in a large volume of brain in mucopolysaccharidosis VII mice. *Exp. Neurol.* **160,** 17–27.

7. Davidson, B. L., Stein, C. S., Heth, J. A., Martins, I., Kotin, R. M., Derksen, T. A., et al. (2000) Recombinant adeno-associated virus type 2, 4, and 5 vectors: transduction of variant cell types and regions in the mammalian central nervous system. *Proc. Natl. Acad. Sci. USA* **97,** 3428–3432.

8. Passini, M. A., Lee, E. B., Heuer, G. G., and Wolfe, J. H. (2002) Distribution of a lysosomal enzyme in the adult brain by axonal transport and by cells of the rostral migratory stream. *J. Neurosci.* **22,** 6437–6446.

9. Kaspar, B. K., Erickson, D., Schaffer, D., Hinh, L., Gage, F. H., and Peterson, D. A. (2002) Targeted retrograde gene delivery for neuronal protection. *Mol. Ther.* **5,** 50–56.

10. Chiorini, J. A., Yang, L., Liu, Y., Safer, B., and Kotin, R. M. (1997) Cloning of adeno-associated virus type-4 (AAV4) and generation of recombinant AAV4 particles. *J. Virol.* **71,** 6823–6833.

11. Cliorini, J. A., Kim, F., Yang, L., and Kotin, R. M. (1999) Cloning and characterizing of adeno-associated virus type-5. *J. Virol.* **73,** 1309–1319.

12. Rutledge, E. A., Halbert, C. L., and Russell, D. W. (1998) Infectious clones and vectors derived from adeno-associated virus (AAV) serotypes other than AAV type 2. *J. Virol.* **72,** 309–319.

13. Bantel-Schaal, U., Delius, H., Schmidt, R., and Van Hausen, H. (1999) Human adeno-associated virus type 5 is only distantly related to other known primate helper-dependent parvovirus. *J. Virol.* **73,** 939–947.

14. Xiao, W., Chirmule, N., Berta, S. C., McCullough, B., Gao, G. P., and Wilson, J. M. (1999) Gene therapy based on adeno-associated virus type 1. *J. Virol.* **73,** 3994–4003.

15. Summerford, C. and Samulski, R. J. (1998) Membrane-associated heparin sulfate proteoglycan is a receptor for adeno-associated virus type 2 virions. *J. Virol.* **72,** 1438–1445.

16. Bartlett, J. S., Wilcher, R., and Samulski, R. J. (2000) Infectious entry pathway of adeno-associated virus and adeno-associated virus vectors. *J. Virol.* **74,** 2777–2785.

17. Walters, R. W., Yi, S. M. P., Keshavjee, S., Brown, K. E., Welsh, M. J., Chiorini, J. A., and Zabner, J. (2001) Binding of adeno-associated virus type 5 to 2,3 sialic acid is required for gene transfer. *J. Biol. Chem.* **276,** 20610–20616.

18. Alisky, J. M., Hughes, S. M., Sauter, S. L., Jolly, D., Dubensky, T. W., Staber, P. D., et al. (2000) Transduction of murine cerebellar neurons with recombinant FIV and AAV5 vectors. *Neuroreport* **1,** 2669–2673.

19. Rabinowitz, J. E., Rolling, F., Li, C., Conrath, H., Xiao, W., Xiao, A., and Samulski, R. J. (2002) Cross-packaging of a single adeno-associated virus (AAV) type 2 vector genome into multiple AAV serotypes enables transduction with broad specificity. *J. Virol.* **76,** 791–801.

20. Zabner, J., Seiler, M., Walters, R., Kotin, R., Fulgeras, W., Davidson, B. L., and Chiorini, J. A. (2000) Adeno-associated virus type 5 (AAV5) but not AAV2 binds to the apical surface of airway epithelia and facilitates gene transfer. *J. Virol.* **74,** 3852–3858.

21. Duan, D., Yan, Z., Yue, Y., Ding, W., and Engelhardt, J. F. (2001) Enhancement of muscle gene delivery with pseudotyped adeno-associated virus type 5 correlates with myoblast differentiation. *J. Virol.* **75,** 7662–7671.

22. McCown, T. J., Xiao, X., Li, J., Breese, G. R., and Samulski, R. J. (1996) Differential and persistent expression patterns of CNS gene transfer by an adeno-associated virus (AAV) vector. *Brain Res.* **713,** 99–107.

23. Klein, R. L., Meyer, E. M., Peel, A. L., Zolotukhin, S., Meyers, C., Muzyczka, N., and King, M. A. (1998) Neuron-specific transduction in the rat septohippocampal or nigrostriatal pathway by recombinant adeno-associated virus vector. *Exp. Neurol.* **150,** 183–194.

24. Klein, R. L., Mandal, R. J., and Muzyczka, N. (2000) Adeno-associated virus vector-mediated gene transfer to somatic cells in the central nervous system. *Adv. Virus Res.* **55,** 507–528.

25. Peel, A. L. and Klein, R. L. (2000) Adeno-associated virus vectors: activity and applications in the CNS. *J. Neurosci. Meth.* **98,** 95–104.

26. Xu, R., Janson, C. G., Mastakov, M., Lawlor, P., Young, D., Mouravlev, A., et al. (2001) Quantitative comparison of expression with adeno-associated virus (AAV-2) brain-specific gene cassettes. *Gene Ther.* **8,** 1323–1332.

27. Prosch, S. J., Stein, K., Staak, C., Liebenthal, H., Volk, H. D., and Kruger, D. H. (1996) Inactivation of the very strong HCMV immediate early promoter by DNA CpG methylation *in vitro. Biol. Chem. Hoppe. Seyler* **377,** 195–201.

28. Peel, A. L., Zolotukhin, S., Schrimsher, G. W., Muzyczka, N., and Reier, P. J. (1997) Efficient transduction of green fluorescent protein in spinal cord neurons using adeno-associated virus vectors containing cell type-specific promoters. *Gene Ther.* **4,** 16–24.

29. Dong, J. Y., Fan, P. D., and Frizzell, R. A. (1996) Quantitative analysis of the packaging capacity of recombinant AAV. *Hum. Gene Ther.* **7,** 2101–2112.

30. Shipley, J. M., Miller, R. D., Wu, B. M., Grubb, J. H., Christensen, S. G., Kyle, J. W., and Sly, W. S. (1991) Analysis of the 5′ flanking region of the human β-glucuronidase gene. *Genomics* **10,** 1009–1018.
31. Wolfe, J. H., Kyle, J. W., Sands, M. S., Sly, W. S., Markowitz, D. G., and Parente, M. K. (1995) High level expression and export of β-glucuronidase from murine mucopolysaccharidosis VII cells corrected by a double-copy retrovirus vector. *Gene Ther.* **2,** 70–78.
32. Passini, M. A. and Wolfe, J. H. (2001) Widespread gene delivery and structure-specific patterns of expression in the brain from an adeno-associated virus vector following intraventricular injections of neonatal mice. *J. Virol.* **75,** 12382–12392.
33. Elliger, S., Elliger, C., Aguilar, C., Raju, N., and Watson, G. (1999) Elimination of lysosomal storage in brains of MPS VII mice treated with intrathecal administration of an adeno-associated virus vector. *Gene Ther.* **6,** 1175–1178.
34. Frisella, W. A., O'Connor, L. H., Vogler, C. A., Roberts, M., Walkley, S., Levy, B., et al. (2001) Intracranial injection of recombinant adeno-associated virus improves cognitive function in a murine model of mucopolysaccharidosis type VII. *Mol. Ther.* **3,** 351–358.
35. Snyder, E. Y., Taylor, R. M., and Wolfe, J. H. (1995) Neural progenitor cell engraftment corrects lysosomal storage throughout the MPS VII mouse brain. *Nature* **374,** 367–370.
36. Fisher, K. J., Jooss, K., Alston, J., Yang, Y., Haecker, S. E., High, K., et al. (1997) Recombinant adeno-associated virus for muscle-directed gene therapy. *Nature Med.* **3,** 306–312.
37. Gao, G. P., Wilson, J. M., and Wivel, N. A. (2000) Production of recombinant adeno-associated virus. *Adv. Virus Res.* **55,** 529–543.
38. Cooley, R. K. and Vanderwold, C. H. (1990) *Stereotaxic Surgery in the Rat: a Photographic Series.* A.J. Kirby Co., London, Ontario.
39. Brooks, A. I., Halterman, M., Chadwick, C., Davidson, B., Haak-Frendscho, M., Radel, C., et al. (1998) Reproducible and efficient murine CNS gene delivery using a microprocessor-controlled injection. *J. Neurosci. Methods* **80,** 137–147.
40. Franklin, K. B. J. and Paxinos, G. (1997) *The Mouse Brain: in Stereotaxic Coordinates.* Academic Press, CA.
41. Watson, D. J., Kobinger, G. P., Passini, M. A., Wilson, J. M., and Wolfe, J. H. (2002) Transduction patterns in the mouse central nervous system by lentivirus vectors pseudotyped with envelope proteins from vesicular stomatitis virus, ebola virus, mikola virus, lymphocytic choriomeningitis virus, or murine leukemia virus. *Mol. Ther.* **5,** 528–537.
42. Wolfe, J. H. and Sands, M. S. (1996) Murine mucopolysaccharidosis type VII: a model for somatic gene therapy of the central nervous system, in *Protocols for Gene Transfer in Neuroscience: Towards Gene Therapy of Neurological Disorders* (Lowenstein, P. R. and Enquist, L.W., eds.) John Wiley & Sons Ltd., London, pp. 263–274.

43. Passini, M. A., Levine, E. M., Canger, A. K., Raymond, P. A., and Schechter, N. (1997) Vsx-1 and Vsx-2: differential expression of two paired-like homeobox genes during zebrafish and goldfish retinogenesis. *J. Comp. Neurol.* **388,** 495–505

44. Barthel, L. K. and Raymond, P. A. (2000) *In situ* hybridization studies of retinal neurons. *Methods Enzymol.* **316,** 579–590.

17

Delivery of DNA to Tumor Cells In Vivo Using Adeno-Associated Virus

Selvarangan Ponnazhagan and Frank Hoover

1. Introduction

The number of published studies on transduction of tumor cells in vivo using recombinant adeno-associated virus (rAAV) vectors is very limited compared with those that have been published on targeting normal cells. A major reason for this can be attributed to the biology of the vector itself. AAV, being a nonpathogenic vector capable of providing transgene integration and long-term expression, is ideally suited for the correction of metabolic defects either to replace a defective protein/enzyme or to elevate their otherwise suboptimal levels in the system. However, increased understanding of both the biology of tumor progression and potential utility of AAV-based vectors suggests that this vector can also be wisely used for cancer gene therapy.

1.1. Possible Approaches for Cancer Gene Therapy Using AAV Vectors

The anti-oncogenic properties of wild-type AAV (wtAAV) have been observed long before the realization of the potential of this vector for gene therapy *(1)*. Until recently, this property was attributed to the nonstructural protein of AAV, which is also toxic to cells *(2)*. However, in a recent report, Raj et al. described selective killing of tumor cells that contain mutant p53 through initiation of apoptotic signals following the introduction of single-stranded AAV genome. This suggested a new approach to gene therapy against tumors that are positive for mutant p53 expression *(3)*. Although stable integration and long-

From: *Methods in Molecular Biology, vol. 246:*
Gene Delivery to Mammalian Cells: Vol. 2: Viral Gene Transfer Techniques
Edited by: W. C. Heiser © Humana Press Inc., Totowa, NJ

term expression capabilities of AAV vectors may not be a desirable/beneficial characteristic for therapy targeting the tumor cells directly, these features can be advantageous for targeting normal cells of the body that can exert influence on the control of tumor growth. Activation of the host immune system against tumor growth or arresting the growth of tumor neovasculature through anti-angiogenesis by targeting endothelial cells of tumor origin are some examples of targeting normal cells for which AAV can be used *(4)*. These strategies can be employed by transducing the vector either directly at the site of tumor or transducing normal tissue such as muscle and systemically expressing the therapeutic protein in secreted form. The main advantage of vector administration to tumor cells is the vicinity of target cells, which not only includes tumor cells themselves but also other target cells such as tumor endothelium in anti-angiogenic therapy or immunomodulation of the host system by cytokine gene transfer or chemoattracting antigen-presenting cells (APCs) to the site of tumor for the uptake, processing, and presentation of tumor antigens for tumor-specific T-cell response.

Prior to the application of intratumoral AAV administration in human patients, it is important to investigate the efficacy, delivery, and therapeutic impact in preclinical animal models. Thus, procedural details given in this chapter may benefit those in this field of study. A vast majority of preclinical evaluation of gene-therapy vectors by intratumoral administration have been done in subcutaneously grown syngeneic or xenograft cell lines. Hence, the present protocol will describe a method of intratumoral rAAV administration in tumors developed as xenografts in rodents. This protocol can, however, be modified to target tumors that are grown in other organs. All animal protocols must be approved by Institutional Animal Care and Use Facility or other appropriate local, state, and federal regulations.

2. Materials

1. rAAV containing a gene of interest, which can be a reporter gene such as β-galactosidase (β-gal), green fluorescent protein (GFP), or luciferase, or a therapeutic gene such as an enzyme, growth factor, or cytokine. Several methods are currently being used to prepare rAAV. The general principle of rAAV preparation involves rescue and packaging of mature virions from a plasmid vector containing the gene of interest cloned within the inverted terminal repeats (ITRs) of AAV. We currently use a discontinuous density-gradient centrifugation followed by heparin-agarose affinity column purification *(5)*. Quantitation of the vector can be done either by slot-blot analysis, by infectious center assay *(5)* or by real-time polymerase chain reaction (PCR) *(6)*. The vector is resuspended in PBS for intratumoral administra-

tion. The vector can be stored at 4°C up to 4 wk. For longer storage, we recommend storage at −80°C.

2. Tumor cell lines (Example: U-87MG, ATCC, Manassas, VA).
3. Microinjection pump Model UMP2-1 (World Precision Instruments Inc., Sarasota, FL).
4. Matrigel (Collaborative Research, Bedford, MA).
5. Nude rats (Rowett, Aberdeen, Scotland).
6. Stereotactic animal frame (Model 900, David Kopf Instruments, Tujunga, CA).
7. Bulldog clamps (Algaier Instruments GmbH, Tuttlingen, Germany).

3. Methods

3.1. Development of Subcutaneous Tumors

Human tumor cell lines can be implanted in 4–6-wk-old nude mice (athymic nude or SCID mice). Depending on the growth characteristics of a particular cell line, the number of cells to be implanted can vary. For example, when we administer between $0.5–1 \times 10^7$ SKOV3.ip1 cells (a human ovarian cancer cell line), palpable tumors appear in approx 10 d. If the growth rate of a tumor cell line is not known, we recommend conducting a pilot study with different amounts of cells before performing an experiment with the vector.

3.1.1. Preparation of Tumor Cells

1. Prepare a fresh culture of the cell line to be implanted using appropriate media and other additives such as serum, antibiotics, cytokines, and so on.
2. Expand cultures by splitting the cells before they reach confluence because overcrowding may retard subsequent growth characteristics.
3. On the day of tumor cell implantation, harvest the cells by using trypsin-ethylenediaminetetraacetic acid (EDTA) to detach the cells; use one-tenth the volume of trypsin-EDTA compared to the total volume of medium used to maintain the culture.
4. Incubate the cells at 37°C until they begin to round up (once again, the time of trypsin treatment varies between cell lines). Tap the vial to detach the cells. Do not incubate the cells in trypsin for an extended period of time because it will decrease the viability.
5. Stop trypsin activity by adding 10 volumes of medium containing 10% serum.
6. Collect the cells in sterile centrifuge tubes and pellet at $500g$ for 5 min.
7. Remove the supernatant and resuspend the pellet in 10–20 mL of PBS.
8. Pellet the cells as before.

9. Resuspend the cells in cold PBS or 50% Matrigel in PBS at a concentration of $1-2 \times 10^8$ cells/mL. Addition of matrigel allows the implanted cells to remain clustered at the site of injection and minimizes dispersion.

3.1.2. Implantation of Tumor Cells

1. Anesthetize the mice by subcutaneous administration of ketamine (100 mg/kg body weight) and xylazine (15 mg/kg body weight), abdominally.
2. Using a sterile alcohol swab, clean the area of the flank region.
3. Using a 1 cc syringe with a 25 G needle, deliver the required number of cells/tumor nodule subcutaneously (sc) in a volume of 50–100 µL: insert the needle carefully approx 1 cm under the skin between the skin and muscle, then inject the cells (*see* **Note 1**).
4. Observe the animal for proper recovery following the injection.
5. Once the tumor starts growing to a palpable size, make measurements at least three times every week. Measure the tumor volume with digital calipers for two-dimensional longest axis (L in mm) and shortest axis (W in mm) and calculate the tumor size using the formula:

$$\text{Volume in mm}^3 = (L \times W^2)/2$$

3.2. Intratumoral Injection of rAAV

The volume of virus to be injected depends on the size of tumor. More care is needed when the vector is delivered to a tumor of less than 100 mm³. Hence, we recommend that the volume of virus suspension be kept as small as possible, ideally around 25 µL for sc developed tumors.

1. Prepare the virus in PBS to be injected in a 0.5-cc insulin syringe with a 26 G. Use a separate syringe for injection into each tumor nodule.
2. Anesthetize the mice with ketamine and xylazine
3. Clean the area of tumor with a sterile alcohol swab.
4. Gently insert the needle into the tumor nodule so that the tip of the needle stays approximately at the center of the tumor. Use the caliper reading from each tumor to empirically determine the required distance to insert the needle.
5. Slowly inject the virus (up to 25 µL) with constant pressure (*see* **Note 2**).
6. Monitor the animals for recovery after the procedure and observe them daily for any complications. If they appear ill or fail to recover properly, they should be humanely euthanized.

3.3. Development of Intracranial Tumors

3.3.1. Implantation of Tumor Cells

A larger animal such as a rat will be easy to handle if the xenografts are developed in the brain. Xenografts into young male and female nude rats can be generated according to a protocol modified from Engebraaten et al. *(7)*. This procedure can be adapted for other cell lines to be grown as xenografts in rodent brain by varying the number of cells injected. Rats weighing 150–200 g should be kept on an ad libitum standard pellet diet with unlimited access to water and caged at constant temperature (21°C) and humidity in rooms with a 12 h light/12 h dark cycle.

1. Anesthetize rats by subcutaneous injection of Midazolam 0.2 g/100 g, Fentanyl Citrate 0.0126 g/100 g, and Fluanizone 0.4 g/100 g into the intraperitoneal cavity and mounted in a stereotactic frame. An additional injection of local anesthesia (1 mL of xylocaine [10 mg/mL] can be delivered sc to the incision area).
2. Make a 2.5-cm mid-sagittal incision on the skull spread the surrounding tissue, held open with bulldog clamps.
3. Position a surgical drill 1 mm posterior to the bregma suture and 3 mm to the right of the midline suture and carefully drill a small burrhole approx 2 mm in diameter. Take care to avoid rupturing the underlying blood vessels.
4. Implant 10–20 tumor spheroids using a 25-μL Hamilton syringe with a 22 G needle 2.5 mm deep into in the cerebral cortex (measured from the dura) using slow, steady hand pressure.
5. Close the wound with thread suture (Ethylon, polyamide 6).
6. Place the animals in a 35°C incubator until they regain consciousness and then return them to their cage. Observe daily.
7. To verify the presence of tumor, 2–4 wk after implantation, perform magnetic resonance imaging (MRI) using Siemens Magnetom Vision Plus 1.5T scanner and a small loop finger coil. Establish the location of the tumor for injection (*see* **Note 3**).

3.3.2. Intratumoral Injection of rAAV

We describe below intratumoral injection of rAAV in brain of nude or athymic rats, which can be modified according to the animal size.

1. Typically, animals should be anesthesized and secured in the stereotaxic frame and viral vector injected into the same coordinates after re-opening the original incision under anesthesia.
2. Inject the virus using a micro-pump connected to a 10-μL Hamilton syringe with a 26 G needle. Inject 10^8–10^9 viral particles over a period of 1 h in a

maximum volume of 8 μL *(8)*. Infusion rates can vary between 33 nL/min and 133 nL/min. It should be emphasized that intra-cranial injections by slow-push hand pressure using a Hamilton syringe give variable and inconsistent results in our experience.

3. As a positive control for infectious virus, inject small amounts of vector (1–2 μL) into the tibialis anterior muscle, in the hind leg of a different animal, through the skin using slow push hand pressure and a 10-μL Hamilton syringe. To aid in identification of the area, measurements can be made from below the knee to the approximate site of injection. Skeletal muscle is chosen because it is easily accessible and is permissive for most AAV serotypes examined (Hoover, unpublished results).

4. Following wound closure, place the animal in a 35°C incubator until it regains consciousness, then return it to its original cage.

5. If animals with tumor implants develop neurological symptoms resulting from tumor burden and tumor phenotype, they should be sacrificed immediately when symptoms such as weight loss, disorientation, lack of movement, or loss of appetite are observed.

6. Following intratumoral delivery of the vector by the aforementioned procedures, the animals can be used for monitoring the effects of gene transfer based on transgene expression analysis if a reporter gene is used, or other end-point measurements such as reduction in tumor size or delayed growth kinetics based on therapeutic efficacy of the transgene used.

3.4. Conclusion

The procedures described can be modified based on the application and growth characteristics of different tumor types and location. Because AAV-based vectors infect a variety of cells, it is possible to get "leakage" of the intratumorally administered vector to other organs as well as through systemic circulation from the tumor site. Thus, depending on the type of study, rAAV may be constructed with tumor-specific expression if transgene expression is deleterious in other organs. An additional concern is the variable infectivity of rAAV among different tumor cells. Thus, prior to conducting intratumoral studies with AAV vectors, it is important to determine the ability of AAV to infect the tumor of interest in vitro.

4. Notes

1. Care should be taken to avoid intra-dermal invasion of the needle. Slow release of the cells minimizes dispersion from the site of injection. Also, the use of matrigel helps in the retention of tumor cells in one region. To determine if implanted cells form tumors, MRI can be performed as described in **Subheading 3.3.1** and **Note 3**.

2. Hold the tumor in one hand so that the needle does not slip and the virus is injected into the tumor. Inject the virus slowly, avoiding any jerking motion.

3. For MRI, anesthetize the rat and place it in a polystyrene immobilizing tube. Obtain coronal T1 (TR 400 ms, TE 14 ms, slice thickness 2 mm, slice center distance 2 mm, totaling 13 coronal slices covering the forebrain) and coronal T2 (TR 4000 ms, TE 96 ms, slice thickness 2 mm, slice center distance 2 mm, totaling 19 coronal slices covering the forebrain) images prior to and 10 min following sc injection of contrast agent (1 mL of 0.5 mM Gadolinium).

Acknowledgments

We thank research support from the American Cancer Society-Institutional Research Grant 60-001-41, Career Development Award from NIH-SPORE grant in ovarian cancer 5 P50CA83591-02, NIH-R01CA90850, RO1-CA98817 grants from the and a Career Development Award from the U.S. Army, Department of Defense BC010494 and PC020372 to S.P. and from the Norwegian Cancer Society to F.H. Work performed by Dr. Per Øyvind Enger and Peter Csaba Huszthy to establish the parameters of the xenograft injections are greatly appreciated.

References

1. Cukor, G., Blacklow, N. R., Kibrick, S., and Swan, I. C. (1975) Effect of adeno-associated virus on cancer expression by herpesvirus-transformed hamster cells. *J. Natl. Cancer Inst.* **55,** 957–959.

2. Schmidt, M., Afione, S., and Kotin, R. M. (2000) Adeno-associated virus type 2 Rep78 induces apoptosis through caspase activation independently of p53. *J. Virol.* **74,** 9441–9450.

3. Raj, K., Ogston, P. and Beard, P. (2001) Virus-mediated killing of cells that lack p53 activity. *Nature* **412,** 914–917.

4. Ponnazhagan S., Curiel D. T., Shaw, D. R., Alvarez, R., and Siegal, G. P. (2001) Adeno-associated viral vectors for cancer gene therapy. *Cancer Res.* **61,** 6313–6321.

5. Zolotukhin, S., Byrne, B., Mason, E., Zolotukhin, I., Potter, M., Chesnut, K., et al. (1999) Recombinant adeno-associated virus purification using novel methods improves infectious titer and yield. *Gene Ther.* **6,** 973–985.

6. Du, B., Wu, P., Boldt-Houle, D. M., and Terwilliger, E. F. (1996) Efficient transduction of human neurons with an adeno-associated virus vector. *Gene Ther.* **3,** 254–261.

7. Engebraaten, O., Hjortland, G. O., Hirschberg, H., and Fodstad, O. (1999) Growth of precultured human glioma specimens in nude rat brain. *J. Neurosurg.* **90,** 125–132.

8. Enger, P. Ø., Thorsen, F., Lonning, P. E., Bjerkvig, R. and Hoover, F. (2002) Adeno-associated viruses penetrate solid human tumor tissue in vivo more effectively than

18

Gene Delivery to Human and Murine Primitive Hematopoietic Stem and Progenitor Cells by AAV2 Vectors

Arun Srivastava

1. Introduction

The adeno-associated virus 2 (AAV2) is known to possess a broad host-range that transcends the species barrier *(1)*. The broad host-range and nonpathogenic nature of AAV, coupled with its site-specificity and stable integration of the proviral genome, have led to the development of recombinant AAV vectors *(2)*. Recombinant AAV vectors have been shown to transduce certain cell types, such as muscle and brain, exceedingly well *(3–5)*. However, controversies exist with regard to the efficacy of AAV vectors in transducing human hematopoietic stem cells *(6)*. Whereas some investigators have concluded that AAV vectors do not transduce human hematopoietic cells at all *(7)*, other reports have documented the need for enormously high multiplicities of infection (MOIs) by AAV vectors for successful transduction *(8–13)*. Several groups, including my own, have reported successful transduction of these cells at relatively low MOIs *(14–17)*. Some of these controversies have been addressed in a recent review article *(18)*.

Here, I summarize our studies in which we have investigated the ability of AAV to infect primary human as well as murine hematopoietic stem and progenitor cells capable of multi-lineage differentiation. We have documented that CD34[+] cells from approx 50% of genetically unrelated donors show unde-

From: *Methods in Molecular Biology, vol. 246:*
Gene Delivery to Mammalian Cells: Vol. 2: Viral Gene Transfer Techniques
Edited by: W. C. Heiser © Humana Press Inc., Totowa, NJ

tectable levels of the transgene expression following infection with recombinant AAV vectors. We also reported that even among the positive donors for AAV infection, there is a wide variation in the transduction efficiency of CD34+ cells. However, significant enhancement in expression of the transduced gene can be achieved upon in vitro expansion of AAV-infected CD34+ cells from both positive and negative donors using recombinant hematopoietic growth factors. The observed level of the transduced gene expression also correlates with the extent of the recombinant AAV entry into these cells. These studies suggest that variability in transduction of human CD34+ hematopoietic cells results from variability in expression of the cellular receptor for AAV, the level of expression of which varies widely in CD34+ cells from different donors *(19)*. Similar results have been reported with human umbilical cord blood-derived CD34+ cells *(16)*.

In studies with mouse transplant model systems, we have also evaluated the efficiency of AAV-mediated transduction of murine hematopoietic stem and progenitor cells ex vivo, and the potential of expression of the transduced gene in vivo following transplantation into recipient mice. We have shown that AAV-mediated high-efficiency transduction of murine hematopoietic progenitor cells and expression of a transduced reporter gene in progeny cells in vivo without any observable cytotoxicity or deleterious effect is indeed possible *(20,21)*. We have also reported successful AAV-mediated stable transduction and high-efficiency, multi-lineage reconstitution, stable proviral integration, long-term, lineage-restricted transgene expression of a human β-globin gene in primary murine hematopoietic stem cells in vivo *(22)*. These results provide further support for the potential use of recombinant AAV vectors in stable transduction of hematopoietic stem cells.

Here, I provide the details of the protocols we have used in our studies. Most of them are previously published techniques with minor modifications.

2. Materials
2.1. Biochemicals, Cytokines, and Antibodies

1. Imagene green $C_{12}FDG$ β-gal substrate (Molecular Probes Inc., Eugene, OR).
2. Ficoll-Hypaque (Pharmacia, Piscataway, NJ).
3. Magnetic separation column, magnetic particles conjugated with anti-CD34 antibodies, and MACS buffer (0.5% bovine serum albumin [BSA] and 5 mM ethlylenediaminetetraacetic acid [EDTA] in 1X phosphate-buffered saline [PBS]) (Miltenyi Biotech, Sunnyvale, CA).
4. Recombinant human cytokines: interleukin (IL)-1, IL-6, IL-7, stem cell factor (SCF), granulocyte-colony stimulating factor (G-CSF), and erythropoietin (Epo) (Stem Cell Technologies, Vancouver, BC, Canada).

5. Rat anti-mouse Gr-1 (granulocytes), B220 (B lymphocytes), TER-119 (erythroid cells), and CD4/CD8 (T lymphocytes) monoclonal antibodies (MAbs) and their FITC conjugates (Pharmingen, San Diego, CA).
6. Goat anti-rat IgG magnetic microbeads (Miltenyi Biotec, Auburn, CA).
7. DNeasy Tissue Isolation Kit (Qiagen, Valencia, CA).
8. Proteinase K buffer: 10 mM Tris-HCl, ph 50 mM KCl, 2.5 mM MgCl$_2$, 0.5% Tween-20, and 100 µg/mL proteinase K.

2.2. Recombinant AAV Vector Stocks

1. Recombinant AAV vector containing either a reporter gene (e.g., bacterial β-galactosidase [β-gal]) or a therapeutic gene (e.g., human β-globin) (*see* **Note 1**).

2.3. Cells

1. Human bone marrow obtained from healthy volunteer donors after obtaining informed consent, approved by the Institutional Review Board for studies involving human subjects.
2. Murine bone marrow cells obtained from each femur and tibia from C57BL6 donor mice

2.4. Mice

1. Transplant donor C57BL6 or B6 Hbb[th/th] homozygous β-thalassemic mice (8–10 wk old) and recipient C57BL6 or B6.c-*kit*W[41/41] mice.

2.5. Equipment

1. Magnetic separation columns (Miltenyi Biotech).
2. Cs[137] irradiator (Nordion, Kanata, Ontario, Canada).

3. Methods
3.1. Isolation of Human Bone Marrow CD34+ Cells

1. Dilute human bone marrow samples immediately after the collection with an equal volume of Iscove's-modified Dulbecco's medium (IMDM) containing 20 U/mL heparin (*see* **Note 2**).
2. Isolate low-density bone marrow (LDBM) cells by Ficoll-Hypaque density centrifugation. Add 15 mL of Ficoll-Hypaque solution in a sterile 50-mL tube and gently layer 30 mL of diluted bone marrow cells on top.
3. Centrifuge at 400g for 30 min. Collect the band at the interface containing the LDBM cells. Wash these cells with sterile PBS and resuspend in 100 µL of IMDM.

4. Incubate 1×10^6 LDBM cells with 10 µg of anti-CD34 antibodies that are conjugated with magnetic particles in 100 µL on ice for 20 min in the dark.

5. Pass the labeled cells through a magnetic separation column. Allow the CD34$^-$ cells to flow through the column then elute magnet-bound CD34$^+$ cells using MACS buffer.

6. To check the purity, incubate 1×10^4 CD34$^+$ cells with 1 µg of fluorescein isothiocyanate (FITC)-labeled anti-CD34 antibodies in 10 µL on ice for 20 min in the dark, and analyze by fluorescence-activated cell sorting (FACS). Between 90 and 95% of the cells should be labeled with FITC.

7. Maintain CD34$^+$ cells in IMDM supplemented with 10% fetal calf serum (FCS) at a cell density of 5×10^4 cells/100 µL. Transduce the cells with AAV (*see* **Subheading 3.3.**) as soon as possible, ideally within 1 h.

3.2. Isolation of Murine Sca1$^+$, lin$^-$ Hematopoietic Cells

Murine stem cell antigen-1-expressing (Sca-1$^+$) and mature blood cell lineage nonexpressing (*lin$^-$*) cells have been shown to contain primitive hematopoietic stem cells capable of engraftment and long-term reconstitution of the hematopoietic system following transplantation in recipient mice. See Yoder et al. *(23,24)* for additional information on purification of Sca-1$^+$, *lin$^-$* cells using flow cytometry.

1. Sacrifice donor mice by CO_2 inhalation followed by cervical dislocation.

2. Using a pair of scissors, remove the large muscles and tendons from the femur and tibia bones. Clean the bones of tiny fragments of muscles using paper towels. Hold the bone with forceps and use a 23 G needle to penetrate both the top and bottom ends of the bone. Use a syringe containing 1 mL of IMDM to flush out the bone from the top and collect the bone marrow cells from the bottom into a sterile 15-mL conical tube.

3. Centrifuge the bone marrow cells at 400*g* for 10 min. Aspirate the supernatant, then add 1 mL of IMDM and gently pipet to dissociate the cells into a single cell suspension.

4. Isolate LDBM cells by centrifugation on Ficoll-Hypaque density gradients as described in **Subheading 3.1.** above. Wash LDBM cells twice with IMDM supplemented with 10% FCS + 100 U/mL penicillin + 10 µg/mL streptomycin + 2 m*M* L-glutamine.

5. Purify Sca1$^+$, *lin$^-$* cells using a FACScan flow cytometer using phycoerythrin-labeled specific antibodies.

   a. Incubate LDBM cells with 1 mg/mL each of rat anti-mouse Gr-1 (granulocytes), B220 (B lymphocytes), TER-119 (erythroid cells), and FITC-conjugated CD4/CD8 (T lymphocytes) MAbs on ice for 20 min.

b. Pellet the cells at 300*g* for 8 min, wash with PBS, and resuspend in 1 mL of IMDM.

c. Add goat anti-rat IgG magnetic microbeads and incubate the cells on ice for 20 min. Isolate the lineage-depleted cell population using a magnetic separation device as described by the manufacturer, and determine the purity as described in **Step 6, Subheading 3.1.** Use these cells for transduction with AAV as described in **Subheading 3.4**.

3.3. Transduction of the Human CD34⁺ Cells With Recombinant AAV Vectors

Perform all experiments with the same recombinant vector stock to minimize any variability.

1. Incubate 5×10^4 CD34⁺ cells with recombinant AAV vectors at an MOI of 50–100 in a volume of 100 µL of serum-free IMDM for 2 h at 4°C (*see* **Note 3**).
2. Following infection, maintain cells in liquid culture at 4°C supplemented with the following recombinant human cytokines: 200 ng/mL of IL-1, 10 ng/mL of IL-6, and 100 ng/mL of SCF.
3. For differentiation into myeloid, erythroid, and lymphoid lineages, incubate cells with 10 ng/mL of G-CSF, 5 U/mL of Epo, and 250 U/mL of IL-7 for 3 d.
4. Use immuno-fluorescence dual-labeling and flow cytometry to determine differentiation of CD34⁺ cells into myeloid and lymphoid lineages using FITC-labeled antibodies against the cell-surface markers as follows: CD15 and CD33 for myeloid differentiation, CD2/3 and CD4/8 for T-cell differentiation, and CD19 for B cell differentiation.

3.4. Ex Vivo Transduction of Murine Sca1⁺, lin⁻ Cells With Recombinant AAV Vectors and Transplantation into Recipient Mice

3.4.1. Short-Term Transduction

1. Incubate murine Sca-1⁺, *lin*⁻ cells (isolated in **Subheading 3.2.**) with different MOIs (10-100) of recombinant AAV vector for 2 h at 37°C as described in **Subheading 3.3.** Prepare mock-infected cells as controls.
2. Irradiate recipient mice with 7 Gy and 4 Gy doses 4 h apart using a Cs¹³⁷ irradiator.
3. Following infection of the Sca-1⁺, *lin*⁻ cells, resuspend the cells in IMDM and inject 500 cells in a volume of 300 µL intravenously (iv) via the tail vein into lethally-irradiated recipient mice.

4. Twelve days posttransplantation, sacrifice the animals, obtain their spleens, and enumerate colony formation.
5. Excise the colonies using a 23 G needle under a dissecting microscope and examine individual spleen colonies for the presence of the transgene by polymerase chain reaction (PCR) analysis as described in **Subheading 3.5**.

3.4.2. Long-Term Transduction

1. Infect the Sca-1$^+$, *lin*$^-$ cells (isolated in **Subheading 3.2.**) with the recombinant AAV vector for 2 h at 37°C at an MOI of 50. Prepare mock-infected cells as controls.
2. Sublethally irradiate the recipient mice with total body irradiation using either 1 Gy or 3 Gy (83 cGy/min) using a Cs^{137} irradiator.
3. Inject 1×10^4 of AAV-infected cells per animal via the tail-vein.
4. Once a month for 12 mo posttransplantation draw 100 µL of peripheral blood from the tail-vein of recipient mice. Lyse the blood in 900 µL of Proteinase K buffer at 55°C overnight.
5. Heat the lysates at 90°C for 10 min to inactivate Proteinase K, and use 5 µL of each sample to perform a 30-cycle PCR amplification with the transgene-specific primer-pair set under conditions described in **Subheading 3.5**.

3.5. Qualitative and Semi-Quantitative PCR Assays for the Presence and Detection of the Transduced Sequences

1. Using a DNeasy Tissue Isolation kit, follow the manufacturer's instructions and prepare DNA samples from spleen colonies isolated in **Subheading 3.4.1.** or from LDBM cells from bone marrow isolated in **Subheading 3.2**.
2. Analyze 100 ng of each DNA sample by PCR amplification for the transduced gene (*see* **Note 4**). Empirically determine that the PCR amplification reaction conditions are in a linear range by assaying the PCR samples after 15, 20, 25, 30, and 35 cycles. For controls, also amplify samples from DNA isolated from cells obtained from mice that did not receive transduced cells and from cells obtained from mock-infected mice. Use 10% of the DNA products from the human β-globin gene and 15-fold dilutions of samples from the mouse β-actin gene amplification reactions for electrophoresis on a 4% polyacrylamide gel at 100 V for 2 h. Also use standard DNA size markers ranging between 100–500 bp. Stain the gel with ethidium bromide and scan using a densitometer. Plot the band intensity against the number of PCR cycles.
3. Transfer the DNA products from the amplification reactions to nitrocellulose membranes by Southern blotting. Use specific DNA probes (human β-globin and mouse β-actin) to detect the amplified DNA products by autoradiography using standard protocols.

4. Determine the relative intensities of the corresponding bands by scanning the autoradiograms using the Photoshop 7.0 program.

5. For semi-quantitative PCR assays, optimize the amplification conditions to be in a linear range by using 15, 20, 25, 30, and 35 cycles as described in **step 2**, and perform the reactions in the presence of 2 μCi of [γ^{32}P] dCTP (sp. act., 800 Ci/mmol) in each reaction mix.

6. Analyze the DNA products from the amplification reactions on 4% polyacrylamide gels followed by autoradiography as described in **step 2**.

7. Determine the relative intensities of the corresponding bands of human β-globin and mouse β-actin DNA by scanning the autoradiograms and analyzing the band intensities with the Photoshop 7.0 program (*see* **Note 5**).

4. Notes

1. The details of construction of a recombinant AAV plasmid containing the bacterial β-galactosidase (*lacZ*) gene under the control of the CMV immediate-early promoter and subsequent packaging into recombinant virions have been described *(19,20)*. Similarly, the details of construction of a recombinant AAV plasmid containing the human globin genes and subsequent packaging into recombinant virions have been also described *(21,22)*. Package and purify all recombinant AAV vector preparations by standard methods *(25)*.

2. It is essential to remove any residual heparin from bone marrow cells because heparin inhibits AAV binding to its receptor. Similarly, serum contains fibroblast growth factor (FGF), which also inhibits AAV binding and entry because AAV uses the FGF receptor as a co-receptor.

3. All experiments should be performed with the same recombinant vector stock to minimize any variability. Infection can also be carried out at 37°C, but in order to maintain cell viability, it is preferable to do it at 4°C. It is also important to note that the level of transduction of CD34$^+$ cells by recombinant AAV vectors treated with Micrococcal nuclease is significantly reduced owing to a marked effect of this nuclease on viability of these cells *(19)*. Thus, it is advisable to use DNase I rather than Micrococcal nuclease during recombinant AAV vector production. It should also be borne in mind that CD34$^+$ cells from approx 50% of normal volunteer donors cannot be infected by AAV because of lack of expression of the AAV receptor *(19)*.

4. Design all primer-pairs such that the GC content, amplimer length, and primer affinity (T$_m$) are not significantly different. For example, for carrying out PCR reactions with an oligonucleotide primer-pair specific for the human β-globin gene, use the forward primer, 5′-TGTCACAGTGCAGCT CACTCAGT-3′, specific for exon 1 coding sequence, and the reverse primer, 5′-TCCTGAGGAGAAGTCTGCCGTT-3′, specific for exon 2 coding se-

quence of human β-globin gene. A specific 402 bp amplified DNA product should be detected following hybridization with the human β-globin DNA probe on Southern blots. To detect the endogenous mouse β-actin gene use the following primer-pair: 5′-TCCTCTTCCTCCCTGGAGAA-3′ and 5′-GCTGATCCACATCTGCTGGA-3′, which should yield a 354 bp amplification product.

5. PCR analysis should indicate the presence of the transduced human β-globin gene sequences in bone marrow cells from the recipient mice. No human β-globin gene sequences should be detected in samples obtained from nontransplanted animals and those transplanted with mock-infected cells. Semi-quantitative PCR assays provide a measure of the transduced human β-globin gene copy number compared with the endogenous mouse β-actin gene.

Acknowledgments

I am grateful to Dr. Mervin C. Yoder for many helpful suggestions. The research in author's laboratory was supported in part by Public Health Service grants (HL-48342, HL-53586, HL-58881, HL-65570, and DK-49218, Centers of Excellence in Molecular Hematology) from the National Institutes of Health, the Phi Beta Psi Sorority, and an Established Investigator Award from the American Heart Association.

References

1. Muzyczka, N. (1992) Use of adeno-associated virus as a general transduction vector for mammalian cells. *Curr. Top. Microbiol. Immunol.* **158,** 97–129.
2. Berns K. I. and Giraud, A. (1996) Biology of adeno-associated virus. *Curr. Top. Microbiol. Immunol.* **218,** 1–23.
3. Xiao, X., Li, J., and Samulski, R. J. (1996) Efficient long-term gene transfer into muscle tissue of immunocompetent mice by adeno-associated virus vector. *J. Virol.* **70,** 8098–8108.
4. Kessler, P. D., Podsakoff, G. M., Chen, X., McQuiston, S. A., Colosi, P. A., Matelis, L.A., et al. (1996) Gene delivery to skeletal muscle results in sustained and expression and systemic delivery of a therapeutic protein. *Proc. Natl. Acad. Sci. USA* **93,** 14082–14087.
5. Kaplitt, M. G., Leone, P., Samulski, R. J., Xiao, X., Pfaff, D. W., O'Malley, K. L., and During, M. J. (1994) Long-term gene expression and phenotypic correction using adeno-associated virus vectors in the mammalian brain. *Nature Genet.* **8,** 148–153.
6. Russell, D. W. and Kay, M. A. (1999) Adeno-associated virus vectors and hematology. *Blood* **94,** 864–874.

7. Alexander, I. A., Russell, D. W., and Miller, A. D. (1997) Transfer of contaminants in adeno-associated virus vector stocks can mimic transduction and lead to artifactual results. *Hum. Gene Ther.* **8,** 1911–1920.

8. Goodman, S., Xiao, X., Donahue, R. E., Moulton, A., Miller, J. L., Walsh, C. E., et al. (1994) Recombinant adeno-associated virus mediated gene transfer into hematopoietic progenitor cells. Blood **84,** 1492–1500.

9. Miller, J. L., Donahue, R. E., Sellers, S. E., Samulski, R. J., Young, N. S., and Nienhuis, A. W. (1994) Recombinant adeno-associated virus (rAAV)-mediated expression of human γ-globin gene in human progenitor-derived erythroid cells. *Proc. Natl. Acad. Sci. USA* **91,** 10183–10187.

10. Walsh, C. E., Nienhuis, A. W., Samulski, R. J., Brown, M. G., Miller, J. L., Young, N. S., Liu, J. M. (1994) Phenotypic correction of Fanconi anemia in human hematopoietic cells with a recombinant adeno-associated virus vector. *J. Clin. Invest.* **94,** 1440–1448.

11. Malik, P., McQuiston, S. A., Yu, X.-J., Pepper, K. A., Krall, W. J., Podsakoff, G. M., et al. (1997) Recombinant adeno-associated virus mediates a high level of gene transfer but less efficient integration in K562 human hematopoietic cell line. *J. Virol.* **71,** 1776–1783.

12. Hargrove, P. W., Vanin, E. F., Kurtzman, G. J., and Nienhuis, A. W. (1997) High-level globin gene expression mediated by a recombinant adeno-associated virus genome that contains the 3′ γ globin gene regulatory element and integrates as tandem copies in erythroid cells. *Blood* **89,** 2167–2175.

13. Nathwani, A. C., Hanawa, H., Vandergriff, J., Kelly, P., Vanin, E. F., and Nienhuis, A. W. (2000) Efficient gene transfer into human cord blood CD34$^+$ cells and the CD34$^+$CD38$^-$ subset using highly purified recombinant adeno-associated viral vector that are free of helper virus and wild-type AAV. *Gene Ther.* **7,** 183–195.

14. Zhou, S. Z., Cooper, S., Kang, L. Y., Ruggieri, L., Heimfeld, S., Srivastava, A., and Broxmeyer, H. E. (1994) Adeno-associated virus 2-mediated high efficiency gene transfer into immature and mature subsets of hematopoietic progenitor cells in human umbilical cord blood. *J. Exp. Med.* **179,** 1867–1875.

15. Fisher-Adams, G., Wong, K. K., Podsakoff, G. M., Forman, S. J., and Chatterjee, S. (1996) Integration of adeno-associated virus vectors in CD34$^+$ human hematopoietic progenitor cells after transduction. *Blood* **88,** 492–504.

16. Chatterjee, S., Li, W., Wong, C. A., Fisher-Adams, G., Lu, D., Guha, M., et al. (1999) Transduction of primitive human marrow and cord blood-derived hematopoietic progenitor cells with adeno-associated virus vectors. *Blood* **93,** 1882–1894.

17. Luhovy, M., McCune, S., Dong, J.-Y., Prchal, J. F., Townes, T. M., and Prchal, J. T. (1996) Stable transduction of recombinant adeno-associated virus into hematopoietic stem cells from normal and sickle cell patients. *Biol. Blood Marrow Transpl.* **2,** 24–30.

18. Srivastava, A. (2002) Obstacles to human hematopoietic stem cell transduction by recombinant adeno-associated virus 2 vectors. *J. Cell. Biochem.* **38,** 39–45.

19. Ponnazhagan, S., Mukherjee, P., Wang, X.-S., Qing, K. Y., Kube, D. M., Mah, C., et al. (1997) Adeno-associated virus type 2-mediated transduction of primary human bone marrow-derived CD34+ hematopoietic progenitor cells: donor variation and correlation of transgene expression with cellular differentiation. *J. Virol.* **71,** 8262–8267.
20. Ponnazhagan, S., Mukherjee, P., Yoder, M. C., Wang, X.-S., Zhou, S. Z. Kaplan, J., et al. (1997) Adeno-associated virus 2-mediated gene transfer in vivo: organ-tropism and expression of transduced sequences in mice. *Gene* **190,** 203–210.
21. Ponnazhagan, S., Yoder, M. C., and Srivastava, A. (1997) Adeno-associated virus type 2-mediated transduction of murine hematopoietic cells with long-term repopulating ability and sustained expression of a human globin gene in vivo. *J. Virol.* **71:** 3098–3104.
22. Tan, M. Q., Qing, K. Y., Zhou, S. Z., Yoder, M. C., and Srivastava, A. (2001) Adeno-associated virus 2-mediated transduction and erythroid lineage-restricted, long-term expression of the normal human β-globin gene in hematopoietic cells from homozygous β-thalassemic mice. *Mol. Ther.* **3,** 940–946.
23. Yoder, M. C., Du, X.-X. and Williams, D. A. (1993) High proliferative potential colony-forming cell heterogeneity identified using counterflow centrifugal elutriation. *Blood* **82,** 385–391.
24. Yoder, M. C., King, B., Hiatt, K., Mukherjee, P., and Williams, D. A. (1995) Growth of bone marrow high-proliferative potential colony-forming cells during in vitro coculture with murine embryonic yolk sac cell lines. *Blood* **86,** 1322–1330.
25. Zolotukhin, S., Byrne, B. J., Zolotukhin, I., Summerford, C., Samulski, R. J., and Muzyczka, N. (1999) Recombinant adeno-associated virus purification using novel methods improves infectious titer and yield. *Gene Ther.* **6,** 973–985.

III

DELIVERY USING HERPES SIMPLEX VIRUSES

19

Delivery Using Herpes Simplex Virus

An Overview

**William F. Goins, Darren Wolfe, David M. Krisky, Qing Bai,
Ed A. Burton, David J. Fink, and Joseph C. Glorioso**

1. Introduction

The human herpesviruses represent excellent candidate viruses for several
types of gene vector applications. As a class, they are large DNA viruses with
the potential to accommodate large or multiple transgene cassettes, and they
have evolved to persist in a lifelong nonintegrated latent state without causing
disease in the immune-competent host. Among the herpesviruses, herpes simplex virus type 1 (HSV-1) is an attractive vehicle because in natural infection,
the virus establishes latency in neurons, a state in which viral genomes may persist for the life of the host as intranuclear episomal elements. The natural lifelong persistence of latent genomes in trigeminal ganglia (TG) without the development of sensory loss or histologic damage to the ganglion attests to the
effectiveness of these natural latency mechanisms. Although the wild-type virus
may be reactivated from latency under the influence of a variety of stresses,
completely replication defective viruses can be constructed that retain the ability to establish persistent quiescent genomes in neurons, but that are unable to
subsequently reactivate in the nervous system. These persistent genomes are devoid of lytic gene expression, but retain the ability to express latency-associated
transcripts (LATs). We have made considerable progress in the construction of
highly defective HSV genomic vectors that are apparently apathogenic for the
nervous system, but are efficiently taken up by neurons to establish persistence

From: *Methods in Molecular Biology, vol. 246:*
Gene Delivery to Mammalian Cells: Vol. 2: Viral Gene Transfer Techniques
Edited by: W. C. Heiser © Humana Press Inc., Totowa, NJ

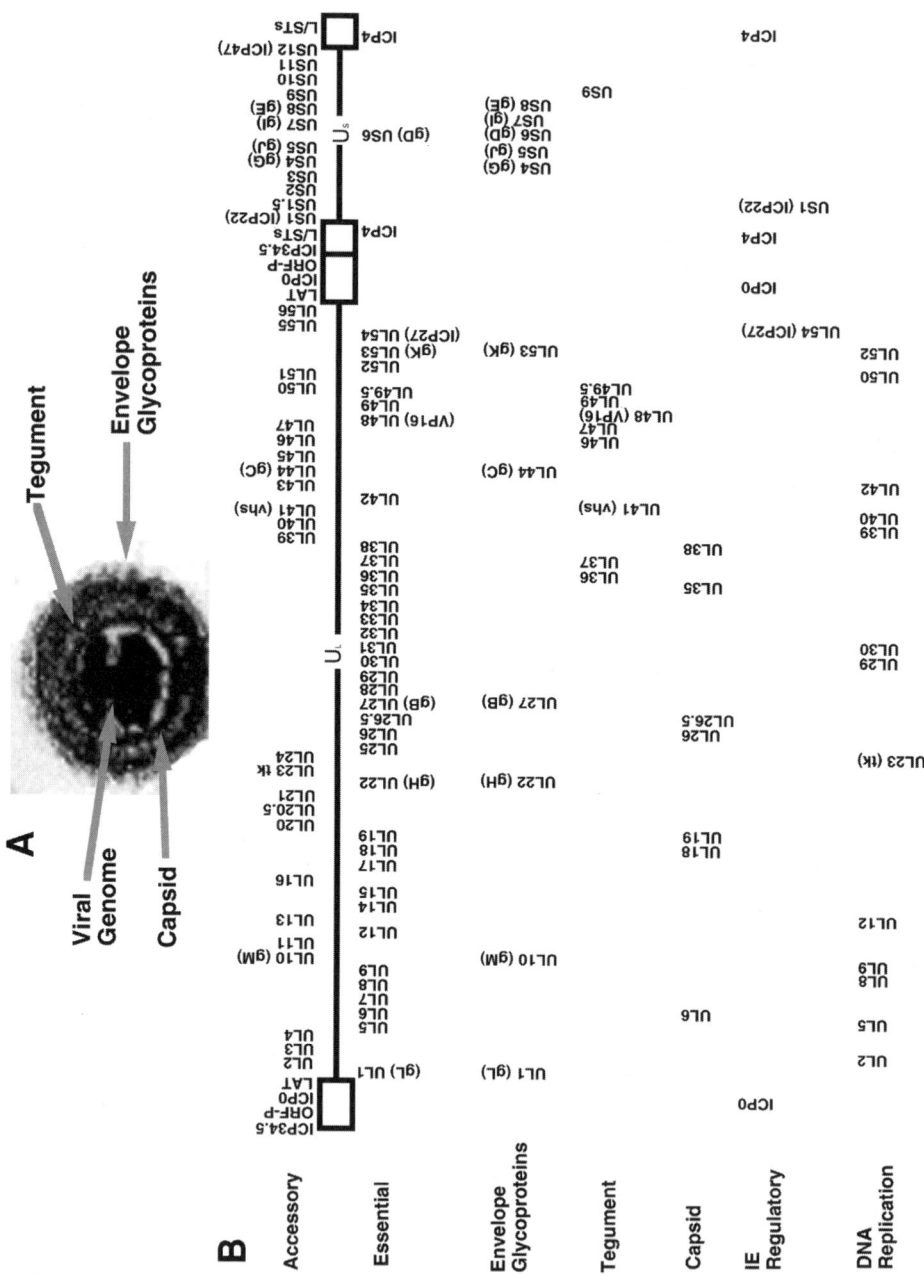

(1). We have identified a promoter element within the latency region of the genome that is capable of expressing transgenes for prolonged periods in the peripheral nervous system (PNS) following delivery in the context of the highly defective viral vector background *(2–4)*.

Our recent studies have determined that HSV can be more broadly used for gene-transfer applications, because replication defective viruses can be constructed for gene delivery without vector toxicity *(1,5–8)* and HSV can persist in a variety of cells and express transgenes long-term *(4)*. HSV-1 has a broad host range and does not require cell division for infection and gene expression. Accordingly, HSV may be generally useful for gene transfer to a variety of normal and disease tissues. However, there are instances where restricted or specifically enhanced vector gene delivery will be very important, such as where systemic or local vector inoculation will require targeting to specific tissues or cells (i.e., motor neurons), where the application dictates that the transgene is cytotoxic (i.e. cancer treatment), or in instances where improved infection of certain cells (i.e., stem cells, endothelial cells, cardiac muscle, or certain classes of neurons) would be highly advantageous. In this chapter, we will highlight the recent developments in the engineering of replication defective HSV genomic vectors with respect to improvements in reducing vector toxicity, regulating transgene expression, and targeting of the vector to specific cells and tissues.

2. The Biology of HSV Infection

2.1. Virus Structure and Membrane Proteins

The virus particle (**Fig. 1A**) is a unique structure composed of at least 34 proteins *(9,10)* many of which are post-translationally modified by cellular and viral enzymes. The virus icosahedral-shaped nucleocapsid *(11)* is packaged with a single linear copy of the double-stranded genome containing approx 85 open reading frames *(12–15)*. The 152 Kb viral genome (**Fig. 1B**) is segmented

---

Fig. 1. HSV-1 particle structure and genome organization. (**A**) Electron micrograph of a typical HSV-1 virus particle displaying the envelope containing virus-specific glycoproteins surrounding the dense tegument layer that encompasses the viral icosahedral-shaped nucleocapsid containing the double-stranded viral genome. (**B**) Schematic representation of the HSV genome, showing the unique long (U$_L$) and unique short (U$_S$) segments, each bounded by inverted repeat elements. The location of the essential genes, which are required for viral replication in vitro, and the nonessential or accessory genes, which contribute to virus replication and spread in vivo, are indicated. Depicted underneath the genome are the 11 HSV-1 glycoprotein genes, the HSV-1 genes encoding the tegument and capsid proteins, the IE transcriptional regulatory genes, as well as the genes, involved in replication of the viral genome.

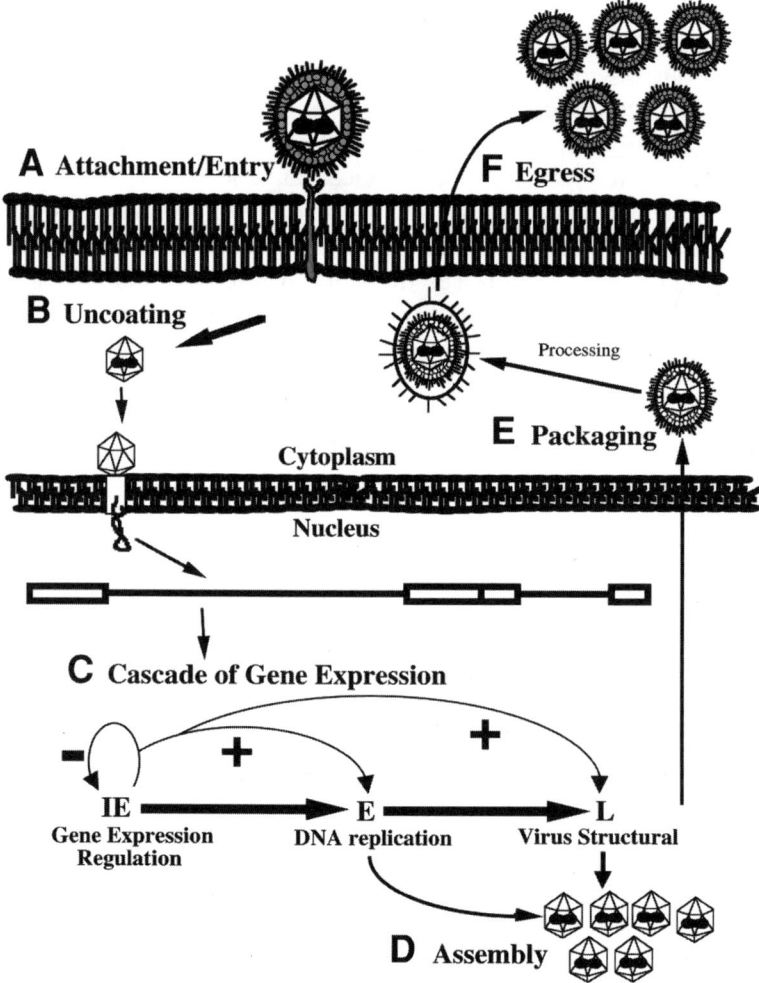

Fig. 2. Schematic representation of the HSV-1 lytic life cycle. **(A)** HSV-1 binds to and enters mucosal or epithelial cells through the binding of HSV-1 glycoproteins to specific cell-surface receptors followed by fusion of the viral envelope with the cell-sur-face membrane. **(B)** Following uncoating, the viral capsid is transported to the nucleus where the linear viral genome is injected through a nuclear pore. **(C)** Temporal cascade of HSV-1 gene expression ensues with the subsequent replication of the viral genome. The immediate early genes (IE or α) are the first genes expressed and do so in the ab-sence of *de novo* protein synthesis. The early (E or β) genes are expressed only when functional IE genes are present, and mainly encode viral proteins involved in replica-tion of the HSV-1 genome. After viral DNA replication has occurred, the late genes (L or γ) that encode the structural gene products composing the viral capsid, tegument, and

with each of its long (U_L) and short (U_S) unique segments flanked by inverted repeats (IRs). The viral functions have been categorized as to whether they are essential for virus replication in cell culture or are accessory (nonessential) functions that contribute to virus replication and spread in vivo. The viral genes are arranged in a manner such that many of the essential functions or accessory genes tend to be clustered within the genome (**Fig. 1B**). Additionally, because very few HSV-1 genes are spliced and the virus possesses a highly evolved recombination system, it is relatively easy to manipulate and engineer the virus for purposes of gene transfer deleting individual genes or blocks of genes that may play a role in vector toxicity.

The capsid is surrounded by a complex matrix of proteins (**Fig. 1A**) referred to as the tegument that is composed of viral structural components involved in shut-off of host protein synthesis *(16–20)*, and activation of immediate early viral gene expression and assembly functions *(21–26)*. The tegument is encased in an envelope containing at least 10 glycoproteins (gB-gE, gG-gJ, gL, and gM) *(10,27,28)*. Of the 10 incorporated glycoproteins, gB, gD, gH and gL are essential for viral infection *(29–32)*, whereas gC, gE, gG, gI, gJ, and gM are dispensable for infection in vitro *(10,27,28)*. Most HSV glycoproteins appear to be functionally independent but closely associate with one another in the particle (i.e., gD and gB) *(33,34)*, which may be important for their function in infection and entry. Some form functional homo-oligomers (i.e., gB) *(35–41)* whereas others form functional hetero-oligomers (gH/gL, gI/gE) *(31,42)*.

2.2. Budding and Envelopment

The assembly of HSV particles occurs in the nucleus where nucleocapsids attach to modified patches of the inner lamella of the nuclear membrane (**Fig. 2**). Budding of DNA-containing capsids occurs at these sites, where enveloped particles are formed. The glycoprotein polypeptides are modified in the endoplasmic reticulum (ER) and require further processing by Golgi enzymes to carry out core trimming and the addition of complex-type glycans, a requirement for virus release *(43–46)*. How these glycoproteins are processed for incorporation into mature enveloped particles remains unclear, but may involve de-envelopment

---

glycoproteins, are expressed. The ICP4 IE gene transactivates the E and L genes, as well as repressing IE expression, while ICP0, ICP22, and ICP27 turn on E and L genes, respectively. (**D**) Particle assembly occurs within the host cell nucleus. (**E**) Immature virion particles bud from the nuclear membrane and viral glycoprotein processing occurs within the Golgi. (**F**) Mature particles exit the cell at the cell surface membrane. The production of progeny virus particles ultimately results in the lysis of the host cell.

and re-envelopment through cytoplasmic membranes *(47–49)*. None of the HSV envelope glycoproteins have been shown to be required for virion assembly *(10,27,28,50)*, and the mechanism for selective incorporation of HSV-1 glycoproteins into mature virions is not well understood. Examination of the incorporation of HSV-1 glycoproteins has shown that shortening of the 109 amino acid gB C-terminus to three or six residues reduced incorporation compared to molecules truncated at residues 22 or 43 *(51)*; however, a recombinant truncated at residue 41 of the gB C-terminus was properly processed, transported to the plasma membrane, and incorporated into mature virus *(52)*. The C-termini of both HSV-1 gC *(53)* and the gC homologue of pseudorabies virus (PRV) *(54,55)* were not required for incorporation of these molecules into mature virions. The transmembrane domain (TMD) was, however, necessary for membrane anchorage and cell-surface expression. Recent studies of HSV-1 *(56)* and bovine herpesvirus type 1 (BHV-1) *(57)* gB have demonstrated that the TMD of this molecule is also required for both membrane anchorage and nuclear envelope localization as well as for transport to the Golgi apparatus and plasma membrane. In our experience, envelope glycoproteins from other viruses can be readily incorporated into the HSV envelope *(58)* provided they are expressed from the HSV genome during infection and undergo similar processing events and with kinetics similar to that of bone fide HSV glycoproteins.

2.3. Viral Infection

The envelope glycoproteins mediate interaction of the virion with the host cell during infection (**Fig. 3**). Two different facets of this process can be identified: attachment to the cell surface and fusion of the viral and cell membranes resulting in viral entry. In cell culture and in animals, the virus is also capable of infecting neighboring cells by moving transcellularly across cell membranes. This process, referred to as cell-to-cell or lateral spread, can be distinguished from infection by exocytosed particles through its ability to occur in the presence of virus neutralizing antibodies as well as by deletion of certain accessory envelope glycoproteins (i.e., gI/gE), which prevents lateral spread but not initial infection *(59,60)*.

2.3.1. Attachment to Heparan Sulfate

Virus attachment is mediated by numerous glycoproteins *(27,28,61)*. Binding to cell surface glucosaminoglycans (GAGs), primarily heparan sulfate (HS) *(62–67)* but also dermatan and chondroitan sulfate *(68,69)*, is mediated by exposed domains of glycoproteins C *(70)* and B *(71,72)* containing positive charged amino acids (**Fig. 3A**). Together this binding represents about 85% of the binding activity to Vero cells with gC contributing the majority of this binding activity *(64,73)*. Deletion of gC (a nonessential gene) and the HS binding domain of gB

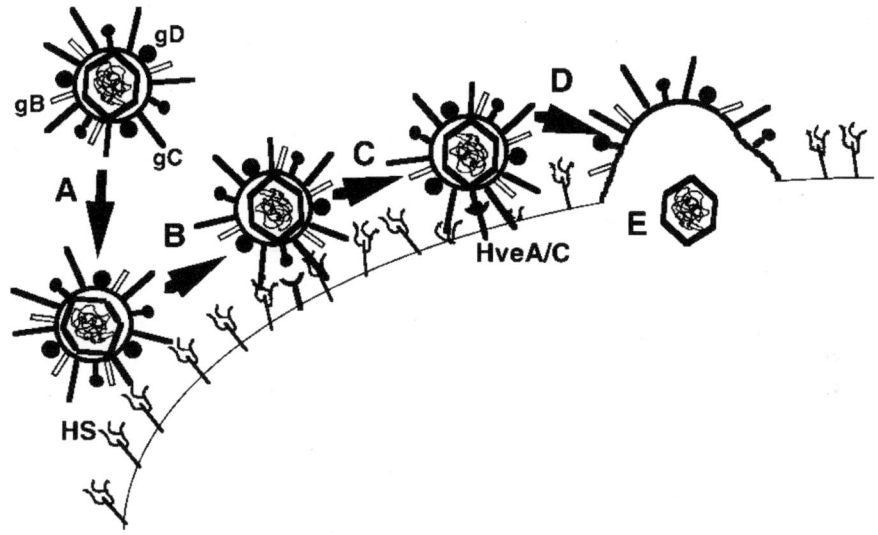

Fig. 3. Mode of HSV-1 entry into susceptible cells. HSV-1 encodes at least 11 glycoproteins, although only 4 are essential for virus entry. Virus entry into susceptible cells involves: (**A**) binding of positive charged amino acid domains of HSV gC and gB to heparan sulfate (HS) moieties present on the cell surface; (**B**) diffusion of the virus on the cell surface until contact with the HveA/HveC receptor; (**C**) binding of gD to its receptor; and (**D**) fusion of the virus envelope and the cell membrane through the action of gB, gD, and the gH/gL complex, that is followed by (**E**) the release of the nucleocapsid into the cytoplasm of the cell.

(an essential gene) in a single mutant virus *(73)* impairs binding (i.e., slower kinetics) to an extent similar to the reduced binding to HS deficient cells by wild-type virus *(63)*, but does not completely eliminate virus adsorption, indicating that other receptors must be involved. Removal of the HS binding domain of gB does not interfere with the ability of gB to participate in virus penetration *(73)*. However, elimination of both HS binding functions severely affects the ability of the virus to form plaques in the presence of neutralizing antibody where the virus is forced to spread to neighboring cells across cell junctions rather than infect neighboring cells with newly exocytosed infectious particles. This phenotype is specific to the HS binding function of gB because deletion of gC alone often results in larger than wild-type plaques *(74,75)*. These data suggest a cooperativity between gC and gB HS binding that is important to both virus attachment and penetration, thereby linking these two processes through the HS receptor interaction. Chemical cross-linking studies of free and bound virus suggest that the relationships between gB, gC, gD, and gH/gL favor the association of the four essential glycoproteins in virus entry (gB, gD, gH/gL) to the exclusion of gC *(76)*. These results are consistent with other evidence that gB is required for cell-to-cell spread of virus *(29,74,77–79)*. HSV mutants selected for escape from soluble HS interference of infection are attributable to mutations in gK and these mutants often acquire a syncytial (syn-) phenotype *(80)*.

2.3.2. Attachment to Herpesvirus Specific Receptors

Initial virus binding to cell surface HS is followed by gD-mediated binding to a second receptor (**Fig. 3C**). Recent work by several labs has identified gD cognate receptors utilized for both virus attachment and penetration. The first herpesvirus entry mediator (HVEM or HveA) was identified by screening a cDNA expression library in HSV-resistant Chinese hamster ovary (CHO) cells for clones that enabled virus infection. HveA was subsequently determined to be a member of the tumor necrosis factor-α (TNF-α)/NGF receptor family *(81)*. Domains of gD that potentially contribute to HveA binding have been identified in virus infection inhibition studies using monoclonal antibodies that require residues 11–19 and 222–252 for binding to gD *(82)*. Consistent with these data, incubation of soluble gD with the host cell prior to infection can block infection by wild-type virus; however, mutants having single amino acid substitutions in the external domain of gD at residues 25 or 27 can circumvent this blocking activity of gD *(83)*. Mutation of residues 1–50 or 222–224 prevent binding of gD to the HveA and a second receptor, HveC/nectin-1, in biochemical and transient transfection assays. Other HSV-1 strains (i.e., ANG and KOS rid-1 and rid-2) altered in these N-terminal residues do not infect CHO cells transduced with the HveA receptor recognized by strain KOS *(84,85)*, but still infect Vero and HeLa cells, indicating the existence of additional receptors. HveA has been detected on a relatively small number of cell types, including activated T-cells *(81)*. Recent work from our laboratory has identified HveA on mouse embryonic stem cells and CD34$^+$ bone marrow and placental-derived cells from both human and mouse origin *(86)*. Together, these studies suggest that HveA is recognized by a folded part of gD that can be affected by mutations spanning a substantial portion of the molecule. A detailed analysis of mutants localized to the N-terminus of gD has pinpointed the binding site for HveA and HveC/nectin-1 (Bai and Glorioso, unpublished) and is in agreement with recent crystal structure data showing that HveA bound to the N-terminus of gD results in a conformational change that may be responsible for inducing virus envelope fusion with the cell surface *(87)*.

Using transfection techniques, a number of other entry mediators have been identified, including HveB (poliovirus receptor-related protein 2 or nectin-2 (Prr-2)) *(88)*, 3-O-sulfated HS *(89)*, and HveC (Prr-1, or nectin-1) *(90)*. HveC/nectin-1 is a member of the immunoglobulin superfamily of genes and has no structural relation to HveA based on binding studies *(90,91)* using soluble forms of

HveC/nectin-1. All alpha herpesvirus gD molecules recognize HveC/nectin-1, including HSV-1, HSV-2, PRV, and BHV-1. Two potentially related but separate HveC/nectin-1 binding sites appear to co-exist within gD. Residues 216–234 have been implicated in the HveC:gD interaction, in as much as monoclonal antibodies (MAbs) specific for an epitope in this region interfere with gD recognition of HveC/nectin-1 but not HveA *(91)*. Recent work from our laboratory (Bai and Glorioso, unpublished) has discovered the first mutants of gD that inactivate binding to HveC/nectin-1 without also preventing recognition of HveA; the reverse is also true, i.e., mutants that attach to cells and enter via HveC/nectin-1 but fail to recognize HveA are available.

2.3.3. Penetration and Lateral Spread

The consequence of the sequential attachment steps in infection is fusion of the virus envelope with the cell-surface membrane and subsequent virus entry (**Fig. 3D**). The events in penetration are not well-understood because multiple glycoproteins are required. A role for gD in virus penetration is supported by evidence that attached virus can be neutralized by anti-gD antibody and virus mutants deleted for gD attach to cells but do not penetrate *(92,93)*. However, the specific binding afforded by gD could trigger penetration without direct involvement of this glycoprotein in the fusion process. Mutants deleted for gH/gL or gB are also blocked in virus penetration but are not defective in attachment. Both gB and gD have been shown to be capable of inducing syncytia if expressed on the cell surface at low pH, supporting a possible role for both molecules in fusion *(94)*. Virus entry may involve a cascade of events in which gD initiates the fusion event *(62)*, in which a fusion bridge is most likely formed by the action of gB followed by extension of the bridge and virus release requiring the activities of gH/gL.

2.4 The HSV-1 Life Cycle

2.4.1. Lytic Infection

Once within the nucleus (**Fig. 2**), the lytic replication cycle of the virus takes place *(50,95,96)* with the viral genes expressed in a highly regulated cascade *(97,98)* of coordinated gene expression consisting of three stages: immediate early (IE or α), early (E or β) and late (L or γ). Three of the viral IE genes are transcriptional activators which induce expression of E and L genes *(99–106)*. Early gene functions participate in viral DNA replication that must proceed in order for late gene expression to occur *(107,108)*. The late genes encode largely structural products comprising the nucleocapsid, tegument, and viral envelope glycoproteins. Viral particles are assembled within the nucleus, bud from the nuclear membrane and particle maturation proceeds during migration through the Golgi followed by egress from the cell.

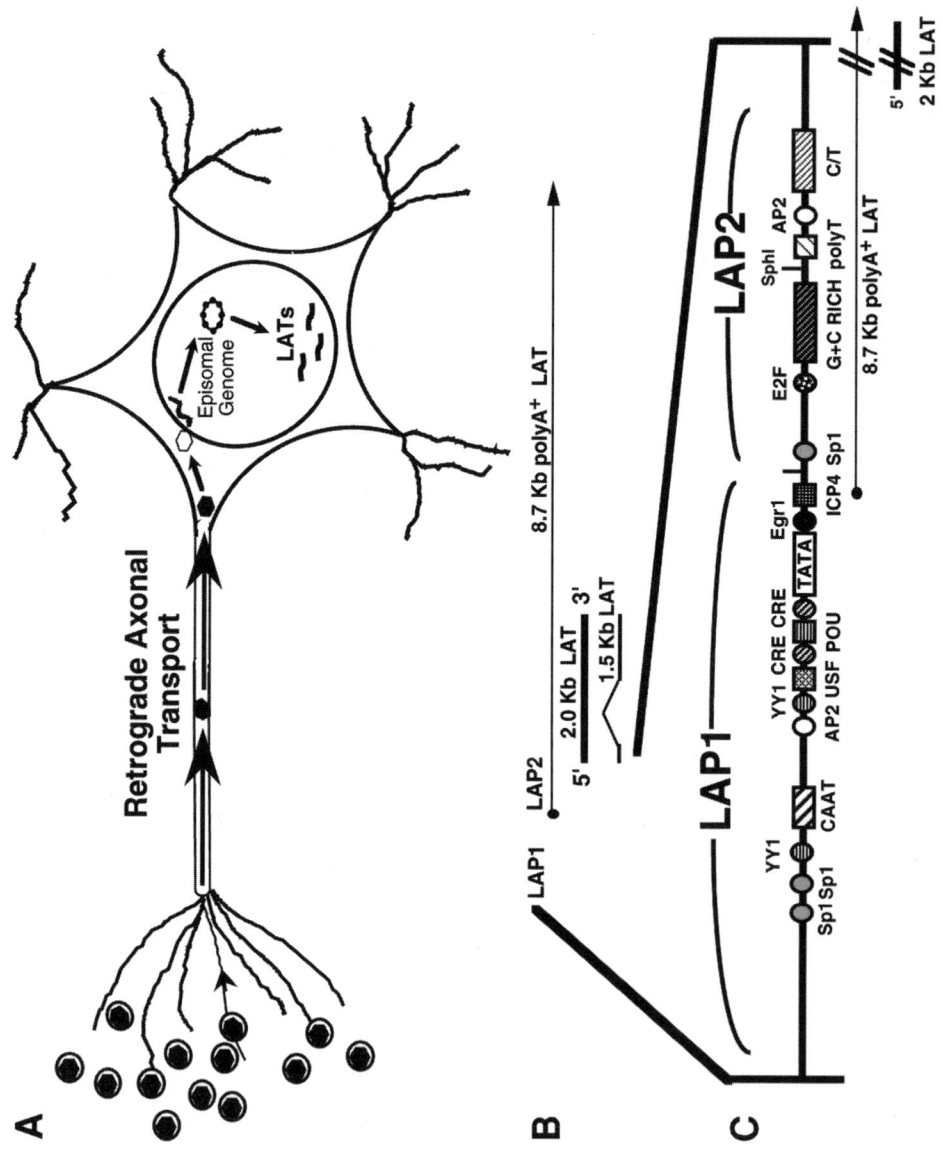

2.4.2. Latency

Following focal replication of the virus in these permissive cell types, the virus invades the nervous system by directly infecting axon terminals of local sensory neurons of the PNS *(109,110)*. The viral nucleocapsids are transported in a retrograde manner back to the neuronal cell body where the virus can either replicate or enter latency (**Fig. 4A**). During latency the linear viral genomes circularizes, becomes methylated and forms a higher-order nucleosomal structure *(111–114)*. At this point the lytic genes become inactive and the viral LATs are readily detected by *in situ* hybridization using LAT-specific riboprobes *(115–118)*. The virus can remain within the latent state for the life-time of the individual or it can be induced to reactivate from latency by a variety of stimuli. Reactivation results in the resumption of the lytic cycle and the subsequent synthesis of progeny virions that may traverse the axon by anterograde transport, establishing an active infection at or near the site of primary infection.

During latency, only a segment of the inverted repeat sequences in U_L, just downstream of and from the strand opposite of the IE ICP0 gene (*see* **Fig. 1B**), remains transcriptionally active producing a family of LATs *(115–118)*. The major LAT species (2.0 and 1.5 Kb) are neither capped nor polyadenylated, and have been shown to be stable introns (**Fig. 4B**) that result from splicing of a large 8.7 Kb primary transcript *(119–123)*. The expression of LATs from these

---

Fig.4. HSV-1 latency. (**A**) HSV-1 latent infection of PNS neurons. During lytic infection the virus enters and replicates within mucosal or epithelial cells and the resulting progeny virus particles encounter peripheral nerve termini, which innervate the site of primary infection. Following fusion of the virus envelope with the nerve cell-surface membrane, the viral nucleocapsid travels via retrograde axonal transport to the neuronal cell body. At this point the virus can either proceed through the highly regulated cascade of lytic gene synthesis, or the virus can enter latency, during which the viral genome becomes methylated and is bound by cellular histones. The latency-associated transcripts (LATs) are the sole viral RNAs expressed from the viral genome during latent infection. (**B**) The LATs are transcribed from the inverted repeat regions flanking the large unique segment of the HSV-1 genome. The transcription products from the LAT region include an 8.7 Kb polyadenylated RNA that is present during lytic infection. The main LAT species of 2.0 and 1.5 Kb represent stable introns that are derived from the 8.7 Kb mLAT. The stable 2 Kb LAT, which is detected in both lytic and latent infection, and the 1.5 Kb LAT species that is spliced from the 2 Kb LAT, which is only detected during latency, are depicted along with the location of the LAT promoters LAP1 and LAP2. (**C**) An enlargement of the latency active promoter (LAP) regions LAP1 and LAP2 shows the position of potentially relevant cis-acting elements in LAP1 such as the CRE, USF, and TATA box as well as the polyT stretch and C/T element in LAP2.

otherwise quiescent genomes demonstrates that with the proper promoter elements, long-term expression of RNA transcripts can be achieved. The LAT promoter/regulatory region has been shown to be composed of two latency-active promoters (**Fig. 4B**), LAP1 *(124–127)* and LAP2 *(125,128,129)*.

2.4.2.1. LAP1

LAP1, which is predominantly responsible for LAT expression during latency *(125,130)*, is located immediately upstream of the 5′ end of the unstable 8.7 Kb LAT (**Fig. 4C**). It contains binding sites for many RNA polymerase class II transcription factors such as a consensus TATA box *(131,132)*, a CAAT box *(133)*, an USF-1 site *(132,134,135)*, and two ATF/CREB elements *(132,136)*. Many of these basal promoter elements, including the TATA box, ATF/CREB and USF-1 sites, are required for LAT expression during latency *(131,132,134, 137)*. Deletion of upstream distal elements that include potential binding sites for neuronal specific transcription factors *(133)* dramatically decreases LAT levels during latency in vivo *(138)*.

2.4.2.2. LAP2

LAP2, located 5′-proximal to the LAT introns (**Fig. 4C**), is the predominant promoter responsible for LAT expression during lytic infection *(125,129)*. However it is also capable of expressing LAT at low levels in the absence of LAP1 in latently infected animals, indicating that LAP2 has independent promoter activity *(125)*. Although LAP2 lacks a TATA box, it contains both a C/T-rich sequence and a polyT element *(128,139)* frequently observed in promoters for cellular housekeeping genes *(140–142)*. We have shown that these elements contribute significantly to LAP2 activity in cell culture and bind transacting factors *(128,139)* such as NSEP-1 *(143,144)* and HMG I(Y) *(145)* respectively. We have mapped an S1 nuclease-sensitive site to this region of LAP2, which perhaps alters the local chromatin conformation to allow long-term expression from LATs in the absence of lytic gene expression. This LAP2 element is moveable and active in expressing transgenes for prolonged periods from both replication-competent and defective vectors *(2,128)*. LAP1 does not possess this capability in the absence of LAP2 *(146–149)*.

3. Design and Engineering of HSV-1 Replication-Defective Vectors

The optimal HSV vector should be: (1) safe and completely devoid of replication competent virus; (2) noncytotoxic; (3) incapable of affecting normal host cell biology; (4) able to persist in the cell nucleus cell body in a non-integrated state; and (5) capable of expressing the therapeutic gene(s) to appropriate levels at the proper time(s). The three major considerations in the design of HSV-

1 vectors concern (1) the elimination of the cytotoxic properties of the virus, (2) the ability to target binding and entry of the virus to the specific cells or tissues of interest, and (3) the development of promoter systems for proper expression of the therapeutic gene. Considerable progress has been made in developing HSV vectors that potentially overcome these obstacles for their potential use in human applications. In this work we will concentrate upon the engineering of HSV-1 vectors deleted for essential gene functions that display reduced toxicity following infection. As discussed in some detail below, these include: (1) elimination of vector toxicity, including direct cytopathic effects; (2) full exploitation of the potential transgene capacity of the vector including rapid methods for the construction of multi-gene vectors; (3) maintenance of the viral vector genome; (4) maintenance of transgene expression; and (5) targeting HSV infection to specific cells and tissues.

3.1. Vector Capacity

Although gene therapy was initially envisioned for the treatment of monogenic diseases, the same methods can be utilized for the delivery of multiple genes for treatment of multifactorial or nongenetic pathologic processes. The large size of the HSV genome makes it ideal for applications requiring the delivery of more than one gene product simultaneously. Using a rapid gene insertion procedure, we have developed novel HSV-1 gene vectors with a background suitable for expression of multiple transgenes *(150,151)*. Together, nine viral genes were deleted, resulting in the removal of 11.6 Kb of viral DNA that was replaced with multiple transgenes under the control of different promoters. We demonstrated that new transgenes could be sequentially added to the multigene vector background with high efficiency using a method involving the insertion and removal of genes flanked by unique *Pac*I restriction enzyme sites engineered into the viral genome. In addition, we have constructed a replication-defective vector that incorporates and expresses the 14 Kb dystrophin full-length cDNA *(152)*. Together, these vectors demonstrate the potential for using HSV-1 vectors for the expression of highly complex sets of transgenes simultaneously, which may be directly applicable to gene-therapy approaches such as wound healing, the destruction of specific tumors, or the expression of large single gene cassettes like those needed to treat Duchenne muscular dystrophy or hemophilia.

3.2. Maintenance of the HSV-1 Vector Genome

In most cell types tested in vitro and in vivo, viral lytic gene promoters and strong promoters like the HCMV IE promoter display transient activity in the background of HSV genomic vectors. Loss of expression from these vectors does not appear to be the result of clearance of the vector genome from the nervous system, because that vector genomes persists in brain for at least 1 yr after inoculation. Moreover, the number of vector genomes remains unchanged at least from 1–8 wk postinoculation *(153)*. Therefore, the loss of expression could be attributed to the sequestering of viral genomes or by the inability of those promoters to remain active in the context of the latent viral genome. Because LATs are continuously expressed from latent genomes in neurons, promoters such as the HCMV IE promoter must lack the cis-elements required for long-term expression from latent viral genomes.

3.3. HSV-1 Vector-Associated Cytotoxicity

Because UV-irradiated virus displays substantially reduced cytotoxicity in vitro *(154,155)* and disruption of viral IE gene expression by interferon also reduces toxicity *(156–160)*, it is presumed that the cytotoxicity of HSV-based vectors results from the expression of HSV-1 gene products. The fundamental approach to designing HSV-1 vectors with reduced toxicity is to remove the essential IE genes of the virus, as well as several nonessential genes, whose products interfere with host-cell metabolism and are part of the virion (tegument) structure (**Fig. 5A**). Deletion of the two essential IE genes which encode the infected cell proteins 4 (ICP4) and 27 (ICP27) blocks early and late gene expression *(99,161,162)*. These deletions require that the missing functions be supplied *in trans* using a complementing cell line *(99,161,163)* to propagate virus. To insure that recombination does not occur between the defective virus and the viral sequences present within the complementing cell line during propagation, it is essential that these sequences do not share homology with sequences present within the viral genome, and that the deletion of the IE genes from the virus exceed the limits of the complementing sequences. Because many of these viral IE genes are toxic to cells, expression of the complementing genes must be inducible upon infection with the defective virus. This is achieved through the use of HSV-1 IE promoters to drive expression of these toxic transactivating genes from the cell, because these promoters have been shown to respond to the HSV-1 transactivator VP16 (Vmw65 or α-TIF) *(22–26)*, a virion tegument component that accompanies the viral DNA molecule into the nucleus of the infected cell. VP16 recognizes an octamer (TAATGARAT) consensus sequence located at various sites in all HSV-1 IE promoters, and together with the cellular transcription factors Oct-1 and HCF *(164–168)*, transactivates the IE gene promoters *(169–173)*. In the absence of virus infection, the complementing IE genes in the cell chromosome are silent. However, upon infection with the replication defective mutant, the VP16/Oct-1/HCF complex transactivates the IE promoters upstream of the complementing viral sequences in the cell line, thereby inducing expression of the necessary products for propagation of the deletion virus.

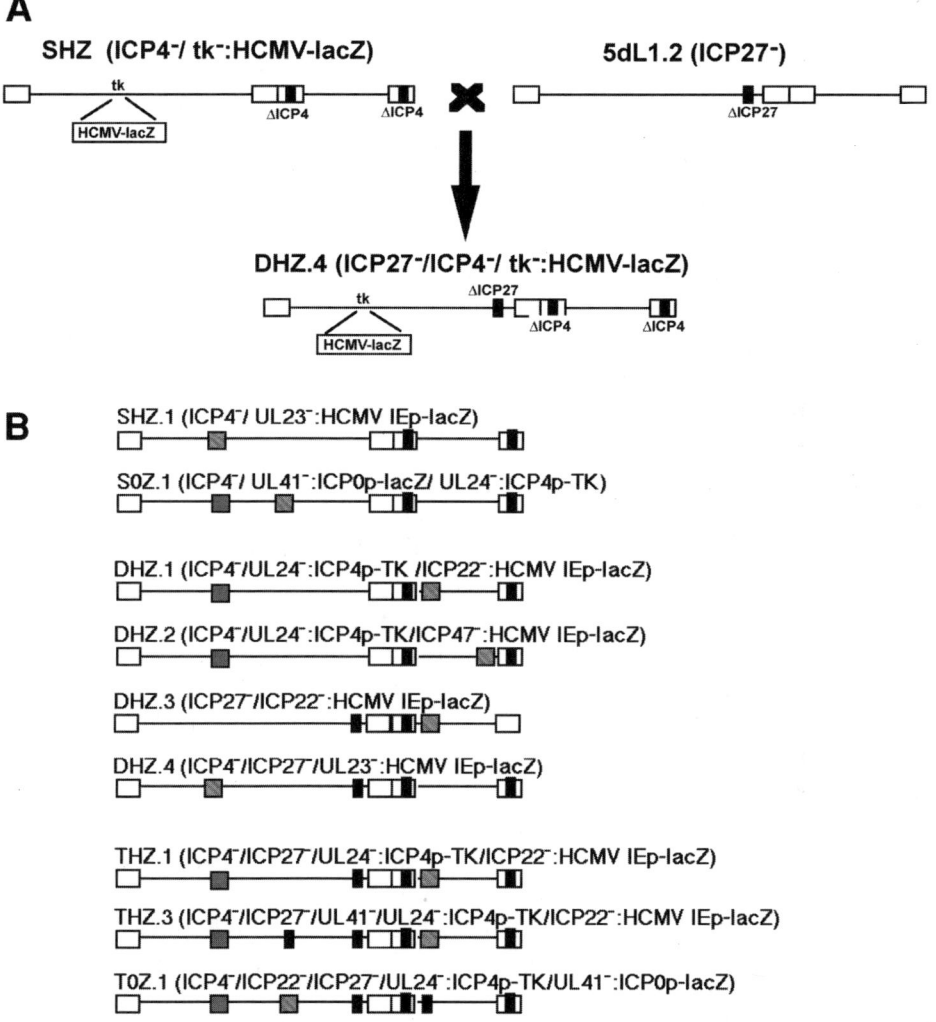

Fig. 5. Construction replication-defective HSV vectors. **(A)** The second generation ICP4⁻/ICP27⁻ double IE gene-deletion mutant (DHZ.4) was constructed by a genetic cross of the ICP4 single IE gene-deletion first-generation vectors SHZ.1 *(211,286)* with the single ICP27 deletion mutant 5dl1.2 *(190)*. Progeny viruses, which were *lacZ* positive and replicated on 7b (ICP4/ICP27) cells but not E5 (ICP4) or N23 (ICP27) cells, were selected and confirmed by Southern blot analysis. **(B)** List of replication-defective HSV-1 vectors containing deletions in HSV-1 genes involved in cytotoxicity. Depicted are general schematics of first-generation single IE gene-deletion vectors, second-generation double IE gene-deletion vectors, and third-generation triple IE gene-deletion vectors showing the location of the various HSV-1 genes involved in cytotoxicity as well as the site of introduction of a reporter gene cassette. Some recombinants contain the ICP4 IE gene promoter driving expression of thymidine kinase (tk) in place of the normal tk early gene promoter

A third IE gene of interest, ICP0, is both cytotoxic and capable of promiscuous transactivation of a variety of cellular genes *(174–176)*, and can also enhance the level of expression of other viral genes *(174)*. ICP0 appears to cooperatively collaborate with ICP4, for example, to increase the activity of this key viral function *(34,177)*. However, ICP0 is not a promoter binding protein and thus appears to stimulate an event prior to direct promoter activation *(178–182)*. Although ICP0 is a nonessential viral protein, deletion of ICP0 results in decreased viral titers *(104,183)*. Therefore, propagation of high titer stocks of virus deleted for ICP0 in conjunction with the essential IE genes will require the production of a cell line capable of complementing ICP0 as well as ICP4 and ICP27. Generation of such a line has been extremely difficult, because even low-level synthesis of ICP0 is toxic to the host cell *(7)*. The two remaining IE genes, ICP22, which affects the phosphorylation of RNA polymerase II *(184,185)* and ICP47, which affects the processing of MHC class I antigens *(186–188)*, may also need to be deleted depending on the specific therapeutic application.

In addition to the viral IE gene functions, an additional gene remains a target of interest for removal to reduce toxicity. The infecting virus carries in with the particle a tegument component, which has a virion-associated host shut-off (vhs) activity *(16,18)*. Vhs appears to interfere nondiscriminately with mRNA stability *(19)*. Removal of the UL41 (vhs) gene does not affect viral replication *(20)* but enhances the health of the cell upon infection with nonreplicating viral mutants *(189)*.

The deletion of a combination of the four IE regulatory genes (ICP4, ICP22, ICP27, and ICP0) along with UL41 (vhs) have yielded vectors (**Fig. 5B**) that display diminished toxicity for a variety of cell types in culture *(1,5,6,8)*. Second-generation recombinants deleted for both ICP4 and ICP27 *(1,5,6,8)* were less toxic than the ICP4/ICP27 first-generation single IE gene-deleted viruses d120 *(161)* or 5dl1.2 *(190)*. The removal of three of the IE gene (ICP4, ICP22, and ICP27) in the third-generation replication-defective vectors dramatically reduced cytotoxicity *(1,6)* and led to increased duration of vector-mediated transgene expression in neurons in culture *(1)*. Recombinants that failed to express any of the five IE genes, including ICP0, essentially shut down viral gene expression upon infection of noncomplementing cells, rendering the virus safe and noncytotoxic *(6)*. However, transgene expression from these recombinants was transient or undetectable in many cell types *(6)*, owing to the deletion of the ICP0 gene *(6,7)*. Thus, it may prove necessary to retain ICP0 to achieve sufficient expression of the therapeutic gene product. For some specific therapeutic applications, the retention of expression of ICP47, which has recently been shown to downregulate MHC class I antigen expression *(186–188)*, should provide an additional level of protection from immune surveillance at the initiation of infection in vivo.

3.4. Expression of Foreign Transgenes from Genomic HSV Vectors

A variety of recombinant HSV genomic vectors have been employed for expression of reporter and therapeutic genes both in cells of the nervous system and other tissues. In most cases, *lacZ* reporter gene expression detected by X-gal staining and immunocytochemistry is generally restricted to the first few days after inoculation, whether reporter gene expression is driven by HSV lytic cycle gene promoters such as ICPO, ICP4, ICP6, tk, gC *(191–204)*, other viral promoters such as retroviral LTRs *(130,147,193,205–212)*, the SV40 promoter *(213–216)* or the human cytomegalovirus immediate early (HCMV IEp) promoter *(217–220)*, RNA polymerase I and III promoters *(221,222)*, or housekeeping gene promoters *(206,208,223–225)* known to be active in the nervous system such as hypoxanthine phosphoribosyltransferase (HPRT), or neuronal-specific promoters such as those for nerve-specific enolase (NSE) or neurofilament (NF). Of interest, the loss of transgene expression from these vectors does not appear to be owing to clearance of the vector genome from the nervous system, because we have found that vector genomes persist in brain for at least 1 yr after inoculation in amounts that remain unchanged from 1–8 wk postinoculation *(153)*. This suggests that these promoters are no longer active as the virus establishes the latent state. However, these promoters have proven useful for driving high-level transient transgene expression *(2,3,152,226–236)*. Recently, we have demonstrated that the HCMV IE promoter remains active long-term in ligament tissue of rabbits and nonhuman primates following injection of the vector into the knee joint *(4)*.

3.4.1. Exploitation of the HSV Latency Promoter System for Long-Term Transgene Expression in Sensory Neurons

Several laboratories have attempted to employ LAPs to drive foreign gene expression from HSV-based vectors. LAP1 alone appears insufficient to provide long-term transgene expression, as a recombinant in which LAP1 was employed to drive expression of β-globin in the native LAT locus expressed β-globin mRNA at 3 wk postinfection *(126)*, but the intensity of the hybridization signal and the number of neurons expressing the transgene decreased 20-fold during latency *(149)*. A wild-type virus with the β-glucuronidase cDNA downstream of LAP1 expressed β-glucuronidase in the trigeminal ganglion (TG) and brainstem, but the number of transgene expressing cells decreased dramatically over time *(237)*. In other studies, vectors in which LAP1 alone was used to drive *lacZ* expression in either the native LAT *(238)* or ectopic gC locus *(147)* expressed β-gal during acute infection but not during latency. All of these recombinant viruses lack LAP2 sequences, suggesting that LAP2 may be required for long-term transgene expression. As proof of this hypothesis, addition of LAP2 sequence to

LAP1-containing reporter gene cassettes in the gC locus of an HSV recombinant restored long-term expression of the transgene product *(148)*. A recombinant with the *lacZ* reporter gene in the native LAT locus downstream of LAP2 expressed β-gal in TG neurons up to 8 wk postinfection *(239)*. To test the ability of LAP to drive transgene expression from the latent HSV genome, we engineered a recombinant vector with the *lacZ* reporter gene inserted downstream of LAP2 in the native LAT locus, interrupting the LAT sequence. This recombinant expressed β-gal in both mouse TG and dorsal root ganglion (DRG) neurons for up to 56 d postinfection as detected by X-gal staining and reverse transcription polymerase chain reaction (RT-PCR) and *lacZ* mRNA was detectable by RT-PCR in brain following intracranial inoculation up to 28 d postinfection. Additionally, we examined the ability of LAP2 to function as a movable genetic element by introducing a LAP2-*lacZ* cassette into the gC locus of the virus *(128)*. LAP2 driven *lacZ* expression from this recombinant was detected in brain by RT-PCR and in TG by X-gal staining and RT-PCR at least 5 mo postinfection *(128)*. These results indicate that the LAP2 sequence, even in an ectopic site in the viral genome, was capable of driving transgene expression during latency. Our studies of LAP2 suggest that it is active and capable of driving transgene expression in the native or ectopic locus in both brain and in the PNS. We have now employed the LAP2 promoter to drive long-term expression of nerve growth factor (NGF) in primary neurons in culture and in vivo in rodents following inoculation of the vector at a peripheral site *(2,3)*. We have recently shown that LAP2 can be active long-term in tissues other than the nervous system. We detected NGF expression in the ligaments of rabbits over 1 yr and in nonhuman primates several months following injection of the LAP2-NGF expression vector into the knee joint *(4)*. This finding opens up the possibility of using the LAP promoter system to obtain long-term transgene expression in cells other than neurons. Recently it has been reported that the inclusion of an internal ribosome entry site preceding the transgene enhanced long-term transgene expression in the native LAT locus *(240)*, presumably by overcoming interference with translation resulting from the numerous upstream start sites between LAP1 and the LAT intron and within the LAT intron itself. This is not a problem with the use of LAP2 to drive transgene expression because the transgene is not preceded by a translation start site or sequences capable of forming complex secondary structures. Moreover, LAP2 placed upstream of other promoters suggests that LAP2 can maintain their activity during latency although the transcriptional start sites were not reported for these chimeric promoter constructs *(241)*.

3.4.2. Development of Transactivation Systems to Increase and Regulate Foreign Gene Expression

It is possible that the level of transgene expression produced by a promoter construct might prove to be insufficient to have the desired biological effect, or in some instances it may be necessary to regulate the timing of therapeutic gene expression. Therefore, we have pursued the development of constitutive (Gal4-VP16) (**Fig. 6A**) and regulatable (RU486) (**Fig. 6B**) transcriptional regulatory elements inserted into HSV-based vectors in an effort to control the duration and level of transgene expression. Gal4/VP16 is a potent transcriptional activator consisting of the DNA binding domain of the yeast transcriptional activator (TA) Gal4 and the strong acidic activating domain of the HSV transactivator VP16 *(242)*. This fusion protein strongly activates promoters to which it is targeted by the placement of the 17 bp Gal4 binding *(242–244)*. Gal4-VP16 appears to function by perturbing the eukaryotic nucleosomal structure of chromatin-bound promoters containing Gal4 sites, thus releasing these promoters from nucleosome repression during transcription activation *(245,246)*. A recombinant virus was created containing the Gal4-VP16 gene under the control of the HCMV IE promoter in the tk locus *(233)*. The vector produced high levels of Gal4-VP16 mRNA and protein in culture and in vivo following stereotactic injection into brain. Infection of a stably transformed cell line with the CAT gene under the control of a Gal4 sensitive promoter (consisting of 5 GAL4 binding sites upstream of the adenovirus E1b TATA box) resulted in a 35-fold increase in the expression of CAT. We then constructed a double recombinant virus containing the Gal4-VP16 construct under the control of the HCMV IE promoter in the gC locus and the *lacZ* gene under the control of the minimal Gal4 TATA promoter in the tk locus of an ICP4-deleted virus. The double recombinant (**Fig. 6A**) produced β-gal in vitro and following intracranial inoculation in vivo, in contrast to the reporter virus containing only the Gal4 TATA promoter driving *lacZ (233)*. These experiments demonstrated the ability of the Gal4-VP16 transactivator to stimulate a promoter containing a Gal4 binding site in the context of the viral genome, although this system lacks the ability to regulate the duration of transgene expression.

We further modified the transactivation system to allow for regulation of transgene expression through the use of an inducible promoter construct. A truncated form of the hormone-binding domain (HBD) of the progesterone receptor, which fails to bind progesterone but is activated by binding of the steroid analogue RU486 was used to form a fusion HBD-Gal4-VP16 transactivator, expressed under the transcriptional control of the HCMV IE promoter *(234)*. With this system, the addition or removal of the drug (RU486) constitutes the on/off regulatory switch that controls the duration of transgene expression. A viral recombinant was constructed containing the transactivator in the gC locus of an ICP4-deleted replication incompetent vector and a reporter cassette consisting of 5 Gal4 binding sites upstream of the adenoviral E1b TATA box driving *lacZ* in the tk locus (**Fig. 6B**). In vitro, β-gal expression was induced 30-fold by treatment of the cultures with RU486 (10^{-7} M). In vivo, following injection of the vector into hippocampus of rodents, treatment with RU486 induced expression

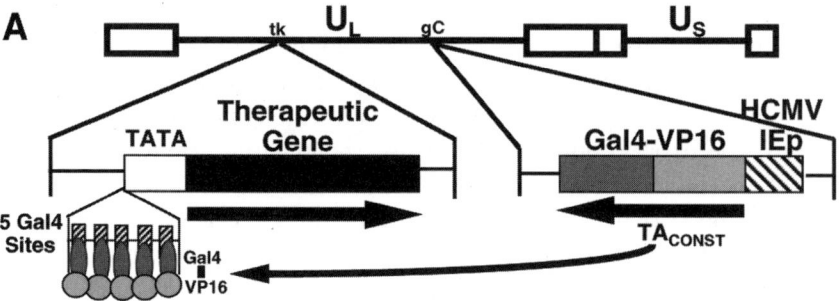

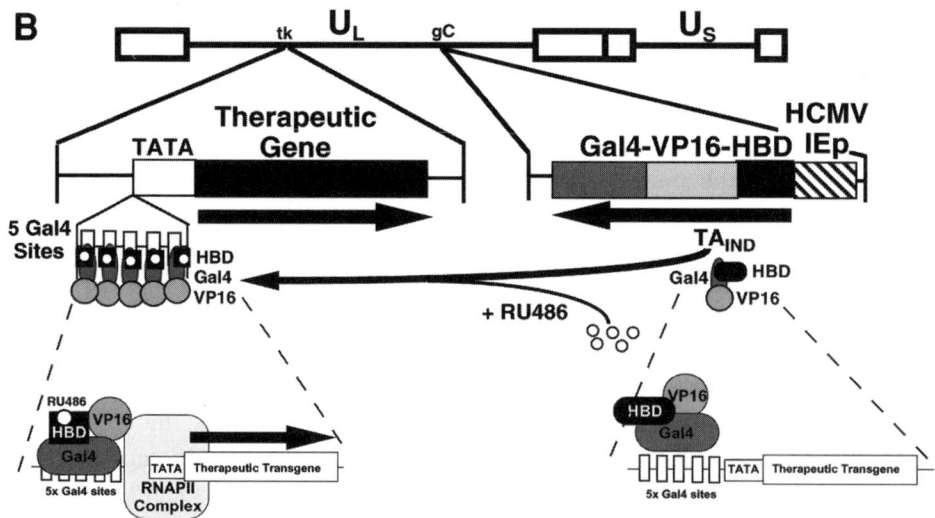

Fig. 6. HSV-1 recombinant vectors containing the constitutive and inducible Gal4-VP16 transactivating function. (**A**) A replication-defective recombinant HSV-1 vector was constructed to contain a minimal promoter with five tandem Gal4 binding sites driving a transgene and a cassette that expresses the Gal4-VP16 constitutive transactivator (TA$_{CONST}$) from the strong HCMV IE gene promoter. The Gal4-VP16 TA$_{CONST}$ was able to boost the level transgene expression from the Gal4 sensitive promoter both in vitro and in vivo. (**B**) A replication-defective recombinant HSV-1 vector was constructed to contain a minimal promoter with five tandem Gal4 binding sites driving a transgene and a cassette that expresses the Gal4-VP16-HBD inducible transactivator (TA$_{IND}$) from the strong HCMV IE gene promoter. Addition of the progesterone receptor hormone-binding domain (HBD) to the constitutive TA creates a molecule that is only transcriptionally active when the drug RU486 is added, as illustrated at the bottom of the figure. The amount of RU486 added and the timing of drug addition controls the level and duration of transgene expression.

of β-gal *(234)*. Thus both constitutive (**Fig. 6A**) and inducible (**Fig. 6B**) transcriptional enhancers function within the context of the HSV genome to regulate transgene expression in the nervous system.

3.5. HSV-1 Vector Targeting

Various attempts to alter the host range of different viruses have been reported. One approach is to construct pseudotype or hybrid vectors by combining elements from different viruses. Examples of this approach include the production of a retroviral (MoMLV) or a lentiviral (HIV-1) nucleocapsid possessing the vesicular stomatitis virus glycoprotein (VSV-G) within the envelope *(247,248)*. The HIV-based vector with VSV tropism demonstrated efficient transfer, integration, and sustained expression of a transgene in adult rat brains. However, the pseudotype approach is limited by the availability of cell type-specific viral-surface proteins. Modification of purified virions by chemical attachment of a ligand capable of binding to a specific cell type has also been explored. Ligands were either chemically attached to viral surface proteins *(249)* or antibodies against viral surface proteins were biochemically linked to a second antibody that recognized a cell-surface receptor *(250)*. Targeting of retroviruses to melanoma cells by the fusion of a single-chain variable fragment antibody directed against a melanoma-specific antigen with the amphotrophic murine leukemia virus envelope separated by a cleavable protease site *(251)* has been demonstrated in vitro. A Sindbis virus displaying a protein A:env chimeric protein was also engineered in a manner that generalizes the applicability of MAb specificity for targeted infection. The recombinant virus had a strong affinity for the F_c region of various mammalian IgGs and cell-specific targeting was accomplished according to the specificity of the antibody *(252,253)*. Although this procedure was highly efficient in vitro, the stability of the protein A-antibody interaction in the blood stream has not been evaluated. A chimeric MoMLV virus with both a wild-type env gene and an env-erythropoietin (EPO) fusion gene displayed infectivity for human cells bearing the EPO receptor *(254)*. Likewise, it has been shown that a single-chain antibody F_{ab} fragment fused to the MoMLV env gene product recognized its target epitope and that a corresponding viral vector could transduce cells that were resistant to infection by the parental MoMLV *(255–257)*. Although these recombinant vectors demonstrated new tropisms, they showed low levels of infectivity. Recently, the in vitro tropism of AAV has been expanded by inclusion of an Arg-Gly-Asp (RGD) motif in the natural AAV viral capsid protein *(251,258)* that resulted in the acquisition of the ability to infect normally refractory cells. Although retroviral targeting has been achieved by the addition of either Ram-1 *(259,260)*, IGF *(261)*, or EGF *(262–264)* ligands as extensions to the native surface (SU) glycoprotein, the efficiency was low because the virus was internalized into endosomes. This was overcome by the addition of protease cleavage sites between the ligand domain and the SU glycoprotein that can be cleaved by matrix metalloprotease addition, which enables the targeting by the novel ligand but fusion through the native SU glycoprotein *(261,265)*. Recombinant adenovirus (Ad)-encoding chimeric proteins have also been engineered *(266)*. One of the two vectors re-directed virus binding to an integrin *(267)*, whereas the other redirected binding to heparan sulfate *(268)* by fusion of the Ad fiber with RGD or polylysine sequences, respectively. Although binding to the fiber receptor still occurred, these viruses demonstrated increased transduction in multiple cell types lacking high levels of Ad fiber receptor *(266)*. Rogers et al. *(269)* cross-linked a F_{ab} fragment from a neutralizing MAb specific for adenoviral knob protein to basic fibroblast growth factor (FGF) *(269)* to allow infection, albeit inefficient, of cell lines expressing the FGF receptor. This approach of using a bi-specific protein has enabled re-targeting of the virus to a variety of cell types *(270–273)*. Recent studies have introduced novel binding peptides into the region of the adenovirus fiber knob to achieve re-targeting *(274)*. In a similar manner, Ad was targeted to the tetanus toxin receptors on motor neurons (MNs) using the $H_C C$ fragment of tetanus toxin tethered to the virus hexon *(275)*. Targeting to primary neurons was demonstrated in cell cultures of spinal cord and to MNs of the mouse brainstem following intramuscular inoculation. In both cases, infectivity was low although muscle cell transduction was substantially reduced and demonstrates the potential feasibility of this approach that may prove feasible for use with HSV vectors.

3.5.1. Targeting HSV-1 Attachment Through gC:EPO

As described earlier, recombinant viruses deleted for gC as well as the polylysine domain of gB (KgBpK$^-$gC$^-$) demonstrated the importance of HS binding for virus entry *(73)*. We have replaced the HS ligand present in gC with erythropoietin (EPO), a molecule that recognizes the EPO receptor (EPO-R) present on a number of cell types, to create KgBpK$^-$gC:EPO$_2$, that was able to bind to and enter EPO-R bearing FD-EPO cells *(276)*, however, the virus was rapidly degraded in endosomal/lysosomal vesicles. Whether infection will occur when a gD-specific receptor is also present on these cells is unknown.

3.5.2. Targeting HSV-1 by Substituting gD With the VSV-G Glycoprotein

Glycoprotein D plays an essential role in the attachment/entry process. Of interest was whether an alternative binding/fusion protein, such as the VSV-G spike glycoprotein, which binds to cell surface receptor and mediates fusion of the VSV envelope and endosomal membrane following endocytosis of VSV-G *(277–280)*, could substitute for gD in the HSV entry process. VSV-G has been employed to pseudotype retroviruses that enter cells via receptor-mediated endocytosis *(281–284)* and measles virus *(285)* that normally bind to and enter cells at the cell-surface membrane similar to HSV. Viruses deleted for Us3 to Us8 (KΔUs3-8) are deficient in virus entry due to the deletion of gD. Using transient complementation assays, the expression of VSV-G from a plasmid was able to complement the gD deletion mutant (KΔUs3-8), which could be neutralized using VSV-G specific antibodies *(58)*. Moreover, we observed large amounts of VSV-G incorporation into purified virus envelopes following transient complementation, which proved to be more efficient than with any VSV-G:gD or VSV-G:gB recombinant glycoprotein composed of shuffled cytoplasmic, transmembrane, and external domains. These experiments demonstrated that (1) a foreign glycoprotein could be incorporated into the virus envelope if expressed during virus replication, and (2) that VSV-G can mediate virus infection in the absence of gD. We are currently addressing whether VSV-G can substitute for HSV-1 gD in the entry process in the background of a virus deleted for the Us3-8 genes, which includes deletion of gD (Us6).

3.5.3. Targeting of HSV-1 Using Soluble Bridging Proteins

In this approach, a region of HveC/nectin-1 capable of binding gD are fused to polypeptides to create bi-specific molecules capable of binding simultaneously to gD and novel cell receptors while blocking gD binding to cell-associated HveC/nectin-1. These bridging molecules are designed to mediate indirect virus interaction with novel receptors triggering fusion of the viral envelope with cell membranes. We have engineered the C-terminal 153 amino acids of HveC/nectin-1 containing the V-loop fused to ligands that bind to cell-specific receptors. We are currently testing binding/entry on CHO-derived cell lines expressing the respective cell-specific receptors using virus pre-mixed with these novel adapter molecules. We are hopeful that this soluble adapter system will prove useful for HSV vector re-targeting.

4. Summary

Over the past several years, we have expended substantial efforts directed at understanding the biology of HSV in order to exploit the natural features of HSV latency to create a vector for the delivery of therapeutic transgenes to the nervous system and other tissues. We now have in hand a number of vectors that demonstrate very low cytotoxicity for neurons and other cells in culture and prolonged expression of a reporter transgene. We have demonstrated the ability of these vectors to express transgenes (i.e., NGF) in peripheral sensory ganglia and have the means to switch the transgene on and off in vivo. We have also dramatically expanded the virus payload and have begun to address the issue of targeting the virus to specific cells. We now propose to use these highly debilitated HSV vectors in animal models, testing the delivery of therapeutic gene products by the vector. The following chapters deal with HSV vector-mediated transduction of neurons and other cell types with the ultimate goal of treating human disease.

References

1. Krisky, D., Wolfe, D., Goins, W., Marconi, P., Ramakrishnan, R., Mata, M., et al. (1998) Deletion of multiple immediate early genes from herpes simplex virus reduces cytotoxicity and permits long-term gene expression in neurons. *Gene Ther.* **5,** 1593–1603.
2. Goins, W. F., Lee, K. A., Cavalcoli, J. D., O'Malley, M. E., DeKosky, S. T., Fink, D. J., and Glorioso, J. C. (1999) Herpes simplex virus type 1 vector-mediated expression of nerve growth factor protects dorsal root ganglia neurons from peroxide toxicity. *J. Virol.* **73,** 519–532.
3. Goins, W. F., Yoshimura, N., Ozawa, H., Yokoyama, T., Phelan, M., Bennet, N., et al. (2001) Herpes simplex virus vector-mediated nerve growth factor expression in bladder and afferent neurons: potential treatment for diabetic bladder dysfunction. *J. Urol.* **165,** 1748–1754.
4. Wolfe, D., Goins, W., Kaplan, T., Capuano, S., Fradette, J., Murphey-Corb, M., et al. (2001) Systemic accumulation of biologically active nerve growth factor following intra-articular herpesvirus gene transfer. *Mol. Ther.* **3,** 61–69.
5. Marconi, P., Krisky, D., Oligino, T., Poliani, P., Ramakrishnan, R., Goins, W., et al. (1996) Replication-defective HSV vectors for gene transfer in vivo. *Proc. Natl. Acad. Sci. USA* **93,** 11319–11320.
6. Samaniego, L., Neiderhiser, L., and DeLuca, N. (1998) Persistence and expression of the herpes simplex virus genome in the absence of immediate-early proteins. *J. Virol.* **72,** 3307–3320.
7. Samaniego, L., Wu, N., and DeLuca, N. (1997) The herpes simplex virus immediate-early protein ICP0 affects transcription from the viral genome and infected-cell survival in the absence of ICP4 and ICP27. *J. Virol.* **71,** 4614–4625.
8. Wu, N., Watkins, S., Schaffer, P., and DeLuca, N. (1996) Prolonged gene expression and cell survival after infection by a herpes simplex virus mutant defective in the immediate-early genes encoding ICP4, ICP27, and ICP22. *J. Virol.* **70,** 6358–6368.
9. Homa, F. L. and Brown, J. C. (1997) Capsid assembly and DNA packaging in herpes simplex virus. *Rev. Med. Virol.* **7,** 107–122.
10. Steven, A. C. and Spear, P. G. (1997) Herpesvirus capsid assembly and envelopment, in *Structural Biology of Viruses* (Chiu, W., Burnett, R., and Garcea, R., eds.), Oxford University Press, New York, NY, pp. 512–533.
11. Newcomb, W. and Brown, J. (1994) Induced extrusion of DNA from the capsid of herpes simplex virus type 1. *J. Virol.* **68,** 443–440.
12. McGeoch, D. J., Cunningham, C., McIntyre, G., and Dolan, A. (1991) Comparative sequence analysis of the long repeat regions and adjoining parts of the long unique regions in the genomes of herpes simplex viruses types 1 and 2. *J. Gen. Virol.* **72,** 3057–3075.

13. McGeoch, D. J., Dalrymple, M. A., Davison, A. J., Dolan, A., Frame, M. C., McNab, D., et al. (1988) The complete DNA sequence of the long unique region in the genome of herpes simplex virus type 1. *J. Gen. Virol.* **69,** 1531–1574.
14. McGeoch, D. J., Dolan, A., Donald, S., and Brauer, D. H. K. (1986) Complete DNA sequence of the short repeat region in the genome of herpes simplex virus type 1. *Nucleic Acids Res.* **14,** 1727–1744.
15. McGeoch, D. J., Dolan, A., Donald, S., and Rixon, F. J. (1985) Sequence determination and genetic content of the short unique region in the genome of herpes simplex virus type 1. *J. Mol. Biol.* **181,** 1–13.
16. Kwong, A. D. and Frenkel, N. (1987) Herpes simplex virus-infected cells contain a function(s) that destablizes both host and viral mRNAs. *Proc. Natl. Acad. Sci. USA* **84,** 1926–1930.
17. Kwong, A. D. and Frenkel, N. (1989) The herpes simplex virus virion host shutoff function. *J. Virol.* **63,** 4834–4839.
18. Kwong, A. D., Kruper, J. A., and Frenkel, N. (1988) Herpes simplex virus virion host shutoff function. *J. Virol.* **62,** 912–921.
19. Oroskar, A. and Read, G. (1989) Control of mRNA stability by the virion host shutoff function of herpes simplex virus. *J. Virol.* **63,** 1897–1906.
20. Read, G. S. and Frenkel, N. (1983) Herpes simplex virus mutants defective in the virion-associated shutoff of host polypeptide synthesis and exhibiting abnormal synthesis of α (immediate early) viral polypeptides. *J. Virol.* **46,** 498–512.
21. Ace, C. I., McKee, T. A., Ryan, J. M., Cameron, J. M., and Preston, C. M. (1989) Construction and characterization of a herpes simplex virus type 1 mutant unable to transinduce immediate-early gene expression. *J. Virol.* **63,** 2260–2269.
22. Batterson, W. and Roizman, B. (1983) Characterization of the herpes simplex virion-associated factor responsible for the induction of alpha-genes. *J. Virol.* **46,** 371–377.
23. Campbell, M. E. M., Palfeyman, J. W., and Preston, C. M. (1984) Identification of herpes simplex virus DNA sequences which encode a trans-acting polypeptide responsible for stimulation of immediate early transcription. *J. Mol. Biol.* **180,** 1–19.
24. Kristie, J. and Roizman, B. (1987) Host cell proteins bind to the cis-acting site required for virion-mediated induction of herpes simplex virus 1 alpha genes. *Proc. Natl. Acad. Sci. USA* **84,** 71–75.
25. McKnight, J. L. C., Kristie, T. M., and Roizman, B. (1987) Binding of the virion protien mediating a gene induction in herpes simplex virus 1-infected cells to its cis site requires cellular proteins. *Proc. Natl. Acad. Sci. USA* **84,** 7061–7065.
26. Post, L., Mackem, S., and Roizman, B. (1981) Regulation of alpha genes of herpes simplex virus: Expression of chimeric genes produced by fusion of thymidine kinase with alpha gene promoters. *Cell* **24,** 555–565.
27. Spear, P. (1993) Membrane fusion induced by herpes simplex virus, in *Viral Fusion Mechanisms* (Bentz, J., eds.), CRC Press, Boca Raton, FL, pp. 201–232.
28. Spear, P. G. (1993) Entry of alphaherpesviruses into cells. *Sem. Virol.* **4,** 167–180.
29. Cai, W., Gu, B., and Person, S. (1988) Role of glycoprotein B of herpes simplex virus type 1 in viral entry and cell fusion. *J. Virol.* **62,** 2596–2604.

30. Desai, P., Schaffer, P., and Minson, A. (1988) Excretion of non-infectious virus particles lacking glycoprotein H by a temperature-sensitive mutant of herpes-simplex virus type 1: evidence that gH is essential for virion infectivity. *J. Gen. Virol.* **69,** 1147–1156.

31. Hutchinson, L., Browne, H., Wargent, V., Davis-Poynter, N., Primorac, S., Goldsmith, K., et al. (1992) A novel herpes simplex virus glycoprotein, gL, forms a complex with glycoprotein H (gH) and affects normal folding and surface expression of gH. *J. Virol.* **66,** 2240–2250.

32. Ligas, M. and Johnson, D. (1988) A herpes simplex virus mutant in which glycoprotein D sequences are replaced by β-galactosidase sequences binds to but is unable to penetrate into cells. *J. Virol.* **62,** 1486–1494.

33. Zhu, Q. and Courtney, R. J. (1988) Chemical crosslinking of glycoproteins on the envelope of herpes simplex virus. *Virology* **167,** 377–384.

34. Zhu, Q. and Courtney, R. J. (1994) Chemical cross-linking of virion envelope and tegument proteins of herpes simplex virus type 1. *Virology* **204,** 590–599.

35. Ali, M. A. (1990) Oligomertization of herpes simplex virus glycoprotein B occurs in the endoplasmic reticulum and a 102 amino acid cytosolic domain is dispensable for dimer assembly. *Virology* **178,** 588–592.

36. Claesson-Welsh, L. and Spear, P. G. (1986) Oligomerization of herpes simplex virus glycoprotein B. *J. Virol.* **60,** 803–806.

37. Highlander, S. L., Goins, W. F., Person, S., Holland, T. C., Levine, M., and Glorioso, J. C. (1991) Oligomer formation of the gB glycoprotein of herpes simplex virus type 1. *J. Virol.* **65,** 4275–4283.

38. Laquerre, S., Argnani, R., Zucchini, S., Manservigi, R., and Glorioso, J. C. (1998) Herpes simplex virus type 1 glycoprotein B requires a cysteine residue at position 633 for folding, processing and incorporation into mature infectious virus particles. *J. Virol.* **72,** 4940–4949.

39. Qadri, I., Gimeno, C., Navarro, D., and Pereira, L. (1991) Mutations in conformation-dependent domains of herpes simplex virus 1 glycoprotein B affect the antigenic properties, dimerization, and transport of the molecule. *Virology* **180,** 135–152.

40. Sarmiento, M., Haffey, M., and Spear, P. G. (1979) Membrane proteins specified by herpes simplex viruses. III. Role of glycoprotein VP7 (B) in virion infectivity. *J. Virol.* **29,** 1149–1158.

41. Sarmiento, M. and Spear, P. G. (1979) Membrane proteins specified by herpes simplex viruses. IV. Conformation of the virion glycoprotein designated VP7(B). *J. Virol.* **29,** 1159–1167.

42. Johnson, D., Frame, M., Ligas, M., Cross, A., and Stow, N. (1988) Herpes simplex virus immunoglobulin G Fc receptor activity depends on a complex of two viral glycoproteins, gE and gI. *J. Virol.* **62,** 1347–1354.

43. Banfield, B. W. and Tufaro, F. (1990) Herpes simplex virus particles are unable to traverse the secretory pathway in the mouse L-cell mutant gro29. *J. Virol.* **64,** 5716–5729.

44. Chatterjee, S. and Sarkar, S. (1992) Studies on endoplasmic reticulum-golgi complex cycling pathway in herpes simplex virus-infected and brefeldin A-treated human fibroblast cells. *Virology* **191,** 327-337.
45. Cheung, P., Banfield, B. W., and Tufaro, F. (1991) Brefeldin A arrests the maturation and egress of herpes simplex virus particles during infection. *J. Virol.* **65,** 1893–1904.
46. Serafini-Cessi, F., Dall'Olio, F., Scannavini, M., and Campadelli-Fiume, G. (1983) Processing of herpes simplex virus-1 glycans in cells defective in glycosyl transferases of the Golgi system: relationship to cell fusion and virion egress. *Virology* **131,** 59–70.
47. Johnson, D. and Spear, P. (1982) Monensin inhibits the processing of herpes simplex virus glycoproteins, their transport to the cell surface, and the egress of virions from infected cells. J. Virol. 43, 1102-1112.
48. Jones, F. and Grose, C. (1988) Role of cytoplasmic vacuoles in varicella-zoster virus glycoprotein trafficking and virion envelopment. *J. Virol.* **62,** 2701–2711.
49. Stackpole, C. W. (1969) Herpes-type virus of the frog renal adenocarcinoma. I. Virus development in tumor transplants maintained at low temperature. *J. Virol.* 4, 75–93.
50. Roizman, B. and Sears, A. (1996) Herpes simplex viruses and their replication, in *Fields Virology* (Fields, B., Knipe, D., Howley, P., Chanock, R., Hirsch, M., Melnick, J., eds.), Lippincott-Raven, Philadelphia, PA, pp. 2231–2295.
51. Raviprakash, K., Rasile, L., Ghosh, K., and Ghosh, H. P. (1990) Shortened cytoplasmic domain affects intracellular transport but not nuclear localization of a viral glycoprotein. *J. Biol. Chem.* **265,** 1777–1782.
52. Huff, V., Cai, W., Glorioso, J. C., and Levine, M. (1988) The carboxy-terminal 41 amino acids of herpes simplex virus type 1 glycoprotein B are not essential for production of infectious virus particles. *J. Virol.* **62,** 4403–4406.
53. Skoff, A. M. and Holland, T. C. (1993) The effect of cytoplasmic domain mutations on membrane anchoring and glycoprotein processing of herpes simplex virus type 1 glycoprotein C. *Virology* **196,** 804–816.
54. Solomon, K. A., Robbins, A. K., and Enquist, L. W. (1991) Mutations in the C-terminal hydrophobic domain of pseudoraabies virus gIII affect both membrane anchoring and protein export. J. Virol. 65, 5952-5960.
55. Solomon, K. A., Robbins, A. K., Whealy, M. E., and Enquist, L. W. (1990) The putative cytoplasmic domain of the pseudorabies virus envelope protein gIII, the herpes simplex virus type 1 glycoprotein C homolog, is not required for normal export and localization. *J. Virol.* **64,** 3516–3521.
56. Gilbert, R., Ghosh, K., Rasile, L., and Ghosh, H. P. (1994) Membrane anchoring domain of herpes simplex virus glycoprotein gB is sufficient for nuclear envelope localization. *J. Virol.* **68,** 2272–2285.
57. Li, Y., Van Drunen Littel-Van Den Hurk, S., Liang, X., and Babiuk, L. A. (1997) Functional analysis of the transmembrane anchor region of bovine herpesvirus 1 glycoprotein gB. *Virology* **228,** 39–54.

58. Anderson, D. B., Laquerre, S., Ghosh, K., Ghosh, H. P., Goins, W. F., Cohen, J. B., and Glorioso, J. C. (2000) Pseudotyping of glycoprotein D (gD) deficient herpes simplex virus type 1 (HSV-1) with vesicular stomatitis virus glycoprotein G (VSV-G) enables mutant virus attachment and entry. *J. Virol.* **74,** 2481–2487.

59. Dingwell, K., Brunetti, C., Hendricks, R., Tang, Q., Tang, M., Rainbow, A., and Johnson, D. (1994) Herpes simplex virus glycoproteins E and I facilitate cell-to-cell spread in vivo and across junctions of cultured cells. *J. Virol.* **68,** 834–845.

60. Dingwell, K. S., Doering, L. C., and Johnson, D. C. (1995) Glycoproteins E and I facilitate neuron-to-neuron spread of herpes simplex virus. *J. Virol.* **69,** 7087–7098.

61. Mettenleiter, T. C. (1994) Initiation and spread of α-herpesvirus infections. *Trends Micro.* **2,** 2–3.

62. Fuller, A. O. and Lee, W. C. (1992) Herpes simplex virus type 1 entry through a cascade of virus-cell interactions requires different roles of gD and gH in penetration. *J. Virol.* **66,** 5002–5012.

63. Gruenheid, S., Gatzke, L., Meadows, H., and Tufaro, F. (1993) Herpes simplex virus infection and propagation in a mouse L cell mutant lacking heparan sulfate proteoglycans. *J. Virol.* **67,** 93–100.

64. Herold, B., Visalli, R., Susmarski, N., Brandt, C., and Spear, P. (1994) Glycoprotein C-independent binding of herpes simplex virus to cells requires cell surface heparan sulfate and glycoprotein B. *J. Gen. Virol.* **75,** 1211–1222.

65. Shih, M., Wudunn, D., Montgomery, R., Esko, J., and Spear, P. (1992) Cell surface receptors for herpes simplex virus are heparan sulfate proteoglycans. *J. Cell Biol.* **116,** 1273–1281.

66. Spear, P. G., Shieh, M. T., Herold, B. C., WuDunn, D., and Koshy, T. I. (1992) Heparan sulfate glycosaminoglycans as primary cell surface receptors for herpes simplex virus. *Adv. Exp. Med. Biol.* **313,** 341–353.

67. Wudunn, D. and Spear, P. (1989) Initial interaction of herpes simplex virus with cells is binding to heparan sulfate. *J. Virol.* **63,** 52–58.

68. Banfield, B., Leduc, Y., Esford, L., Visalli, R., Brandt, C., and Tufaro, F. (1995) Evidence for an interaction of herpes simplex virus with chondroitin sulfate proteoglycans during infection. *Virology* **208,** 531–539.

69. Williams, R. K. and Straus, S. E. (1997) Specificity and affinity of binding of herpes simplex virus type 2 glycoprotein B to glycosaminoglycans. *J. Virol.* **71,** 1375–1380.

70. Tal-Singer, R., Peng, C., Ponce de Leon, M., Abrams, W. R., Banfield, B. W., Tufaro, F., et al. (1995) Interaction of herpes simplex virus glycoprotein gC with mammalian cell surface molecules. *J. Virol.* 69, 4471–4483.

71. Herold, B. C., Gerber, S. I., Polonsky, T., Belval, B. J., Shaklee, P. N., and Holme, K. (1995) Identification of structural features of heparin required for inhibition of herpes simplex virus type 1 binding. *Virology* **206,** 1108–1116.

72. Li, Y., van Drunen Littel-van den Hurk, S., Babiuk, L. A., and Liang, X. (1995) Characterization of cell-binding properties of bovine herpesvirus 1 glycoproteins B, C, and D: identification of a dual cell-binding function of gB. *J. Virol.* **69,** 4758–4768.

73. Laquerre, S., Argnani, R., Anderson, D. B., Zucchini, S., Manservigi, R., and Glorioso, J. C. (1998) Heparan sulfate proteoglycan binding by herpes simplex virus type 1 glycoproteins B and C which differ in their contribution to virus attachment, penetration, and cell-to-cell spread. *J. Virol.* **72,** 6119–6130.

74. Manservigi, R., Spear, P. G., and Buchan, A. (1977) Cell fusion induced by herpes simplex virus is promoted and suppressed by different viral glycoproteins. *Proc. Natl. Acad. Sci. USA* **74,** 3913–3917.

75. Pogue-Geile, K. L., Lee, G. T., Shapira, S. K., and Spear, P. G. (1984) Fine mapping of mutations in the fusion-inducing MP strain of herpes simplex virus type 1. *Virology* **136,** 100–109.

76. Handler, C. G., Cohen, G. H., and Eisenberg, R. J. (1996) Cross-linking of glycoprotein oligomers during herpes simplex virus type 1 entry. *J. Virol.* **70,** 6076–6082.

77. Gage, P. J., Levine, M., and Glorioso, J. C. (1993) Syncytium-inducing mutations localize to two discrete regions within the cytoplasmic domain of herpes simplex virus type 1 glycoprotein B. *J. Virol.* **67,** 2191–2201.

78. Hutchinson, L., Graham, F. L., Cai, W., Debroy, C., Person, S., and Johnson, D. C. (1993) Herpes simplex virus (HSV) glycoproteins B and K inhibit cell fusion induced by HSV syncytial mutants. *Virology* **196,** 514–531.

79. Navarro, D., Paz, P., and Pereira, L. (1992) Domains of herpes simplex virus 1 glycoprotein B that function in virus penetration, cell-to-cell spread, and cell fusion. *Virology* **186,** 99–112.

80. Pertel, P. E. and Spear, P. G. (1996) Modified entry and syncytium formation by herpes simplex virus type 1 mutants selected for resistance to heparin inhibition. *Virology* **226,** 22–33.

81. Montgomery, R., Warner, M., Lum, B., and Spear, P. (1996) Herpes simplex virus 1 entry into cells mediated by a novel member of the TNF/NGF receptor family. *Cell* **87,** 427–436.

82. Nicola, A. V., Ponce de Leon, M., Xu, R., Hou, W., Whitbeck, J. C., Krummenacher, C., et al. (1998) Monoclonal antibodies to distinct sites on herpes simplex virus (HSV) glycoprotein D block HSV binding to HVEM.J. *J. Virol.* **72,** 3595–3601.

83. Dean, H. J., Terhune, S. S., Shieh, M. T., Susmarski, N., and Spear, P. G. (1994) Single amino acid substitutions in gD of herpes simplex virus 1 confer resistance to gD-mediated interference and cause cell-type-dependent alterations in infectivity. *Virology* **199,** 67–80.

84. Nicola, A. V., Peng, C., Lou, H., Cohen, G. H., and Eisenberg, R. J. (1997) Antigenic structure of soluble herpes simplex virus (HSV) glycoprotein D correlates with inhibition of HSV infection. *J. Virol.* **71,** 2940–2946.

85. Whitbeck, J., Peng, C., Lou, H., Xu, R., Willis, S., Ponce de Leon, M., et al. (1997) Glycoprotein D of herpes simplex virus (HSV) binds directly to HVEM, a mediator of HSV entry. *J. Virol.* **71,** 6083–6093.

86. Wechuck, J., Wolfe, D., Goins, W., Wendell, S., Goff, J., Greenberger, J., Ataai, M., and Glorioso, J. 2003 Gene transfer to hematopoietic stem cells using replication defective herpes simplex virus type 1 (HSV-1) vectors. *J. Virol.*(submitted).

87. Carfi, A., Willis, S. H., Whitbeck, J. C., Krummenacher, C., Cohen, G. H., Eisenberg, R. J., and Wiley, D. C. (2001) Herpes simplex virus glycoprotein D bound to the human receptor HveA. *Mol. Cell* **8,** 169–179.

88. Warner, M. S., Geraghty, R. J., Martinez, W. M., Montgomery, R. E., Whitbeck, J. C., Xu, R., et al. (1998) A cell surface protein with herpesvirus entry activity (HveB) confers susceptibility to infection by mutants of herpes simplex virus type 1, herpes simplex virus type 2, and pseudorabies virus. *Virology* **246,** 179–189.

89. Shukla, D., Liu, J., Blaiklock, P., Shworak, N. W., Bai, X., Esko, J. D., Cohen, G. H., et al. (1999) A novel role for 3-O-sulfated heparan sulfate in herpes simplex virus 1 entry. *Cell* 99, 13–22.

90. Geraghty, R. J., Krummenacher, C., Cohen, G. H., Eisenberg, R. J., and Spear, P. G. (1998) Entry of alphaherpesviruses mediated by poliovirus receptor-related protein 1 and poliovirus receptor. *Science* **280,** 1618–1620.

91. Krummenacher, C., Nicola, A., Whitbeck, J., Lou, H., Hou, W., Lambris, J., et al. (1998) Herpes simplex virus glycoprotein D can bind to poliovirus receptor-related protein 1 or herpesvirus entry mediator, two structurally unrelated mediators of virus entry. *J. Virol.* **72,** 7064–7074.

92. Fuller, A. O. and Spear, P. G. (1985) Specificities of monoclonal and polyclonal antibodies that inhibit adsorption of herpes simplex virus to cells and lack of inhibition by potent neutralizing antibodies. *J. Virol.* **55,** 475-482.

93. Highlander, S., Sutherland, S., Gage, P., Johnson, D., Levine, M., and Glorioso, J. (1987) Neutralizing monoclonal antibodies specific for herpes simplex virus glycoprotein D inhibit virus penetration. *J. Virol.* **61,** 3356–3364.

94. Butcher, M., Raviprakash, K., and Ghosh, H. P. (1990) Acid pH-induced fusion of cells by herpes simplex virus glycoproteins gB and gD. *J. Biol. Chem.* **265,** 5862–5868.

95. Roizman, B. and Sears, A. (1990) Herpes simplex viruses and their replication, in *Virology* (Fields, B., Knipe, D., Chanock, R., Hirsch, M., Melnick, J. Monath, T., and Roizman, B., eds.), Raven Press, Ltd., New York, NY, pp. 1795–1841.

96. Roizman, B. and Sears, A. E. (1993) Herpes simplex viruses and their replication, in *The Human Herpesviruses* (Roizman, B., Whitley, R. J., and Lopez, C., eds.), Raven Press, New York, NY, pp. 11–68.

97. Honess, R. and Roizman, B. (1974) Regulation of herpes simplex virus macromolecular synthesis. I. Cascade regulation of the synthesis of three groups of viral proteins. *J. Virol.* 14, 8–19.

98. Honess, R. W. and Roizman, B. (1975) Regulation of herpes virus macromolecular synthesis: sequential transition of polypeptide synthesis requires functional viral polypeptides. *Proc. Natl. Acad. Sci. USA* **72,** 1276–1280.

99. DeLuca, N. A. and Schaffer, P. A. (1985) Activation of immediate-early, early, and late promoters by temperature-sensitive and wild-type forms of herpes simplex virus type 1 protein ICP4. *Mol. Cell. Biol.* **5,** 1997–2008.

100. Dixon, R. A. F. and Schaffer, P. A. (1980) Fine-structure mapping and functional analysis of temperature-sensitive mutants in the gene encoding the herpes simplex virus type 1 immediate early protein VP175. *J. Virol.* **36,** 189–203.

101. O'Hare, P. and Hayward, G. (1985) Three trans-acting regulatory proteins of herpes simplex virus modulate immediate-early gene expression in a pathway involving positive and negative feed regulation. *J. Virol.* **56,** 723–733.

102. O'Hare, P. and Hayward, G. S. (1985) Evidence for a direct role for both the 175,000 and 110,000-molecular-weight immediate-early protein of herpes simplex vius in transactivation of delayed-early promoters. *J. Virol.* **53,** 751–760.

103. Preston, C. (1979) Abnormal properties of an immediate early polypeptide in cells infected with the herpes simplex virus type 1 mutant tsK. *J. Virol.* **32,** 357–369.

104. Sacks, W. R. and Schaffer, P. A. (1987) Deletion mutants in the gene encoding the herpes simplex virus type 1 immediate-early protein ICP0 exhibit impaired growth in cell culture. *J. Virol.* **61,** 829–839.

105. Stow, N. and Stow, E. (1986) Isolation and characterization of a herpes simplex virus type 1 mutant containing a deletion within the gene encoding the immediate early polypeptide Vmw 110. *J. Gen. Virol.* **67,** 2571–2585.

106. Watson, R. and Clements, J. (1980) A herpes simplex virus type 1 function continuously required for early and late virus RNA synthesis. *Nature* **285,** 329–330.

107. Holland, L. E., Anderson, K. P., Shipman, C., and Wagner, E. K. (1980) Viral DNA synthesis is required for efficient expression of specific herpes simplex virus type 1 mRNA. *Virology* **101,** 10–24.

108. Mavromara-Nazos, P. and Roizman, B. (1987) Activation of herpes simplex virus 1 γ2 genes by viral DNA replication. *Virology* **161,** 593–598.

109. Cook, M. L. and Stevens, J. G. (1973) Pathogenesis of herpetic neuritis and ganglionitis in mice: evidence of intra-axonal transport of infection. *Infect. Immun.* **7,** 272–288.

110. Stevens, J. G. (1989) Human herpesviruses: a consideration of the latent state. *Microbiol. Rev.* **53,** 318–332.

111. Deshmane, S. L. and Fraser, N. W. (1989) During latency, herpes simplex virus type 1 DNA is associated with nucleosomes in a chromatin structure. *J. Virol.* **63,** 943–947.

112. Dressler, G., Rock, D., and Fraser, N. (1987) Latent herpes simplex virus type 1 DNA is not extensively methylated in vivo. *J. Gen. Virol.* **68,** 1761–1765.

113. Mellerick, D. M. and Fraser, N. (1987) Physical state of the latent herpes simplex virus genome in a mouse model system: evidence suggesting an episomal state. *Virology* **158,** 265–275.

114. Rock, D. and Fraser, N. (1985) Latent herpes simplex virus type 1 DNA contains two copies of the virion DNA joint region. *J. Virol.* **55,** 849–852.

115. Gordon, Y. J., Johnson, B., Romanonski, E., and Araullo-Cruz, T. (1988) RNA complementary to herpes simplex virus type 1 ICP0 gene demonstrated in neurons of human trigeminal ganglia. *J. Virol.* **62,** 1832–1835.

116. Rock, D. L., Nesburn, A. B., Ghiasi, H., Ong, J., Lewis, T. L., Lokensgard, J. R., and Wechsler, S. (1987) Detection of latency-related viral RNAs in trigeminal ganglia of rabbits latently infected with herpes simplex virus type 1. *J. Virol.* **61,** 3820–3826.

117. Spivack, J. G. and Fraser, N. W. (1987) Detection of herpes simplex virus type 1 transcripts during latent infection in mice. *J. Virol.* **61**, 3841–3847.
118. Stevens, J. G., Wagner, E. K., Devi-Rao, G. B., Cook, M. L., and Feldman, L. T. (1987) RNA complementary to a herpesviruses α gene mRNA is prominent in latently infected neurons. *Science* **255**, 1056–1059.
119. Alvira, M. R., Cohen, J. B., Goins, W. F., and Glorioso, J. C. (1999) Genetic studies exposing the splicing events involved in HSV-1 latency associated transcript (LAT) production during lytic and latent infection. *J. Virol.* **73**, 3866–3876.
120. Farrell, M. J., Dobson, A. T., and Feldman, L. T. (1991) Herpes simplex virus latency-associated transcript is a stable intron. *Proc. Natl. Acad. Sci. USA* **88**, 790–794.
121. Krummenacher, C., Zabolotny, J., and Fraser, N. (1997) Selection of a nonconsensus branch point is influenced by an RNA stem-loop structure and is important to confer stability to the herpes simplex virus 2-kilobase latency-associated transcript. *J. Virol.* **71**, 5849–5860.
122. Rodahl, E. and Haarr, L. (1997) Analysis of the 2-kilobase latency-associated transcript expressed in PC12 cells productively infected with herpes simplex virus type 1: evidence for a stable, nonlinear structure. *J. Virol.* **71**, 1703–1707.
123. Zabolotny, J., Krummenacher, C., and Fraser, N. W. (1997) The herpes simplex virus type 1 2.0-kilobase latency-associated transcript is a stable intron which branches at a guanosine. *J. Virol.* **71**, 4199–4208.
124. Batchelor, A. H. and O'Hare, P. O. (1990) Regulation and cell-type-specific activity of a promoter located upstream of the latency-associated transcript of herpes simplex virus type 1. *J. Virol.* **64**, 3269–3279.
125. Chen, X., Schmidt, M. C., Goins, W. F., and Glorioso, J. C. (1995) Two herpes simplex virus type-1 latency active promoters differ in their contribution to latency-associated transcript expression during lytic and latent infection. *J. Virol.* **69**, 7899–7908.
126. Dobson, A. T., Sederati, F., Devi-Rao, G., Flanagan, W. M., Farrell, M. J., Stevens, J. G., et al. (1989) Identification of the latency-associated transcript promoter by expression of rabbit β-globin mRNA in mouse sensory nerve ganglia latently infected with a recombinant herpes simplex virus. *J. Virol.* **63**, 3844–3851.
127. Zwaagstra, J., Ghiasi, H., Nesburn, A. B., and Wechsler, S. L. (1989) In vitro promoter activity associated with the latency-associated transcript gene of herpes simplex virus type 1. *J. Gen. Virol.* **70**, 2163–2169.
128. Goins, W. F., Sternberg, L. R., Croen, K. D., Krause, P. R., Hendricks, R. L., Fink, D. J., et al. (1994) A novel latency-active promoter is contained within the herpes simplex virus type 1 UL flanking repeats. *J. Virol.* **68**, 2239–2252.
129. Nicosia, M., Deshmane, S. L., Zabolotny, J. M., Valyi-Nagy, T., and Fraser, N. W. (1993) Herpes simplex virus type 1 Latency-Associated Transcript (LAT) promoter deletion mutants can express a 2-kilobase transcript mapping to the LAT region. *J. Virol.* **67**, 7276–7283.

130. Dobson, A., Margolis, T., Sederati, F., Stevens, J., and Feldman, L. (1990) A latent, nonpathogenic HSV-1-derived vector stably expresses beta-galactosidase in mouse neurons. *Neuron* **5,** 353–360.

131. Rader, K. A., Ackland-Berglund, C. E., Miller, J. K., Pepose, J. S., and Leib, D. A. (1993) In vivo characterization of site-directed mutations in the promoter of the herpes simplex virus type 1 latency-associated transcripts. *J. Gen. Virol.* **74,** 1859–1869.

132. Soares, M. K., Hwang, D.-Y., Schmidt, M. C., Fink, D. J., and Glorioso, J. C. (1996) Cis-acting elements involved in transcriptional regulation of the herpes simplex virus type-1 latency-associated promoter 1 (LAP1) in vitro and in vivo. *J. Virol.* **70,** 5384–5394.

133. Batchelor, A. H. and O'Hare, P. O. (1992) Localization of cis-acting sequence requirements in the promoter of the latency-associated transcript of herpes simplex virus type 1 required for cell-type-specific activity. *J. Virol.* **66,** 3573–3582.

134. Kenny, J., Millhouse, S., Wotring, M., and Wigdahl, B. (1997) Upstream stimulatory factor family binds to the herpes simplex virus type 1 latency-associated transcript promoter. *Virology* **230,** 381–391.

135. Zwaagstra, J. C., Ghiasi, H., Nesburn, A. B. and Wechsler, S. L. (1991) Identification of a major regulatory sequence in the latency-associated transcript (LAT) promoter of herpes simplex virus type 1 (HSV-1). *Virology* **182,** 287–297.

136. Kenny, J. I., Krebs, F. C., Hartle, H. T., Gartner, A. E., Chatton, B., Leiden, J. M., et al. (1994) Identification of a second ATF/CREB-like element in the herpes simplex virus type 1 (HSV-1) latency-associated transcript (LAT) promoter. *Virology* **200,** 220–235.

137. Leib, D. A., Bogard, C. L., Kosz-Vnenchak, M., Hicks, K. A., Coen, D. M., Knipe, D. M., and Schaffer, P. A. (1989b) A deletion mutant of the latency-associated transcript of herpes simplex virus type 1 reactivates from the latent infection. *J. Virol.* **63,** 2893–2900.

138. Wang, K., Pesniacak, L., and Straus, S. E. (1997) Mutations in the 5' end of the herpes simplex virus type 2 latency-associated transcript (LAT) promoter affect LAT expression in vivo but not the rate of spontaneous reactivation of genital herpes. *J. Virol.* **71,** 7903–7910.

139. French, S. W., Schmidt, M. C., and Glorioso, J. C. (1996) Involvement of an HMG protein in the transcriptional activity of the herpes simplex virus latency active promoter 2. *Mol. Cell. Biol.* **16,** 5393–5399.

140. Kasai, Y., Chen, H., and Flint, S. J. (1992) Anatomy of an unusual RNA polymerase II promoter containing a downstream TATA element. *Mol. Cell. Biol.* **12,** 2884–2897.

141. Koller, E., Hayman, A. R., and Trueb, B. (1991) The promoter of the chicken a2(VI) collagen gene has features characteristic of housekeeping genes and of proto-oncogenes. *Nucleic Acids Res.* **19,** 485–491.

142. McDermott, J. B., Peterson, C. A., and Piatigorsky, J. (1992) Structure and lens expression of the gene encoding chicken βA3/A1-crystallin. *Gene* **117,** 193–200.

143. Kolluri, R., Torrey, T. A., and Kinniburgh, A. J. (1992) A CT promoter element binding protein: definition of a double-strand and a novel single-strand DNA binding motif. *Nucleic Acids Res.* **20,** 111–116.

144. Postel, E. H., Mango, S. E., and Flint, S. J. (1989) A nuclease-hypersensitive element of the human c-myc promoter interacts with a transcription initiation factor. *Mol. Cell. Biol.* **9,** 5123–5133.

145. Thanos, D. and Maniatis, T. (1992) The high mobility group protein HMG I(Y) is required for NF-κB-dependent virus induction of the human IFN-β gene. *Cell* 71, 777–789.

146. Berthomme, H., Lokensgard, J., Yang, L., Margolis, T., and Feldman, L. T. (2000) Evidence for a bi-directional element located downstream from the herpes simplex virus type 1 latency-associated promoter that increases its activity during latency. *J. Virol.* **74,** 3613–3622.

147. Lokensgard, J. R., Bloom, D. C., Dobson, A. T., and Feldman, L. T. (1994) Long-term promoter activity during herpes simplex virus latency. *J. Virol.* **68,** 7148-7158.

148. Lokensgard, J. R., Feldman, L. T., and Berthomme, H. (1997) The latency-associated promoter of herpes simplex virus type 1 requires a region downstream of the transcription start site for long-term expression during latency. *J. Virol.* **71,** 6714–6719.

149. Margolis, T. P., Bloom, D. C., Dobson, A. T., Feldman, L. T., and Stevens, J. G. (1993) Decreased reporter gene expression during latent infection with HSV LAT promoter constructs. *Virology* **197,** 585–592.

150. Krisky, D., Marconi, P., Oligino, T., Rouse, R., Fink, D., Cohen, J., et al. (1998) Development of herpes simplex virus replication-defective multigene vectors for combination gene therapy applications. *Gene Ther.* **5,** 1517–1530.

151. Krisky, D., Marconi, P., Oligino, T., Rouse, R., Fink, D., and Glorioso, J. (1997) Rapid method for construction of recombinant HSV gene transfer vectors. *Gene Ther.* **4,** 1120–1125.

152. Akkaraju, G. R., Huard, J., Hoffman, E. P., Goins, W. F., Pruchnic, R., Watkins, S. C., et al. (1999) Herpes simplex virus vector-mediated dystrophin gene transfer and expression in MDX mouse skeletal muscle. *J. Gene Med.* **1,** 280–289.

153. Ramakrishnan, R., Fink, D. J., Guihua, J., Desai, P., Glorioso, J. C., and Levine, M. (1994) Competitive quantitative polymerase chain reaction (PCR) analysis of herpes simplex virus type 1 DNA and LAT RNA in latently infected cells of the rat brain. *J. Virol.* **68,** 1864–1870.

154. Johnson, P. A., Miyanohara, A., Levine, F., Cahill, T., and Friedmann, T. (1992) Cytotoxicity of a replication-defective mutant herpes simplex virus type 1. *J. Virol.* **66,** 2952–2965.

155. Leiden, J., Frenkel, N., and Rapp, F. (1980) Identification of the herpes simplex virus DNA sequences present in six herpes simplex virus thymidine kinase-transformed mouse cell lines. *J. Virol.* **33,** 272–285.

156. DeStasio, P. R. and Taylor, M. W. (1989) Trans-activation of herpes simplex virus type 1 immediate early genes is specifically inhibited by human recombinant interferons. Biochem. Biophys. *Res. Commun.* **159,** 439–444.

157. DeStasio, P. R. and Taylor, M. W. (1990) Specific effect of interferon on the herpes simplex virus type 1 transactivation event. *J. Virol.* **64,** 2588–2593.

158. Klotzbucher, A., Mittnacht, S., Kirchner, H., and Jacobsen, H. (1990) Different effects of IFN gamma and IFN alpha/beta on immediate early gene expression of HSV-1. *Virology* **79,** 487–491.

159. Mittnacht, S., Straub, P., Kirchner, H., and Jacobsen, H. (1988) Interferon treatment inhibits onset of herpes simplex virus immediate-early transcription. *Virology* **164,** 201–210.

160. Nicholl, M. J. and Preston, C. M. (1996) Inhibition of herpes simplex virus type 1 immediate early gene expression by alpha interferon is not VP16 specific. *J. Virol.* **70,** 6336–6339.

161. DeLuca, N., McCarthy, A., and Schaffer, P. (1985) Isolation and characterization of deletion mutants of Herpes simplex virus type 1 in the gene encoding immediate-early regulatory protein ICP4. *J Virol* **56,** 558–570.

162. Sacks, W., Greene, C., Aschman, D., and Schaffer, P. (1985) Herpes simplex virus type 1 ICP27 is essential regulatory protein. *J. Virol.* **55,** 796–805.

163. Samaniego, L., Webb, A., and DeLuca, N. (1995) Functional interaction between herpes simplex virus immediate-early proteins during infection: gene expression as a consequence of ICP27 and different domains of ICP4. *J. Virol.* **69,** 5705–5715.

164. Katan, M., Haigh, A., Verrijzer, C., Vliet, P. v. d., and O'Hare, P. (1990) Characterization of a cellular factor which interacts functionally with Oct-1 in the assembly of a multicomponent transcription complex. *Nucleic Acids Res.* **18,** 6871–6880.

165. Kristie, T. and Sharp, P. (1993) Purification of the cellular C1 factor required for the stable recognition of the Oct-1 homeodomain by herpes simplex virus α-transinduction factor (VP16). *J. Biol. Chem.* **268,** 6525–6534.

166. Werstuck, G. and Capone, J. (1993) An unusual cellular factor potentiates protein-DNA complex assembly Oct-1 and Vmw65. *J. Biol. Chem.* **268,** 1272–1278.

167. Wilson, A., LaMarco, K., Peterson, M., and Herr, W. (1993) The VP16 accessory protein HCF is a family of polypeptides processed from a large precursor protein. *Cell* **74,** 115–125.

168. Xiao, P. and Capone, J. (1990) A cellular factor binds to the herpes simplex virus type 1 transactivator Vmw65 and is required for Vmw65-dependent protein-DNA complex assembly with Oct-1. *Mol. Cell. Biol.* **10,** 4974–4977.

169. Gerster, T. and Roeder, R. (1988) A herpesvirus trans-activating protein interacts with transcription factor OTF-1 and other cellular proteins. *Proc. Natl. Acad. Sci. USA* **85,** 6347–6351.

170. O'Hare, P. and Goding, C. (1988a) Herpes simplex virus regulatory elements and the immunoglobulin octamer domain bind a common factor and are both targets for virion transactivation. *Cell* **52,** 435–445.

171. O'Hare, P., Goding, C., and Haigh, A. (1988b) Direct combinational interaction between a herpes simplex virus regulatory protein and a cellular octamer binding factor mediates specific induction of virus immediate-early gene expression. *EMBO J.* **7,** 4231–4238.

172. Preston, C., Frame, M., and Campbell, M. (1988) A complex formed between cell components and an HSV structural polypeptide binds to a viral immediate early gene regulatory DNA sequence. *Cell* **52**, 425–434.

173. Stern, S., Tanaka, M., and Herr, W. (1989) The Oct-1 homeodomain directs formation of a multiprotein-DNA complex with the HSV transactivator VP16. *Nature* **341**, 624–630.

174. Cai, W. and Schaffer, P. A. (1992) Herpes simplex virus type 1 ICP0 regulates expression of immediate-early, early, and late genes in productively infected cells. *J. Virol.* 66, 2904–2915.

175. Everett, R. D. (1987) The regulation of transcription of viral and cellular genes by herpesvirus immediate-early gene products. *Anticancer Res.* **7**, 589–604.

176. Gelman, I. H. and Silverstein, S. (1985) Identification of immediate-early genes from herpes simplex virus that transactivate the virus thymidine kinase gene. *Proc. Natl. Acad. Sci. USA* **82**, 5265–5269.

177. Quinlan, M. P. and Knipe, D. M. (1985) Stimulation of expression of a herpes simplex virus DNA-binding protein by two viral factors. *Mol. Cell. Biol.* **5**, 957–963.

178. Everett, R. and Maul, G. (1994) HSV-1 IE protein Vmw110 causes redistribution of PML. *EMBO J.* **13**, 5062–5069.

179. Everett, R., O'Hare, P., O'Rourke, D., Barlow, P., and Orr, A. (1995) Point mutations in the herpes simplex virus type 1 Vwm110 RING finger helix affect activation of gene expression, viral growth, and interaction with PML-containing nuclear structures. *J. Virol.* **69**, 7339–7344.

180. Maul, G. and Everett, R. (1994) The nuclear location of PML, a cellular member of the C3HC4 zinc-binding domain protein family, is rearranged during herpes simplex virus infection by the C3HC4 viral protein ICP0. *J. Gen. Virol.* **75**, 1223–1233.

181. Maul, G. G., Guldner, H. H., and Spivack, J. G. (1993) Modification of discrete nuclear domains induced by herpes simplex virus type1 immediate early gene 1 product (ICP0). *J. Gen. Virol.* **74**, 2679–2690.

182. Maul, G. G., Ishov, A. M., and Everett, R. D. (1996) Nuclear domain 10 as preexisting potential replication start sites of herpes simplex virus type-1. *Virology* **217**, 67–75.

183. Chen, J. and Silverstein, S. (1992) Herpes simplex viruses with mutations in the gene encoding ICP0 are defective in gene expression. *J. Virol.* **66**, 2916–2927.

184. Rice, S., Long, M., Lam, V., and Spencer, C. (1994) RNA polymerase II is aberrantly phosphorylated and localized to viral replication compartments following herpes simplex virus infection. *J. Virol.* **68**, 988–1001.

185. Rice, S. A., Long, M. C., Lam, V., Schaffer, P. A., and Spencer, C. A. (1995) Herpes simplex virus immediate-early protein ICP22 is required for viral modification of host RNA polymerase II and establishment of the normal viral transcription program. *J. Virol.* **69**, 5550–5559.

186. Hill, A., Jugovic, P., York, I., Russ, G., Bennink, J., Yewdell, J., et al. (1995) Herpes simplex virus turns off the TAP to evade host immunity. *Nature* **375**, 411–415.

187. Hill, A. and Ploegh, H. (1995) Getting the inside out: the transporter associated with antigen processing (TAP) and the presentation of viral antigen. *Proc. Natl. Acad. Sci. USA* **92,** 341–343.

188. York, I., Roop, C., Andrews, D., Riddell, S., Graham, F., and Johnson, D. (1994) A cytosolic herpes simplex virus protein inhibits antigen presentation to CD8+ T lymphocytes. *Cell* **77,** 525–535.

189. Johnson, P., Wang, M., and Friedmann, T. (1994) Improved cell survival by the reduction of immediate-early gene expression in replication-defective mutants of herpes simplex virus type 1 but not by mutation of the viron host shutoff function. *J. Virol.* **68,** 6347–6362.

190. McCarthy, A., McMahan, L., and Schaffer, P. (1989) Herpes simplex virus type 1 ICP27 deletion mutants exhibit altered patterns of transcription and are DNA deficient. *J. Virol.* **63,** 18–27.

191. Chiocca, E. A., Choi, B. B., Cai, W. Z., DeLuca, N. A., Schaffer, P. A., DiFiglia, M., et al. (1990) Transfer and expression of the lacZ gene in rat brain neurons mediated by herpes simplex virus mutants. *New Biol.* **2,** 739–746.

192. Desai, P., Ramakrishnan, R., Lin, Z., Osak, B., Glorioso, J., and Levine, M. (1993) The RR1 gene of herpes simplex virus type 1 is uniquely trans activated by ICP0 during infection (published erratum appears in *J Virol* 1994 Feb;**68(2):**1264). *J. Virol.* **67,** 6125–6135.

193. Ecob-Prince, M., Hassan, K., Denheen, M., and Preston, C. (1995) Expression of β-galactosidase in neurons of dorsal root ganglia which are latently infected with herpes simplex virus type 1. *J. Gen. Virol.* **76,** 1527–1532.

194. Fink, D. J., Sternberg, L. R., Weber, P. C., Mata, M., Goins, W. F., and Glorioso, J. C. (1992) In vivo expression of β-galactosidase in hippocampal neurons by HSV-mediated gene transfer. *Hum. Gene Ther.* **3,** 11–19.

195. Hendricks, R. L., Weber, P. C., Taylor, J. L., Koumbis, A., Tumpey, T. M., and Glorioso, J. C. (1991) Endogenously produced interferon α protects mice from herpes simplex virus type 1 corneal disease. *J. Gen. Vir.* **72,** 1601–1610.

196. Ho, D. and Mocarski, E. (1988) β-galactosidase as a marker in the peripheral and neural tissues of the herpes simplex virus-infected mouse. *Virology* **167,** 279–283.

197. Huang, Q., Vonsattel, J.-P., Schaffer, P., Martuza, R., Breakefield, X., and DiFiglia, M. (1992) Introduction of a foreign gene (Escherichia coli lacZ) into rat neostriatal neurons using herpes simplex virus mutants: a light and electron microscopic study. *Exp. Neurol.* **115,** 303–316.

198. Kramm, C. M., Rainov, N. G., Sena-Esteves, M., Chase, M., Pechan, P. A., Chiocca, E. A., and Breakefield, X. O. (1996) Herpes vector-mediated delivery of marker genes to disseminated central nervous system tumors. *Hum. Gene Ther.* **7,** 291–300.

199. Nilaver, G., Muldoon, L., Kroll, R., Pagel, M., Breakefield, X., Davidson, B., and Neuwelt, E. (1995) Delivery of herpesvirus and adenovirus to nude rat intracerebral tumors after osmotic blood-brain barrier disruption. *Proc. Natl. Acad. Sci. USA* **92,** 9829–9833.

200. Oligino, T., Ghivizzani, S. C., Wolfe, D., Lechman, E. R., Krisky, D., Mi, Z., et al. (1999) Intra-articular delivery of a herpes simplex virus IL-1Ra gene vector reduces inflammation in a rabbit model of arthritis. *Gene Ther.* **6,** 1713–1720.

201. Palella, T., Silverman, L., Schroll, C., Homa, F., Levine, M., and Kelley, W. (1988) Herpes simplex virus-mediated human hypoxanthine-guanine phosphoribosyl-transferase gene transfer into neuronal cells. *Mol. Cell. Biol.* **8,** 457–460.

202. Ramakrishnan, R., Levine, M., and Fink, D. (1994) PCR-based analysis of herpes simplex virus type 1 latency in the rat trigeminal ganglion established with a ribonucleotide reductase-deficient mutant. *J. Virol.* **68,** 7083–7091.

203. Weir, J. and Dacquel, E. (1995) Plasmid insertion vectors that facilitate construction of herpes simplex virus gene delivery vectors. *Gene* **154,** 123–128.

204. Weir, J. P. and Elkins, K. L. (1993) Replication-incompetent herpesvirus vector delivery of an interferon alpha gene inhibits human immunodeficiency virus replication in human monocytes. *Proc. Natl. Acad. Sci. USA* **90,** 9140–9144.

205. Bloom, D., Maidment, N., Tan, A., Dissette, V., Feldman, L., and Stevens, J. (1995) Long-term expression of a reporter gene from latent herpes simplex virus in the rat hippocampus. *Mol. Brain Res.* **31,** 48–60.

206. Carpenter, D. E. and Stevens, J. G. (1996) Long-term expression of a foreign gene from a unique position in the latent herpes simplex virus genome. *Hum. Gene Ther.* **7,** 1447–1454.

207. Coffin, R. S., Howard, M. K., Cumming, D. V., Dollery, C. M., McEwan, J., Yellon, D. M., et al. (1996) Gene delivery to the heart in vivo and to cardiac myocytes and vascular smooth muscle cells in vitro using herpes virus vectors. *Gene Ther.* **3,** 560–566.

208. Davar, G., Kramer, M., Garber, D., Roca, A., Andersen, J., Bebrin, W., et al. (1994) Comparative efficacy of expression of genes delivered to mouse sensory neurons with herpes virus vectors. *J. Comp. Neurol.* **339,** 3-11.

209. Keir, S. D., Mitchell, W. J., Feldman, L. T., and Martin, J. R. (1995) Targeting and gene expression in spinal cord motor neurons following intramuscular inoculation of an HSV-1 vector. *J. NeuroVirol.* **1,** 259–267.

210. Maidment, N. T., Tan, A. M., Bloom, D. C., Anton, B., Feldman, L. T., and Stevens, J. G. (1996) Expression of the lacZ reporter gene in the rat basal forebrain, hippocampus, and nigrostriatal pathway using a nonreplicating herpes simplex vector. *Exp. Neurol.* **139,** 107–114.

211. Mester, J. C., Pitha, P., and Glorioso, J. C. (1995) Anti-viral activity of herpes simplex virus vectors expressing alpha-interferon. *Gene Ther.* **3,** 187–196.

212. Warden, M. P. and Weir, J. P. (1996) Inducible gene expression of the human immunodeficiency virus LTR in a replication-incompetent herpes simplex virus vector. *Virology* **226,** 127–131.

213. Dilloo, D., Rill, D., Entwistle, C., Boursnell, M., Zhong, W., Holden, W., et al. (1997) A novel herpes vector for the high-efficiency transduction of normal and malignant human hematopoietic cells. *Blood* **89,** 119–127.

214. Pyles, R. B., Warnick, R. E., Chalk, C. L., Szanti, B. E., and Parysek, L. M. (1997) A novel multiply-mutated HSV-1 strain for the treatment of human brain tumors. *Hum. Gene Ther.* **8,** 533–544.

215. Rasty, S., Thatikunta, P., Gordon, J., Khalili, K., Amini, S., and Glorioso, J. (1996) Human immunodeficiency virus tat gene transfer to the murine central nervous

system using a replication-defective herpes simplex virus vector stimulates transforming growth factor beta 1 gene expression. *Proc. Natl. Acad. Sci. USA* **93,** 6073–6078.

216. Roemer, K., Johnson, P. A., and Friedmann, T. (1991) Activity of the simian virus 40 early promoter-enhancer in herpes simplex virus type 1 vectors is dependent on its position, the infected cell type, and the presence of Vmw175. *J. Virol.* **65,** 6900–6912.

217. Huard, J., Akkaraju, G., Watkins, S. C., Pike-Cavalcoli, M., and Glorioso, J. C. (1997) LacZ gene transfer to skeletal mucsle using a replication-defective herpes simplex virus type 1 mutant vector. *Hum. Gene Ther.* **8,** 439–452.

218. Huard, J., Goins, W. F., and Glorioso, J. C. (1995) Herpes simplex virus type 1 vector mediated gene transfer to muscle. *Gene Ther.* **2,** 385–393.

219. Johnson, P. A., Yoshida, K., Gage, F. H., and Friedmann, T. (1992) Effects of gene transfer into cultured CNS neurons with a replication-defective herpes simplex virus type 1 vector. *Brain Res. Mol. Brain Res.* **12,** 95–102.

220. Shering, A. F., Bain, D., Stewart, K., Epstein, A. L., Castro, M. G., Wilkinson, G. W., and Lowenstein, P. R. (1997) Cell type-specific expression in brain cell cultures from a short human cytomegalovirus major immediate early promoter depends on whether it is inserted into herpesvirus or adenovirus vectors. *J. Gen. Virol.* **78,** 445–459.

221. Glorioso, J., Goins, W., Meaney, C., Fink, D., and DeLuca, N. (1994) Gene transfer to brain using herpes simplex virus vectors. *Ann. Neurol.* **35,** S28–S34.

222. Lachmann, R. H., Brown, C., and Efstathiou, S. (1996) A murine RNA polymerase I promoter inserted into the herpes simplex virus type 1 genome is functional during lytic, but not latent, infection. *J. Gen. Virol.* **77,** 2575–2582.

223. Andersen, J., Frim, D., Isacson, O., and Breakefield, X. (1993) Herpesvirus-mediated gene delivery into the rat brain: specificity and efficiency of the neuron-specif enolase promoter. *Cell Mol. Neurobiol.* **13,** 503–515.

224. Andersen, J., Garber, D., Meaney, C., and Breakefield, X. (1992) Gene transfer into mammalian central nervous system using herpes virus vectors: extended expression of bacterial lacZ in neurons using the neuron-specific enolase promoter. *Hum. Gene Ther.* **3,** 487–499.

225. Kennedy, P. and Steiner, I. (1993) The use of herpes simplex virus vectors for gene therapy in neurological diseases. *Q. J. Med.* **86,** 697–702.

226. Goss, J. R., Mata, M., Goins, W. F., Wu, H. H., Glorioso, J. C., and Fink, D. J. (2001) Antinociceptive effect of genomic herpes simplex virus-based vector expressing human proenkephalin in rat dorsal root gangkion. *Gene Ther.* **8,** 551–556.

227. Marconi, P., Simonato, M., Zucchini, S., Bregola, G., Argnani, R., Krisky, D., et al. (1999) Replication-defective herpes simplex virus vectors for neurotrophic factor gene transfer in vitro and in vivo. *Gene Ther.* **6,** 904–912.

228. Marconi, P., Tamura, M., Moriuchi, S., Krisky, D., Goins, W., Cohen, J., and Glorioso, J. (2000) Connexin43-enhanced suicide gene therapy using herpesviral vectors. *Mol. Ther.* **1,** 71–81.

229. Moriuchi, S., Krisky, D., Marconi, P., Tamura, M., Shimizu, K., Yoshimine, T., et al. (2000) HSV vector cytotoxicity is inversely correlated with effective TK/GCV suicide gene therapy of rat gliosarcoma. *Gene Ther.* **7,** 1483–1490.

230. Moriuchi, S., Oligino, T., Krisky, D., Marconi, P., Fink, D., Cohen, J., and Glorioso, J. (1998) Enhanced tumor-cell killing in the presence of ganciclovir by HSV-1 vector-directed co-expression of human TNF-α and HSV thymidine kinase. *Cancer Res.* **58,** 5731–5737.

231. Natsume, A., Mata, M., Wolfe, D., Oligino, T., Goss, J., Huang, S., et al. (2002) Bcl-2 and GDNF delivered by HSV-mediated gene transfer after spinal root avulsion provide a synergistic effect. *J. Neurotrauma* **19,** 61–68.

232. Niranjan, A., Moriuchi, S., Lunsford, L., Kondziolka, D., Flickinger, J., Fellows, W., et al. (2000) Effective treatment of experimental glioblastoma by HSV vector-mediated TNF-α and HSV-tk gene transfer in combination with radiosurgery and ganciclovir administration. *Mol. Ther.* **2,** 114–120.

233. Oligino, T., Poliani, P. L., Marconi, P., Bender, M. A., Schmidt, M. C., Fink, D. J., and Glorioso, J. C. (1996) In vivo transgene activation from an HSV-based gene vector by GAL4:VP16. *Gene Ther.* **3,** 892–899.

234. Oligino, T., Poliani, P. L., Wang, Y., Tsai, S. Y., O'Malley, B. W., Fink, D. J., and Glorioso, J. C. (1998) Drug inducible transgene expression in brain using a herpes simplex virus vector. *Gene Ther.* **5,** 491–496.

235. Wilson, S. P., Yeomans, D. C., Bender, M. A., Lu, Y., Goins, W. F., and Glorioso, J. C. (1999) Antihyperalgesic effects of infection with a preproenkephalin-encoding herpes virus. *Proc. Natl. Acad. Sci. USA* **96,** 3211–3216.

236. Yamada, M., Oligino, T., Mata, M., Goss, J. R., Glorioso, J. C., and Fink, D. J. (1999) HSV vector-mediated expression of Bcl-2 prevents 6-hydroxydopamine induced degeneration of neurons in the substantia nigra in vivo. *Proc. Natl. Acad. Sci. USA* **96,** 4078–4083.

237. Wolfe, J. H., Deshmane, S. L., and Fraser, N. W. (1992) Herpesvirus vector gene transfer and expression of β-glucuronidase in the central nervous system of MPS VII mice. *Nat. Genet.* **1,** 379–384.

238. Margolis, T., Sedarati, F., Dobson, A., Feldman, L., and Stevens, J. (1992) Pathways of viral gene expression during acute neuronal infection with HSV-1. *Virology* **189,** 150–160.

239. Ho, D. Y. and Mocarski, E. S. (1989) Herpes simplex virus latent RNA (LAT) is not required for latent infection in the mouse. *Proc. Natl. Acad. Sci. USA* **86,** 7596–7600.

240. Lachmann, R. H. and Efstathiou, S. (1997) Utilization of the herpes simplex virus type 1 latency-associated regulatory region to drive stable reporter gene expression in the nervous system. *J. Virol.* **71,** 3197–3207.

241. Coffin, R. S., Thomas, S. K., Thomas, N. S. B., Lilley, C. E., Pizzey, A. R., Griffiths, C. H., et al. (1998) Pure populations of transduced primary human cells can be produced using GFP expressing herpes virus vectors and flow cytometry. *Gene Ther.* **5,** 718–722.

242. Sadowski, I., Ma, J., Triezenberg, S., and Ptashne, M. (1988) GAL4/VP16 is an unusually potent transcriptional activator. *Nature* **335,** 563–564.

243. Carey, M., Leatherwood, J., and Ptashne, M. (1990) A potent GAL4 derivative activates transcription at a distance in vitro. *Science* **247,** 710–712.

244. Chasman, D. I., Leatherwood, M., Carey, M., Ptashne, M., and Kornberg, R. D. (1989) Activation of yeast polymerase II transcription by herpesvirus VP16 and GAL4 derivative in vitro. *Mol. Cell. Biol.* **9,** 4746–4749.

245. Axelrod, J. D., Reagan, M. S., and Majors, J. (1993) GAL4 disrupts a repressing nucleosome during activation of GAL 1 transcription in vivo. *Genes Dev.* **7,** 857–869.

246. Xu, L., Schaffner, W., and Rungger, D. (1993) Transcription activation by recombinant GAL4/VP16 in the *Xenopus* oocyte. *Nucleic Acids Res.* **21,** 2775.

247. Emi, N., Friedmann, T., and Yee, J. (1991) Pseudotype formation of murine leukemia virus with the G protein of vesicular stomatitis virus. *J. Virol.* **65,** 1202–1207.

248. Naldini, L., Blomer, U., Gallay, P., Mulligan, R., Gage, F. H., Verma, I. M., and Trono, D. (1996) In vivo gene delivery and stable transduction of nondividing cells by a lentiviral vector. *Science* **272,** 263–267.

249. Neda, H., Wu, C. H., and Wu, G. Y. (1991) Chemical modification of an ecotropic murine leukemia virus results in redirection of its target cell specificity. *J. Biol. Chem.* **266,** 14143–14146.

250. Roux, P., Jeanteur, P., and Piechaczyk, M. (1989) A versatile and potentially general approach to the targeting of specific cell types by retroviruses: application to the infection of human cells by means of major histocompatibility complex class I and class II antigens by mouse ecotropic murine leukemia virus-derived viruses. *Proc. Natl. Acad. Sci. USA* **86,** 9079–9083.

251. Martin, F., Neil, S., Kupsch, J., Maurice, M., Cosset, F., and Collins, M. (1999) Retrovirus targeting by tropism restriction to melanoma cells. *J. Virol.* **73,** 6923–6929.

252. Iijima, Y., Ohno, K., Ikeda, H., Saawai, K., Levin, B., and Meruelo, D. (1999) Cell-specific targeting of a thymidine kinase/ganciclovir gene therapy system using a recombinant Sindbis virus vector. *Int. J. Cancer* **80,** 110–118.

253. Ohno, K., Sawai, K., Iijima, Y., Levin, B., and Meruelo, D. (1997) Cell-specific targeting of Sindbis virus vectors displaying IgG-binding domains of protein A. *Nat. Biotech.* **15,** 763–767.

254. Kasahara, N., Dozy, M., and Kan, Y. W. (1994) Tissue-specific targeting of retroviral vectors through ligand-receptor interactions. *Science* **255,** 1373–1376.

255. Marin, M., Noel, D., Valsesia-Wittman, S., Brockly, F., Etienne-Julan, M., Russel, S., et al. (1996) Targeted infection of human cells via major histocompatibility complex class I molecules by moloney murine leukemia virus-derived viruses displaying single-chain antibody fragment-envelope fusion proteins. *J. Virol.* **70,** 2957–2962.

256. Russell, S. J., Hawkins, R. E. and Winter, G. (1993) Retroviral vectors displaying functional antibody fragments. *Nucleic Acids Res.* **21,** 1081–1085.

257. Somia, N. V., Zoppe, M., and Verma, I. M. (1995) Generation of targeted retroviral vectors by using single-chain variable fragment: an approach to in vivo gene delivery. *Proc. Natl. Acad. Sci. USA* **92**, 7570–7574.

258. Girod, A., Ried, M., Wobus, C., Lahm, H., Leike, K., Kleinschmidt, J., et al. (1999) Genetic capsid modifications allow efficient re-targeting of adeno-associated virus type 2. *Nat. Med.* **5**, 1052–1056.

259. Morling, F. J., Peng, K. W., Cosset, F. L., and Russell, S. J. (1997) Masking of retroviral envelope functions by oligomerizing polypeptide adaptors. *Virology* **234**, 51–61.

260. Valesia-Wittmann, S., Morling, F. J., Hatziioannou, T., Russell, S. J., and Cosset, F. L. (1997) Receptor co-operation in retrovirus entry: recruitment of an auxiliary entry mechanism after retargeted binding. *EMBO J.* **16**, 1214–1223.

261. Chadwick, M., Morling, F., Cosset, F., and Russell, S. (1999) Modification of retroviral tropism by display of IGF-I. *J. Mol. Biol.* **285**, 485–494.

262. Cosset, F. L., Morling, F. J., Takeuchi, Y., Weiss, R. A., Collins, M. K., and Russell, S. J. (1995) Retroviral retargeting by envelopes expressing an N-terminal binding domain. *J. Virol.* **69**, 6314–6322.

263. Fielding, A. K., Maurice, M., Morling, F. J., Cosset, F. L., and Russell, S. J. (1998) Inverse targeting of retroviral vectors: selective gene transfer in a mixed population of hematopoietic and nonhematopoietic cells. *Blood* **91**, 1802–1809.

264. Nilson, B., Morling, F., Cosset, F.-L., and Russell, R. (1996) Targeting of retroviral vectors through protease-substrate interactions. *Gene Ther.* **3**, 280–286.

265. Peng, K. W., Morling, F. J., Cosset, F. L., Murphy, G., and Russell, S. J. (1997) A gene delivery system activatable by disease-associated matrix metalloproteinases. *Hum. Gene Ther.* **8**, 729–738.

266. Wickham, T. J., Tzeng, E., Shears II, L. L., Roelvink, P. W., Li, Y., Lee, G. M., et al. (1997) Increased in vitro and in vivo gene transfer by adenovirus vectors containing chimeric fiber proteins. *J. Virol.* **71**, 8221–8229.

267. Wickham, T. J., Carrion, M. E., and Koveski, I. (1995) Targeting of adenovirus penton base to new receptors through replacement of its RGD motif with other receptor-specific peptide motifs. *Gene Ther.* **2**, 750–756.

268. Wickham, T. J., Roelvink, P. W., Brough, D. E., and Kovesdi, I. (1996) Adenovirus targeted to heparan-containing receptors increases its gene delivery efficiency to multiple cell types. *Nat. Biotechnol.* **14**, 1570–1573.

269. Rogers, B. E., Douglas, J. T., Ahlem, C., Buchsbaum, D. J., Frincke, J., and Curiel, D. T. (1997) Use of a novel cross-linking method to modify adenovirus tropism. *Gene Ther.* **4**, 1387–1392.

270. Grill, J., Van Beusechem, V., Van Der Valk, P., Dirven, C., Leonhart, A., Pherai, D., et al. (2001) Combined targeting of adenoviruses to integrins and epidermal growth factor receptors increases gene transfer into primary glioma cells and spheroids. *Clin. Cancer Res.* **7**, 641–650.

271. Li, L., Wickham, T., and Keegan, A. (2001) Efficient transduction of murine B lymphocytes and B lymphoma lines by modified adenoviral vectors: enhancement via targeting to FcR and heparan-containing proteins. *Gene Ther.* **8**, 938–945.

272. Nettlebeck, D., Miller, D., Jerome, V., Zuzzarte, M., Watkins, S., Hawkins, R., et al. (2001) Targeting of adenovvirus to endothelial cells by a bispecific single-chain diabody directed against the adenovirus fiber knob domain and human endoglin (CD105). *Mol. Ther.* **3**, 882–891.

273. Wesseling, J., Bosma, P., Krasnykh, V., Kashentseva, E., Blackwell, J., Reynolds, P., et al. (2001) Improved gene transfer efficiency to primary and established human pancreatic carcinoma target cells via epidermal growth factor receptor and integrin-targeted adenoviral vectors. *Gene Ther.* **8**, 969–976.

274. Mizuguchi, H., Koizumi, N., Hosono, T., Utoguchi, N., Watanabe, Y., Kay, M., and Hayakawa, T. (2001) A simplified system for constructing recombinant adenoviral vectors containing heterologous peptides in the HI loop of their fiber knob. *Gene Ther.* **8**, 730–735.

275. Schneider, H., Groves, M., Muhle, C., Reynolds, P. N., Knight, A., Themis, M., et al. (2000) Retargeting of adenoviral vectors to neurons using the HC fragment of tetanus toxin. *Gene Ther.* **7**, 1584–1592.

276. Laquerre, S., Anderson, D. B., Stolz, D. B., and Glorioso, J. C. (1998) Recombinant herpes simplex virus type 1 engineered for targeted binding to erythropoietin receptor-bearing cells. *J. Virol.* **72**, 9683–9697.

277. Fan, D. P. and Sefton, B. M. (1978) The entry into host cells of Sindbis virus, vesicular stomatitis virus and Sendai virus. *Cell* **15**, 985–992.

278. Florkiewicz, R. Z. and Rose, J. K. (1984) A cell line expressing vesicular stomatitis virus glycoprotein fuses at low pH. *Science* **225**, 721–723.

279. Matlin, K. S., Reggio, H., Helenius, A., and Simons, K. (1982) Pathway of vesicular stomatitis virus entry leading to infection. *J. Mol. Biol.* **156**, 609–631.

280. Riedel, H., Kondor-Koch, C., and Garoff, H. (1984) Cell surface expression of fusogenic vesicular stomatitis virus G protein from cloned cDNA. *EMBO J.* **3**, 1477–1483.

281. Abe, A., Chen, S. T., Miyanohara, A., and Friedmann, T. (1998) In vitro cell-free conversion of noninfectious Moloney retrovirus particles to an infectious form by the addition of the vesicular stomatitis virus surrogate envelope G protein. *J. Virol.* **72**, 6356–6361.

282. Luo, T., Douglas, J. L., Livingston, R. L., and Garcia, J. V. (1998) Infectivity enhancement by HIV-1 Nef is dependent on the pathway of virus entry: implications for HIV-based gene transfer systems. *Virology* **241**, 224–233.

283. Manning, W. C., Murphy, J. E., Jolly, D. J., Mento, S. J., and Ralston, R. O. (1998) Use of a recombinant murine cytomegalovirus expressing vesicular stomatitis virus G protein to pseudotype retroviral vectors. *J. Virol. Methods* **73**, 31–39.

284. Ory, D. S., Neugeboren, B. A., and Mulligan, R. C. (1996) A stable human-derived packaging cell line for production of high titer retrovirus/vesicular stomatitis virus G pseudotypes. *Proc. Natl. Acad. Sci. USA* **93**, 11400–11406.

285. Spielhofer, P., Bachi, T., Fehr, T., Christiansen, G., Cattaneo, R., Kaelin, K., et al. (1998) Chimeric measles viruses with a foreign envelope. *J. Virol.* **72**, 2150–2159.

286. Rasty, S., Goins, W., and Glorioso, J. (1995) Site-specific integration of multigenic shuttle plasmids into the herpes simplex virus type 1 (HSV-1) genome using a cell-free Cre-*lox* recombination system, in *Methods in Molecular Genetics* (Adolph, K., ed.), Academic Press, San Diego, CA, pp. 114–130.

20

Gene Transfer to Skeletal Muscle Using Herpes Simplex Virus-Based Vectors

Baohong Cao and Johnny Huard

1. Introduction

Type 1 herpes simplex virus (HSV-1)-based vectors, which are naturally capable of carrying large DNA fragments like the 14 kb dystrophin cDNA, have been studied for their ability to transduce muscle cells *(1–5)*. These vectors can persist in the host cell in a nonintegrated state and can be prepared at adequately high titers (10^7–10^9 PFU/mL). They also infect myoblasts, myotubes, and immature myofibers efficiently *(1–5)*. The major disadvantage of the first-generation HSV vectors is their relatively high cytotoxicity, which hampers long-term transgene expression. Second-generation mutants defective for multiple immediate early (IE) genes (e.g., *ICP4, ICP22,* and *ICP27*) display substantially reduced cytotoxicity in vitro, which improves the duration of transgene expression *(6–11)*. In this chapter, we describe a new method of gene delivery using second-generation HSV-1 vectors. This procedure should enable an investigator to transduce normal mouse muscle cells, both in vitro and in vivo. We explain the conditions for muscle cell isolation, transduction in vitro and in vivo, and the technique for evaluating transduction efficiency (β-galactosidase; β-gal) using histology or the β-gal assay (ONPG) method.

From: *Methods in Molecular Biology, vol. 246:*
Gene Delivery to Mammalian Cells: Vol. 2: Viral Gene Transfer Techniques
Edited by: W. C. Heiser © Humana Press Inc., Totowa, NJ

2. Material

2.1. Sources of Enzymes

1. Collagenase XI: obtained from Clostridium histolyticcum (> 1200 collagen digestion units per mg) (Sigma, St Louis, MO);
2. Trypsin: 10x 0.5% trypsin-ethylenediaminetetraacetic acid (EDTA) (Invitrogen, Carlsbad, CA);
3. Dispase: 1–12 U/mg (Invitrogen);
4. *Escherichia coli* β-gal: this enzyme is commercially available (e.g., Sigma).

2.2. Virus

1. The HSV-1 referred to in the following protocol is THZ.4 with the IE genes *ICP4*, *ICP22*, and *ICP27* deleted; the virus carries a *LacZ* gene under the control of the human cytomegalovirus immediate-early promoter (HCMV) *(11,12)*;
2. Examples of HSV used in muscle transduction:
   a. Replication-defective HSV-1 vector, which contains an expression cassette with the HCMV promoter driving *LacZ* in the thymidine kinase (tk) locus of an *ICP4⁻* virus *(13–15)*.
   b. THZ.3 with the cytotoxicicity genes *ICP4*, *ICP22*, *ICP27*, and *UL41* deleted *(7)*.

2.3. Solutions and Culture Medium

1. DMEM is available from Invitrogen. Complete DMEM: DMEM supplemented with 10% horse serum, 10% fetal calf serum (FCS), 50 U/mL penicillin, 50 μg/mL streptomycin, and 1% chick embryo extract; Opti-MEM is available from Invitrogen.
2. Muscle cell or section fixation reagent: 1.5% glutaraldehyde in PBS. This solution is always freshly prepared.
3. Phosphate-buffered saline (PBS): 1 L contains 0.144 g KH_2PO_4, 9 g NaCl, and 0.795 g $Na_2HPO_4 \cdot 7H_2O$.
4. 100X Mg solution: 0.1 M $MgCl_2$ and 4.5 M β-mercaptoethanol.
5. Na_2CO_3 (1 M).
6. 1X ONPG: 4 mg/mL o-nitrophenyl-β-D-galactosidase dissolved in 0.1 M sodium phosphate, pH 7.5.
7. Tris-HCl: 1 M, pH 7.8.
8. X-Gal substrate: 0.4 mg/mL 5-bromo-chloro-3-indolyl-β-D-galactoside, 1 mM $MgCl_2$, and 5 mM $K_4Fe(CN)_6/K_3Fe(CN)_6$ in PBS.
9. Hanks' Balanced Salt Solution (HBSS) is available from Invitrogen.

3. Methods

3.1. Transducing Muscle Cells In Vitro

3.1.1. Preparation of Muscle Cells

1. Anesthetize newborn or adult CD57 BL/10J mice (usually, $n = 3$ or 4) with intraperitoneal (ip) injection of 0.1 mL anesthetic cocktail composed of katamine, xylazine, and PBS (2:1:9, volume). Sacrifice mice by cervical dislocation.
2. Peel off the skin of the four legs, remove the whole legs, and incubate them in 20 mL of HBSS in a Petri dish inside a tissue-culture hood; dissect and discard the bone and fat tissue.
3. Mince the muscle into small pieces with scissors.
4. Collect the pieces of muscle by centrifuging at 2500g for 5 min.
5. Enzymatically dissociate the muscle cells by adding 0.2% collagenase-type XI for 1 h at 37°C, followed by 0.125% dispase for 30 min then 0.1% trypsin for 30 min, all at 37°C.
6. Collect the cells by centrifuging at 2500g for 5 min.
7. Resuspend the cells in 20 mL complete DMEM.
8. Pass cells serially through 18, 27, and 30 G needles.
9. Plate cells into collagen-coated T75 flasks and incubate at 37°C in a humidified, 5% CO_2 incubator. Use one T75 flask for an isolation from newborn mice and two T75 flasks for an isolation from adult mice.
10. Two h after plating, some cells will have adhered to the flask (PP1). Collect the supernatant containing the nonadherent cells and transfer to a second flask (PP2).
11. Twenty-four hours after the second plating, transfer the nonadherent cells to a third flask (PP3).
12. Repeat **step 11** until PP6 is obtained. The adherent PP6 cells are mostly myogenic cells, which will be used for the following viral transduction (*see* **Note 1**). The entire muscle preparation process requires 5 d to complete. The cells obtained by this method can go through 30 passages without major phenotype change. Cells can be frozen for future use.

3.1.2. HSV-1 Transduction of Muscle Cells

1. Infect monolayer muscle cells ($1\text{X }10^5$) with HSV-1 in 6-well plates at a multiplicity of infection (MOI) of 3 (*see* **Note 2**). Perform the infection in 500 µL of Dulbecco's modified Eagle's medium (DMEM) without serum. After 3 h, add 1 mL of complete medium. Incubate the culture for 24–48 h at 37°C in a humidified, 5% CO_2 incubator.

3.1.3. Assay for β-Galactosidase Expression in Muscle Cells

3.1.3.1. HISTOCHEMISTRY

1. Fix cells in 1.5% glutaraldehyde for 10 min at room temperature.
2. Rinse cells twice with PBS for 5 min each time.
3. Overlay cells with X-Gal substrate overnight at 37°C (*see* **Note 3**).
4. Remove the X-Gal substrate, wash the cells with PBS, and observe the staining under a light microscope to determine β-gal expression.

3.1.3.2. BIOCHEMISTRY

1. Use gentle aspiration to remove the medium from the transduced mono-layers of cells growing in the tissue-culture plates. Wash the monolayers three times using 1 mL of PBS without calcium and magnesium salts. Stand the plates at an angle for 2–3 min to allow the last traces of PBS to drain to one side. Remove any remaining PBS by aspiration.
2. Add 1 mL of PBS to each well and use a rubber policeman to scrape the cells into microfuge tubes. Store the tubes in ice until all of the plates have been processed. Recover the cells by microfugation at maximum speed for 10 s at room temperature.
3. Gently resuspend the cell pellets in 1 mL of ice-cold PBS and, again, re-cover the cells by microfugation. Remove any remaining PBS from the cell pellets and from the sides of the tubes. Store the cell pellets at −20°C for future analysis, or prepare cell extracts by the method in **step 4.**
4. Lyse the cells via three cycles of freezing and thawing in 200 μL of 1 *M* Tris-Cl buffer.
5. For each sample of cell lysate to be assayed, mix 3 μL of 100X Mg solu-tion, 66 μL of ONPG, and 201 μL of 0.1 *M* sodium phosphate, pH 7.5. Add this 300 μL ONPG mixture to 30 μL of cell lysate (*see* **Note 4**). If a heat treatment is to be used to deactivate endogenous β-gal, incubate the cell lysates for 45–60 min at 50°C prior to assay *(16–18)*.
6. Incubate the reactions for 30 min at 37°C or until a faint yellow color has developed. In most cases, the background of endogenous β-gal activity is very low, allowing incubation times as long as 4–6 h to be used.
7. Stop the reactions by adding 500 μL of 1 *M* Na_2CO_3 to each tube. Read the optical density of the solutions at a wavelength of 420 nm in a spectropho-tometer (*see* **Note 5**).

3.2. In Vivo Gene Delivery to Skeletal Muscle

1. Anesthetize adult or newborn mice with an ip injection of 0.1 mL of an anesthetic cocktail comprising katamine, xylazine, and PBS. Inject 20 μL HSV-1 viral suspension (1×10^9 pfu/mL) percutaneously into the muscle

of interest, e.g., the *gastrocnemius*, to an approximate depth of 2.0 mm using a 50 µL Hamilton syringe with a 30 G needle (*see* **Note 6**).

2. Following injection, hold the perforated skin closed with forceps for 10–15 s. A good injection should result in no exudation of the injected material.

3. Sacrifice the mice at different time points postinjection. Peel off the skin and cut the tendo calcaneus. Starting from this position, use a blade to separate the hindlimb muscle along the tibia. Cut the muscle off after separation to the knee. Remove and freeze the injected muscle in 2-methylbutane precooled in liquid nitrogen, then store it in an Eppendorf tube at −80°C.

3.2.1. Histological Examination of the Gene-Delivery Efficiency

1. Cryosection the muscle at 10-µm thickness. Collect sections on glass slides.

2. Air-dry the sections, fix them in 1.5% glutaraldehyde for 1 min, then rinse them twice with PBS for 2 min each time.

3. Incubate the sections overnight in the X-Gal substrate solution at 37°C (*see* **Note 3**).

4. Remove the X-Gal solution, rinse the slides with PBS, and observe the staining under light microscope to determine the efficiency of gene delivery.

3.2.2. Examination of the Gene-Delivery Efficiency Using ONPG Assay

Homogenize the muscle with Tris-Cl, pH 7.8, using a homogenizer (e.g., Tissue-Tearor homogenizer). This procedure can be performed at room temperature, although it is better to keep the sample on ice. Proceed with the steps described in **Subheading 3.1.3.2., Steps 5–7**.

4. Notes

1. There are many methods by which to isolate muscle cells. The method described here was used to enrich for myogenic cells *(19,20)*. Examination of the desmin staining showed that PP1 contained only 7% desmin-positive cells, whereas the subsequent preplates contained increasing fractions of desmin-positive cells (PP2 = 14%; PP3 = 25%; PP4 = 72%; PP5 = 75%; PP6 = 78%). We believe PP6 represents a relatively pure population of myogenic cells.

2. Multiplicity of infection with HSV-1 can vary from 0.1 to 5. In our experience, 3 is the optimal condition. Viral titer can be determined as follows: serially dilute the viral stocks in Opti-MEM media, and place them onto confluent E5 cell (a HSV helper cell) monolayers in 6-well plates. Allow the virus to adsorb to the cells for 2 h at 37°C. Aspirate the virus solutions and wash the cells with HBSS. Overlay the cells with 2 mL DMEM containing 5% FBS and 0.3% methylcellulose, and then incubate them for an

additional 3 d. Fix the cells and visualize plaques by staining with 0.5% crystal violet (Sigma) for 10 min. The titers are defined as plaque forming unit per mL (PFU/mL). Transduction time of muscle cell can vary from 24–48 h. In our experiments the results using 48-h transduction were not superior to those obtained using a 24-h transduction time.

3. LacZ staining time also can vary from 2 h to overnight. We suggest observation every 2–3 h to ensure optimal staining and reduce background staining owing to endogenous β-gal.

4. The exact amount of extract required for ONPG assay will depend on the strength of the promoter driving the expression of the β-gal gene, the efficiency of transduction, and the incubation time of the assay.

5. It is essential to include positive and negative controls. These assays check for the presence of endogenous inhibitors and β-gal, respectively. All of the controls should contain 30 μL of cell extract from mock-transfected cells. In addition, the positive controls should include 1 μL of a commercial preparation of *E. coli* β-gal (50 U/mL). The commercial enzyme preparation should be dissolved at a concentration of 3,000 U/mL in 0.1 *M* sodium phosphate, pH 7.5. Just prior to use, transfer 1 μL of the stock solution of β-gal into 60 μL of 0.1 *M* sodium phosphate, pH 7.5, to make a working stock of the enzyme containing 50 U/mL. One unit of *E. coli* β-gal is defined as the amount of enzyme that will hydrolyze 1 μmole of ONPG substrate in 1 min at 37°C.

6. The choice of muscle depends on the researcher's interest. The *gastrocnemius* muscle is generally selected; it is the largest muscle in the hindlimb and is located in the middle of the muscle belly of the leg, making it the easiest muscle to access.

References

1. Huard, J., Akkaraju, G., Watkins, S. C., Pike-Cavalcoli, M., and Glorioso, J. C. (1997) LacZ gene transfer to skeletal muscle using a replication-defective herpes simplex virus type 1 mutant vector. *Hum. Gene Ther.* **8,** 439–452.
2. Huard, J., Feero, W. G., Watkins, S. C., Hoffman, E. P., Rosenblatt, D. J., and Glorioso, J. C. (1996) The basal lamina is a physical barrier to herpes simplex virus-mediated gene delivery to mature muscle fibers. *J. Virol.* **70,** 8117–8123.
3. Falk, T., Kilani, R. K., Yool, A. J., and Sherman, S. J. (2001) Viral vector-mediated expression of K+ channels regulates electrical excitability in skeletal muscle. *Gene Ther.* **8,** 1372–1379.
4. Wright, M. J., Wightman, L. M., Lilley, C., de Alwis, M., Hart, S. L., Miller, A., et al. (2001) In vivo myocardial gene transfer: optimization, evaluation and direct comparison of gene transfer vectors. *Basic Res. Cardiol.* **96,** 227–236.

5. Akkaraju, G. R., Huard, J., Hoffman, E. P., Goins, W. F., Pruchnic, R., Watkins, S. C., et al. (1999) Herpes simplex virus vector-mediated dystrophin gene transfer and expression in MDX mouse skeletal muscle. *J. Gene Med.* **1,** 280–289.
6. Krisky, D. M., Marconi, P. C., Oligino, T. J., Rouse, R. J., Fink, D. J., Cohen, J. B., et al. (1998) Development of herpes simplex virus replication-defective multigene vectors for combination gene therapy applications. *Gene Ther.* **5,** 1517–1530.
7. Krisky, D. M., Wolfe, D., Goins, W. F., Marconi, P. C., Ramakrishnan, R., Mata, M., et al. (1998) Deletion of multiple immediate-early genes from herpes simplex virus reduces cytotoxicity and permits long-term gene expression in neurons. *Gene Ther.* **5,** 1593–1603.
8. Samaniego, L. A., Webb, A. L., and DeLuca, N. A. (1995) Functional interactions between herpes simplex virus immediate-early proteins during infection: gene expression as a consequence of ICP27 and different domains of ICP4. *J. Virol.* **69,** 5705–5715.
9. Samaniego, L. A., Wu, N., and DeLuca, N. A. (1997) The herpes simplex virus immediate-early protein ICP0 affects transcription from the viral genome and infected-cell survival in the absence of ICP4 and ICP27. *J. Virol.* 71, 4614–4625.
10. Wu, N., Watkins, S. C., Schaffer, P. A., and DeLuca, N. A. (1996) Prolonged gene expression and cell survival after infection by a herpes simplex virus mutant defective in the immediate-early genes encoding ICP4, ICP27, and ICP22. *J. Virol.* **70,** 6358–6369.
11. Marconi, P., Krisky, D., Oligino, T., Poliani, P. L., Ramakrishnan, R., Goins, W. F., et al. (1996) Replication-defective herpes simplex virus vectors for gene transfer in vivo. *Proc. Natl. Acad. Sci. USA* **93,** 11319–11320.
12. van Deutekom, J. C., Floyd, S. S., Booth, D. K., Oligino, T., Krisky, D., Marconi, P., et al. (1998) Implications of maturation for viral gene delivery to skeletal muscle. *Neuromuscul. Disord.* **8,** 135–148.
13. Mester, J. C., Pitha, P. M., and Glorioso, J. C. (1995) Antiviral activity of herpes simplex virus vectors expressing murine alpha 1-interferon. *Gene Ther.* **2,** 187–196.
14. Huard, J., Goins, W. F., Akkaraju, G. R., Krisky, D., Oligino, T., Marconi, P., et al. (1998) Gene transfer to muscle and spinal cord using herpes simplex virus-based vectors, in: *Stem Cell Biology and Gene Therapy* (Quesenberry, P.J., Stein, G.S., Forget, B., and Weissman, S., eds.), John Wiley & Sons, Inc., Indianapolis, IN pp. 179–200.
15. Gage, P. J., Sauer, B., Levine, M., and Glorioso, J. C. (1992) A cell-free recombination system for site-specific integration of multigenic shuttle plasmids into the herpes simplex virus type 1 genome. *J. Virol.* 66, 5509–5515.
16. Hall, C. V., Jacob, P. E., Ringold, G. M., and Lee, F. (1983) Expression and regulation of Escherichia coli lacZ gene fusions in mammalian cells. *J. Mol. Appl. Genet.* **2,** 101–109.
17. Norton, P. A. and Coffin, J. M. (1985) Bacterial beta-galactosidase as a marker of Rous sarcoma virus gene expression and replication. *Mol. Cell Biol.* **5,** 281–290.

18. Young, D. C., Kingsley, S. D., Ryan, K. A., and Dutko, F. J. (1993) Selective inactivation of eukaryotic beta-galactosidase in assays for inhibitors of HIV-1 TAT using bacterial beta-galactosidase as a reporter enzyme. *Anal. Biochem.* **215,** 24–30.

19. Qu, Z., Balkir, L., van Deutekom, J. C., Robbins, P. D., Pruchnic, R., and Huard, J. (1998) Development of approaches to improve cell survival in myoblast transfer therapy. *J. Cell Biol.* **142,** 1257–1267.

20. Lee, J. Y., Qu-Petersen, Z., Cao, B., Kimura, S., Jankowski, R., Cummins, J., et al. (2000) Clonal isolation of muscle-derived cells capable of enhancing muscle regeneration and bone healing. *J. Cell Biol.* **150,** 1085–1100.

21

Delivery of Herpes Simplex Virus-Based Vectors to the Nervous System

James R. Goss, Atsushi Natsume, Darren Wolfe, Marina Mata, Joseph C. Glorioso, and David J. Fink

1. Introduction

Gene transfer to the nervous system is an attractive option to treat a wide variety of neurological insults *(1–3)*. The expression of trophic factor and/or anti-apoptotic genes may be beneficial in halting the slow neurodegeneration in such conditions as Parkinson's disease *(4,5)*, the rapid neuronal cell death following trauma to the brain or spinal cord *(6,7)*, or in treating peripheral neuropathies associated with diabetes or use of chemotherapeutic agents *(8,9)*. Introduction of dominant-negative mutant genes or antisense RNA to treat diseases such as Huntington's disease *(10)*, or transfer of genes to replace lost or mutated endogenous proteins to treat disorders such as lysosomal storage diseases *(11)*, may prove useful. In addition, gene transfer to overexpress endogenous antinociceptive proteins has great potential in pain management *(12)*. The problem faced by all of these applications is finding a suitable methodology that will facilitate the transfer of exogenous genes to the appropriate nerve cells; virus-based vectors have proven quite efficient in transferring genes to many different cell types *(13)*.

A viral vector candidate useful to treat neurological disorders must have the following characteristics: first, it must be able to infect and have high or preferential tropism to neurons. Second, it must have a large capacity to accommo-

From: *Methods in Molecular Biology, vol. 246:*
Gene Delivery to Mammalian Cells: Vol. 2: Viral Gene Transfer Techniques
Edited by: W. C. Heiser © Humana Press Inc., Totowa, NJ

date more than one transgene because many neurological disorders may best be treated with a multivariate approach requiring more than one gene product. In addition, clinically useful vectors will most likely have to be designed to allow regulation of the transgene(s). In either case, the vector must be able to accommodate large transgenes. Third, in many cases, the vector must be capable of expressing the transgene(s) for a prolonged period. Finally, as an economically feasible treatment option, gene-therapy vectors must be able to be grown to high titers, be able to be purified for human use, and be stable enough to be stored for long periods of time until applied by the end user. The biological features of HSV-1 (as reviewed in Chapter 19) make this a very attractive candidate to deliver exogenous gene products to the nervous system.

This chapter will outline the use of modified genomic herpes simplex virus (HSV) vectors to transfer exogenous genes to nerve cells including the purification procedures necessary to ensure removal of cellular debris from the harvesting of the vectors, which can have toxic and/or immunogenic affects when injected directly into central nervous system (CNS) tissue, and the methods utilized to infect neurons both in vitro and in vivo.

2. Materials

2.1. Purification of HSV Vectors for Use in Transfecting Neurons

2.1.1. Solutions

1. 7b Culture Media: this is a buffered media required to stabilize media pH in sealed bottles; it is used for all procedures with roller bottles. Dulbecco's modified Eagle's medium (DMEM), 10% fetal bovine serum (FBS), 250 mM HEPES.
2. RSB Buffer: 10 mM NaCl, 10 mM Tris-HCl, pH 7.4, 3 mM MgCl$_2$.
3. 50% OptiPrep: 5 volumes of 60% OptiPrep (stock) plus 1 volume of PBS.
4. 20% OptiPrep: 2 volumes of 50% OptiPrep plus 3 volumes of PBS.

2.1.2. Reagents and Materials

1. DMEM and fetal bovine serum (FBS) are available from GIBCO/Invitrogen (Carlsbad, CA).
2. Ethylenediaminetetraacetic acid (EDTA), HEPES, MgCl$_2$, NaCl, phosphate buffered saline (PBS) (10X stock), Tris-HCl, and trypsin, are available from Sigma (St. Louis, MO).
3. Roller bottles (850 cm^2) are available from Corning (Corning, NY).
4. Falcon tissue-culture flasks (T150) and 15-mL conical tubes are available from BD PharMingen (La Jolla, CA).
5. Cell Production Roller Apparatus and 15-mL dounce homogenizer ("B" or "tight" pestle) are available from Bellco Biotechnology (Vineland, NJ).

6. Beckman ultra-clear centrifuge tubes (Cat. no. 344058) and Beckman 13-mL polyallomer quick-seal tubes (Cat. no. 342143) are available from Beckman (Fullerton, CA).
7. OptiPrep is available from Greiner Bio-One (Longwood, FL). Note that this comes as a 60% solution when making 50% and 20% working solutions.
8. The 7B complementing cell line is proprietary. Contact Dr. Joseph Glorioso at the University of Pittsburgh, Department of Molecular Genetics and Biochemistry, for permission to use.

2.1.3. Centrifuges and Rotors

1. GS-6R refrigerated tabletop centrifuge and GH-3.8 rotor are used in these protocols; similar models are available from Beckman Coulter (Fullerton, CA).
2. J2-MC floor model centrifuge and JLA-10.5 and JA-20 rotors are used in these protocols; similar models are available from Beckman Coulter.
3. XL90 ultracentrifuge and SW-28 and NVT-65 rotors are used in these protocols; similar models are available from Beckman Coulter.

2.2. Infecting Neurons In Vitro

1. Defined Neurobasal Media Neurobasal medium, B-27 (1 X concentration), GlutaMAX I (0.5 mL/50 mL), Albumax II (50 µL/50 mL), 100 U/mL penicillin, 100 ng/mL 7.0s NGF.
2. Lysis buffer: 1 X PBS, 1 % NP-40, 0.5% sodium deoxycholate, 100 µg/mL PMSF, 20 µg/mL leupeptin.
3. EDTA, 5-fluoro-2'-deoxyuridine, leupeptin, 7.0s NGF, paraformaldehyde, penicillin, NP-40, PMSF, poly-D-lysine (MW 70,000–150,000), sodium deoxycholate, trypsin, and uridine are available from Sigma.
4. Albumax II, B-27, GlutaMAX I, and NeuroBasal medium are available from GIBCO/Invitrogen.
5. 24-well plates are available from Falcon (Lincoln Park, NJ)
6. Protein assay kit is available from Bio-Rad (Hercules, CA).
7. Histochoice is available from Amresco (Solon, OH).

2.3. Infecting PC12 Cells In Vitro

1. DMEM is available from GIBCO/Invitrogen.
2. 2.5s NGF is available from Sigma.

2.4. Infecting Neurons In Vivo

1. Glass bead sterilizer, stereotaxic device, and electronically controlled injector are available from Stoelting (Wood Dale, IL).

2. Ketamine and xylazine are available from Sigma. Note that ketamine is a controlled substance.

3. Isoflurane is available from Fort Dodge Animal Health (Overland Park, KS).

4. A rechargeable high-speed micro-drill is available from Fine Science Tools (Foster City, CA).

3. Methods

3.1 Purification of HSV Vectors for Use in Transfecting Neurons

HSV viral vectors must be grown in various complimenting cell lines depending on what genes have been deleted from the construct. Our group most often utilizes the 7B complementing cell line *(14)*. This cell line has been engineered to express ICP27 and ICP4 upon viral infection allowing propagation of HSV based viral vectors deleted for the essential gene functions for both ICP4 and ICP27. Standard tissue-culture practices are used to propagate and expand the 7B cell line in T150 flasks. Subculture cell lines (1:10) every 3 d. Incubate at 37°C in a humidified incubator with 5% CO_2.

3.1.1. Seeding Roller Bottles

1. Seed two roller bottles with at least 1×10^7 7B cells each (approx one trypsinized T150 flask at confluence) with 100 mL of 7b culture media (*see* **Note 1**). Rotate roller bottles on a Cell Production Roller Apparatus at 1 rpm for 48 to 72 h or until confluent by visual inspection. It is important that the cells do not become over-confluent because this negatively affects vector production.

2. When the cells become confluent, aspirate the media, add 10 mL of trypsin/ EDTA, and incubate for 8 to 10 min while rotating.

3. Harvest cells by washing the inside of each roller bottle with 25 mL of media containing HEPES.

4. Seed 10 roller bottles from the initial two roller bottles (7 mL; 1:5 dilution). Add 100 mL of 7b culture medium to each flask and rotate at 37°C for an additional 48 h. At this point the cells should be approx 70–80% confluent; this is the optimal time for proceeding with infection.

3.1.2. Infection

1. To infect 10 roller bottles, place 200 mL of fresh 7b medium into a sterile 250-mL container.

2. Calculate the volume of viral stock required to achieve an MOI of 0.05. The number of cells must be estimated. At 75% confluence, 10 roller bottles (each 850 cm²) contain 7.5×10^8 7B cells. Add the appropriate amount of titered vector stock (about 3.8×10^6 PFU/roller bottle) to the 200 mL of media reserved earlier and mix. This will serve as the inoculum for infection.

3. Aspirate the media from each roller bottle and add 20 mL of vector inoculum to each.
4. Incubate the roller bottles at 37°C for 2 h.
5. Add 80 mL of medium, bringing the volume of each roller bottle to 100 mL.
6. Incubate for 24 h at 37°C while rotating.
7. Move the roller bottles to a 34°C incubator roller and allow the infection to proceed for an additional 24–48 h or until approx 70% of the cells are detached (*see* **Note 2**).

3.1.3. Harvesting Virus

For all steps in harvesting and centrifuging the virus (**Subheadings 3.13.–3.16.**), be sure to keep the tubes containing virus on ice and to carry out the centrifugations at 4°C.

1. To harvest the virus, swirl the media around the inside of the roller bottle to dislodge the cells. This leaves any uninfected cells attached, thereby cutting down on cell debris.
2. Collect media and cells in 250-mL conical tubes.
3. Spin at 2300*g* for 10 min (3000 rpm in a GS-6R centrifuge).
4. Transfer and reserve the supernatant in 500-mL centrifuge tubes and place on ice.
5. Resuspend the cell pellets in 30 mL of RSB buffer and centrifuge at 1000*g* (2000 rpm in a GH3.8 rotor in a GS-6R centrifuge) for 10 min.
6. Transfer the supernatant to reserved supernatant from **step 4**. Keep the cell pellet on ice.
7. Spin the combined supernatants at 18,500*g* for 1 h (10,000 rpm in a JLA-10.5 rotor in a J2-MC centrifuge). At this point you need to decide if you want a viral preparation with a lower titer and very little cellular debris or with a higher titer but with more cellular debris.
   a. If the vector is to be used in the brain or for other extremely sensitive experiments, then the viral preparation must be as clean as possible. In this is case, resuspend the pellet in 3 mL of PBS plus 3 mL of RSB and continue with the procedure in **Subheading 3.1.4.**
   b. To prepare vector with a higher titer but containing more cellular debris, do not resuspend the pellet yet but keep it on ice and continue with **steps 8–12**.
8. Add 3 mL of RSB buffer to the cell pellet from **step 6** and incubate on ice for more than 10 min to swell the cells. Place a sterile dounce homogenizer (with a B or tight pestle) on ice (*see* **Note 3**).
9. Transfer the cells to the homogenizer and dounce more than 30 times.

10. Transfer the lysed cells to a 15-mL conical centrifuge tube and spin at 1000*g* (2000 rpm in a GH-3.8 rotor in a GS-6R centrifuge) at 4°C for 10 min.
11. Transfer and reserve the supernatant, add an equal volume of PBS to the supernatant, and centrifuge at 1000*g* for 10 min (2000 rpm in a GS-6R centrifuge); discard the pellet.
12. Resuspend the pellet from **step 7** with the supernatant from **step 11**. Be sure to completely but gently resuspend the pellet. Vortex on low speed and/or pass through a small gauge needle if necessary. Titer the vector as described in **Subheading 3.1.7.**

3.1.4. Cushioning

1. Add 7.5 mL of PBS to a Beckman ultra-clear 25 × 89 mm centrifuge tube and mark the level on the outside of the tube.
2. Add the virus from **Subheading 3.1.3.**, **step 7a** to the tube.
3. With a Pasteur pipet, carefully layer 3 mL of 50% OptiPrep on the bottom of the tube.
4. Carefully top off the tube with PBS. Add the PBS very slowly as not to disturb the layer of OptiPrep on the bottom. It is important to fill the tube as much as possible so that it does not collapse during centrifugation. If a balance tube is made, be sure to add 3 mL of 50% OptiPrep.
5. Centrifuge at 112,700*g* for 30 min (25,000 rpm in a Beckman SW-28 rotor and a Beckman XL-90 ultracentrifuge with acceleration set on max and break set on 9).
6. Carefully aspirate down to the line marked on the tube before the spin. Reserve the bottom 7.5 mL (3 mL of 50% OptiPrep, 4.5 mL of PBS, and virus). Mix these to produce a concentrated solution of virus in 20% OptiPrep.

3.1.5. OptiPrep Gradient

1. Prewarm a Beckman tube sealer.
2. Transfer the virus/OptiPrep solution to a Beckman 13-mL polyallomer quick-seal tube using a 5-cc syringe with a 20 G needle.
3. Prepare a 10-cc syringe with a 20 G needle and 15 mL of 20% OptiPrep. Use the syringe to top off the quick-seal tube and aspirate any bubbles.
4. Seal the tube with the Beckman tube sealer.
5. Spin at 402,000*g* for 4.5 h (65,000 rpm in a Beckman NVT-65 rotor and a Beckman XL-90 ultracentrifuge with acceleration set on max and break set on 9).

3.1.6. Harvesting Bands

1. To harvest the virus from the gradient, insert an 18 G needle into the top of the ultracentrifuge tube. Using a 3-cc syringe with a 20G needle, puncture

the tube just above the large band near the bottom of the tube (approx 1/4 from the bottom) and remove 3 mL of virus. Remove the virus slowly, avoiding the thick band of debris banded just below the virus (*see* **Note 4**).

2. Place the collected bands and 25 mL of PBS into an oak ridge tube and centrifuge at 48,400g (20,000 rpm in a JA-20 rotor) for 30 min at 4°C.

3. Discard the supernatant and resuspend virus pellet in an appropriate volume of PBS, usually between 250 and 1000 µL. To avoid damaging the virus, the best way to resuspend it is to pipet the virus off of the wall and then allow it to shake at 4°C overnight. If there are any large chunks remaining, they can be easily broken up by pipetting.

4. Aliquot 10 µL samples of the virus into microfuge tubes and store at −80°C.

3.1.7. Assay of Viral Titer

1. Reserve 10 µL of virus from the completed vector preparation in a 1.5-mL microfuge tube to determine the titer.

2. Add 990 µL of media. This makes the dilution 1×10^{-2}.

3. Prepare another seven microfuge tubes and fill each with 0.9 mL of media for virus dilutions (*see* **Note 5**).

4. Using a micropipet, remove 100 µL from the 1×10^{-2} tube and add it to the next tube, giving a 1×10^{-3} dilution. Repeat this until all of the tubes contain virus. It is important to vortex each tube before moving on to the next dilution and to change tips between each pipetting.

5. Trypsinize one or two flasks of the complementing cell line for the vector to be titered. (For 7B cells, one confluent 75 cm^2 flask contains about 8.25 $\times$ 10^6 cells.) If the virus is not wild-type, it is also important to infect Vero cells to confirm that the vector does not contain wild type virus. If wild type virus is present, plaques will form on the Vero cells.

6. For each virus, prepare 4.5 $\times$ 10^6 cells in 6 mL of medium (7.5 $\times$ 10^5 cells for each well of a 6-well plate).

7. Add 1 mL (7.5 $\times$ 10^5 cells) to each of six microfuge tubes and label them 1×10^{-5} to 1×10^{-10}.

8. With the dilutions from **step 4**, add 100 µL of the 1×10^{-4} tube to the tube containing cells labeled 1×10^{-5}. Remove 100 µL from the dilution tube labeled 1×10^{-5} and add it to the tube containing cells labeled 1×10^{-6}. Repeat this up to the last tube, making sure to vortex the dilution tube before removing the diluted virus and to change tips between pipettings.

9. Incubate the virus and cells at 37°C for 1 h.

10. Plate the cells in a 6-well plate and incubate at 37°C in a humidified incubator containing 5% CO_2.

11. After no more than 20 h, carefully aspirate the media and add 2 mL of media containing 0.5% methycellulose (*see* **Note 6**).
12. Incubate the plate for 4–5 d at 37°C.
13. Depending on the virus, develop the plate by one of the following methods.
    a. If the virus contains no reporter genes (β-galactosidase [β-gal] or green fluorescent protein [GFP]), aspirate the media. Quickly fix the cells with 95% ethanol. Remove the ethanol after a few seconds. Add 1–2 mL of 0.4% crystal violet stain to cover the bottom of the plate. Allow the plate to stain for 20–30 min. Wash the plate carefully with water. The plaques will appear clear against the purple background.
    b. If the virus contains GFP, view the plate under a fluorescent microscope and count green plaques. Stain the plate with crystal violet afterwards to determine if there are any viral plaques present that do not express GFP.
    c. If the virus contains β-gal, fix the cells with 95% ethanol, remove the ethanol, then add 1–2 mL of 0.2% 5-bromo-4-chloro-3-indolyl-β-D-galactopyranoside (X-gal). Incubate 3–4 h to stain the plate then count blue plaques.

3.2. Infecting Neurons In Vitro

Often, it is desirable to test the expression of a transgene from an HSV vector by infecting primary cell cultures. We generally use cell cultures derived from E17 rat dorsal root ganglia (DRG), trigeminal ganglia (TG), or cortex. In addition we have used differentiated and nondifferentiated PC12 cells to examine neuron-specific characteristics of our HSV vectors *(15)*. It should be pointed out that the transgene expression characteristics (amount and length of time) of any given HSV vector might not be the same in vitro as in vivo. Still, in vitro testing of HSV vectors is an important and vital step when characterizing any potential therapeutic use.

3.2.1. Infecting Neurons In Vitro

1. Dissect and remove DRG, TG, and/or cortex from 17-d rat embryos (*see* **Note 7**).
2. Dissociate cells with 0.25% trypsin, 1 mM EDTA for 45 min at 37°C with constant shaking. Gently centrifuge the cells and remove the trypsin-EDTA.
3. Resuspend the cells in a small volume of Defined Neurobasal medium and plate them on poly-D-lysine coated coverslips at 1 × 10^5 cells/well in a 24-well plate; bring the volume to 500 μL with Defined Neurobasal medium.
4. Incubate the cells at 37°C in an humidified incubator containing 5% CO_2.
5. On d 4, add 5 μM 5-fluoro-2′-deoxyuridine and 5 μM uridine to cultures to inhibit the growth of dividing cells (Schwann and satellite cells).

6. Cells can be infected with vector from d 6 onward. Remove the medium and add 250 µL of fresh medium to the cells in each well to be infected. Add the appropriate amount of vector (generally infect at an MOI of 1). Incubate for 1 h at 37°C with gentle agitation every 10 min. Remove the infection medium and add 500 µL of fresh medium to the cells.

7. Incubate the cells for the length of time desired. Usually 24–48 h is sufficient to observe transgene expression by histochemistry or enzyme-linked immunosorbent assay (ELISA). To isolate protein from cells, remove the medium and lyse the cells in Lysis Buffer for 20 min at room temperature (RT). Determine total protein using a Bio-Rad Protein Assay Kit as described by the manufacturer; aliquots can be stored at −80°C until used. Transgene products, such as trophic factors, which are released into the medium, can be measured by ELISA. Collect cell culture medium at regular times and remove cells and virus by centrifugation in a microfuge at 16,000*g* for 10 min. Store samples at −80°C until assayed. We have found that most commercially available ELISA kits work well.

8. To perform immunohistochemistry on the cells, aspirate the medium, then, using a forcep, remove the coverslip containing the cells. Fix the cells in Histochoice or cold 4% paraformaldehyde for 10 min. Wash the cells in PBS and incubate with the primary antibody (plus 1% of the appropriate serum) for 2 h at room temperature or overnight at 4°C.

3.2.2. Infecting PC12 Cells In Vitro

1. Plate PC12 cells on poly-D-lysine coated coverslips at 1×10^5 cells/well in a 24 well plate in 500 µL of DMEM supplemented with 7.5% fetal calf serum (FCS) and 7.5% horse serum.

2. Incubate cells at 37°C in an humidified incubator containing 5% CO_2.

3. PC12 cells can be used as is or differentiated to assume a neuronal phenotype. To infect undifferentiated PC12 cells, see **Subheading 3.2.1., step 6.**

4. To differentiate PC12 cells, replate the cells at less than 2×10^4 cells/cm^2 in DMEM containing 1% serum and 100 ng/mL 2.5s NGF; incubate for 7 d at 37°C in an humidified incubator containing 10% CO_2. To infect differentiated PC12 cells, see **Subheading 3.2.1., step 6**.

3.3. Infecting Neurons In Vivo

The use of HSV-based vectors in the nervous system can be divided between delivery to the central nervous system (CNS, brain, and spinal cord) or the sensory ganglia of the peripheral nervous system (PNS). Delivery of HSV vectors to the CNS is accomplished by stereotaxic inoculation as with other viral vectors, liposomes, plasmids, or naked DNA constructs, and requires invasive surgical procedures. On the other hand, delivery into the PNS can take advantage of the unique HSV neurotropism to sensory nerves. A specific sensory ganglion can be infected with HSV vectors by simple, nonsurgical application of the vectors to the epithelial tissue corresponding to the receptive field of the sensory ganglion. Thus the TG can be infected by placing the vector on the surface of the eye following corneal abrasion (which facilitates uptake of the vector by the sensory neurons) or the lumbar DRG can be targeted by subcutaneous (sc) inoculation of the vector to the plantar surface of the hind feet in rodents.

3.3.1. Infecting Neurons in the CNS

All surgical procedures should be performed in as sterile an environment as possible. Make sure that all surgical instruments are sterilized. If vectors are to be injected into multiple animals then multiple sets of sterilized instruments should be available, or some method of sterilizing the instruments between surgeries should be available. We have found that glass bead sterilizers work well for this purpose.

3.3.1.1. INJECTING VECTORS INTO THE BRAIN

1. Anesthetize the rat or mouse with a mixture of kctamine/xylazine (80/10 mg/kg). Proceed when the animal no longer responds to a noxious stimulus (foot or tail pinch).
2. Clean the hair overlying the scalp with 70% ethanol and then shave off the hair using electric clippers.
3. Disinfect the scalp with betadine solution.
4. Using a sterile scalpel, make a midline incision through the scalp. The size of the incision should be about 2–2.5 cm in a rat and about 1 cm in a mouse. Cut and scrape the connective tissue overlying the skull to expose the skull. Use sterile gauze to stop any bleeding.
5. Place the rat or mouse in a small animal stereotaxic device with ear and incisor bars. Adjust the incisor bar so that the lambda and bregma sutures lie on a flat plane (flat skull position *[16]*).
6. Locate the desired coordinates on the skull using the stereotaxic device and a brain atlas. Mark the spot with a pen. Use a micro-drill to drill through the skull. Be careful not to drill into the underlying brain tissue.
7. Inject the vector into the brain at the desired coordinates. The amount of vector injected will depend on the application, but we usually inject 1–2 µL containing 1×10^6 PFU/µL. It is important to inject the vector slowly (<0.2 µL/min) to avoid causing any tissue damage. Owing to the small volumes and slow rate of injection, we currently use either a Hamilton syringe with a 30 G needle attached directly to an electronically controlled injec-

tor, or a glass micropipet attached, via plastic tubing, to a syringe that is attached to an electronically controlled injector.
8. Fill the burr hole with bone wax and suture the scalp closed.

3.3.1.2. INJECTING VECTORS INTO THE SPINAL CORD

1. Anesthetize rats or mice with a mixture of ketamine/xylazine (80/10 mg/kg). Proceed when they no longer respond to a noxious stimulus (foot or tail pinch).
2. Clean the hair on the back with 70% ethanol and then shave off the hair using electric clippers.
3. Disinfect the skin with betadine solution.
4. Make a 2–3 cm incision in the skin along the area of the spinal cord where the injection will be. Blunt dissect any tissue overlying the vertebrate column.
5. If a single injection of vector is required, drill a burr hole through the lamina and pierce the dura with a sterile hypodermic needle. If you wish to inject multiple sites along the spinal cord, then perform a laminectomy (or hemilaminectomy if injecting only one side) to reveal the spinal cord (*see* **Note 8**).
6. Following the injection, cover the laminectomy by muscle and close the wound by suturing the muscle and skin layer by layer.

3.3.2. Infecting Neurons in Sensory Ganglia

3.3.2.1. INFECTING TRIGEMINAL GANGLION NEURONS

The TG is easy to infect with HSV vectors and is easy to dissect, thus it is a convenient tissue in which to examine in vivo vector transfection efficiencies and transgene expression characteristics. Approximately one-third of the neurons in the trigeminal ganglion project to the cornea. Cornea scarification is a simple method to transfect these neurons with an HSV vector.

1. Anesthetize rats or mice with a mixture of ketamine/xylazine (80/10 mg/kg). Proceed when the animals no longer respond to a noxious stimulus (foot or tail pinch).
2. Draw a syringe needle across the surface of the cornea about 15 times using enough pressure to scratch the surface but being careful not compromise the integrity of the cornea.
3. Place 5 μL (mouse) to 15 μL (rat) of the vector (1×10^9 PFU/mL) directly onto the surface of the eye and let it sit there for 15 min.
4. Liberally wash the eye and surrounding tissues with sterile saline to remove any vector.
5. Allow the animal to recover from the anesthesia. Because corneal scarification can cause pain, a nonsteroidal anti-inflammatory agent such as ke-

toprofen should be given to the animals (5 mg/kg sc twice per day) for 5–7 d following the corneal abrasions.

3.3.2.2. Infecting Dorsal Root Ganglion Neurons

Sensory neurons in DRG project to specific body dermatomes and can be infected with HSV vectors by applying the vector to their respective dermatome. In our studies of peripheral neuropathy, we have focused on examining the nerves innervating the hindlimbs *(9)*. The cell bodies of these sensory nerves are found predominantly in the fourth, fifth, and sixth lumbar DRGs. The plantar surface of the hind foot in rats and mice are innervated by neurons from each of these three DRGs making them easy to infect.

1. Anesthetize rats or mice with isoflurane or another short-lasting anesthetic.
2. Wash the plantar surface of the foot with 70% alcohol.
3. Inject 10 µL (mouse) or 30 µL (rat) of vector (1 × 10⁹ PFU/mL) sc into the plantar surface. When injecting into the rat foot, it is best to inject three 10 µL volumes into three separate areas of the foot in order to spread the vector over as wide an area as possible. Conversely, instead of an sc injection, the plantar surface can be scarified with sandpaper and the vector placed on the surface and allowed to sit for at least 15 min (in this case, a longer-lasting anesthetic must be used).
4. Allow the animal to recover from the anesthesia. There is usually very little pain or discomfort associated with this procedure and analgesics are not necessary.

4. Notes

1. Roller bottles are utilized to increase the density of cells and to simplify production and subsequent purification.
2. This takes practice to determine because each infection proceeds at slightly different rates. Do not allow the infection to go longer than 72 h as virus will begin to degenerate. If the infection has not gone to completion in 72 h, discard the rollers bottles. Failed infections are usually due to inaccurate titers of vector stocks, using overconfluent cells, or cells over passage 50.
3. Using a dounce homogenizer is advantageous over sonication and freeze/thaw because it only disrupts the cell membrane, leaving nuclei intact, and does not rupture the viral envelope.
4. If you inadvertently remove some debris along with the virus, place the contents in a 50 ml conical tube and centrifuge at 3k for 10 min at 4°C. Remove the supernatant and continue with **step 2** in **Subheading 3.1.6.**

5. Depending on the type and volume of vector, the number of dilutions required will vary. For the average large preparation (five roller bottles), it is best to take the dilution out to 1×10^{-10}.

6. Media with methylcellulose can be added as soon as the cells have adhered to the plate. This special media is more viscous than standard DMEM, which prevents secondary plaque formation. Although this step can be omitted in cases where a titer is needed in a shorter amount of time, the addition of it will give a much more accurate result.

7. Euthanize a pregnant female rat (17-d gestation) with an overdose of ketamine/xylazine (200/25 mg/kg) or with carbon dioxide. Make a midline incision through the abdomen; remove the placenta then the embryos using a surgical microscope. To collect cortex and TG, use a scalpel to sever the top of the head by making a complete rostral to caudal cut just above the eyes. The upper portion contains cortical cells. Remove this tissue mass using Vanna-style micro scissors and forceps (Fine Science Tools, Foster City, CA); transfer the tissue to 1X Leibovitz's L-15 Medium (Invitrogen/ GIBCO). To prepare the tissue for culture, transfer all of the collected tissue to 1X Trypsin-EDTA (Invitrogen/GIBCO) and proceed as described in **Subheading 3.2.1., step 2**. The TG are strands of tissue attached to each eye in the bottom portion of the head of the embryo and should be visible under the dissecting microscope. Remove the TG using Vanna-style micro scissors and forceps; prepare the tissue for cell culture as described above for cortex. To collect the DRG, make a ventral midline incision and remove all of the body organs revealing the ventral aspect of the spinal column. Carefully cut along the vertebrate to reveal the spinal cord and remove the spinal cord. The DRG are attached to the spinal cord and can be easily removed using Vanna-style micro scissors and forceps. Prepare the tissue for culture as described **Subheading 3.2.1.**

8. The same stereotaxic device used to inject into specific brain coordinates can also be used to inject into the spinal cord. This assures accurate dorsal–ventral and medial–lateral placement of the needle in the spinal cord. As with brain injections, the amount of vector injected will depend on the application. Generally, for a single point injection, we inject a volume of no more than 1 µL. Transduction of motor neurons is usually best achieved by injecting smaller volumes (200 nL) along a given length of the spinal cord.

References

1. Fink, D. J., DeLuca, N. A., Goins, W. F., and Glorioso, J. C. (1996) Gene transfer to neurons using herpes simplex virus-based vectors. *Annu. Rev. Neurosci.* **19,** 265–287.

2. Zlokovic, B. V. and Apuzzo, M. L. J. (1997) Cellular and molecular neurosurgery: pathways from concept to reality—Part 1: Target disorders and concept approaches to gene therapy of the central nervous system. *Neurosurgery* **40,** 789–804.

3. Fink, D. J., DeLuca, N. A., Yamada, M., Wolfe, D. P., and Glorioso, J. C. (2000) Design and application of HSV vectors for neuroprotection. *Gene Ther.* 7, 115–119.

4. Eberhardt, O., Coelln, R. V., Kugler, S., Lindenau, J., Rathke-Hartlieb, S., Gerhardt, E., et al. (2000) Protection by synergistic effects of adenovirus-mediated X-chromosome-linked inhibitor of apoptosis and glial cell line-derived neurotrophic factor gene transfer in the 1-methyl-4-phenyl-1,2,3,6-tetrahydropyridine model of Parkinson's disease. *J. Neurosci.* **20,** 9126–9134.

5. Natsume, A., Mata, M., Goss, J. R., Huang, S., Wolfe, D., Oligino, T., et al. (2001) Bcl-2 and GDNF delivered by HSV-mediated gene transfer act additively to protect dopaminergic neurons from 6-OHDA induced degeneration. *Exp. Neurol.* **169,** 231–238.

6. Romero, M. I., Rangappa, N., Garry, M. G., and Smith, G. M. (2001) Functional regeneration of chronically injured sensory afferents into adult spinal cord after neurotrophin gene therapy. *J. Neurosci.* **21,** 8408–8416.

7. Philips, M. F., Mattiasson, G., Wieloch, T., Bjorklund, A., Johansson, B. B., Tomasevic, G., et al. (2001) Neuroprotective and behavioral efficacy of nerve growth factor-transfected hippocampal progenitor cell transplants after experimental traumatic brain injury. *J. Neurosurg* **94,** 765–774.

8. Schratzberger, P., Walter, D. H., Rittig, K., Bahlmann, F. H., Pola, R., Curry, C., et al. (2001). Reversal of experimental diabetic neuropathy by VEGF gene transfer. *J. Clin. Invest.* **107,** 1083–1092.

9. Chattopadhyay, M., Wolfe, D., Huang, S., Goss, J., Glorioso, J. C., Mata, M., and Fink, D. J. (2002). In vivo gene therapy of pyridoxine-induced neuropathy by HSV-mediated gene transfer of neurotrophin-3. *Ann. Neurol.* **51,** 19–27.

10. Haque, N. and Isacson, O. (1997) Antisense gene therapy for neurodegenerative disease? *Exp. Neurol.* **144,** 139–146.

11. Poenaru, L. (2001) From gene transfer to gene therapy in lysosomal storage diseases affecting the central nervous system. *Ann. Med.* **33,** 28–36.

12. Goss, J. R., Mata, M., Goins, W. F., Wu, H. H., Glorioso, J. C., and Fink, D. J. (2001) Antinociceptive effect of a genomic herpes simplex virus-based vector expressing human proenkephalin in rat dorsal root ganglion. *Gene Ther.* **8,** 551–556.

13. Hermens, W. T., and Verhaagen, J. (1998) Viral vectors, tools for gene transfer in the nervous system. *Prog. Neurobiol.* **55,** 399–432.

14. Krisky, D. M., Wolfe, D., Goins, W. F., Marconi, P. C., Ramakrishnan, R., Mata, M., et al. (1998) Deletion of multiple immediate early genes from herpes simplex virus reduces cytotoxicity and permits long-term gene expression in neurons. *Gene Ther.* **5,** 1593–1603.

15. Chen, X., Li, J., Mata, M., Goss, J., Wolfe, D., Glorioso, J. C., and Fink, D. J. (2000) Herpes simplex virus type 1 ICP0 protein does not accumulate in the nucleus of primary neurons in culture. *J Virol* **74,** 10132–10141.

16. Paxinos, G., and Watson, C. (1998) *The Rat Brain in Stereotaxic Coordinates*, 4th ed. Academic Press, New York, NY.

22

Gene Transfer to Glial Tumors Using Herpes Simplex Virus

Ajay Niranjan, Darren Wolfe, Wendy Fellows, William F. Goins, Joseph C. Glorioso, Douglas Kondziolka, and L. Dade Lunsford

1. Introduction

Glial tumors occur as intraaxial masses in the brain and are uniformly fatal due to lack of effective therapy. Resection combined with radiation and chemotherapy fails to eradicate malignant cells infiltrating into normal brain, and recurrence at the original site is ultimately fatal. Gene transfer offers the potential to enhance tumor cell killing while sparing surrounding normal brain. Several approaches have been developed to deliver genes to tumor cells in order to kill these cells. The first strategy involves the use of viral vectors that are replication-competent, but depend on attributes unique to the tumor cell to support viral growth. Both replication-competent adenovirus and herpes simplex virus (HSV) vectors have been employed in pre-clinical studies and most recently in human clinical trials (1–3). For this purpose, HSV vectors have been engineered that replicate in dividing cells, such as tumor cells, but not in normal neurons. The use of conditional replication competent viruses could allow for their spread in tumor tissue while minimizing damage to normal brain, thus increasing the specificity and effectiveness. Such mutants include those lacking the viral thymidine kinase (tk) gene (4–7), ribonucleotide reductase gene (8,9), a protein kinase gene (10), or a gene (γ34.5) required for growth specifically in neurons (11–13). Deleting these genes in combination creates viruses that are highly compromised for their ability to replicate in and kill neuronal cells, yet

From: *Methods in Molecular Biology, vol. 246:*
Gene Delivery to Mammalian Cells: Vol. 2: Viral Gene Transfer Techniques
Edited by: W. C. Heiser © Humana Press Inc., Totowa, NJ

retain the ability to replicate in and kill dividing tumor cells. Although these vectors are compromised for replication in nondividing cells, they have been shown to replicate and kill ependymal cells lining the ventricles following inoculation of vector into rodent brain. These vectors have shown no gross signs of toxicity to date in two Phase I/II human clinical trials *(14,15)* and have shown some preliminary indication of efficacy. Although these recombinant viruses are replication competent, they remain difficult to grow and purify.

A second strategy utilizes replication defective vectors to induce destruction of tumor cells mediated by a variety of mechanisms, including the recruitment of natural killer (NK) cells using the appropriate cytokines, the activation of anti-cancer drugs at the tumor site that kill multiple tumor cells in addition to those transduced by the vector, and the use of antigens and cytokine expressing genes to elicit specific anti-tumor immunity. Although each of these strategies suffers from limitations, approaches of this nature are under evaluation for efficacy in patients with brain tumors. Appropriate animal models of brain tumors are necessary to study gene transfer to glial tumors. Rodent models of intracranial tumors are reproducible and have been widely used for therapeutic investigations. Most of our current studies were performed using the U87 malignant glioma nude mouse model and the 9L gliosarcoma rat model.

Effective gene transfer for the treatment of cancer will likely require the simultaneous delivery of multiple transgenes using complex vectors capable of delivery of multiple transgenes to an individual tumor cell. HSV is very well-suited for this purpose because of its considerable genome size (152Kb) and because it is possible to engineer vectors that are safe and, by themselves, incapable of causing disease *(16–19)*. HSV efficiently infects many different cell types and a single virus particle can deliver and express multiple transgenes at high levels *(20–24)*. We have used HSV-1 multigene vectors for treatment of brain tumors by direct injection. The multigene vector NUREL-C2 expresses the HSV ICP0 gene, the viral tk gene for activation of ganciclovir (GCV), the rat connexin 43 gene that reduces the metastatic growth properties of glioblastoma cells by the formation of gap junctions and enhances the transmission of activated GCV to neighboring cells, and the human tumor-necrosis factor-α (TNF-α) gene that is toxic to some tumor cells as a soluble mediator, but is also cytotoxic to TNF-α resistant tumor cells when expressed intracellularly *(20)*. TNF-α also activates NK cells to attack tumor tissue and destroy tumor blood vessel endothelium, particularly in combination with radiation.

1.1. Development of HSV Vectors Suitable for Cancer Gene Therapy

Herpes simplex virus type 1 (HSV-1) is a nuclear-replicating DNA virus that infects a wide variety of cells and tissues, and grows to relatively high titers. These characteristics, along with well-established methods to insert large amounts of exogenous DNA and to manipulate the genotype and phenotype of the virus, make HSV an attractive gene-transfer vector. Wild-type HSV replicates efficiently, is cytolytic, and is pathogenic, factors of the natural biology of HSV that have been used in the construction of the replication competent oncolytic HSV vectors. In contrast, our strategy relies on the use of virus vector backbones that do not replicate in any natural cell type. Specifically, the attributes of this system are: (1) the vectors are replication-defective in any cell type and have been engineered so that replication competent viruses are not produced in the process of generating virus stocks; (2) both tumor and nontumor cells are growth-arrested by infection with the vector; and, (3) infected cells are programmed to express very high levels of the therapeutic transgenes, which in turn result in the destruction of infected as well as uninfected neighboring tumor cells. The details of HSV infection, replication, and vector construction are presented in the overview chapter within this section. Vectors defective for expression of the immediate-early regulatory genes *ICP4*, *ICP27*, *ICP47*, and *ICP22* genes do not grow or appreciably express any of the approx 80 early and late proteins in noncomplementary cells. However, engineered transgene expression from strong viral promoters from these genomes is sufficient to produce protein detectable on sodium dodecyl sulfate (SDS) gels of total infected cell lysate. The expression of ICP0 from the virus results in the growth arrest of tumor cells. Neurons regulate and process ICP0 differently than replicating cells such as fibroblasts *(21)* and are not affected by ICP0 expression *(22)*. These effects, including promyelocytic leukemia oncogenic domains (PODs) disruption and replication arrest, in addition to the ability of ICP0 to promote high levels of transgene expression, provide the basis for the choice of vector backbone in this study. Additional published studies have demonstrated that cell cycle arrest is a function of ICP0 *(18,19)*. Subsequently, published studies have demonstrated that the expression of ICP0 from these defective viruses results in a cellular gene-expression profile consistent with cell-cycle arrest, which is independent of the tumor status of the cell *(23)*. Moreover, these replication-defective vectors can be grown to relatively high titer. Replication-competent virus (RCV) is not detected in virus stocks demonstrating the safety profile of this type of vector backbone. In summary, we have utilized a defective HSV backbone engineered by N. DeLuca *(19,24)* with attributes that are optimal for the safe application to the destruction of tumors, including retention of viral mechanisms to promote high levels of transgene expression and to induce growth arrest of infected tumor cells, and the ability to obtain high yields of virus devoid of RCV to elicit the greatest possible effect on the tumor without the possibility of pathogenesis.

1.2. Treatment of Glial Tumors Using Multigene HSV Vectors

The HSV tk gene is commonly employed as a suicide gene in experimental and human clinical protocols *(25–30)* to kill tumor cells. Transfer of the tk gene into tumor cells results in tumor cell death when these cells are exposed to the antiviral drug GCV as the TK enzyme processes the prodrug into a toxic nucleotide analog, which, upon incorporation into nascent DNA, results in strand termination during DNA replication in actively dividing cells such as tumor cells. A powerful characteristic observed in TK-GCV treatments is that transduction of a small fraction of the tumor cells results in significant destruction of noninfected neighboring tumor cells, a phenomenon that has been termed the "bystander effect" *(25,28,30–36)*. The bystander effect results from cell-to-cell transfer of phosphorylated GCV via gap junctions between TK-transduced tumor cells and neighboring unmodified cells *(33–36)*. We have previously tested the ability of TK overexpressing HSV vectors to act as a treatment for established tumors in rodent glioma models and demonstrated significant increases in survival with the HSV-TK vector *(20)*.

In order to augment the cell killing seen in suicide gene therapy, we have taken two approaches that can be used in combination with surgery and radiation. In the first, we created a replication defective HSV-1-based vector that expresses the human TNF-α gene product in conjunction with the HSV-tk gene (TH:TNF) *(20)*. In the second approach, we introduced the connexin 43 gene into the HSV-TK expression vector to increase the GCV-mediated bystander effect in tumor cells in which connexin was poorly expressed *(37)*.

1.2.1. Rationale for Using TK in Combination With TNF-α

TNF-α has been demonstrated to possess an array of anti-tumor activities, including potent cytotoxicity exerted directly on tumor cells *(38)*, enhancement of the expression of HLA antigens *(39)* and ICAM-1 *(40)* on tumor cell surfaces, enhancement of interleukin-2 (IL-2) receptors on lymphocytes *(41)*, and promotion of the activation of such effector cells as NK cells, lymphokine-activated killer (LAK) cells, and cytotoxic T lymphocytes (CTL) *(41–45)*. However, despite this promising anti-tumor profile, the clinical use of TNF-α has been constrained by the toxicity of systemic TNF-α delivery *(44,45)*. The possibility exists, though, that local production of TNF-α at the site of tumor growth may allow for effective use of this cytokine as an anti-tumor agent. Furthermore, the radio-sensitizing ability of TNF-α should optimize the anti-tumor effects of this cytokine when combined with a radiotherapy approach. In an effort to augment the effectiveness of HSV-TK-mediated suicide gene therapy, we created an HSV vector that expresses HSV-TK and TNF-α.

1.2.2. The Rationale for Using TK in Combination With Connexin 43

Connexins are the components of gap junctions *(46)* that play a major role in intercellular communication to control homeostasis and cell proliferation. Reduced intercellular communication through gap junctions has been regularly observed in transformed cells *(47–49)* and may be due to reduced connexin expression. It has been shown that retrovirus-mediated introduction of the connexin 43 (Cx43) gene resulted in growth reduction of transformed cells that shared the characteristic of reduced gap-junctional activity *(50)*. Various reports confirmed these findings and suggested that connexin possess tumor-suppressor activity *(51–54)*. Gap junctions play a critical role in the HSV-TK/GCV bystander effect by enabling the transfer of activated GCV from TK-positive vector-transduced cells to neighboring TK-negative cells *(33,54–58)*. Tumor cells having reduced gap-junction formation are less susceptible to the bystander effect *(58)*, suggesting that transfer of connexin genes into these cells to restore or augment intercellular communication would improve the effectiveness of HSV-TK/GCV therapy. In vitro and in vivo experiments have demonstrated that the bystander effect can be potentiated by the expression of connexin genes *(37,57,59,60)*. The bystander effect requires connexin expression not only by the TK-positive GCV-activating cell, but also by the TK-negative recipient cell *(58)*, suggesting that gene-therapy approaches aimed at co-delivery of connexin with HSV-TK to mediate an enhanced bystander effect may not be effective against tumors that express no connexin at all. However, it may be predicted from a study of 17 cell lines measuring bystander effects and gap junctional activity *(56)* that this will be a rather small or limited group. Vast overexpression of connexin in the transduced cells will increase transfer of activated GCV providing even limited connexin expression in neighboring cells.

1.2.3. Rationale for Using Radiosurgery in Combination With HSV-Based TNF-α Gene Transfer

Radiation has been shown to enhance TNF-α production by tumor cells and macrophages *(61)*. Radiation can also prime astrocytes and microglial cells for TNF-α production *(62)*. The synergistic or additive effect of radiation and TNF-α on a variety of tumors has been demonstrated by several investigators *(63–67)*. Both radiation and TNF-α induce free radicals, which mediate many of their biological effects. Additionally, a high concentration of TNF-α induces apoptosis in endothelial cells *(68)*, which can cause loss of endothelial cell anticoagulant surface, ultimately leading to a pro-coagulant microenvironment *(69)*. Similarly, radiation has been associated with a pro-coagulant state as well as endothelial apoptosis. Free-radical production following radiation causes thrombosis of tumor vasculature. The combination of radiation therapy and TNF-α can lead to enhanced tumor vessel thrombosis. Staba et al. in a study on human malignant glioma (D54) xenografts in a nude mice model, observed increased tumor thrombosis and necrosis in animals treated with adenoviral TNF-α gene therapy and radiation *(70)*. We have shown enhanced tumor-cell killing with intratumoral TNF-α and radiosurgery *(71)*.

2. Materials

1. 9L Gliosarcoma cells (ATCC, Manassas, VA).
2. U87 MG cells (ATCC).
3. Fischer 344 rats (Harlan Sprague Dawley, Indianapolis, IN).
4. Nude Mice (Harlan Sprague Dawley).
5. Ketamine (100 mg/mL) (Phoenix Pharmaceutical Inc., St Joseph, MO).
6. Acepromazine (10 mg/mL) (Phoenix Pharmaceutical Inc.).
7. Betadine solution (Purdue Frederick Company, Norwalk, CT).
8. Shaver/razor (Oster Company, McMinnville, TN).
9. Electric drill (Foredom Company, Bethel, CT).
10. Bone Wax (Ethicon Inc, Somerville, NJ).
11. Small animal stereotactic (Frame Model 900, David Kopf Instruments, Tujunga, CA).
12. Surgical blade no. 15 (Bard Parker, Becton Dickinson, Franklin Lakes, NJ).
13. Surgical instruments (Codman Instruments, Randolph, MA).
14. Suture, 3-0 silk (USSC Sutures, Norwalk, CT).
15. Ten microliter syringe (Cat. no. 80360, Hamilton Company, Reno, NV).
16. Transpore adhesive tape (3M, St. Paul, MN).
17. 25 G needle (Becton Dickinson, Franklin Lakes, NJ).
18. Autoclips, 9-mm (Clay Adams, Becton Dickinson, Sparks, MD).
19. Microinjection unit (Cat. no. 1208, David Kopf Instruments).
20. Microtome (Leitz Company, Austria).
21. X-gal staining solution: 1 μg/mL X-gal, 10 mM K_3Fe (CN)$_6$, 10 mM K_4Fe (CN)$_6$, 1 mM MgCl$_2$ in PBS.
22. Crystal Violet Solution: 1% cytstal violet (Sigma, St. Louis, MO) (w/v) in 50% methanol.
23. OCT (Tissue Tek, Torrance, CA).
24. Paraformaldehyde (4%) (Sigma).
25. PBS (1X, Sigma).
26. Gel Mount, (Biomeda, Foster City, CA).

3. Methods

3.1. Tumor Cell Inoculation Technique

We have used 6- to 8-wk-old Fischer 344 rats for the 9L rat gliosarcoma model and Balb/C nude mice for the U87 malignant glioma model in our experiments. Any rodent model can be used when employing the following technique.

1. Anesthetize the animal with an intramuscular (im) injection of Ketamine (100 mg/mL) and Acepromazine (10 mg/mL) at a ratio of 9:1 and a dose of 44 mg/kg.

2. Remove the hair from the top of the skull with an animal shaver and swab the surgical site with Betadine solution. Place the anesthetized animal in a small animal stereotactic frame.
3. Align the animal in the ear bars first. Keep one ear bar stationary (fixed in the frame) and place it into the ear canal on the first side. Place the second ear bar into the other ear canal. Check that the head is fixed properly. Place two top incisor teeth in the mouthpiece and immobilize the nose with a nose clip (*see* **Note 1**).
4. Make a 1-cm incision in the skin atop the skull with a no. 15 surgical blade and identify the bregma. Bregma is the landmark at which the two coronal sutures meet the sagittal suture.
5. Select a site on the cranium 2 mm to the right of bregma and 1 mm anterior to the coronal suture. Drill a small craniotomy using an electric drill, taking care not to open the dura.
6. Implant 1×10^5 U87 malignant glioma cells or 9L glioma cells in a 3 µL volume stereotactically in the right frontal lobe region 3 mm below the dura mater using a 10-µL syringe with stationary needle. Perform the injection over 4 min using a microinjection unit to ensure that the solution does not leak up the needle tract when the needle is removed.
7. Leave the needle in place for 1 min after injection is complete and then retract the needle slowly.
8. Seal the craniotomy with bone wax after removal of the injection needle.
9. Close the scalp using Autoclip skin staples. Mark the animals and divide them into different experimental groups (*see* **Note 2**).
10. Observe the animals until they are fully awake (*see* **Note 3**).

3.2. Intratumoral Vector Injection Technique

Tumor bearing animals receive intratumoral injections of 5 µL of vector (on third day following tumor implantation in our experiments). The following technique is employed for intratumoral vector injection.

1. Remove the vector from $-80°C$, thaw it on ice, and dilute to the appropriate concentration such that the number of particles for each vector and controls being compared is the same (*see* **Note 4**).
2. Transport the vector stocks to the animal facility on ice (*see* **Note 5**).
3. Anesthetize the animal with Ketamine and Acepromazine. Immobilize the animal head in the small animal stereotactic frame as described in **Subheading 3.1.**
4. Using a scalpel, remove the skin staple and open the skin incision used to inoculate the tumor cells (*see* **Note 6**).

5. Identify the previous burr hole and remove the bone wax using fine forceps (*see* **Note 7**).
6. Inject 5 µL of vector stereotactically through the previous craniotomy using the same coordinates as used for tumor cell implantation (3 mm below the dura). Complete the injection over a period of 4 min using a microinjection unit attached to the stereotactic frame.
7. Leave the needle in place for 1 min following completion of injection.
8. Retract the needle slowly and again seal the burr hole with bone wax.
9. If radiosurgery is planned at a later stage, put a 2 mm long section of a 25 G needle in the bone wax in the craniotomy site. This needle marker is used in subsequent stereotactic targeting of the tumor for gamma knife radiation. The tumor will be 3 mm below the marker, which can be seen on plain X-ray.
10. Close the scalp with a 3-0 silk suture. Alternatively the scalp can be closed using metal wound clips, but if a subsequent radiological procedure is planned, the metal clip will cause radiation scatter and artifacts in the imaging.
11. Mark the animals for future identification (*see* **Note 2**).
12. Observe the animals are until they are fully awake. Once animals are awake and mobile transfer them back to the animal facility.
13. Examine the animals twice daily for the development of signs of increased intracranial pressure from developing brain tumor. Initial signs of brain tumor progression are unwillingness to groom and black discoloration around the eyes. Several days later, hemiparesis will begin to be exhibited. Sacrifice the animals at this point. This time point is well before animals are no longer able to feed themselves.

3.3. Assessment of Gene-Transfer Efficiency

Gene-transfer efficiency is a critical parameter in the development of effective treatment for tumors.

1. Three days after LacZ vector delivery to brain tumor, sacrifice anesthetized animals by decapitation (*see* **Note 8**).
2. Using small surgical scissors, carefully cut the skull starting from the foramen magnum proceeding toward each eye socket.
3. Remove the top of the skull and expose the brain.
4. Remove the brain and locate the needle injection site.
5. Using a disposable scalpel, cut a 5 mm × 5 mm block surrounding the injection site and transfer to freshly prepared 4% paraformaldehyde.
6. After 20 min incubation at room temperature, imbed the brain sample in OCT, freeze on dry ice, and cut 10–20 micron sections on a microtome (*see* **Note 9**).

7. Rinse sections in PBS and incubate with X-gal staining solution.
8. Visually inspect sections until blue color is apparent (1–4 h).
9. Rinse sections in PBS and mount cover slip with Gel Mount.
10. Count LacZ$^+$ (blue) cells and total cells to obtain the efficiency of gene transfer to the tumor cells and confirm delivery into tumor.

4. Notes

1. If the animal starts moving while in the frame, give 0.10 mL of Ketamine/Acepromazine im and wait 5 min. Intracranial injection should not be performed when the animal is moving or appears in pain because increased intracranial pressure could cause the injected material to leak out through the needle tract.
2. We mark rat tails with a series of different colors in line or dot formation. Alternatively ear tags can be used. However, if future treatment requires X-ray of the animal's head, the ear tags can cause scatter of the X-rays.
3. For pain control, give 160 mg of acetominophen for 3 d during the postoperative period if needed. Children's chewable Tylenol can also be given as the rats readily consume it.
4. The vector concentration delivered is usually 1×10^6 plaque forming units (PFU) in a volume of 5 µL per animal. Determine the vector concentration on complementing cells. Trypsinize the cells and dilute to 1×10^6/mL. Aliquot 1 mL of cells into microfuge tubes and add 100 µL of serially diluted (1×10^{-4} to 1×10^{-8}) vector stock to each tube. Incubate at 37°C for 1 h with frequent agitation to prevent the cells from settling. Plate the contents of each tube into a well of a 6-well cell-culture dish with 2 mL of media and incubate for 2 d. Remove the media and stain the cells with crystal violet. Count the number of plaques and multiply by the dilution factor to obtain the concentration of the vector stock.
5. Take care not to warm the vector stock. Always keep the vial containing the vector stock on ice until the vector injection procedure is finished.
6. Sterilize all surgical instruments by autoclaving and maintain a sterile field during surgery.
7. If bleeding from the dura mater is encountered during the process of bone wax removal, apply light pressure with a cotton swab on the drill hole area for 2 min.
8. Routine investigation must be performed to confirm gene delivery to the tumor.
9. An experienced technician should be consulted for use of a microtome.

References

1. Bischoff, J. R., Kirn, D. H., Williams A., Heise, C., Horn, S., Muna, M., et al. (1996) An adenovirus mutant that replicates selectively in p53-deficient human tumor cells. *Science* 274, 373–376.
2. Boviatsis, E. J., Chase, M., Wei, M. X., Tamiya, T., Hurford, R. K., Kowall. N. W., et al. (1994) Gene transfer into experimental brain tumors mediated by adenovirus, herpes simplex virus, and retrovirus vectors. *Hum. Gene Ther.* 5, 183–191.
3. Rampling, R., Cruickshank, G., Papanastassiou, V., Nicoll, J., Hadley, D., Brennan, D., et al. (2000) Toxicity evaluation of replication-competent herpes simplex virus (ICP 34.5 null mutant 1716) in patients with recurrent malignant glioma. *Gene Ther.* 7, 859–866.
4. Boviatsis, E. J., Scharf, J. M., Chase, M., Harrington, K., Kowall, N. W., Breake-field, X. O., and Chiocca, E. A. (1994) Antitumor activity and reporter gene transfer into rat brain neoplasms inoculated with herpes simplex virus vectors defective in thymidine kinase or ribonucleotide reductase. *Gene Ther.* 1, 323–331.
5. Martuza, R., Malick, A., Markert, J., Ruffner, K., and Coen, D. (1991) Experimental therapy of human glioma by means of a genetically engineered virus mutant. *Science* 252, 854–856.
6. Bak, I. J., Markhan, C. H., and Cook, M. L. (1997) Intra-axonal transport of herpes simplex virus in the rat central nervous system. *Brain Res.* 136, 415–429.
7. Kosz-Vnenchak, M., Coen, D., and Knipe, D. (1990) Restricted expression of herpes simplex virus lytic genes during establishment of latent infection by thymidine kinase-negative mutant viruses. *J. Virol.* 64, 5396–5402.
8. Mineta, T., Rabkin, S., and Martuza, R. (1994) Treatment of malignant gliomas using ganciclovir-hypersensitive, ribonucleotide reductase-deficient herpes simplex viral mutant. *Cancer Res.* 54, 3963–3966.
9. Yamada, Y., Kimura, H., Morishima, T., Daikoku, T., Maeno, K., and Nishiyama, K. (1991) The pathogenicity of ribonucleotide reductase-null mutants of herpes simplex virus type 1 in mice. *J. Infect. Dis.* 164, 1091–1097.
10. Chambers, R., Gillespie, G. Y., Soroceanu, L., Andreansky, S., Chatterjee, S., Chou, J., et al. (1995) Comparison of genetically engineered herpes simplex viruses for the treatment of brain tumors in a scid mouse model of human malignant glioma. *Proc. Natl. Acad. Sci. USA* 92, 1411–1415.
11. MacLean, A., ul-Fareed, M., Robertson, L. Harland, J., Brown, S. (1991) Herpes simplex virus type 1 deletion variants 1714 and 1716 pinpoint neurovirulenece-related sequences in Glasgow strain 17+ between immediate early gene 1 and the 'a' sequence. *J. Gen. Virol.* 72, 631–639.
12. Markert, J., Malick, A., Coen, D., and Martuza, R. (1993) Reduction of elimination of encephalitis in experimental glioma therapy model with attenuated herpes simplex mutantsw that retain susceptibility to acyclovir. *Neurosurgery* 32, 597–603.
13. Whitley, R., Kern, E., Chatterjee, S., Chou, J., and Roizman, B. (1993) Replication establishment of latency, and induced reactivation of herpes simplex virus gl 34.5 deletion mutants in rodent models. *J. Clin. Invest.* 91, 2837–2843.

14. Markert, J. M., Medlock, M. D., Rabkin, S. D., Gillespie, G. Y., Todo, T., Hunter, W. D., et al. (2000) Conditionally replicating herpes simplex virus mutant, G207 for the treatment of malignant glioma: results of a phase I trial. *Gene Ther.* 7, 867–874.

15. Varghese, S. Rabkin, SD (2002). Oncolytic herpes simplex virus vectors for cancer virotherapy. *Cancer Gene Ther.* 9, 967–68.

16. Krisky, D., Wolfe, D., Goins, W., Marconi, P., Ramakrishnan, R., Mata, M., et al. (1998) Deletion of multiple immediate early genes from herpes simplex virus reduces cytotoxicity and permits long-term gene expression in neurons. *Gene Ther.* 5, 1593–1603.

17. Krisky, D., Marconi, P., Oligino, T., Rouse, R., Fink, D., Cohen, J., et al. (1998) Development of herpes simplex virus replication-defective multigene vectors for combination gene therapy applications. *Gene Ther.* 5, 1517–1530.

18. Samaniego, L., Wu, N., and DeLuca, N. (1997) The herpes simplex virus immediate-early protein ICP0 affects transcription from the viral genome and infected-cell survival in the absence of ICP4 and ICP27. *J. Virol.* 71, 4614–4625.

19. Samaniego, L., Neiderhiser, L., and DeLuca, N. (1998) Persistence and expression of the herpes simplex virus genome in the absence of immediate-early proteins. *J. Virol.* 72, 3307–3320.

20. Moriuchi, S., Oligino, T., Krisky, D., Marconi, P., Fink, D., Cohen, J., and Glorioso, J. (1998) Enhanced tumor-cell killing in the presence of ganciclovir by HSV-1 vector-directed co-expression of human TNF-α and HSV thymidine kinase. *Cancer Res.* 58, 5731–5737.

21. Chen, X., Li, J., Mata, M., Goss, J., Wolfe, D., Glorioso, J., and Fink, D. (2000) Herpes simplex virus type 1 ICP0 protein does not accumulate in the nucleus of primary neurons in culture. *J. Virol.* 74, 10132–10141.

22. Moriuchi, S., Krisky, D., Marconi, P., Tamura, M., Shimizu, K., Yoshimine, T., et al. (2000) HSV vector cytotoxicity is inversely correlated with effective TK/GCV suicide gene therapy of rat gliosarcoma. *Gene Ther.* 7, 1483–1490.

23. Hobbs, W. and DeLuca, N. (1999) Perturbation of cell cycle progression and cellular gene expression as a function of herpes simplex virus ICP0. *J. Virol.* 73, 8245–8255.

24. Samaniego, L., Webb, A., and DeLuca, N. (1995) Functional interaction between herpes simplex virus immediate-early proteins during infection: gene expression as a consequence of ICP27 and different domains of ICP4. *J. Virol.* 69, 5705–5715.

25. Barba, D., Hardin, J., Ray, J., and Gage, F. H. (1993) Thymidine kinase-mediated killing of rat brain tumors. *J. Neurosurg.* 79, 729–735.

26. Caruso, M., Panis, Y., Gagandeep, S., Houssin, D., Salzmann, J. L., and Klatzmann, D. (1993) Regression of established macroscopic liver metastases after in situ transduction of a suicide gene. *Proc. Natl. Acad. Sci. USA* 90, 7024–7028.

27. Culver, K., Ram, Z., Walbridge, S., Ishii, H., Oldfield, E., and Blaese, R. (1992) In vivo gene transfer with retroviral vector-producer cells for treatment of experimental brain tumors. *Science* 256, 1550–1552.

28. Ezzeddine, Z. D., Martuza, R. L., Platika, D., Short, M. P., Malick, A., Choi, B., and Breakefield, X. O. (1991) Selective killing of glioma cells in culture and in vivo

by retrovirus transfer of the herpes simplex virus thymidine kinase gene. *New Biol.* 3, 608–614.

29. Moolten, F. L. (1986) Tumor chemosensitivity conferred by inserted herpes thymidine kinase genes: paradigm for a prospective cancer control strategy. *Cancer Res.* 46, 5276–5281.

30. Ram, Z., Culver, K. W., Walbridge, S., Blaese, R. M., and Oldfield, E. H., (1986) In situ retroviral-mediated gene transfer for the treatment of brain tumors in rats. *Cancer Res.* 46, 5276–5281.

31. van Dillen, I.J., Mulder, N.H., Vaalburg, W., de Vries, E.F. Hospers, G.A. (2002) Influence of the bystander effect HSV-tk/GCV gene therapy. A review. *Current Gene Ther.* 2, 307–22.

32. Short, M. P., Choi, B. C., Lee, J. K., Malick, A., Breakefield, X. O., and Martuza, R. L. (1990). Gene delivery to glioma cells in rat brain by grafting of a retrovirus packaging cell line. *J. Neurosci. Res.* 27, 427–439.

33. Bi, W. L., Parysek, L. M. Warnick, R., and Stambrook, P. J. (1993) In vitro evidence that metabolic cooperation is responsible for the bystander effect observed with HCV Tk retroviral gene therapy. *Hum. Gene. Ther.* 4, 725–731.

34. Freeman, S., Abboud, C., Whartenby, K., Packman, C., Koeplin, D., Moolten, F., and Abraham, G. (1993) The "Bystander Effect": tumor regression when a fraction of the tumor mass is genetically modified. *Cancer Res.* 53, 5274–5283.

35. Kato, K., Yoshida, J., Mizuno, M., Sugita, K., and Emi, N. (1994) Retroviral transfer of herpes simplex thymidine kinase into glioma cells targeting of gancyclovir cytotoxic effect. *Neurol. Med. Chir.* (Tokyo). 34, 339–344.

36. Wu, J. K., Cano, W. G., Meylaerts, S. A., Qi, P., Vrionis, F., and Cherington, V. (1994). Bystander tumoricidal effect in the treatment of experimental brain tumors. *Neurosurgery* 35, 1094–1102.

37. Marconi, P., Tamura, M., Moriuchi, S., Krisky, D. M., Goins, W. F., Cohen, J. B., and Glorioso, J. C. (2000) Connexin43-enhanced suicide gene therapy using herpesviral vectors. *Mol. Ther.* 1, 71–81.

38. Han, S. K., Brody, S. L., and Crystal, R. G. (1994) Suppression of in vivo tumorigenicity of human lung cancer cells by retrovirus-mediated transfer of the human tumor necrosis factor-alpha cDNA. *Am. J. Resp. Cell. Mol. Biol.* 11, 270–278.

39. Pfizenmaier, K., Pfizenmaier, K., Scheurich, P., Schluter, C., and Kronke, M. (1987). Tumor necrosis factor enhances HLA-A, B, C and HLA-DR gene expression in human tumor cells. *J. Immunol.* 138, 975–980.

40. Watanabe, Y., Kuribayashi, K., Miyatake, S., Nishihara, K., Nakayama, E., Taniyama, T., and Sakata, T. (1989) Exogenous expression of mouse interferon cDNA in mouse neuroblastoma C1300 cells results in reduced tumorigenicity by augmented anti-tumor immunity. *Proc. Nat. Acad. Sci. USA* 86, 9456–9460.

41. Ostensen, M. E., Thiele, D. L., and Lipsky, P. E. (1987) Enhancement of human natural killer cell function by the combined effects of tumor necrosis factor alpha or interleukin-1 and interferon-alpha or interleukin-2. *J. Biol. Res. Mod.* 8, 53–61.

42. Plaetinck, G., Declercq, W., Tavernier, J., Jabholz, M., and Fiers, W. (1987) Re-

combinant tumor necrosis factor can induce interleukin 2 receptor expression and cytolytic activity in a rat x mouse T cell hybrid. *Eur. J. Immunol.* 17, 1835–1838.

43. Rothlein, R., Czajkowski, M., O'Neill, M. M., Marlin, S. D., and Merluzzi, V. J. (1988) Induction of intercellular adhesion molecule 1 on primary and continuous cell lines by pro-inflammatory cytokines. Regulation by pharmacologic agents and neutralizing antibodies. *J. Immunol.* 141, 1665–1669.

44. Ranges, G. E., Figari, I. S., Espevik, T., and Palladino Jr., M. A. (1987) Inhibition of cytotoxic T cell development by transforming growth factor beta and reversal by recombinant tumor necrosis factor alpha. *J. Exp. Med.* 166, 991–998.

45. Owen-Schaub, L. B., Gutterman, J. U., and Grimm, E. A. (1988). Synergy of tumor necrosis factor and interleukin 2 in the activation of human cytotoxic lymphocytes: effect of tumor necrosis factor alpha and interleukin 2 in the generation of human lymphokine-activated killer cell cytotoxicity. *Cancer Res.* 48, 788–792.

46. Bennett, M. and Verselis, V. (1992) Biophysics of gap junctions. *Semin Cell Biol.* 3, 29–47.

47. Colombo, B. M., Benedetti, S., Ottolenghi, S., Mora, M., Pollo, B., Poli, G., and Finocchiaro, G. (1995) The "bystander effect": association of U-87 cell death with ganciclovir-mediated apoptosis of nearby cells and lack of effect in athymic mice. *Hum. Gene Ther.* 6, 763–772.

48. Naus, C. C., Bechberger, J. F., and Paul, D. L. (1991) Gap junction gene expression in human seizure disorder. *Exp. Neurol.* 111, 198–203.

49. Yamasaki, H. (1996) Role of disrupted gap junctional intercellular communication in detection and characterization of carcinogens. *Mutat. Res.* 365, 91–105.

50. Mehta, P. P., Hotz-Wagenblatt, A., Rose, B., Shalloway, D., and Loewenstein, W. R. (1991) Incorporation of the gene for a cell-cell channel protein into transformed cells leads to normalization of growth. *J. Membr. Biol.* 124, 207–225.

51. Loewenstein, W. R. and Kanno, Y. (1966) Intercellular communication and the control of tissue growth: lack of communication between cancer cells. *Nature* 209, 1248–1249.

52. Mehta, P. P., Bertram, J. S., and Loewenstein, W. R. (1986) Growth inhibition of transformed cells correlates with their junctional communication with normal cells. *Cell* 44, 187–196.

53. Shinoura, N., Chen, L., Wani, M. A., Kim, Y. G., Larson, J. J., Warnick, R. E., et al. (1996) Protein and messenger RNA expression of connexin43 in astrocytomas: implications in brain tumor gene therapy. *J. Neurosurg.* 84, 839–845.

54. Zhu, D., Caveney, S., Kidder, G. M., and Naus, C. C. (1991) Transfection of C6 glioma cells with connexin 43 cDNA: analysis of expression, intercellular coupling, and cell proliferation. *Proc. Natl. Acad. Sci. USA* 88, 1883–1887.

55. Elshami, A. A., Saavedra, A., Zhang, H., Kucharczuk, J. C., Spray, D. C., Fishman, G. I., et al. (1996) Gap junctions play a role in the 'bystander effect' of the herpes simplex virus thymidine kinase/glnciclovir system in vitro. *Gene Ther.* 3, 85–92.

56. Fick, J., Barker, F. G. N., Dazin, P., Westphale, E. M., Beyer, E. C., and Israel, M. A.

(1995). The extent of heterocellular communication mediated by gap junctions is predictive of bystander tumor cytotoxicity in vitro. *Proc. Natl. Acad. Sci. USA* 92, 11071–11075.

57. Mesnil, M., Piccoli, C., Tiraby, G., Willecke, K., and Yamasaki, H., (1996) By- stander killing of cancer cells by herpes simplex virus thymidine kinase gene is me- diated by connexins. *Proc. Natl. Acad. Sci. USA* 93, 1831–1835.

58. Touraine, R. L., Vahanian, N., Ramsey, W. J., and Blaese, R. M. (1998). Enhance- ment of the herpes simplex virus thymidine kinase/ganciclovir bystander effect and its antitumor efficacy in vivo by pharmacologic manipulation of gap junctions. *Hum. Gene Ther.* 9, 2385–2391.

59. Dilber, M. S., Abedi, M. R., Christensson, B., Bjorkstrand, B., Kidder, G. M., Naus, C. C., et al. (1997) Gap junctions promote the bystander effect of herpes simplex virus thymidine kinase in vivo. *Cancer Res.* 57, 1523–1528.

60. Vrionis, F. D., Wu, J. K., Qi, P., Waltzman, M., Cherington, V., and Spray, D. C. (1997) The bystander effect exerted by tumor cells expressing the herpes simplex virus thymidine kinase (HSVtk) gene is dependent on connexin expression and cell communication via gap junctions. *Gene Ther.* 4, 577–585.

61. Hallahan, D. E., Spriggs, D. R., Beckett, M. A., Kufe, D. W., and Weichselbaum, R. R. (1989) Increased tumor necrosis factor alpha mRNA after cellular exposure to ionizing radiation. *Proc. Natl. Acad. Sci. USA* 86, 10104–10107.

62. Chiang, C. S. and McBride, W. H. (1991) Radiation enhances tumor necrosis fac- tor alpha production by murine brain cells. *Brain Res.* 566, 265–269.

63. Baher, A. G., Andres, M. L., Folz-Holbeck, J., Cao, J. D., and Gridley, D. S. (1999) A model using radiation and plasmid-mediated tumor necrosis factor-alpha gene therapy for treatment of glioblastomas. *Anticancer Res.* 19, 2917–2924.

64. Sersa, G., Willingham, V., and Milas, L. (1988) Anti-tumor effects of tumor necro- sis factor alone or combined with radiotherapy. *Intl. J. Cancer.* 42, 129–134.

65. Hallahan, D. E., Beckett, M. A., Kufe, D., and Weichselbaum, R. R. (1990) The in- teraction between recombinant human tumor necrosis factor and radiation in 13 human tumor cell lines. *Intl. J. Rad. Oncol. Biol. Physics* 19, 69–74

66. Gridley, D. S., Archambeau, J. O., Andres, M. A., Mao, X. W., Wright, K., and Slater, J. M. (1997) Tumor necrosis factor-alpha enhances antitumor effects of ra- diation against glioma xenografts. *Oncol. Res.* 9, 217–227.

67. Gridley, D. S., Li, J., Kajioka, E. H., Andres, M. L., Moyers, M. F., and Slater, J. M. (2000). Combination of pGL1-TNF-alpha gene and radiation (proton and gamma- ray) therapy against brain tumor. *Anticancer Res.* 20, 4195–4203.

68. Fajardo, L. F., Kwan, H. H., Kowalski, J., Prionas, S. D., and Allison, A. C. (1992) Dual role of tumor necrosis factor-alpha in angiogenesis. *Am. J. Pathol.* 140, 539–544.

69. Bombeli, T., Karsan, A., Tait, J. F., and Harlan, J. M. (1997) Apoptotic vascular en- dothelial cells become procoagulant. *Blood* 89, 2429–2942.

70. Staba, M. J, Mauceri, H. J., Kufe, D. W., Hallahan, D. E., and Weichselbaum, R. R.

(1998) Adenoviral TNF-alpha gene therapy and radiation damage tumor vasculature in a human malignant glioma xenograft. *Gene Ther.* 5, 293–300.

71. Niranjan, A., Moriuchi, S., Lunsford, L., Kondziolka, D., Flickinger, J., Fellows, W., et al. (2000) Effective treatment of experimental glioblastoma by HSV vector-mediated TNF-α and HSV-tk gene transfer in combination with radiosurgery and ganciclovir administration. *Mol Ther.* 2, 114–120.

23

Delivery of Herpes Simplex Virus-Based Vectors to Stem Cells

Darren Wolfe, James B. Wechuck, David M. Krisky, Julie P. Goff, William F. Goins, Ali Ozuer, Michael E. Epperly, Joel S. Greenberger, David J. Fink, and Joseph C. Glorioso

1. Introduction

In contrast to traditional drugs that generally act by altering existing gene product function, gene therapy aims to target the root cause of the disease by altering the genetic makeup of the cell to treat the disease. Researchers have adapted several classes of viruses as gene-transfer vectors, taking advantage of natural viral mechanisms designed to efficiently and effectively deliver DNA to the host-cell nucleus. Among these, the human herpesviruses are excellent candidate vectors for a variety of applications. Herpes simplex virus type 1 (HSV-1) is a particularly attractive gene-transfer vehicle because natural infection in humans includes a latent state in which the viral genome persists in a nonintegrated form without causing disease in an immune-competent host *(1–3)*. HSV-1 is a large DNA virus with a broad host range that can be engineered to accommodate multiple or large therapeutic transgenes *(4)*. HSV vectors may be generally useful for gene transfer to a variety of tissues in which short-term or extended transgene expression of therapeutic transgenes achieve a therapeutic effect. We have used therapeutic vectors to successfully treat human disease models in animals, including cancer, Parkinson's disease, and nerve damage *(5–10)*.

From: *Methods in Molecular Biology, vol. 246:*
Gene Delivery to Mammalian Cells: Vol. 2: Viral Gene Transfer Techniques
Edited by: W. C. Heiser © Humana Press Inc., Totowa, NJ

1.1. Biology of HSV-1

The HSV-1 particle contains a condensed core of DNA surrounded by the capsid and tegument, all of which is contained within a glycoprotein studded membrane (reviewed in Chapter 19). The genome structure of HSV can be divided into viral genes that are either essential or accessory for replication in cell culture. Following virus attachment by both nonspecific and ligand/receptor specific interactions with viral glycoproteins, the capsid penetrates the cell membranes and is transported to the nuclear membrane, where the viral DNA enters through a nuclear pore. Once inside the nucleus, early (E) genes primarily specify enzyme functions required for viral DNA synthesis, and late (L) genes primarily produce virion structural components. Following assembly of viral structural components, unit lengths of viral DNA are packaged into the newly assembled capsid. Tegument proteins accumulate and the immature particle buds through the inner nuclear membrane where the viral glycoproteins are acquired. Enveloped virions are modified by enzymes within the Golgi apparatus and fuse with the cell membrane, forming a mature, extracellular virus particle. Once this infectious particle is released from the infected cell, neighboring cells can be infected and the lytic replication cycle is reinitiated, resulting in viral proliferation.

1.2. Vector Technology

HSV genes are expressed in a sequential cascade during lytic infection with viral immediate early (IE) genes initiating the sequential cascade of lytic gene synthesis *(11)*. Removal of an essential IE gene prevents expression of subsequent E and L genes, resulting in a defective vector incapable of replication in normal cells or tissues. The IE genes ICP4 and ICP27 encode products that are essential for viral replication and expression of the E and L genes *(12,13)*. Our vectors are minimally defective for both ICP4 and ICP27, providing a considerable margin of safety against generation of replication competent virus. Elimination of multiple IE genes reduces the cytotoxicity of HSV-based vectors for cell lines and primary neuronal cell cultures *(2,3)*. Essential IE gene functions can be complemented in stable cell lines, allowing production of replication-defective HSV-based vectors *(14)*. We have not detected replication-competent virus (RCV) in vector stocks minimally defective for the two essential genes ICP4 and ICP27. The lack of expression of viral lytic genes enhances vector safety and reduces host responses that may result in elimination of vector transduced cells. We have achieved expression of five transgene products from a single HSV-1 vector deleted for nine viral functions, including the IE genes ICP4, ICP22, ICP27, and ICP47 *(4)*. Considerable technical progress has been made in developing HSV vectors through: (1) elimination of vector toxicity, including direct cytopathic effects and the potential inflammatory or immune responses; (2) full exploitation of the potential transgene capacity of the vector; (3) maintenance of transgene expression, including the efficiency of cellular transduction, the level and duration of expression, and expression regulation; and (4) targeting of transgene expression to specific cell populations. These points are fully covered in the accompanying Overview chapter to this section of this book (*see* Chapter 19).

1.3. Stem Cells as a Therapeutic Target for Gene Transfer

The nomenclature of stem cells is based on a theoretical hierarchy on differentiation, proliferation, and the self-renewal potential. As all stem cells proliferate, mature, and differentiate, their daughter cells approach a terminal state restricting their differentiation capacity to more defined cell types. Totipotent stem cells, exemplified by embryonic tissue or zygotes, are capable of giving rise to every lineage of cells required to form an adult animal. Pluripotent stem cells of adult tissues are able to give rise to cells of one or more particular tissue types with some cross-tissue capabilities. Pluripotent muscle-derived stem cells are able to repopulate the hematopoietic system; mesenchymal stem cells can differentiate into osteoblasts, fibroblasts, myoblasts, and other cell types; and hematopoietic cells can incorporate into muscle or endothelium *(15–18)*. Multipotent stem cells are committed to becoming one cell type of a particular tissue such as the white blood cell lineage of the hematopoietic system. Certain cells may be able to dedifferentiate or trans-differentiate, although this has not been formally proven and contaminating cell populations must be addressed *(19,20)*. The features of all three classes of stem cells that make them attractive as cell and gene-based therapy applications are their self-renewal and differentiation capabilities. These features permit therapies to be developed to replace functions missing in genetic diseases or to provide additional functions to treat acquired diseases.

The phenotypic potential of hematopoietic stem cells has been extensively studied and provides a basis for comparison of more recently identified and purified stem cells. One criterion for classifying primitive stem cells is their capacity to reconstitute the hematopoietic system by transplantation into lethally irradiated mice *(21,22)*. Experiments of this nature have been performed with stem cells derived from umbilical cord blood, bone marrow, peripheral blood, and muscle *(23)*. Stem cells are also routinely classified by the presence of cell-surface markers identified by antibody labeling *(24,25)*. One of the first markers for hematopoietic cells, CD34, is present on a wide population of primitive and lineage committed cells. Other surface markers have been discovered that can be used in further classifying subpopulations within CD34$^+$ populations *(26)*.

The absence of lineage specific cell surface markers on CD34$^+$ cells (Lin$^-$) has been shown, in general, to indicate a more primitive cell with higher proliferative potential. Searching for primitive hematopoietic stem cells, subsets of stem cells are being studied including CD34+Thy-1+/CD38− and CD34+/Lin− *(27,28)*. Two subsets of CD34+ cells have been found, CD34+/KDR+ and CD34+/AC133+, that when transplanted in limiting numbers to severe combined immunodeficiency (SCID) mice, verify the presence of pluripotent stem cells *(29–31)*.

1.4. Gene Transfer to Stem Cells With Viral Vectors

Manipulation of the genetic make-up of stem cells could be used to treat disease by infusing the patient's own transduced (corrected) stem cells back into the afflicted individual, thus avoiding the rejection of foreign donor cells. Subsequent cellular homing and/or differentiation and expression of the introduced transgene would correct the genetic deficit. This approach has been used to correct the SCID-X1 disorder in human patients by transducing CD34+ bone marrow cells with a Moloney murine leukemia virus (MMLV) vector carrying the γc cytokine receptor gene *(32)*. Conceptually, growth factor or morphogen gene therapy combined with stem-cell transplantation may provide a method to direct the differentiation process, enhancing the recovery of particular cellular functions by stimulating the production of specific cell lineages. The ability to efficiently transduce unstimulated human stem cells with the vector of choice must precede experiments directed toward this gene therapy application.

Both viral and nonviral vectors have been employed for stem-cell transduction, with each having distinct benefits and disadvantages. For stem cells, issues relevant to the choice of vector include: (1) efficiency of transduction, (2) mitotic state of the target cell, (3) maintenance of transduced DNA, (4) length of transgene expression, (5) vector capacity, and (6) unwanted host immune recognition of transduced cells.

1.4.1. Retrovirus Vectors

Initially, for stem cell-transduction experiments, retroviral vectors were utilized for cell-marking experiments because they integrate into the host genome. Several reports have shown that transduction levels can be increased by inclusion of cytokines during the transduction process. Although this stimulation may be especially important for viral integration into nondividing cells, stem-cell properties may be altered after cytokine stimulation. Several groups have transduced stimulated human cord-blood cells with VSV-G pseudotyped lentiviral vectors with an average transduction efficiency of 59%, whereas a MMLV-based vector showed fewer than 4% GFP positive cells; these results indicate that lentivirus is a more efficient vector than MMLV for the transduction of a variety of stem-cell types *(33–35)*.

1.4.2. AAV Vectors

AAV vectors have also been used for gene transfer into human stem cells *(36,37)*. Cytokine stimulated CD34$^+$ cells were 23% positive for transgene expression. Additional stimulation with tumor necrosis factor-α (TNF-α) increased the transduction efficiency of CD38$^-$ cells from 33 to 60% *(38,39)*.

1.4.3. Adenoviral Vectors

As many as 21% of serum-stimulated human CD34$^+$ and CD34$^+$/CD38$^-$ cells were transduced by adenoviral vectors; however, colony assays showed deleterious effects at higher MOI *(40)*. In a similar report, 30–45% of cytokine stimulated human CD34$^+$/CD38$^-$ cells were transduced by adenoviral vectors *(41)*. The level of adenoviral transduction appears to be limited by the presence of the major coxsackie and adenoviral receptor (CAR) on just 6–15% of CD34$^+$ cells *(42,43)*.

1.4.4. HSV Vectors

Because HSV does not integrate into the host-cell chromosome, applications for long-term gene therapy in CD34$^+$ cells with the goal of repopulating and expanding the number of transgene expressing cells is not feasible. However, applications using HSV to deliver genes that direct stem cells into specific therapeutic lineages could produce populations of untransduced cells that could replace cellular functions lost in disease. Alternatively, HSV-transduced stem cells could be used to deliver suicide genes such as HSV-tk to tumor vasculature prior to activating drug delivery, resulting in tumor starvation.

Disabled infectious single-cycle (DISC) HSV-2 efficiently infects human bone marrow CD34$^+$ stem cells, cultured on stromal support cultures, when exposed at an MOI of 2 for 2 h *(44)*. CD34$^+$ cells, derived from human peripheral blood, exposed to a replication attenuated HSV vector deficient in ICP34.5 and U$_L$43, resulted in minimal transduction *(45)*. These reports have marked differences in transduction efficiency that may be owing to the specific cell population utilized, the transduction protocol, the vector backbone (HSV-1 vs HSV-2), or the culture conditions employed.

Stimulated CD34$^+$ cells from human peripheral blood or monkey bone marrow cells were transduced with our replication defective HSV vector (TOZ.1) that is deficient in two essential genes, contains the *lacZ* reporter gene, and expresses thymidine kinase *(46)*. FACS analysis for *lacZ* expression by fluorescein di-β-D-galactopyranoside (FDG) staining demonstrated 99.9% cells positive at an MOI of 3. These transduced CD34$^+$ cells were transplanted into primates and skin autograft biopsies were analyzed 5 d later. Treatment with Ganciclovir to induce cell killing of HSV-TK transduced cells resulted in a progressive necrotic process producing detachment of the autograft, demonstrating the potential for eliminating newly forming vasculature such as that of a growing tumor.

In the aforementioned transduction protocols, cytokines, serum, and/or growth factors were used to stimulate stem cells; however, these factors alter the pleiotropic phenotype of the cells and limit their use to specific applications. We are pursuing transduction of minimally stimulated human stem cells with HSV-based replication defective vectors. When previous protocols were utilized for transduction, dramatic reductions in transduction efficiency were seen when compared to transduction of cytokine-stimulated stem cells. Here we detail the requirements for transduction of minimally stimulated CD34$^+$ cells derived from human umbilical cord blood with an HSV-based vector defective in multiple essential genes. Only thrombopoietin and stem cell factor are included in the transduction protocol that serves to preserve the viability and pluripotent phenotype of the isolated stem cells *(47)*.

2. Materials

2.1. CD34$^+$ Stem Cell Purification

1. Ficol-Paque Research Grade is available from Amersham Pharmacia Biotech (Uppsala, Sweden).
2. Fifty-mL tubes containing 0.6% anticoagulants citrate dextrose-formula A (ACD-A) (Cytosol Labs, Braintree, MA).
3. CD34$^+$ Progenitor Cell Isolation Kit (includes FcR Blocking Reagent, Hapten Antibody QBEND/10, and Anti-Hapten MicroBeads), Magnetic Cell Separator, Columns, and Pre-Separation Filters from Miltenyi Biotec (Auburn, CA).
4. Iscove's modified Dulbecco's medium (IMDM) from Invitrogen/GIBCO (Carlsbad, CA).
5. BIT9500 Medium from StemCell Technologies (Vancouver, BC, Canada).
6. Stem-cell factor (SCF) (50 ng/mL) and Thrombopoietin (10 U/mL) from StemCell Technologies (Vancouver, BC, Canada).
7. Ca^{+2}/Mg^{+2} free phosphate-buffered saline (PBS), available as a 10X stock solution from Roche (Indianapolis, IN).

2.2. Stem Cell Transduction

1. 24- and 96-well tissue-culture plates, available from Falcon/Becton-Dickinson (Franklin Lakes, NJ).
2. Safe-Lock Microfuge tubes available from Brinkman Instruments (Westbury, NY).
3. Human Erythopoietin (EPO) enzyme-linked immunosorbent assay (ELISA) kit available from (R&D Systems, Inc., Minneapolis, MN).

4. Parafilm (American National Can Co., Menasha, WI).
5. Nutator Mixer (VWR International, West Chester, PA).

2.3. Flow Cytometry and ELISA Analysis of Transduced Stem Cells

1. Five-mL round-bottom polystyrene tubes (Falcon/Becton-Dickinson).
2. Flow cytometer (FACSCalibur, Becton-Dickinson, San Diego, CA) or comparable.
3. Phosphate Buffered Saline (PBS), 10X stock solution (Roche).
4. 1% Paraformaldehyde (Sigma, St. Louis, MO) in PBS, prepared just before use.

3. Methods

3.1. Purification of CD34+ Stem Cells from Human Umbilical Cord Blood

3.1.1. Cord Blood

1. Immediately upon delivery, in accordance with institutional guidelines, transfer 50–75 mL of human umbilical cord blood (CB) to 50-mL tubes containing 0.6% anticoagulants citrate dextrose-formula A (ACD-A).
2. Transport stabilized cord blood to lab on ice using approved transport container (contact hospital and University officials for transport guidelines)

3.1.2. Mononuclear Cell Isolation

1. Dilute cord blood 1:2 with Ca^{+2} and Mg^{+2} free PBS + 0.6% ACD-A.
2. Transfer 15 mL of Ficoll-Paque (1.077 g/mL) to a 50-mL tube.
3. Slowly layer 35 mL of diluted cord blood onto Ficoll-Paque and centrifuge at $400g$ for 35 min at room temperature.
4. Aspirate the upper layer taking care not to disturb the interphase containing the mononuclear cells (MNCs).
5. Transfer the MNCs (~5 mL) to a new tube containing 45 mL of PBS/ACD-A and centrifuge at $200g$ for 10 min at room temperature.
6. Wash the cell pellet with PBS/ACD-A and resuspend in 300 µL of PBS/ACD-A.

3.1.3. CD34+ Cell Isolation

1. Add 100 µL of FcR blocking reagent to the MNCs and mix well.
2. Add 100 µL of hapten-antibody QBEND/10 CD34 and incubate for 15 min at 4°C.
3. Wash labeled cells with 10 mL of PBS/ACD-A and centrifuge at $200g$ for 10 min at room temperature.

4. Aspirate the supernatant and add 400 µL of PBS/ACD-A and 100 µL of Anti-Hapten anti-CD34 magnetic MicroBeads for 15 min at 4°C.
5. Wash the cells and beads with 10 mL of PBS/ACD-A and centrifuge at 200g for 10 min at room temperature.
6. Aspirate and resuspend the cells in 500 µL of PBS/ACD-A.
7. Place a Magnetic Cell Separator column in a magnetic holder and rinse it with 500 µL of PBS/ACD-A.
8. Pass the cells through a pre-wetted Pre-Filter and apply the eluate to the column.
9. Wash the column three times with 500 µL of PBS/ACD-A.
10. Remove the column from the magnetic holder and place it in a 50-mL tube.
11. Apply 1 mL of elution buffer and elute the cells with the supplied plunger.
12. Determine the percentage of CD34$^+$ cells by flow cytometric analysis (**Fig. 1**).

3.2. Transduction of Stem Cells With HSV Vectors

1. Centrifuge freshly isolated CD34$^+$ cells at 800g and resuspend at a concentration of 1×10^6 cells/mL in prewarmed IMDM containing BIT9500 (20%), SCF (50 ng/mL), and thrombopoietin (10 U/mL).
2. Transfer 0.5×10^6 (500 µL) cells to microfuge tubes (*see* **Note 1**).
3. Thaw the vector stock at room temperature; add to the cells and seal the tube with Parafilm (*see* **Note 2**).
4. Rock cells on a Nutator Mixer at 37°C for 1 h (*see* **Note 3**).
5. Remove the cells from the microfuge tube and transfer them to tissue-culture plates (0.5×10^6 cells/well in 12-well tissue-culture plates) (*see* **Note 4**).
6. Incubate cells at 37°C with 5% CO_2 for at least 18 h (*see* **Note 5**).

3.3. Analysis of Transduced Stem Cells

Subheading 3.3.1. briefly describes a procedure for assaying transduction efficiency when using a vector containing a green fluorescent protein (GFP) reporter gene. **Subheading 3.3.2.** describes a procedure for assaying transduction efficiency when using a vector containing a reporter EPO gene.

3.3.1. Flow Cytometry Analysis of GFP

1. Wash the transduced cells twice with 1X PBS.
2. Resuspend the cells at 1×10^6/mL in 1% paraformaldehyde (at a minimum, use 1×10^5 cells) and transfer to a 5-mL Falcon round-bottom polypropylene tube.
3. Set up the Flow Cytometer by setting forward and side scatter with mock transduced cells to gate for cell size and granularity ranges. Set fluoren-

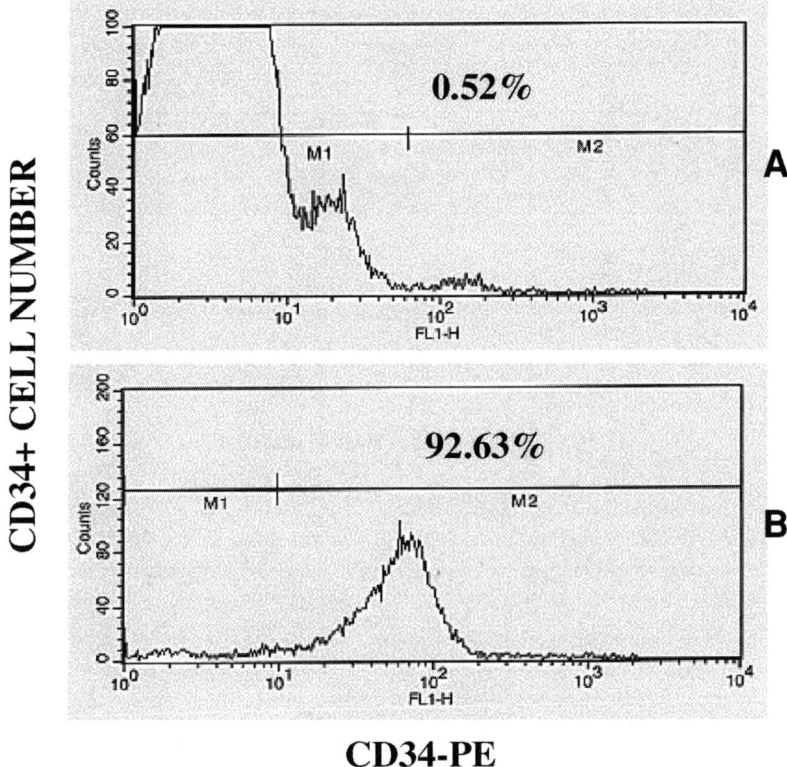

CD34-PE

Fig. 1. Analysis of CD34+ progenitor cell-purification efficiency from human umbilical cord blood. Mononuclear cells (MNC) were purified by Ficoll-Paque density gradient centrifugation. CD34+ progenitor cells were purified from the mononuclear fraction by immunomagnetic separation using a CD34+ Progenitor Cell Isolation Kit from Miltenyi Biotec. **(A)** Analysis of MNC for CD34+ by flow cytometry demonstrates 0.5% of purified mononuclear cells are CD34+. **(B)** Analysis of purified CD34+ cells demonstrate that greater than 92% are CD34+. This population of cells represents our starting material and is generally between 92% and 97% CD34+.

scence gate to eliminate background autofluorescence (>99%) and analyze transduced stem-cells populations (*see* **Note 6**).

3.3.2. ELISA for rhEPO

1. From cultures of stem cells transduced using a recombinant vector that expresses human erythropoietin (e.g., DHE), freeze 20 µL samples of supernatant in duplicate at −80°C every 24 h.
2. Thaw samples at room temperature and centrifuge at maximum speed for 1 min in a microfuge (*see* **Note 7**).

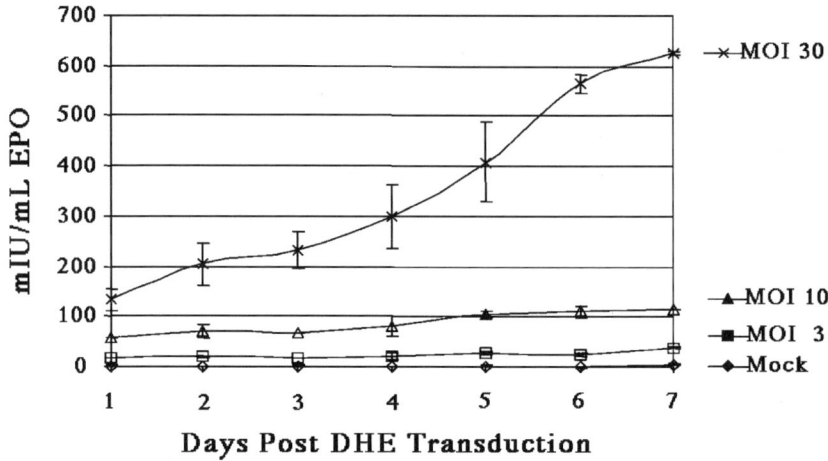

Fig. 2. Production of human erythopoietin (EPO) from CD34+ stem cells transduced with vector DHE. Vector DHE is deleted for the essential genes ICP4 and ICP27 and carries the human erythopoietin cDNA driven by the HCMV IE promoter in the U_L41 locus. Cell supernatant was collected at 24-h intervals and frozen until analyzed by ELISA for human EPO. This demonstrates the ability to deliver genes with morphogenic activity into CD34+ cells with the aim of stimulating the production of specific cell lineages from noncommitted stem cells.

3. Prepare 1:10 serial dilutions of the samples. Prepare 1:2 dilutions of the standards ranging from 2.5–200 mIU/mL. Use 50 μL for rhEPO ELISA according to manufacturer's instructions.
4. Read ELISA microtiter plates using a microplate reader with a 495-nm filter.
5. Calculate EPO concentrations in the samples from the standard curve (**Fig. 2**).

4. Notes
1. Optimal transduction occurs at 1×10^6 cells/mL using 500 μL/tube. Higher or lower cell concentration decreases transduction efficiency.
2. We have found that increasing the multiplicity of infection (MOI) up to 300 plaque forming units (PFU)/cell increases transduction rates. We routinely achieve greater than 70% transduction with an MOI of 30.
3. Incubation in microfuge tubes for longer than 1 h decreases cell viability.
4. Washing the cells after rocking for 1 h decreases transduction efficiency.
5. Removing the inoculum prior to 18 h posttransduction reduces gene transfer efficiency.

6. Because each flow cytometer is different, consult the equipment manual or an experienced technician for details on use of the particular flow cytometer being used. The setup of the flow cytomoter gates can dramatically influence the reported results. We utilize very stringent paramaters for size and autofluorescence gating that may *underestimate* the actual transduction efficiency. We strongly suggest consulting experienced technicians prior to flow cytometry analysis.
7. For analysis by ELISA, include controls such as media from mock or empty vector transduced stem cells to account for background EPO production.

References

1. Fink, D., DeLuca, N., Yamada, M., Wolfe, D. and Glorioso, J. (2000) Design and application of HSV vectors for neuroprotection. *Gene Ther.* **7,** 115–120.
2. Krisky, D., Wolfe, D., Goins, W., Marconi, P., Ramakrishnan, Mata, R., et al. (1998) Deletion of multiple immediate early genes from herpes simplex virus reduces cytotoxicity and permits long-term gene expression in neurons. *Gene Ther.* **5,** 1593–1603.
3. Samaniego, L., Neiderhiser, L., and DeLuca, N. (1998) Persistence and expression of the herpes simplex virus genome in the absence of immediate-early proteins. *J Virol.* **72,** 3307–3320.
4. Krisky, D., Marconi, P., Oligino, T., Rouse, R., Fink, D., Cohen, J., Watkins, S., and Glorioso, J. (1998) Development of herpes simplex virus replication-defective multigene vectors for combination gene therapy applications. *Gene Ther.* **5,** 1517–1530.
5. Chattopadhayay, M., Wolfe, D., Huang, S., Goss, J., Glorioso, J., Mata, M., and Fink, D. (2002) In vivo gene therapy of pyridoxine-induced neuropathy by HSV-mediated gene transfer of neurotrophin-3. *Ann. Neurol.* **51,** 19–27.
6. Marconi, P., Tamura, M., Moriuchi, S., Krisky, D., Goins, W., Cohen, J. and Glorioso, J. (2000) Connexin43-enhanced suicide gene therapy using herpesviral vectors. *Mol. Ther.* **1,** 71–81.
7. Moriuchi, S., Oligino, T., Krisky, D., Marconi, P., Fink, D., Cohen, J., and Glorioso, J. (1998) Enhanced tumor-cell killing in the presence of ganciclovir by HSV-1 vector-directed co-expression of human TNF-α and HSV thymidine kinase. *Cancer Res.* **58,** 5731–5737.
8. Natsume, A., Mata, M., Goss, J., Huang, S., wolfe, D., Oligino, T., Glorioso, J., and Fink, D. (2001) Bcl-2 and GDNF delivered by HSV-mediated gene transfer act additively to protect dopaminergic neurons from 6-ohda-induced degeneration. *Exp Neurol.* **169,** 231–238.
9. Oligino, T., Ghivizzani, S. C., Wolfe, D., Lechman, E. R., Krisky, D., Mi, Z., et al. (1999) Intra-articular delivery of a herpes simplex virus IL-1Ra gene vector reduces inflammation in a rabbit model of arthritis. *Gene Ther.* **6,** 1713–1720.
10. Wolfe, D., Goins, W., Yamada, M., Moriuchi, S., Krisky, D., Oligino, T., Marconi, P.,

Fink, D., and Glorioso, J. (1999) Engineering herpes simplex virus vectors for CNS applications. *Exp. Neurol.* **159,** 34–46.

11. Honess, R. and B. Roizman, (1974) Regulation of herpes simplex virus macromolecular synthesis. I. Cascade regulation of the synthesis of three groups of viral proteins. *J. Virol.* **14,** 8–19.

12. DeLuca, N., McCarthy, A., and Schaffer, P. (1985) Isolation and characterization of deletion mutants of Herpes simplex virus type 1 in the gene encoding immediate-early regulatory protein ICP4. *J. Virol.* **56,** 558–570.

13. Sacks, W., Greene, C., Aschman, D., and Schaffer, P. (1985) Herpes simplex virus type 1 ICP27 is essential regulatory protein. *J. Virol.* **55,** 796–805.

14. Marconi, P., Krisky, D., Oligino, T., Poliani, P., Ramakrishnan, R., Goins, W., et al. (1996) Replication-defective HSV vectors for gene transfer in vivo. *Proc. Natl. Acad. Sci. USA* **93,** 11319–11320.

15. Ferrari, G., Cusella-De, Angelis, G., Coletta, M., et al. (1998) Muscle regenration by bone marrow-derived myogenic progenitors. *Science* **279,** 1528–1530.

16. Gussoni, E., Suneoka, Y., Strickland, C. D., et al. (1999) Dystrophin expression in the mdx mouse restored by stem cell transplantation. *Nature* **401,** 390–394.

17. Jackson, K., T. Mi, and Goodell, M. (1999) Hematopoietic potential of stem cells isolated from murine skeletal muscle. *Proc. Natl. Acad. Sci. USA* **96,** 14482–14486.

18. Prockop, D. (1997) Marrow stromal cells as stem cells for nonhematopoietic tissues. *Science* **276,** 71–74.

19. Fuchs, E. and Segre, J. (2000) Stem cells: a new lease on life. *Cell* **100,** 143–155.

20. Weissman, I. (2000) Stem cells: Units of development, units of regeneration, and units in evolution. *Cell* **100,** 157–168.

21. Greiner, D., Hesselton, R., and Shultz, L. (1998) SCID mouse models of human stem cell engraftment. *Stem Cells* **16,** 166–177.

22. Piacibello, W., Sanavio, F., Severino, A., Dane, A., Gammaitoni, L., Fagioli, F., et al. (1999) Engraftment in nonobese diabetic severe combined immunodeficient mice of human CD34+ cord blood cells after ex vivo expansion: evidence for the amplification and self-renewal of repopulating stem cells. *Blood* **93,** 3736–3749.

23. Wang, J., Doedens, M., and Dick, J. (1997) Primitive human hematopoietic cells are enriched in cord blood compared with adult bone marrow or mobilized peripheral blood as measured by the quantitative in vivo SCID-repopulating assay. *Blood* **89,** 3919–3924.

24. Bhatia, M., Wang, J., Kapp, U., Bonnett, D., and Dick, J. (1997) Purification of primative human hematopoietic cells capable of repopulating immune-deficient mice. *Proc. Natl. Acad. Sci. USA* **94,** 5320–5325.

25. Burt, R. (1999) Clinical utility in maximizing CD34+ cell count in stem cell grafts. *Stem Cells* **17,** 373–376.

26. Buhring, H., Seiffert, M., Bock, T., Scheding, S., Thiel, A., Scheffold, A., et al. (1999) Expression of novel surface antigens on early hematopoietic cells. *Ann. NY Acad. Sci.* **872,** 25–38.

27. Baum, C., Weissman, I., Tsukamoto, A., Buckle, A., and Peault, B. (1999) Isolation

of a candidate human hematopoietic stem cell population. *Proc. Natl. Acad. Sci. USA* **89,** 2804–2808.

28. Terstappen, L., Huang, S., Safford, M., Lansdorp, P., and Loken, M. (1991) Sequential generations of hematopoietic colonies derived from single nonLineage-committed CD34+ CD38- progenitor cells. *Blood* **77,** 1218–1227.
29. Gehling, U., Ergun, S., Schumacher, C., Wagener, K. Pantel, and Otte, M. (2000) In vitro differentiation of endothelial cells from AC133-positive progenitor cells. *Blood* **95,** 3106–3112.
30. Yin, A., Miraglia, S., Zanjani, E., Almeida-Porada, G., Ogawa, M., Leary, A., et al. (1997) AC133, a novel marker for human hematopoietic stem and progenitor cells. *Blood* **90,** 5002–5012.
31. Ziegler, B., Valtieri, M., Porada, G., DeMaria, R., Masella, B., Gabbianelli, M., et al. (1999) KDR Receptor: a key marker defining hematopoietic stem cells. *Science* **285,** 1553–1558.
32. Cavazzana-Calvo, M., Hacein-Bey, S., de Saint Basile, G., Gross, F., Yvon, E., Nusbaum, P., et al. (2000) Gene therapy of human severe combined immunodeficiency (SCID)-X1 disease. *Science* **288,** 669–672.
33. Case, S., Price, M., and Jordan, C. (1999) Stable transduction of quiescent CD34(+)CD38(-) human hematopoietic cells by HIV-1-based lentiviral vectors. *Proc. Natl. Acad. Sci. USA* **96,** 2988–2993.
34. Evans, J., Kelly, P., O'Neill, E., and Garcia, J. (1999) Human cord blood CD34+ CD38- cell transduction via lentivirus-based gene transfer vectors. *Hum. Gene Ther.* **10,** 1479–1489.
35. Haas, D., Case, S., Crooks, G., and Kohn, D. (2000) Critical factors influencing stable transduction of human CD34(+) cells with HIV-1-derived lentiviral vectors. *Mol. Ther.* **2,** 71–80.
36. Fisher-Adams, G., Wong, K. Jr., Prodsakoff, G., Forman, S., and Chatterjee, S. (1996) Integration of adeno-associated virus vectors in CD34+ human hematopoietic progenitor cells after transduction. *Blood* **88,** 492–504.
37. Goodman, S., Xiao, X., Donahue, R., Moulton, A., Miller, J., Walsh, C., et al. (1994) Recombinant adeno-associated virus-mediated gene transfer into hematopoietic progenitor cells. *Blood* **84,** 1492–1500.
38. Nathwani, A., Hanawa, H., Vandergriff, J., Kelly, P., Vanin, E., and Nienhuis, A. (2000) Efficient gene transfer into human cord blood CD34+ cells and the CD34+ CD38- subset using highly purified recombinant adeno-associated viral vector preparations thata are free of helper virus and wild-type AAV. *Gene Ther.* **7,** 183–195.
39. Zhou, S., Cooper, S., Kang, L., Ruggieri, L., Heimfeld, S., Srivastava, A., and Broxmeyer, H. (1994) Adeno-associated virus 2-mediated high efficiency gene transfer into immature and mature subsets of hematopoietic progenitor cells in human umbilical cord blood. *J. Exp. Med.* (1994) **179,** 1867–1875.
40. Watanabe, T., Kuszynski, C., Ino, K., Heimann, D., Shephard, H., Yasui, Y., et al. (1996) Gene transfer into human bone marrow hematopoietic cells mediated by adenovirus vectors. *Blood* **87,** 5032–5039.

41. Neering, S., Hardy, S., Minamoto, D., Spratt, S., and Jordan, C. (1996) Transduction of primitive human hematopoietic cells with recombinant adenovirus vectors. *Blood* **88,** 1147–1155.

42. Rebel, V., Hartnett, S., and Denham, J. (2000) Maturation and lineage-specific expression of the coxsackie and adenovirus receptor in hematopoietic cells. *Stem Cells* **18,** 176–182.

43. Shayakhmetov, D., Papayannopoulou, T., Stamatoyannopoulos, G., and Lieber, A. (2000) Efficient gene transfer into human CD34(+) cells by a retargeted adenovirus vector. *J. Virol.* **74,** 2567–2583.

44. Dilloo, D., Rill, D., Entwistle, C., Boursnell, M., Zhong, W., Holden, W., et al. (1997) AC133, a novel marker for human hematopoietic stem and progenitor cells. *Blood* **90,** 5002–5012.

45. Coffin, R., Thomas, S., Thomas, N., Lilley, C., Pizzey, A., Griffiths, C., et al. (1998) Pure populations of transduced primary human cells can be produced using GFP expressing herpes virus vectors and flow cytometry. *Gene Ther.* **5,** 718–722.

46. Gomez-Navarro, J., Contreras, J. L., Bilbao, G., Krisky, D. M., Oligino, T., Marconi, P., et al. (2000) Genetically modified CD34+ cells as cellular vehicles for gene delivery into areas of angiogenesis in a Rhesus model. *Gene Ther.* **7,** 43–52

47. Goff, J., Shields, D., and Greenberger, J. (1998) Influence of cytokines on the growth kinetics and immunophenotype of daughter cells resulting from the first division of single CD34+Thy-1+lin− cells. *Blood* **92,** 4098–4107.

48. Miltenyi, S., Guth, S., and Radbruch, A. (1994) Isolation of CD34+ hematopoietic progenitor cells by high-gradient magnetic sorting, in *Hematopoietic Stem Cells*, (Wunder, E., ed.) Alpha MedPress, Dayton, Ohio pp. 201–213.

IV

DELIVERY USING BACULOVIRUSES

24

Baculovirus-Mediated Gene Delivery into Mammalian Cells

Raymond V. Merrihew, Thomas A. Kost, and J. Patrick Condreay

1. Introduction

A relatively recent advance in the use of recombinant baculoviruses is their use for delivery of genes and genetic elements into mammalian cells. Baculovirus vectors retrofitted with mammalian gene promoters have been shown to efficiently deliver and express genes in a broad assortment of cell types. These baculovirus transductions are simple to perform, reproducible, and demonstrate no overt cell toxicity. Baculovirus-mediated gene delivery is particularly useful for repetitive or moderately high-throughput procedures such as cell-based assays, or for situations where transfection procedures are inadequate.

Recombinant baculoviruses are most commonly used to express high quantities of proteins for purification from insect cells *(1,2)*. Numerous genetically engineered derivatives of the *Autographa californica* nucleopolyhedrovirus (AcMNPV), an enveloped virus with a double-stranded, circular genome of approx 130 kilobases (kb), are commercially available for this purpose. The use of this insect virus as a tool for efficient gene delivery into mammalian cells was first described for cell lines of hepatic origin *(3–5)*. Subsequent investigations demonstrated baculovirus transduction with a broad array of mammalian cell types *(6–8)*. The list of mammalian cells successfully transduced includes primary cells *(8–10)*, nondividing cells *(11)*, and other cells traditionally difficult to transfect, such as the Saos-2 osteosarcoma cell line *(12)*. In our laboratory, we observed 70–90% transient transduction efficiencies with a majority of

From: *Methods in Molecular Biology, vol. 246:*
Gene Delivery to Mammalian Cells: Vol. 2: Viral Gene Transfer Techniques
Edited by: W. C. Heiser © Humana Press Inc., Totowa, NJ

commonly used cell lines, such as CV-1, HeLa, HEK 293, BHK, and Chinese hamster ovary (CHO). (Notable exceptions were cells of hemopoietic origin, which demonstrated transient transduction efficiencies well below 10%.) These frequencies were achieved using a viral multiplicity of 200 plaque-forming units (PFU) per cell. Despite the moderately high multiplicity, we observed no appreciable cell toxicity. Overall expression levels could be enhanced by the addition of histone deacetylase inhibitors to the cell-culture medium following baculovirus transduction (8). Additionally, by including a dominant selectable marker on the baculovirus vector, we were able to select for stable clones at a frequency of approx 1 of every 50–100 transduced cells. Detailed analysis of four clones demonstrated randomly integrated, single-copy fragments ranging in size from 5–18 kb (13). Thus, recombinant baculoviruses can serve both as transient and stable gene delivery vehicles.

Researchers seeking alternative modes of gene delivery into mammalian cells are finding increasingly wide applications for recombinant baculoviruses. Chimeric baculoviruses have been employed to overcome specific problems associated with launching viral infections. Hepatitis B (HBV) and hepatitis C (HCV) suffer from an inability to infect cultured cells. By generating baculovirus hybrids with HBV (14) and HCV (15), investigators were able to deliver their genomes into hepatoma cells. Others have reported improved efficiencies of generating recombinant adeno-associated virus (AAV) (16) and fully deleted adenovirus vectors (17) by using recombinant baculovirus-mediated gene delivery instead of standard plasmid-transfection procedures. Despite these and other successful applications, the use of baculovirus as a mammalian gene-delivery system remains relatively uncharacterized. As such, many genetic modifications for improving the overall utility of this system may be envisioned.

In this chapter, we will provide detailed steps for generating recombinant baculoviruses for use as mammalian gene-delivery tools. As an illustrative example, a baculovirus vector carrying a β-galactosidase (β-gal) reporter gene driven by the constitutive cytomegalovirus (CMV) promoter will be described. In the course of outlining the procedures, we will highlight some of the capabilities of the baculovirus system for transient and stable mammalian gene delivery.

2. Materials

2.1. Plasmids

1. Plasmid pFastBac1 from Invitrogen (Carlsbad, CA) is a shuttle vector used to construct recombinant baculoviruses.
2. Derivatives of pFastBac1 include plasmids pFastBac-CMV βGal (**Fig. 1**) and pFastBacMam1GFP (8). Plasmid pFastBac-CMV βGal carries the β-gal reporter gene driven by the CMV immediate early promoter/enhancer.

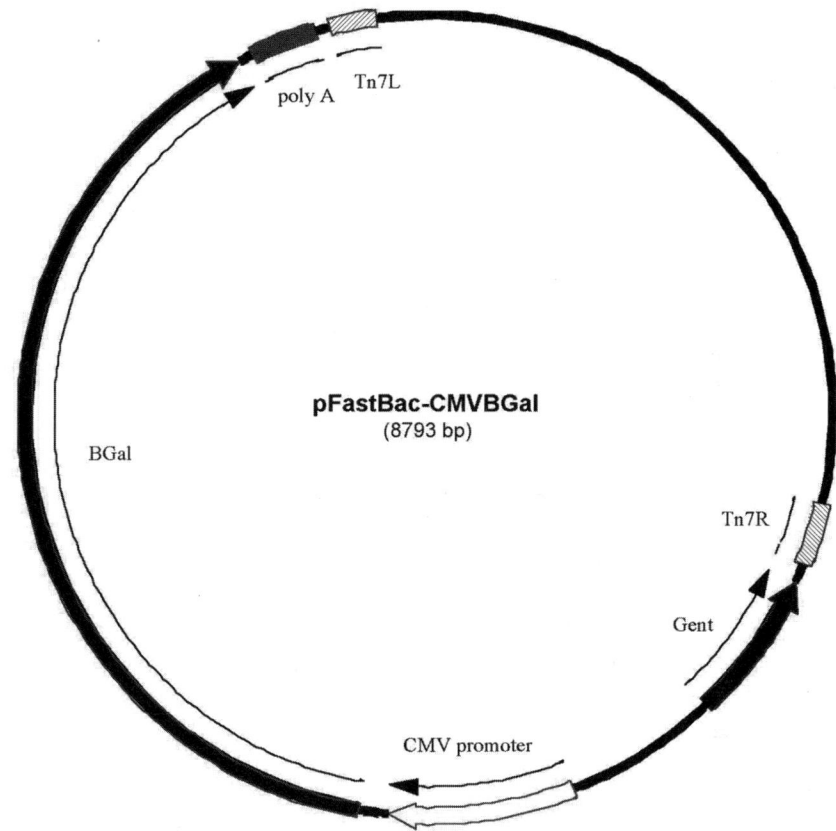

Fig. 1. Shuttle plasmid used to generate BacMam-CMVβGal baculovirus. Details of the construction are in the text.

Plasmid pFastBacMam1GFP carries two marker genes: (1) the dominant selectable marker Neo (encoding the neomycin phosphotransferase II protein) driven by the simian virus 40 (SV40) promoter, and (2) the gene for green fluorescent protein (GFP) driven by the CMV immediate early (IE) promoter/enhancer.

2.2. Chemicals

1. Geneticin (G418), gentamicin, Bluo-gal, IPTG, and Cellfectin are available from Invitrogen.
2. Kanamycin, tetracycline, ethylenediaminetetraacetic acid (EDTA), Tris, sodium hydroxide, sodium dodecyl sulfate (SDS), ammonium acetate, sodium butyrate, dimethylsulfoxide (DMSO), formaldehyde, glutaraldehyde,

5-bromo-4-chloro-3-indolyl-β-D-galactopyranoside (X-gal), ferricyanide, ferrocyanide, and magnesium chloride are available from Sigma Chemical (St. Louis, MO).
3. Luria-Bertuni (LB) agar is available from KD Medical (Columbia, MD).

2.3. Cells

1. Competent DH5α cells and competent DH10Bac cells are available from Invitrogen.
2. The *Spodoptera frugiperda* cell line Sf9 may be obtained from the American Type Culture Collection (ATCC, Manassas, VA). Culture Sf9 cells at 27°C in Grace's Insect Medium supplemented with 10% fetal calf serum (FCS) and 0.1% pluronic F-68.
3. Human embryonic kidney (HEK) 293EBNA cells may be obtained from the ATCC. Culture cells in DMEM/F-12 media supplemented with 10% FCS at 37°C in a 5% CO_2 humidified environment.

2.4. Culture Medium and Reagents

1. SOC media, LB broth, Grace's Insect Medium, pluronic F-68, Grace's Insect Plaquing Medium (2×), 4% agarose gel, DMEM/F-12, Dulbecco's phosphate-buffered saline (PBS), and trypsin are available from Invitrogen.
2. FCS is available from Hyclone (Logan, UT).

3. Methods
3.1. Production of Baculovirus Stock Cultures

Recombinant baculovirus DNA is generated in DH10Bac *Escherichia coli* according to the Bac-to-Bac protocol (Invitrogen) originally described by Luckow and co-workers *(18)*. DH10Bac has been engineered to carry the baculovirus genome as a low copy vector called a bacmid. In addition, DH10Bac carries a separate plasmid expressing the Tn7 transposase. The first step to generating a recombinant baculovirus is the transformation of DH10Bac with a pFastBac1-derived shuttle plasmid (*see* **Note 1**). Recombinant virus DNA is generated by site-specific transposition of shuttle plasmid sequences flanked by left and right ends of a Tn7 site into its target site on the bacmid. This site-specific transposition event disrupts a β-gal α-peptide cassette on the bacmid, allowing for blue/white screening of colonies. Recombinant virus DNA from a white colony is then purified and transfected into insect cells. Because baculovirus is a budding virus, a stock of virus simply represents the growth-medium supernatant.

1. Transform 100 μL of competent DH10Bac cells with a pFastBac-derived shuttle plasmid (10 ng to 1 μg) on ice for 20 min (*see* **Note 2**).
2. Heat-shock cells at 42°C for 60 s, add 1 mL of SOC media, and incubate with shaking at 37°C for 4 h. A considerably shorter incubation time than

4 h is inadequate for sufficient expression of the drug resistance genes for gentamicin (carried on pFastBac1 and its derivatives), kanamycin (carried on the bacmid), and tetracycline (carried on the Tn7 transposase expression plasmid).

3. Dilute transformation mixes 10^{-1} to 10^{-4} and spread on LB agar plates supplemented with 50 µg/mL kanamycin, 7 µg/mL gentamicin, 10 µg/mL tetracycline, 100 µg/mL Bluo-gal, and 40 µg/mL IPTG. Incubate at 37°C for 2–3 d.
4. Pick white colonies and inoculate into LB medium, then replate for independent colony isolation and confirmation of a white phenotype.
5. Pick an independent white colony and inoculate into a tube containing 5 mL of LB broth supplemented with 50 µg/mL kanamycin, 7 µg/mL gentamicin, and 10 µg/mL tetracycline. Shake overnight at 37°C.
6. Perform a standard alkaline lysis miniprep *(19)* to isolate recombinant bacmid DNA.

    a. Pellet 1.5 mL of culture in a microcentrifuge and resuspend in 100 µL of solution I (10 m*M* EDTA, 25 m*M* Tris, pH 8.0).
    b. Add 200 µL of solution II (0.2 *N* sodium hydroxide, 1% SDS), mix by inversion, and place on ice for 5 min.
    c. Add 150 µL of solution III (7.5 *M* ammonium acetate) and mix by inversion.
    d. Spin tube for 5 min in a microcentrifuge at maximum speed and transfer the supernatant to a new 1.5-mL tube.
    e. Add 1 mL of ethanol and pellet DNA for 5 min.
    f. Gently resuspend nearly dry DNA in 40 µL of water.

7. Seed Sf9 cells in a 6-well plate at 9×10^5 cells per well and allow the cells to attach for 30 min at room temperature.
8. Prepare a transfection mix by combining 100 µL of serum-free Grace's Insect Medium with 7 µL of CellFectin reagent and 5 µL of bacmid DNA. Incubate at room temperature for 15 min and then dilute to 1 mL with serum-free Grace's Insect Medium.
9. Rinse Sf9 cells once with serum-free Grace's Insect Medium and add the transfection mix. Incubate for 5 h at room temperature.
10. Remove the transfection mix from the cells and add 2 mL of growth medium (Grace's Insect Medium supplemented with 10% FCS and 0.1% pluronic F-68). Incubate the 6-well dish in a humidified chamber at 27°C for 3 d (*see* **Note 3**).
11. To scale-up the baculovirus, plate 4×10^6 Sf9 cells in 5 mL of medium in a T-25 flask and add the 2 mL of the culture supernatant from the primary transfection well. Incubate at 27°C for 3 d.

12. To generate the final virus stock, seed Sf9 cells at 1×10^6 cells/mL in 150 mL of growth medium in a 500 mL shaker flask with the cap loosened to allow for aeration. Add the culture supernatant from the T-25 flask. Incubate with shaking at 130 rpm at 27°C for 3 d.

13. To remove the Sf9 cells, centrifuge the culture and collect the supernatant. Virus may be sterile filtered through a 0.2-μm filter, although this is not strictly necessary.

14. Determine the virus titer by plaque assay *(20)* on Sf9 cells.

   a. Seed ten 60-mm dishes each with 2×10^6 Sf9 cells. Allow cells to attach at room temperature for 30 min.

   b. Make 10-fold serial dilutions of virus stock from 10^{-1} to 10^{-8} in growth medium. Make sufficient volume (3 mL) of the 10^{-4} to 10^{-8} dilutions to add to the cells.

   c. Remove medium from 60-mm dishes. Overlay duplicate plates each with 1 mL of 10^{-4} to 10^{-8} virus dilutions. Incubate for 1 h at room temperature.

   d. Prepare plaquing overlay. Combine 25 mL of Grace's Insect Plaquing Medium (2×) supplemented with 20% FBS, 12.5 mL of melted 4% agarose, and 12.5 mL of sterile water. Place in a 40°C water bath until ready for use.

   e. Remove virus dilutions and add to each dish 4 mL of plaquing overlay. Allow gel to harden for 15 min.

   f. Incubate dishes in a humidified chamber at 27°C for 4–10 d. Monitor dishes daily until the number of visible plaques stabilizes.

   g. Select dishes with fewer than 20 plaques for scoring. Determine the titer of the original stock by the following calculation: PFU/mL = 1/dilution factor × number of plaques.

15. Store virus stocks at 4°C, protected from light. Under these conditions, the virus is highly stable.

3.2. Transient Expression of β-Gal in HEK 293EBNA Cells

For the majority of commonly used cell lines, we previously reported 70–90% transient transduction efficiencies with a GFP-containing virus (BacMam1GFP). The addition of butyrate significantly enhanced the levels of total GFP protein expression. Similar increases in expression were observed using the deacetylase inhibitor trichostatin A *(8)*. Here, we report the transduction of 293EBNA cells with a virus carrying a β-gal reporter gene. Transduction of 293EBNA cells with recombinant baculovirus BacMam-CMVβGal (BM-CMVβGal) was achieved simply by replacing cell culture medium with virus inoculum, incubating, and then replacing virus inoculum with fresh growth medium. Expression of β-gal is visualized by fixing and staining the cells (**Fig. 2**).

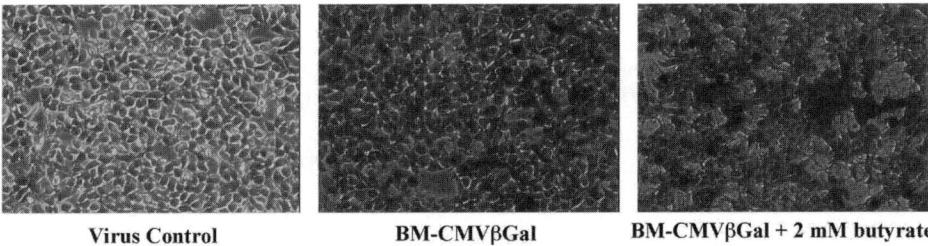

Virus Control   **BM-CMVβGal**   **BM-CMVβGal + 2 mM butyrate**

Fig. 2. Transduction of HEK 293EBNA cells with virus carrying the β-galactosidase cassette driven by the CMV promoter. Left, cells transduced with control virus lacking the CMVβGal cassette. Middle, cells transduced with BacMam-CMVβGal. Right, cells transduced with BacMam-CMVβGal and then treated with 2 mM butyrate. In each case, cells were transduced with 250 PFU/cell. Cells were stained for β-gal activity 16 h following transduction.

1. Grow HEK 293EBNA cells in D-MEM/F-12 media supplemented with 10% FCS. Plate cells in 6-well plates at 2×10^5 cells/well and allow the cells to attach overnight at 37°C (*see* **Note 4**).
2. Remove medium and add 1 mL of virus inoculum at 1×10^8 PFU/mL to the cells (*see* **Note 5**). Incubate cells at 37°C for 1–2 h.
3. Remove virus inoculum and replace with 2 mL of growth medium or 2 mL of medium supplemented with 2 m*M* sodium butyrate (*see* **Note 6**). Incubate plates at 37°C overnight.
4. Rinse cells once with PBS and then fix for 5 min at 4°C with a solution of 2% formaldehyde and 0.2% glutaraldehyde in PBS.
5. Rinse cells twice with PBS.
6. Cover cells with X-gal staining solution (**Table 1**). Incuate overnight at room temperature.
7. Rinse cells with PBS and photograph.

Table 1
Cytochemical Stain for β-Galactosidase

β-gal stain	Stock	Volume stock (10 mL of stain)
0.1% (1 mg/mL) X-gal	50 mg/mL[a]	0.2 mL
5 m*M* potassium ferricyanide	100 m*M*	0.5 mL
5 m*M* potassium ferrocyanide	100 m*M*	0.5 mL
2 m*M* magnesium chloride	1 *M*	20 μL
PBS		8.8 mL

[a]Prepare X-gal in dimethyl sulfoxide (DMSO).

3.3. Selection of Stable Clones

We previously reported the generation of stable clones of BacMam1GFP-transduced CHO cells *(13)*. BacMam1GFP was generated using the Bac-to-Bac system with shuttle plasmid pFastBacMam1GFP, which carries the Neo cassette. Detailed analysis of four G418-resistant, GFP-expressing stable clones demonstrated single-copy integration of 5- to 18-kb fragments. Here, we provide a procedure for selecting stable clones expressing a dominant selectable marker (Neo) and a reporter gene (GFP).

1. Seed CHO cells at approx 2×10^5 cells/well into six wells of a 6-well plate. Allow cells to attach overnight at 37°C.
2. Remove the medium and transduce the cells with 1 mL of virus at 1×10^8 PFU/mL exactly as described above for transient expression (*see* **Subheading 3.2.**); however, as a control for proper selection, do not add virus to one of the wells. Incubate for 1 h at 37°C.
3. Replace the virus inoculum with 2 mL of fresh growth medium. Incubate overnight at 37°C.
4. Select stable cells by replacing the growth medium with 2 mL of G418-containing medium (*see* **Note 7**). Incubate cells at 37°C with medium changes every 3 d for two weeks. Periodically observe the control to monitor the complete loss of viable cells in this well.
5. Pool the selected five wells of cells by trypsinizing and transfering to a T-75 flask for further expansion in G418-containing medium.
6. When the cells are near confluent, collect cells by trypsinization. Half of the cells may be resuspended in freezing medium (growth medium plus 10% DMSO), frozen in a styrofoam container at −80°C, and then stored in liquid nitrogen. Wash the remaining cells and resuspend in PBS with 1% FCS.
7. Single-cell sort GFP-expressing cells by flow cytometry into 96-well plates using a Becton Dickinson FACStar Plus flow cytometer (*see* **Note 8**).
8. Propogate and cryopreserve several clones that maintain G418 resistance and GFP expression.
9. To characterize the integrated DNA, clones may be expanded for genomic DNA extraction and Southern blotting *(19)*. Use of the shuttle plasmid as a probe is generally sufficient for analysis.

An overview of the procedures described above for baculovirus-mediated gene delivery into mammalian cells is presented in **Fig. 3**.

4. Notes

1. To generate suitable shuttle plasmids, we have used standard subcloning procedures to replace the baculovirus polyhedrin promoter sequences of pFastBac1 with a mammalian promoter and gene of interest.

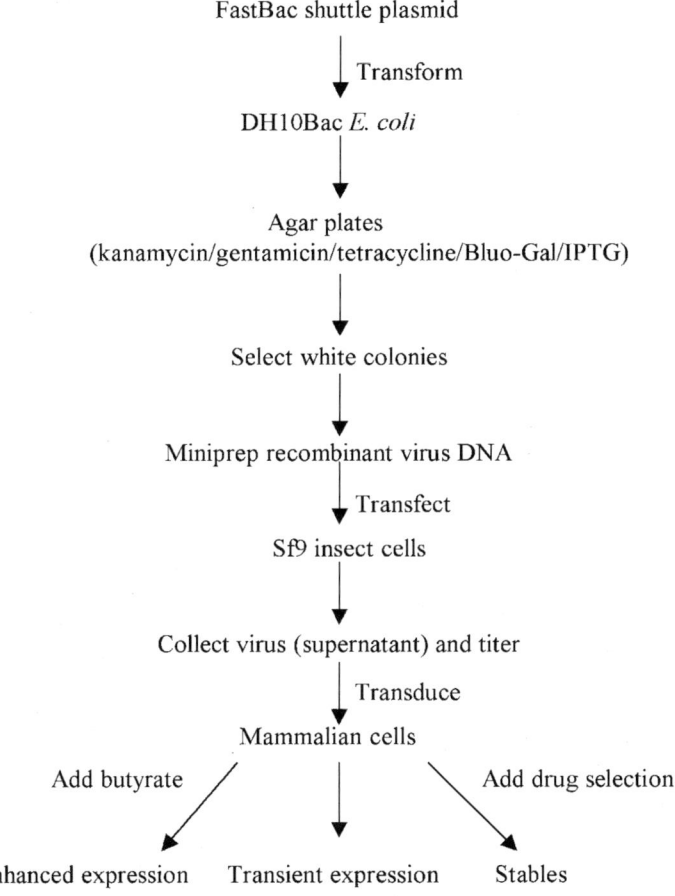

Fig. 3. Summary of procedures for construction and use of recombinant baculovirus as a mammalian gene-delivery vehicle.

2. The manufacturer of the Bac-to-Bac system recommends using approx 1 ng of shuttle plasmid DNA in transformations of DH10Bac. Given that several dilutions are plated, we have found it acceptable to add a greater amount of shuttle plasmid to ensure the generation of colonies with recombinant bacmids. This is especially prudent when the exact concentration of the plasmid has not been determined.

3. Following transfection of Sf9 cells, 6-well plates are incubated in a humidified environment to prevent evaporation of the growth medium. This may be achieved by placing the plates on towels soaked with water within a lidded box.

4. We have found that transduction frequencies are best when the cells are subconfluent. Although each cell line should be tested for optimal densities, we recommend 50% to 80% confluency at the time of transduction as a general rule.

5. The optimal multiplicity of virus for a particular application with a given cell line should be determined empirically. For maximal gene expression, a moderately high multiplicity such as 200 PFU/cell is recommended. Higher multiplicities may require concentration of the virus, which can be achieved by ultracentrifugation (80,000g for 30 min at 4°C) or by chromatography *(21)*. Low multiplicities may be used for procedures that require gene expression closer to physiological levels.

6. We have observed that sodium butyrate has a cytostatic effect on cells.

7. We have found that CHO cells are effectively selected in 1 mg/mL G418 for the stable integration of the Neo cassette. For other cell lines, the proper dose of G418 should be determined experimentally by testing a range of concentrations and selecting the lowest concentration that effectively kills nontransduced cells.

8. We have used flow cytometry to isolate individual clones carrying the GFP reporter gene. However, standard methods for clonal isolation, such as the use of cloning rings, may also be employed.

References

1. Luckow, V. A. (1993) Baculovirus systems for the expression of human gene products. *Curr. Opin. Biotechnol.* **4,** 564–572.
2. Jones, I. and Morikawa, Y. (1996) Baculovirus vectors for expression in insect cells. *Curr. Opin. Biotechnol.* **7,** 512–516.
3. Hofmann, C., Sandig, V., Jennings, G., Rudolph, M., Schlag, P. and Strauss, M. (1995) Efficient gene transfer into human hepatocytes by baculovirus vectors. *Proc. Natl. Acad. Sci. USA* **92,** 10099–10103.
4. Boyce, F. M. and Bucher, N. L. (1996) Baculovirus-mediated gene transfer into mammalian cells. *Proc. Natl. Acad. Sci. USA* **93,** 2348–2352.
5. Sandig, V., Hofmann, C., Steinert, S., Jennings, G., Schlag, P., and Strauss, M. (1996) Gene transfer into hepatocytes and human liver tissue by baculovirus vectors. *Hum. Gene Ther.* **7,** 1937–1945.
6. Yap, C.-C., Ishii, K., Aoki, Y., Aizaki, H., Tani, H., Shimizu, H., et al. (1997) A hybrid baculovirus-T7 RNA polmerase system for recovery of an infectious virus from cDNA. *Virology* **231,** 192–200.
7. Shoji, I., Aizaki, H., Tani, H., Ishii, K., Chiba, T., Saito, I., et al. (1997) Efficient gene transfer into various mammalian cells, including non-hepatic cells, by baculovirus vectors. *J. Gen. Virol.* **78,** 2657–2664.
8. Condreay, J. P., Witherspoon, S. M., Clay, W. C., and Kost, T. A. (1999) Transient and stable gene expression in mammalian cells transduced with a recombinant baculovirus vector. *Proc. Natl. Acad. Sci. USA* **96,** 127–132.

9. Ma, L., Tamarina, N., Wang, Y., Kuznetsov, A., Patel, N., Kending, C., et al. (2000) Baculovirus-mediated gene transfer into pancreatic islet cells. *Diabetes* **49,** 1986–1991.

10. Sarkis, C., Serguera, C., Petres, S., Buchet, D., Ridet, J.-L., Edelman, L., and Mallet, J. (2000) Efficient transduction of neural cells in vitro and in vivo by a baculovirus-derived vector. *Proc. Natl. Acad. Sci. USA* **97,** 14638–14643.

11. van Loo, N.-D., Fortunati, E., Ehlert, E., Rabelink, M., Grosveld, F., and Scholte, B. J. (2001) Baculovirus infection of nondividing mammalian cells: mechanisms of entry and nuclear transport of capsids. *J. Virol.* **75,** 961–970.

12. Song, S. U. and Boyce, F. M. (2001) Combination treatment for osteosarcoma with baculoviral vector mediated gene therapy (p53) and chemotherapy (adriamycin). *Exp. Mol. Med.* **33,** 46–53.

13. Merrihew, R. V., Clay, W. C., Condreay, J. P., Witherspoon, S. M., Dallas, W. S., and Kost, T. A. (2001) Chromosomal integration of transduced recombinant baculovirus DNA in mammalian cells. *J. Virol.* **75,** 903–909.

14. Delaney, W. E. and Isom, H. C. (1998) Hepatitis B virus replication in human HepG2 cells mediated by hepatitis B virus recombinant baculovirus. *Hepatology* **28,** 1134–1146.

15. Fipaldini, C., Bellei, B., and La Monica, N. (1999) Expression of hepatitis C virus cDNA in human hepatoma cell line mediated by a hybrid baculovirus-HCV vector. *Virology* **255,** 302–311.

16. Sollerbrant, K., Elmen, J., Wahlestedt, C., Acker, J., Leblois-Prehaud, H., Latta-Mahieu, M., et al. (2001) A novel method using baculovirus-mediated gene transfer for production of recombinant adeno-associated virus vectors. *J. Gen. Virol.* **82,** 2051–2060.

17. Cheshenko, N., Krougliak, N., Eisensmith, R. C., and Krougliak, V. A. (2001) A novel system for the production of fully deleted adenovirus vectors that does not require helper adenovirus. *Gene Ther.* **8,** 846–854.

18. Luckow, V. A., Lee, S. C., Barry, G. F., and Olins, P. O. (1993) Efficient generation of infectious recombinant baculoviruses by site-specific transposon-mediated insertion of foreign genes into a baculovirus genome propagated in *Escherichia coli. J. Virol.* **67,** 4566–4579.

19. Sambrook, J., Fritsch, E. F., and Maniatis, T. (1989) *Molecular Cloning: A Laboratory Manual.* Cold Spring Harbor Laboratory, Cold Spring Harbor, NY.

20. O'Reilly, D. R., Miller, L. K., and Luckow, V. A. (1994) *Baculovirus Expression Vectors: A Laboratory Manual.* Oxford University Press, Inc., New York, NY.

21. Barsoum, J. (1999) Concentration of recombinant baculovirus by cation-exchange chromatography. *Biotechniques* **26,** 834–840.

V

DELIVERY USING LENTIVIRUSES

25

Gene Delivery by Lentivirus Vectors

An Overview

Tal Kafri

1. Introduction

For more than two decades, retroviral biology has been the most intensely studied field in virology. The retroviral genome is encoded by a 7–11 kb positive-sense single-stranded RNA molecule, two of which homodimerize and package in lipid-enveloped viral particles. Following attachment and receptor-mediated entry into host cells, viral reverse transcriptase and integrase enzymes mediate reverse transcription and integration of the virus genome into the host-cell chromatin. The ability of a replication competent retrovirus to incorporate a herpes simplex virus thymidine kinase (*tk*) gene into the genome of a mouse cell and to convert NIH-3T3 TK$^-$ cells into TK$^+$ transformants was first described in 1981 *(1,2)*. These studies established the basis of using retroviruses as vehicles for efficient therapeutic gene delivery into mammalian cells. Twenty years of extensive research of retrovirus-vector biology resulted in major improvements in vector design and retrovirus-vector production. High-titer concentrated retrovirus vectors ($>10^9$ infectious units [IU]/mL) can be generated by several retrovirus-vector stable producer lines. The ability to pseudotype retrovirus vectors with a variety of envelope proteins, including the vesicular stomatitis virus G glycoprotein (VSV-G), significantly broadens the tropism of replication-defective retrovirus vectors. In addition, combinations of synthetic and tissue-specific promoters, which were incorporated into retrovirus vectors, allowed long-term

From: *Methods in Molecular Biology, vol. 246:*
Gene Delivery to Mammalian Cells: Vol. 2: Viral Gene Transfer Techniques
Edited by: W. C. Heiser © Humana Press Inc., Totowa, NJ

and regulated gene expression in vector transduced cells. However, in spite of these improvements, all retrovirus vector-based clinical trials (excluding a recently published study *[3]*) failed to demonstrate clinical efficacy. The major disadvantage of simple retrovirus vectors is their inability to transduce nondividing cells, which renders them extremely inefficient as an in vivo gene-delivery system. In contrast with simple retroviruses, lentiviruses have evolved the ability to infect nondividing cells *(4)*. The lentiviradea genus includes the human immunodeficiency viruses, HIV-1 and HIV-2; the simian immunodeficiency virus, SIV; and the nonprimate lentiviruses, such as visna virus, equine infectious anemia virus (EIAV), caprine arthritis-encephalitis virus (CAEV), and the feline and bovine immunodeficiency viruses (FIV and BIV). The mechanism by which lentiviruses infect nondividing cells has not been completely elucidated. However, several studies have indicated that by using the host-cell nuclear-import machinery, the *gag*, *vpr*, and *pol* gene products can mediate an active transfer of HIV-1 preintegration complexes into nuclei of nondividing cells *(5–7)*. The first lentiviral vectors developed were based on HIV-1. These were found to be efficient at transducing nondividing cells and yet retained the ability of simple retrovirus vectors to integrate transgenes into the target cell genome, without triggering an inflammatory response *(8–11)*.

Similar to simple retrovirus vectors, the design of replication defective HIV-1 vectors is based on the strategy of segregation of the *cis*-acting elements in the HIV-1 genome (which are required for vector RNA synthesis, packaging, reverse transcription, and integration) from protein-encoding sequences *(9)*. As an additional measurement of safety, envelope-encoding sequences are usually separated from the rest of the HIV-1 packaging cassette. Based on this approach, the components required for the generation of HIV-1 based vectors are supplied in producer cells from three separate expression cassettes: envelope, packaging, and vector cassette.

2. The Envelope Cassette

Substituting the parental HIV-1 envelope with the vesicular stomatitis virus G protein (VSV-G) was a major breakthrough in lentivirus vector development *(9,11)*. The VSV-G envelope confers three new features on lentivirus vector particles: (1) it stabilizes vector particles and therefore allows vector concentration by ultracentrifugation; (2) it mediates attachment of vector particles to phosphatidyl serine molecules on target cells, thus dramatically broadening vector tropism; and (3) it directs lentivirus vector entry to an endocytic pathway, which reduces the requirements for viral accessory proteins for full infectivity *(12)*. Although VSV-G pseudotyped lentivirus vectors were found to be efficient at transducing nondividing cells in various animal models, results from two studies have indicated that the use of lentivirus vectors in in vivo human clinical trials can be hampered by complement- and antibody-mediated immune response directed against the VSV-G envelope *(13,14)*. Several studies demonstrated that lentivirus vectors could be successfully pseudotyped with a variety of simple retrovirus envelope proteins, including: the xenotropic, polytropic, amphotropic 4070A, the10A1 subtype envelope protein; the glycoprotein of the lymphocytic choriomeningitis virus (LCMV); and the hemagglutinin (HA) of the avian influenza virus *(9,15,16)*. The gibbon ape leukemia virus envelope protein was shown to facilitate transduction of hematopoietic stem cells by simple retrovirus vectors. However, efficient pseudotyping of HIV-1 vector particles with this envelope protein required modification of its cytoplasmic tail *(16)*. In general, titers of VSV-G pseudotyped lentivirus vectors are more than 10-fold higher than the typical titers of non-VSV-G pseudotyped lentivirus vectors. Interestingly, in vivo transduction of airway epithelia with VSV-G pseudotyped lentivirus vectors was found inefficient, presumably owing to the lack of the VSV-G receptor on the apical surface of well-differentiated airway epithelia cells *(17)*. Kubinger overcame this obstacle by pseudotyping lentivirus vectors with the envelope protein from the Zaire strain of the Ebola virus, thus opening a new way of gene delivery to small and large airways, which are major targets of cystic fibrosis (CF) gene therapy *(18)*.

Clearly, further studies are required to optimize envelope protein usage in lentivirus vector-based gene-therapy models. In this regard, the ability to specifically direct lentivirus vectors to unique target organs in vivo is most desirable.

3. The Packaging Cassette

Excluding the envelope protein, the packaging cassette expresses all HIV-1 proteins required for vector particles production and efficient transduction of target cells. To avoid transfer of HIV-1 gene-coding sequences to target cells, the packaging cassette should be deleted of HIV-1 *cis*-acting sequences, such as the packaging signal and the LTRs, yet it should retain the Rev response element (RRE) sequence and the parental splice donor site *(9)*. Hetrologous potent promoters and polyadenylation signals, such as cytomegalovirus (CMV) and the Rous sarcoma virus (RSV) promoter sequences and the insulin gene polyadenylation signal, were incorporated into the packaging cassette in place of the parental LTRs to maximize full-length mRNA synthesis and stabilization. Unlike simple retroviruses, lentiviruses express a further six proteins in addition to the gag, pol, and env genes. These genes are termed accessory genes and include the *Vpu, Vpr, Vif, Nef,* and the virus regulatory proteins Tat and Rev, which control HIV-1 gene expression at transcriptional and posttranscriptional levels. The accessory proteins are essential for high-rate HIV-1 replication in vivo, and determine the pathogenic features of the virus. The long-term survival of some AIDS patients maybe explained by their infection with HIV-1 strains carrying defective accessory genes *(19,20)*. Interestingly, the *Vpr*, *Vpu*, *Nef*, and *Vif* accessory genes are not required for viral propagation in vitro. Thus, to improve the biosafety of the HIV-1 vectors, several research groups have developed packaging cassettes from which the HIV-1 accessory genes were systematically deleted *(8,10,21,22)*. The initial packaging construct, which was developed by Naldini et al. *(9)* and termed first-generation packaging cassette, contains all of the HIV-1 accessory genes. A later packaging construct, which is devoid of all of the accessory genes excluding the *Tat* and the *Rev*, is termed second-generation packaging cassette *(21)*. Further deletion of the Tat gene and separating the *Gag/Pol* and the *Rev* genes into two expression cassettes resulted in the development of the third-generation packaging system *(22)*. These modifications significantly reduced the likelihood of generating replication-competent retroviruses (RCRs). Furthermore, should RCRs be inadvertently generated by the new packaging system, they will lack all of the accessory proteins and the pathogenic features of HIV-1. The improvements in biosafety did not reduce vector titers, nor did they hamper the ability of HIV-1 vectors to transduce nondividing cells, including hematopoietic stem cells, monocyte-derived macrophages, terminally differentiated neurons, and murine hepatocytes. In contrast with that, two research groups reported on dramatic reduction in the ability of HIV-1 vectors generated in the absence of the accessory proteins to transduce nonactivated human lymphocytes *(23,24)*. These effects of HIV-1 accessory proteins on transduction efficiency were determined for VSV-G pseudotyped vectors. Other studies indicated that HIV-1 vector particles pseudotyped with envelope proteins other than VSV-G may exhibit significant reduction in transduction efficiency in the absence of the accessory protein *(12,25)*. Recent developments in the vector packaging system uncovered additional ways to improve the biosafety of HIV-1 vectors. These include the humanization of the *gag/pol* codon usage and the separation of the *gag/pol* gene sequence into two independent expression cassettes.

3.1. Humanization of the Gag/Pol Codon Usage

The exceptionally high content of AU bases in the HIV genome results in a biased codon usage and the generation of unstable transcripts. This reduces *gag-pol* translation efficiency, destabilizes gag-pol mRNA, and renders HIV-1 Gag/Pol protein expression Rev dependent. Following humanization of the HIV-1 *gag-pol* sequence, Kotsopoulou et al. could report on highly efficient, Rev-independent *gag-pol* gene expression, which was not hampered by a deletion of the RRE sequences in the packaging cassette *(26)*. Because the novel humanized and RRE deleted packaging cassette share minimal sequence homology with the vector construct, it significantly reduces the likelihood of generating RCRs, and thus improves the biosafety of HIV-1 vectors. Furthermore, it allows efficient production of vectors expressing anti-HIV-1 gene products, such as trans-dominant proteins, ribozymes, and antisense sequences directed against Rev function and *gag/pol* sequences.

3.2. Splitting the Gag/Pol Expression Cassette

A novel HIV-1 vector packaging construct was developed by Wu et al. in which the Gag/Pol coding region is split into two expression cassettes: the Gag-Protease cassette and the Vpr/Pol fusion cassette *(27)*. Vpr binding to the Gag precursor protein p6 mediates efficient incorporation of the Vpr/Pol fusion protein to vector particles. Using the new packaging system, it was possible to completely prevent vector-mediated transfer of functional *gag-pol* sequences into target cells. Thus, it is theoretically possible to improve the biosafety of the lentivirus vector system by further splitting it into five constructs from which the vector, envelope, Gag/Protease, Vpr/Pol and Rev are independently expressed.

4. The Vector Cassette

The vector plasmid expresses the full-length vector RNA, containing all of the HIV-1 *cis*-acting elements (required for efficient packaging, reverse transcription, and integration) and the transgene-expression cassette (the internal promoter and the transgene sequence). Initially, the HIV-1 sequences in the vector included the 5' LTR and the 5' leader sequence, followed by 360 bp of the *gag* gene and 700 bp of the envelope gene *(9)*. These sequences contain important *cis*-elements including the primer binding site (PBS), the splice donor site, the packaging signal, the RRE, and the splice-acceptor site. The internal promoter and the transgene sequences were located in the middle of the vector and were followed by 800 bp from the 3' end of the HIV-1 genome, which contains the 3' end of the Nef gene, a polypurine tract (PPT) and the 3' LTR. Intensive studies by several research groups resulted in some significant modifications in HIV-1 vector design, which substantially improved vector biosafety, raised vector titers, and increased transgene expression. Replacement of the 3' U3 region with a potent heterologous promoter rendered the vector Tat-independent and thus enabled vector production with a third-generation packaging construct *(28)*. A further increase in vector titer was obtained by substituting the bovine growth hormone polyadenylation signal for the 3' U5 *(29)*. Incorporating various deletions (133–400 bp) in the 3' U3 region generated self-inactivating (SIN) HIV-1 vectors *(28–30)*. The deleted regions contained the HIV-1 enhancer/promoter sequences, including the TATA box. Because during reverse transcription the modified 3' U3 region is transferred to the 5' LTR, integrated SIN vectors are completely devoid of their parental enhancer/promoter sequences, and thus lack the ability to transcribe full-length vector RNA that may be packaged. This reduces the likelihood of generating RCR and minimizes the ability to mobilize integrated vectors by replication-competent viruses. In addition, the lack of enhancer/promoter sequences in the SIN vector LTR reduces the risk of inadvertently activating silent host-cell promoters or interfering with the activity of an internal promoter controlling a transgene.

The incorporation of the woodchuck hepatitis virus posttranscriptional regulatory element (WPRE) *(31)* and the central polypurine tract (cPPT) *(32)* sequences into the HIV-1 vector cassette significantly improved transgene expression and transduction efficiency, respectively. The mechanism by which 600 bp of the WPRE sequence increases transgene expression has not been completely resolved. However, when placed in the sense orientation in the 3′ untranslated region of a transgene, the WPRE sequence increases overall transgene expression by more than fivefold.

In addition to the 3′ end polypurine tract, the HIV-1 genome contains a cPPT and a central termination sequence (cTS). Thus, in the process of reverse transcription, synthesis of the HIV-1 plus strand DNA is initiated at two origins, which results in the displacement of the plus DNA strand and the formation of a DNA flap. Although studies by Charneau et al. demonstrated the importance of the cPPT/cTS for efficient HIV-1 replication *(33,34)*, the initial HIV-1 vectors in which these sequences were not incorporated were able to efficiently transduce nondividing cells in vitro and in vivo *(9,10)*. Later studies showed that the incorporation of the cPPT and the cTS sequences into HIV-1 vectors accelerated their transduction kinetics. It appears that the DNA flap improves nuclear importation of preintegration complexes of HIV-1 vectors in dividing and nondividing cells. More importantly, incorporation of the 150 bp sequence containing the cPPT/cTS in the sense orientation 5′ to the vector internal promoter increased transduction efficiency by 3–10-fold in a variety of cells, including terminally differentiated neurons, hematopietic stem cells (HCS), hepatocytes and freshly isolated T-cells *(32,35–39)*.

Incorporation of the tetracycline inducible system into HIV-1 vectors added a desirable feature to the newly developed lentivirus vectors, because it conferred the ability to control transgene expression in vivo *(40,41)*. This increased the feasibility of using lentivirus vectors in clinical trials in which regulation of transgene expression is required. The tetracycline inducible system is based on two components: (1) the tetracycline regulated transactivator (tTA), which is a fusion protein of the tet repressor, and the transactivator domain of the herpes simplex virus protein VP16; and (2) the inducible promoter, which contains a minimal promoter and seven copies of the Tet operon. In the absence of tetracycline, the tTA binds and activates the inducible promoter. Binding of tetracycline (or its more potent analog, doxycyline) to the tTA induces conformational changes that render the tTA incapable of binding to the Tet operon and thus abolish transgene expression *(42)*. Initially, the two regulatory components of the tetracycline-inducible system were contained in a single inducible HIV-1 vector. Based on this design, a CMV promoter located in the middle of the vector genome constitutively expressed the tTA, while the inducible promoter at the 3′ end of the vector regulated the expression of the transgene of interest *(40)*. Although first-generation inducible HIV-1 vectors could be generated efficiently by transient transfection with three plasmids (*see* **Subheading 5.**), their titers (up to 10^9 IU/mL) were 5–10-fold lower than the titers obtained with conventional noninducible vectors. The first-generation inducible HIV-1 vector efficiently transduced dividing cells in culture and terminally differentiated neurons in vivo. Induction of transgene expression by doxycyline withdrawal resulted in more than a 500-fold increase in the level of a reporter green fluorescent protein (GFP), yet basal GFP expression could be detected by fluorescent activated cell sorting (FACS) in doxycyline-treated cells *(40)*. The mechanism responsible for the leakiness of the inducible promoter in the lentivirus vector was not characterized. A binary-inducible lentivirus vector system in which the CMV promoter/tTA expression cassette and the inducible promoter/ GFP cassette were separated into two lentivirus vectors demonstrated some reduction in doxycyline-independent basal-promoter activity. Interestingly, the binary-vector system demonstrated a higher level of maximal induced expression (Kafri, unpublished data). However the requirement to transduce a single target cell with two independent vectors may limit the use of the binary system to in vitro applications. The development of second-generation inducible lentivirus vectors was based on placing the inducible promoter in the vector U3 region. This significantly reduced basal-gene expression, but the modified vectors showed lower levels of maximal induced expression *(43)*.

5. Production of Lentivirus Vectors

To date, the common method of HIV-1 and other lentivirus vector production is based on a transient transfection of three to four plasmids into 293T cells (**Fig. 1**). In addition, several lentivirus vector-packaging cell lines were recently developed as an alternative method of vector production. The following section briefly summarizes the methods for producing lentiviral vectors by these two techniques.

5.1. Vector Production by Transient Transfection

5.1.1. The Cells

Lentivirus vectors are usually generated in human embryonic kidney (HEK) 293T cells, which are SV40 large-T antigen-expressing 293 cells. These cells are highly transfectable, thus at least 60% of the cells should be successfully transfected by the calcium-phosphate transfection method. Prior to transfection,

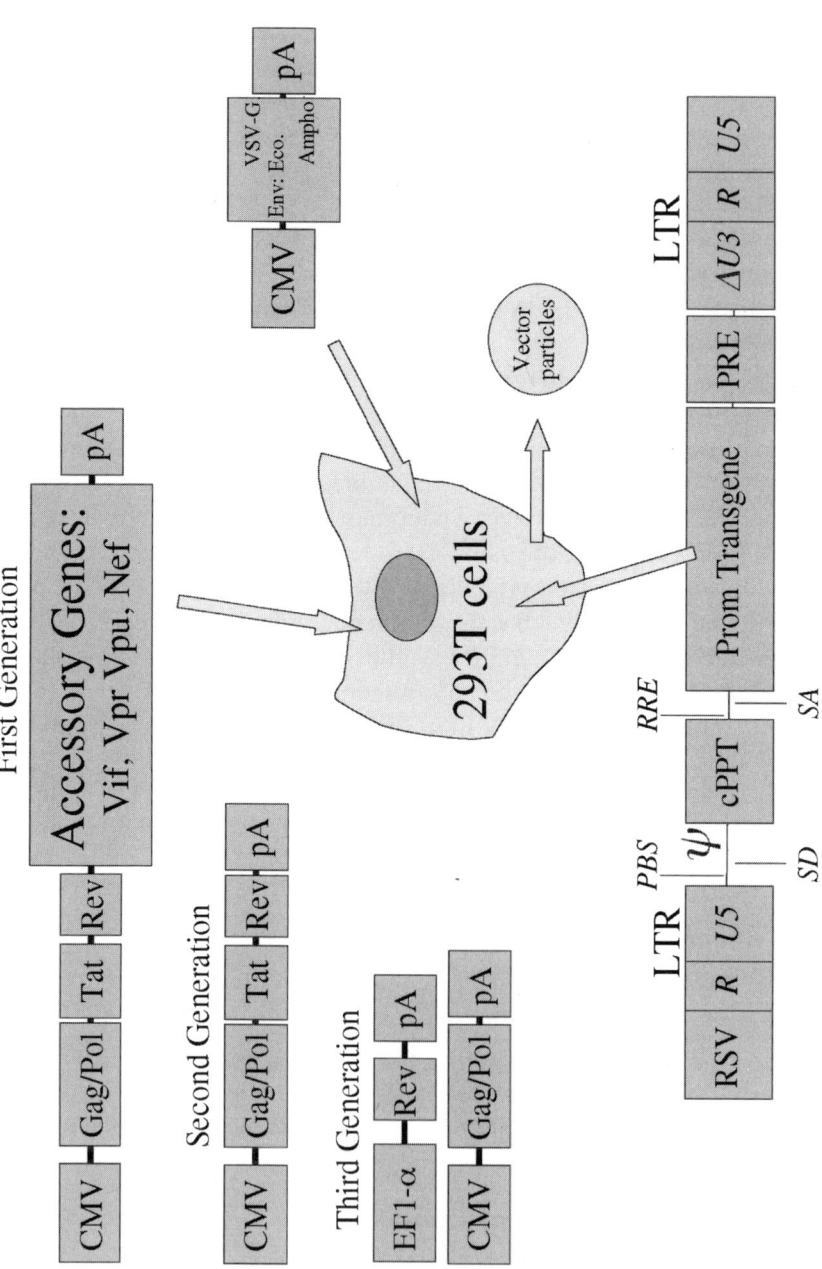

Fig. 1. Production of lentivirus vectors transient transfection of 3–4 plasmids into 293T cells. The plasmids should include the vector cassette, the envelope cassette, and a packaging construct. The third-generation packaging cassette was spilt into two plasmids, which express the *gag/pol* and the *rev* gene. At 48-h posttransfection, lentivirus vectors can be collected from conditioned media and concentrated further by ultracentrifugation.

plate the cells on 10 cm poly-L-lysine-coated plates at a density such that they are 60–70% confluent at the time of transfection. Prepare poly-L-lysine-coated plates by covering plates with 0.001% poly-L-lysine (Cat. no. P-4707, Sigma, St. Louis, MO) in PBS for 5 min, then removing the solution.

5.1.2. The DNA

Use 30–40 μg of endotoxin free DNA to transfect 4–5×10^6 cells in a 10-cm plate. This should include 15, 10, and 5 μg of DNA of the vector, the packaging cassette (either first- or second-generation), and the VSV-G envelope expression cassette, respectively. If a non-VSV-G envelope expression cassette is used, increase the amount of envelope DNA to 10 μg. If a third-generation HIV-1 packaging cassette is used, use 10 μg of the Gag/Pol expression cassette DNA and 3μ g of the Rev expression cassette DNA *(22)*.

5.1.3. The Transfection

The conventional calcium phosphate transient transfection method (using either the HEPES-based system or the BES-based system) is the most efficient technique for HIV-1 vector production. Eight to twelve h after the addition of DNA, wash the transfected cells and add fresh media. Collect vector particles by harvesting the media 60 h posttransfection. To separate vector particles from cell debris, centrifuge the media at $4000g$ for 5 min then filter it through a 0.45-micron filter. Concentrate VSV-G pseudotyped particles by two rounds of ultracenrifugation at $50,000g$ for 2 h each time. Following centrifugation, resuspend the vector pellets in either phosphate-buffered saline (PBS), Hank's balanced salt solution (HBSS), or Dulbecco's modified Eagle's medium (DMEM), and freeze at $-80°C$.

5.1.4. Vector Titers

Typical titers of VSV-G pseudotyped HIV-1 vectors are in the range of 10^6–10^7 IU/mL prior to concentration and 10^9–10^{10} IU/mL following concentration. Vectors expressing reporter genes (*GFP* or *LacZ*) or selection markers, such as the gene conferring resistance to puromycin, can be titered by serial dilutions on target cells (HEK293, HeLa cells). The titer of vectors, which do not express reporter genes, can be estimated by assays that measure either the amount or activity of proteins incorporated into the vector particles. These include the p24^{gag} ELISA Kit (Cat. no. NEK050B, Perkin Elmer Life Science, Boston, MA) and the reverse-transcription assay (Cat. no. 1828657, Roche, Indianapolis, IN). Extrapolating from the p24^{gag} concentration to vector titers is not accurate, and the ratio between p24^{gag} and the number of infectious particles may vary significantly between different vectors, laboratories, and vector stocks. Typically, 4–60×10^3 infectious vector particles contain 1 ng of p24^{gag}. A real-time polymerase chain reaction (PCR)-based method to determine vector titers has been recently described. The PCR primers in this system were designed to allow amplification of the minus strong-stop cDNA fragment that is present in all infectious vector particles, regardless of the transgene. Thus, this procedure does not require a reverse transcription step and can be applied to all HIV-1-based vectors (44).

5.2. Vector Production by Stable Packaging Cell Lines

Several drawbacks to producing lentiviral vectors by the transient transfection method include: higher chances of generating replication competent viruses, variability in the quality of vector stocks, and difficulties in scaling up vector production. For these reasons several HIV-1 vector packaging cell lines were developed. The cytotoxic effects of the VSV-G envelope and the HIV-1 gene products (protease, Vpr) dictated the use of Tetracycline or the ecdysone inducible gene expression systems (45–49).

Some of these cell lines can generate up to 10^6 infectious vector particles/mL. In general, vector titers obtained from the currently used packaging cell lines are 10–50-fold lower than the titers of HIV-1 vectors generated by transient transfection. Only three of these cell lines were tested and demonstrated their ability to generate HIV-1 vectors capable of transducing nondividing cells in vivo (terminally differentiated neurons) (45,48,49). Furthermore, it is not yet clear whether HIV-1 vectors generated by these cell lines can efficiently transduce hematopoietic stem cells. Transduction of a packaging line with the transfer vector is the last step in the generation of a retro/lentivirus vector producer line.

However, this approach is neither applicable for third-generation packaging cell lines, which are devoid of the Tat gene, nor can it be used with SIN vectors. Owing to these limitations, cotransfecting vector DNA with a selection marker and screening for stable high producer clones has been used to generate a stable HIV-1 vector producer line (46). Earlier studies indicated that this time-consuming approach results in tandem integration of genetically unstable copies of the vector DNA, which are often subjected to transcription shutoff. Thus, it is not surprising that a producer cell line that was generated by this approach could generate merely 1.8×10^5 IU/mL (46). In another approach, a CMV promoter was placed in the vector 3′U3 region. Because the process of reverse transcription results in the transfer of the vector 3′ U3 region to the 5′ LTR, the CMV promoter supported efficient production of vector RNA in the transduced packaging cells (47). This approach generated a self-activating vector, which increases the likelihood of generating replication-competent viruses, and may inadvertently activate neighboring endogenous promoters. Recently, a conditional SIN (cSIN) vector has been developed (48). The vector 3′ U3 region has been replaced with the tetracycline-inducible promoter which, following reverse transcription and integration, is transferred to the vector 5′ LTR. This allows the newly developed vector to retain its SIN features in normal target cells, yet be able to be produced at high titers in cells expressing the tTA. It is clear that a new generation of packaging cell lines should be developed before we can consider employing HIV-1 vectors in human clinical trials. Based on earlier studies, one can expect that in these novel packaging cell lines, the *gag* and the *pol* coding sequences will be separated from each other and from the *rev* gene. Ideally, the codon usage of the *gag, pol,* and *rev* genes will be humanized, and the vectors will be self-inactivating.

5.2.1. Using the SODk1 Packaging Cell Line

In general, the procedure of vector production by stable packaging cell lines can be divided into two stages: the induction stage and the vector collection/ concentration stage. Both stages vary slightly between the different packaging cell lines. Here we describe a method of producing first-generation HIV-1 vectors pseudotyped with the VSV-G envelope by the SODk-1 packaging cell line. Several research groups *(45,48,50)* have successfully used this method. The SODk1 cell line is derived from HEK293 cells, which express all of the HIV-1 genes (excluding the envelope gene) under the control of a tetracycline-inducible promoter. The cells also express the VSV-G envelope gene and the GFP gene from an inducible bi-directional promoter. SODk1 cells are resistant to G418 and blasticidin. Culture SODk1 cells in medium containing DMEM (high glucose), 10% fetal calf serum (FCS) and 1 μg/mL doxycyline. Transfer the cells at 1:5 dilution every 3–4 d, during which time they will reach 80–90% confluence. To generate a stable vector producer line, introduce an HIV-1 vector into the SODk1 packaging cell line either by stable transfection and clonal selection using a selection marker such as Hygromycin B, or by transduction with rescuable vectors (non-SIN and cSIN vectors).

5.2.2. Induction of Vector Production

Remove doxycyline from the SODk1 cells by washing the cells with PBS. Split the cells at 1:5 dilution into medium containing DMEM and 10% FCS (tetracycline-free FCS, BD Biosciences Clontech, Palo Alto, CA). Change the medium daily. When the cells reach 90% confluence, transfer them at 1:4 dilution into poly-L-lysine-coated plates prepared as described in **Subheading 5.1.1.** Twelve hours after transferring the cells, add sodium butyrate (Cat. no. B-5887, Sigma) to a final concentration of 5 m*M*. The tetracycline-free medium containing 5 m*M* butyrate should be changed daily. Induction of gene expression from the bi-directional tetracycline-inducible promoter should result in GFP and VSV-G gene expression. Determine induction efficiency by fluorescence microscopy. One d after the addition of sodium butyrate more than 85% of the cells should express GFP.

5.2.3. Vector Collection and Concentration

Harvest medium containing vector particles daily starting from d 2 to 4 after the addition of sodium butyrate; maximal vector titer (up to 10^6 IU/mL) occurs at d 3 following addition of sodium butyrate. To separate vector particles from cell debris, centrifuge the collected media at 4000g for 5 min and filter through a 0.45-micron filter. Concentrate the vector by ultracentifugation as described in **Subheading 5.1.3.**

6. Biosafety

The risks associated with lentivirus vector-based gene therapy are: the generation of replication competent viruses, insertional mutagenesis, vector transduction of germ cells, and vector mobilization by wild-type HIV-1. Recent improvements in the lentivirus vector system include: the development SIN vectors, the minimization and splitting of the packaging cassette, and the generation of packaging cell lines. These alleviated some of the biosafety concerns regarding the HIV-1 vector. However, the ability to reliably screen vector stocks for replication-competent retroviruses (RCRs) remains a prerequisite for using retrovirus vectors in clinical trials. Thus, it is not surprising that several assays have been developed to probe vector stocks for the emergence of RCRs and the transfer of viral genes into target cells' genome. Most of the currently used assays are based on cell lines, which were genetically engineered to report on the transfer of functional viral genes *(9)*. The HeLa P4.2 cell line constitutively expresses CD4 and contains the *LacZ* gene under the control of the HIV-1 LTR *(34)*. Transduction of the P4.2 cells with the non-SIN vector HR′CGFP resulted in the generation of the HeLa4G reporter cell line, which allows detection of RCRs by three independent methods: (1) the tat transfer assay, (2) the HIV-1 gag transfer assay, (3) the GFP marker rescue assay *(45)*.

6.1. The tat Transfer Assay

The tat transfer assay is relevant for screening vector stocks generated with a first- or second-generation packaging cassette (which contains the *tat* gene). This assay is based on the fact that the HIV-1 LTR, which controls the *LacZ* gene in HeLa4G cells, is inactive in the absence of the tat gene. Therefore, the appearance of X-Gal staining HeLa4G cells 3–4 wk following transduction of a vector sample serves as an indicator of the HIV-1 *tat* gene in the vector sample. Studies have demonstrated that the sensitivity of this assay is 20 tat-transducing units per mL of test medium *(9,34)*.

6.2. The HIV-1 gag Transfer Assay

The HIV-1 *gag* transfer assay is based on an HIV-1 p24^{gag} ELISA (p24 ELISA Kit, Perkin Elmer Life Science). Similar to the *tat* transfer assay, 10^5

HeLa4G cells are transduced with a vector sample at a multiplicity of infection (MOI) of 10 and cultured for 3 wk, after which time the p24^{gag} concentration in the culture medium is determined by an ELISA assay. The detection limit of this method is greater than 1 pg/mL, which is about 1–2 IU/mL.

Recently, an assay has been developed by Farson et al. *(49)* to detect recombination between sequences containing functional *gag/pol* genes and a non-SIN transfer vector. In this assay, HEK 293 cells expressing the VSV-G gene under a tetracycline-inducible promoter are transduced with vector. Induction of VSV-G envelope production could support amplification of a non-SIN recombinant vector expressing the *gag/pol* genes, resulting in a continuous increase in p24^{gag} in the culture media. This assay can detect an increase in p24^{gag} following initial transduction with a replication competent virus sample containing only 6 fg of p24^{gag}. A major drawback to this assay is its limited ability to detect recombination events between functional *gag/pol* genes and SIN vectors, which cannot replicate even in the presence of functional *gag/pol* and envelope genes.

6.3. The GFP Marker Rescue Assay

The GFP marker rescue assay tests for recombination events resulting in the incorporation of functional *gag/pol* and *envelope* sequences into an HIV-1 transfer vector. Transduction of HeLa4G cells with a vector sample expressing the *gag/pol* and *envelope* genes will result in the rescue and the release of the integrated HR'CGFP vectors into the culture media. Productive transduction of naïve 293T cells with HR'CGFP vectors rescued from HeLa4G cells can be detected by the presence of GFP-expressing cells and indicates that the tested vector stocks are unsafe for research use *(45)*.

To screen for recombination between functional *gag/pol* and Tat/Rev sequences and a non-SIN vector, Wu et al. developed a sensitive rescue assay *(27)*. Their approach was based on the ability of rescued HIV-1 vectors containing *gag/pol* and *tat/rev* genes to activate the transcription of the puromycin-resistance gene from HIV-1 LTR in HEK 293 reporter cells. However, this assay cannot detect recombination between sequences of the packaging cassette and SIN vectors, and its sensitivity has not been determined.

7. Target Organs

During the last 5 years, lentivirus vectors have become increasingly popular as a growing number of laboratories demonstrated their efficiency at delivery and maintaining long term transgene expression in a wide spectrum of target organs, some of which were resistant to transduction by nonlentivirus vectors. The rat central nervous system (CNS) was the first target organ to be efficiently transduced by lentivirus vectors in vivo *(9)*. Long-term constitutive and inducible transgene expression in rat brain have been reported following a single injection of lentivirus vectors (40,51). Furthermore, no evidence of cellular or structural pathology could be found in any of the transduced tissues. Interestingly, the majority (>88%) of the transduced cells were neurons (52). The permissiveness of the CNS to lentivirus vector transduction was also demonstrated in murine and primate animal models, in which lentivirus vectors were found efficient at delivering biologically active genes that changed the pathologic course of CNS diseases (28,53–61).

The spectrum of hematological diseases and nonhematological disorders (adrenoleukodystrophy, hemophilia), which can be treated by bone-marrow transplantation (62,63), underscores the importance of developing an efficient gene-delivery system to HCS. The ability of lentivirus vectors to transduce nondividing cells marks them as the vector of choice for HCS gene delivery. A growing number of research reports have demonstrated long-term expression of reporter and functional transgenes in all hematopoietic lineages, following primary and secondary transplantion with HIV-1 vector transduced HCS (37,64–76). Other hematopoietic cells which were transduced efficiently by HIV-1 vectors include primary nonactivated human lymphocytes, acute myeloid leukemia cells, and human dendritic cells (DCs) derived from either CD34(+) or human monocytes. Efficient tranduction of these cells opens new avenues in anti HIV-1 and cancer therapy (23,24,67,77–84).

Another promising target organ for lentivirus vector gene delivery is the eye, as several studies demonstrated efficient gene delivery to different cell types in the eye, including photoreceptors, pigmented epithelium, and corneal cells (85–88). Several publications have demonstrated that lentivirus vectors are efficient at in vitro transgene delivery to pancreatic islet cells (89–91). Furthermore, transduction of murine islet cells with lentivirus vector-expressing IL-4 abrogated their destruction following implantation into diabetes mellitus prone NOD/SCID mice (89).

The muscle tissue is a major target organ for gene therapy. Interestingly, it appears that the ability of lentivirus vectors to transduce muscle tissue is species-dependent. Murine muscle tissue was found resistant to lentivirus-vector transduction, whereas relatively efficient transduction of muscle tissue was reported in hamster and rat animal models (10,92).

Recently, successful lentivirus vector-transgene delivery into Rhesus monkey fetal tissues in vivo and into mouse embryonic stem cells in vitro has been reported (93–95). These studies are the first steps in establishing the feasibility of in utero and embryonic gene therapy, using lentivirus vectors. Clearly, additional studies are required to establish the safety of this approach, and to determine the risk of germ-cell transduction.

The liver is a site for a number of genetic and acquired diseases associated with high morbidity and mortality. It is not surprising, therefore, that the ability of lentivirus vectors to deliver and express transgenes in liver tissue was studied by several research groups. An initial study, in which lentivirus vectors were injected into the rat liver parenchyma, demonstrated efficient Vpr/Vif-dependent transduction of rat hepatocytes *(10)*. A study by Park et al. *(96)* reported on a limited success at transducing murine hepatocytes (<2%) by intraportal administration of lentivirus vectors and indicated that efficient lentivirus transduction of hepatocytes in vivo is cell-cycle-dependent. This study also reported on vector-associated liver injury and animal mortality, which could be attributed to either direct toxic effects of viral particles or to contaminants in the lentiviral preparation. In contrast to the Park study, a later study by Pfeifer et al. *(38)* demonstrated improved hepatocyte transduction (>5%) following either intraportal, intravenous, or intraperitoneal administration of lentivirus vectors. These results indicated that progression through the cell cycle is not required for in vivo hepatocyte transduction by HIV-1 based vectors. In addition, no evidence to vector-induced liver damage was found in this research.

The ability to transduce nondividing cells will determine the feasibility of using lentivirus vectors as in vivo gene-delivery systems. Although lentivirus vectors proved efficient at transducing terminally differentiated neurons in a normal rat brain, several studies indicated that a block at the reverse transcription step may hamper their ability to transduce quiescent G_0 cells *(97,98)*. Although the mechanism responsible for blocking lentivirus vector transduction of G_0 cells may by cell-type specific, recent studies revealed some common characteristics in lentivirus vector transduction of different primary cells, such as HCS and pancreatic islet cells *(64,71,90,98)*. In these in vitro systems, the transduction rate could be improved either by cytokine pretreatment of the target HCS or by adding hepatic growth factor (HGF) to the islet cells' culture media. Also, although the transduction rate in these studies was multiplicity of infection (MOI)-dependent, it was not proportional to the increase in MOI. It appears that maximal in vitro transduction of HCS and islet cells can be reached between an MOI of 60–100. One study demonstrated that vector concentration, rather than the actual MOI, is a better parameter by which transduction efficiency should be determined *(71)*. However, regardless of vector concentration and the addition of HGF/cytokines, the transduction rate of HCS and pancreatic islet cells did not exceed 45–60%. Evidently, additional studies are required to better understand the mechanism by which lentivirus vectors transduce primary nondividing cells.

8. Non-HIV-1 Lentivirus Vectors

Concerns regarding the safety of the HIV-1 vectors prompted several research groups to develop nonhuman lentivirus-vector systems. This approach was based on the idea that a lentivirus vector, which is derived from a virus that cannot replicate in human cells, will retain its ability to transduce nondividing human cells, yet will pose less biosafety risks when used in human clinical trials. To date, several nonhuman lentivirus gene-delivery systems, including the SIV-, EIAV-, FIV-, and BIV-based vectors, have been developed and were found efficient at transducing dividing and nondividing cells in vitro and in vivo (17,92,99–106). However, the safety advantages of these newly developed vectors over the HIV-1 based vectors have not yet been tested. In the process of their development, the nonhuman lentivirus vectors have been genetically engineered to efficiently transduce primary human cells, and to allow their production in human cell lines. More worrisome is the fact that FIV and SIV full-length RNA was shown to be cross-packaged by HIV-1 particles (107,108), which raises the risks of generating a chimeric virus with the ability to traverse between humans and domestic animals. It is important to directly compare the effectiveness of HIV-1 and nonhuman lentivirus vectors at transducing human cells. A recent study by Hofmann et al. (109) indicated that transduction efficiency of human cells by vectors originating from nonhuman lentiviruses may be hampered by the requirement for some species-specific factors during early stages of lentivirus transduction. In this regard, one can expect that the efficacy of the HIV-1 based vector as a gene-delivery system in human clinical trials will surpass its performances in animal models. Clearly, additional studies are required in order to determine the advantages of using one lentivirus vector system over the others.

Acknowledgment

The author of this review wishes to offer his apology to investigators whose work was not described in this manuscript or was inadequately presented.

References

1. Shimotohno, K. and Temin, H. M. (1981) Formation of infectious progeny virus after insertion of herpes simplex thymidine kinase gene into DNA of an avian retrovirus. *Cell* **26,** 67–77.
2. Wei, C. M., Gibson, M., Spear, P. G., and Scolnick, E. M. (1981) Construction and isolation of a transmissible retrovirus containing the src gene of Harvey murine sarcoma virus and the thymidine kinase gene of herpes simplex virus type 1. *J. Virol.* **39,** 935–944.
3. Cavazzana-Calvo, M., Hacein-Bey, S., de Saint Basile, G., et al. (2000) Gene therapy of human severe combined immunodeficiency (SCID)-X1 disease. *Science* **288,** 669–672.
4. Lewis, P., Hensel, M., Emerman, M. (1992) Human immunodeficiency virus infection of cells arrested in the cell cycle. *EMBO J.* **11,** 3053–3058.
5. Gallay, P., Swingler, S., Song, J., Bushman, F., and Trono, D. (1995) HIV nuclear

import is governed by the phosphotyrosine-mediated binding of matrix to the core domain of integrase. *Cell* **83,** 569–576.

6. Gallay, P, Stitt, V., Mundy, C., Oettinger, M., and Trono, D. (1996) Role of the karyopherin pathway in human immunodeficiency virus type 1 nuclear import. *J. Virol.* **70,** 1027–1032.

7. Gallay, P., Hope, T., Chin, D., and Trono, D. (1997) HIV-1 infection of nondividing cells through the recognition of integrase by the importin/karyopherin pathway. *Proc. Natl. Acad. Sci. USA* **94,** 9825–9830.

8. Kim, V. N., Mitrophanous, K., Kingsman, S. M., and Kingsman, A. J. (1998) Minimal requirement for a lentivirus vector based on human immunodeficiency virus type 1. *J. Virol.* **72,** 811–816.

9. Naldini, L., Blomer, U., Gallay, P., et al. (1996) In vivo gene delivery and stable transduction of nondividing cells by a lentiviral vector [see comments]. *Science* **272,** 263–267.

10. Kafri, T., Blomer, U., Peterson, D. A., Gage, F. H., and Verma, I. M. (1997) Sustained expression of genes delivered directly into liver and muscle by lentiviral vectors. *Nat. Genet.* **17,** 314–317.

11. Akkina, R. K., Walton, R. M., Chen, M. L., Li, Q. X., Planelles, V., and Chen, I. S. (1996) High-efficiency gene transfer into CD34+ cells with a human immunodeficiency virus type 1-based retroviral vector pseudotyped with vesicular stomatitis virus envelope glycoprotein G. *J. Virol.* **70,** 2581–2585.

12. Aiken C. (1997) Pseudotyping human immunodeficiency virus type 1 (HIV-1) by the glycoprotein of vesicular stomatitis virus targets HIV-1 entry to an endocytic pathway and suppresses both the requirement for Nef and the sensitivity to cyclosporin A. *J. Virol.* **71,** 5871–5877.

13. DePolo, N. J, Reed, J. D., Sheridan, P. L., et al. (2000) VSV-G pseudotyped lentiviral vector particles produced in human cells are inactivated by human serum. *Mol. Ther.* **2,** 218–222.

14. Higashikawa, F. and Chang, L. (2001) Kinetic analyses of stability of simple and complex retroviral vectors. *Virology* **280,** 124–131.

15. Christodoulopoulos, I. and Cannon, P. M. (2001) Sequences in the cytoplasmic tail of the gibbon ape leukemia virus envelope protein that prevent its incorporation into lentivirus vectors. *J. Virol.* **75,** 4129–4138.

16. Stitz, J., Buchholz, C. J., Engelstadter, M., et al. (2000) Lentiviral vectors pseudotyped with envelope glycoproteins derived from gibbon ape leukemia virus and murine leukemia virus 10A1. *Virology* **273,** 16–20.

17. Olsen, J. C., Kelly, M., Patel, M., and Johnson, L. G. (1999) Gene delivery to murine trachea by an equine lentiviral vector. *Pediatr. Pulmonol.* Suppl. **19:222** (Abstr.).

18. Kobinger, G. P., Weiner, D. J., Yu, Q. C., and Wilson, J. M. (2001) Filovirus-pseudotyped lentiviral vector can efficiently and stably transduce airway epithelia in vivo. *Nat. Biotechnol.* **19,** 225–230.

19. Wang, B., Ge, Y. C., Palasanthiran, P., et al. (1996) Gene defects clustered at the C-terminus of the vpr gene of HIV-1 in long-term nonprogressing mother and child

pair: in vivo evolution of vpr quasispecies in blood and plasma. *Virology* **223,** 224–232.

20. Kirchhoff, F., Greenough, T. C., Brettler, D. B., Sullivan, J. L., and Desrosiers, R.C. (1995) Brief report: absence of intact nef sequences in a long-term survivor with nonprogressive HIV-1 infection. *N. Engl. J. Med.* **332,** 228–232.

21. Zufferey, R., Nagy, D., Mandel, R. J., Naldini, L., and Trono, D. (1997) Multiply attenuated lentiviral vector acheives efficient gene delivery in vivo. *Nat. Biotechnol.* **15,** 871–875.

22. Dull, T., Zufferey, R., Kelly, M., et al. (1998) A third-generation lentivirus vector with a conditional packaging system. *J. Virol.* **72,** 8463–8471.

23. Chinnasamy, D., Chinnasamy, N., Enriquez, M. J., Otsu, M., Morgan, R. A., and Candotti, F. (2000) Lentiviral-mediated gene transfer into human lymphocytes: role of HIV-1 accessory proteins. *Blood* **96,** 1309–1316.

24. Costello, E., Munoz, M., Buetti, E., Meylan, P. R., Diggelmann, H., and Thali, M. (2000) Gene transfer into stimulated and unstimulated T lymphocytes by HIV-1-derived lentiviral vectors. *Gene Ther.* **7,** 596–604.

25. Shen, H., Cheng, T., Preffer, F. I., et al. (1999) Intrinsic human immunodeficiency virus type 1 resistance of hematopoietic stem cells despite coreceptor expression. *J. Virol.* **73,** 728–737.

26. Kotsopoulou, E., Kim, V. N., Kingsman, A. J., Kingsman, S. M., and Mitrophanous, K. A. (2000) A Rev-independent human immunodeficiency virus type 1 (HIV-1)-based vector that exploits a codon-optimized HIV-1 gag-pol gene. *J. Virol.* **74,** 4839–4852.

27. Wu, X., Wakefield, J. K., Liu, H., et al. (2000) Development of a novel trans-lentiviral vector that affords predictable safety. *Mol. Ther.* **2,** 47–55.

28. Miyoshi, H., Blomer, U., Takahashi, M., Gage, F. H., and Verma, I. M. (1998) Development of a self-inactivating lentivirus vector. *J. Virol.* **72,** 8150–8157.

29. Iwakuma, T., Cui, Y., and Chang, L. J. (1999) Self-inactivating lentiviral vectors with U3 and U5 modifications. *Virology* **261,** 120–132.

30. Zufferey, R., Dull, T., Mandel, R. J., et al. (1998) Self-inactivating lentivirus vector for safe and efficient in vivo gene delivery. *J. Virol.* **72,** 9873–9880.

31. Zufferey, R., Donello, J. E., Trono, D., and Hope, T. J. (1999) Woodchuck hepatitis virus posttranscriptional regulatory element enhances expression of transgenes delivered by retroviral vectors. *J. Virol.* **73,** 2886–2892.

32. Zennou, V., Petit, C., Guetard, D., Nerhbass, U., Montagnier, L., and Charneau, P. (2000) HIV-1 genome nuclear import is mediated by a central DNA flap. *Cell* **101,** 173–185.

33. Charneau, P., Alizon, M., and Clavel, F. (1992) A second origin of DNA plus-strand synthesis is required for optimal human immunodeficiency virus replication. *J. Virol.* **66,** 2814–2820.

34. Charneau, P., Mirambeau, G., Roux, P., Paulous, S., Buc, H., and Clavel, F. (1994) HIV-1 reverse transcription. A termination step at the center of the genome. *J. Mol. Biol.* **241,** 651–662.

35. Zennou, V., Serguera, C., Sarkis, C., et al. (2001) The HIV-1 DNA flap stimulates HIV vector-mediated cell transduction in the brain. *Nat. Biotechnol.* **19,** 446–450.

36. Sirven, A., Pflumio, F., Zennou, V., et al. (2000) The human immunodeficiency virus type-1 central DNA flap is a crucial determinant for lentiviral vector nuclear import and gene transduction of human hematopoietic stem cells. *Blood* **96,** 4103–4110.

37. Sirven, A., Ravet, E., Charneau, P., et al. (2001) Enhanced transgene expression in cord blood cd34(+)-derived hematopoietic cells, including developing t cells and nod/scid mouse repopulating cells, following transduction with modified trip lentiviral vectors. *Mol. Ther.* **3,** 438–448.

38. Pfeifer, A., Kessler, T., Yang, M., et al. (2001) Transduction of liver cells by lentiviral vectors: analysis in living animals by fluorescence imaging. *Mol. Ther.* **3,** 319–322.

39. Follenzi, A., Ailles, L. E., Bakovic, S , Geuna, M., and Naldini, L. (2000) Gene transfer by lentiviral vectors is limited by nuclear translocation and rescued by HIV-1 pol sequences. *Nat. Genet.* **25,** 217–222.

40. Kafri, T., van Praag, H., Gage, F. H., and Verma, I. M. (2000) Lentiviral vectors: regulated gene expression. *Mol. Ther.* **1,** 516–521.

41. Reiser, J., Lai, Z., Zhang, X. Y., and Brady, R. O. (2000) Development of multigene and regulated lentivirus vectors. *J. Virol.* **74,** 10589–10599.

42. Gossen, M. and Bujard, H. (1992) Tight control of gene expression in mammalian cells by tetracycline- responsive promoters. *Proc. Natl. Acad. Sci. USA* **89,** 5547–5551.

43. Haack, K. and Kafri, T. (2001) Development of second generation inducible lentiviral vectors. *Mol. Ther.* **3,** S2.

44. Scherr, M., Battmer, K., Blomer, U., Ganser, A., and Grez, M. (2001) Quantitative determination of lentiviral vector particle numbers by real-time PCR. *Biotechniques* **31,** 520, 522, 524, passim.

45. Kafri, T., van Praag, H., Ouyang, L., Gage, F. H., and Verma, I. M. (1999) A packaging cell line for lentivirus vectors. *J. Virol.* **73,** 576–584.

46. Pacchia, A. L., Adelson, M. E., Kaul, M., Ron, Y., and Dougherty, J. P. (2001) An inducible packaging cell system for safe, efficient lentiviral vector production in the absence of HIV-1 accessory proteins. *Virology* **282,** 77–86.

47. Klages, N., Zufferey, R., and Trono, D. (2000) A stable system for the high-titer production of multiply attenuated lentiviral vectors. *Mol. Ther.* **2,** 170–176.

48. Xu, K., Ma, H., McCown, T. J., Verma, I. M., and Kafri, T. (2001) Generation of a stable cell line producing high-titer self-inactivating lentiviral vectors. *Mol. Ther.* **3,** 97–104.

49. Farson, D., Witt, R., McGuinness, R., et al. (2001) A new-generation stable inducible packaging cell line for lentiviral vectors. *Hum. Gene Ther.* **12,** 981–997.

50. Hansen, M. S., Smith, G. J., 3rd, Kafri, T., Molteni, V., Siegel, J. S., and Bushman, F. D. (1999) Integration complexes derived from HIV vectors for rapid assays in vitro. *Nat. Biotechnol.* **17,** 578–582.

51. Naldini, L., Blomer, U., Gage, F. H., Trono, D., and Verma, I. M. (1996) Efficient gene transfer, integration, and sustained long-term expression in adult rat brains injected with a lentiviral vector. *Proc. Natl. Acad. Sci. USA* **93,** 11382–11388.

52. Blomer, U., Naldini, L., Kafri, T., Trono, D., Verma, I. M., and Gage, F. H. (1997) Highly efficient and sustained gene transfer in adult neurons with a lentivirus vector. *J. Virol.* **71,** 6641–6649.

53. Bensadoun, J. C., Deglon, N., Tseng, J. L., Ridet, J. L., Zurn, A. D., and Aebischer, P. (2000) Lentiviral vectors as a gene delivery system in the mouse midbrain: cellular and behavioral improvements in a 6-OHDA model of Parkinson's disease using GDNF. *Exp. Neurol.* **164,** 15–24.

54. Kordower, J. H., Bloch, J., Ma, S. Y., et al. (1999) Lentiviral gene transfer to the nonhuman primate brain. *Exp. Neurol.* **160,** 1–16.

55. Deglon, N., Tseng, J. L., Bensadoun, J. C., et al. (2000) Self-inactivating lentiviral vectors with enhanced transgene expression as potential gene transfer system in Parkinson's disease. *Hum. Gene Ther.* **11,** 179–190.

56. Lai, Z., Han, I., Zirzow, G., Brady, R. O., and Reiser, J. (2000) Intercellular delivery of a herpes simplex virus VP22 fusion protein from cells infected with lentiviral vectors. *Proc. Natl. Acad. Sci. USA* **97,** 11297–11302.

57. Bosch, A., Perret, E., Desmaris, N., Trono, D., and Heard, J. M. (2000) Reversal of pathology in the entire brain of mucopolysaccharidosis type VII mice after lentivirus-mediated gene transfer. *Hum. Gene Ther.* **11,** 1139–1150.

58. Consiglio, A., Quattrini, A., Martino, S., et al. (2001) In vivo gene therapy of metachromatic leukodystrophy by lentiviral vectors: correction of neuropathology and protection against learning impairments in affected mice. *Nat. Med.* **7,** 310–316.

59. Blomer, U., Kafri, T., Randolph-Moore, L., Verma, I. M, and Gage, F. H. (1998) Bcl-xL protects adult septal cholinergic neurons from axotomized cell death. *Proc. Natl. Acad. Sci. USA* **95,** 2603–2608.

60. Bjorklund, A., Kirik, D., Rosenblad, C., Georgievska, B., Lundberg, C., and Mandel, R. J. (2000) Towards a neuroprotective gene therapy for Parkinson's disease: use of adenovirus, AAV and lentivirus vectors for gene transfer of GDNF to the nigrostriatal system in the rat Parkinson model. *Brain Res.* **886,** 82–98.

61. Kordower, J. H., Emborg, M. E., Bloch, J., et al. (2000) Neurodegeneration prevented by lentiviral vector delivery of GDNF in primate models of Parkinson's disease. *Science* **290,** 767–773.

62. Benhamida, S., Pflumio, F., Fichelson, S., et al. (2001) HIV-flap mediated gene therapy for adrenoleukodystrophy: efficient correction of hematopoietic stem cells using short clinically applicable protocole. *Mol. Ther.* **3,** S327.

63. Kootstra, N., Matsumura, R., and Verma, I. (2001) Gene therapy for hemophilia A using lentiviral vectors. *Mol. Ther.* **3,** S221.

64. Miyoshi, H., Smith, K. A., Mosier, D. E., Verma, I. M., and Torbett, B. E. (1999) Transduction of human CD34+ cells that mediate long-term engraftment of NOD/SCID mice by HIV vectors. *Science* **283,** 682–686.

65. Guenechea, G., Gan, O. I., Inamitsu, T., et al. (2000) Transduction of human CD34+ CD38- bone marrow and cord blood-derived SCID-repopulating cells with third-generation lentiviral vectors. *Mol. Ther.* **1,** 566–573.

66. An, D. S., Wersto, R. P., Agricola, B. A., et al. (2000) Marking and gene expression by a lentivirus vector in transplanted human and nonhuman primate CD34(+) cells. *J. Virol.* **74,** 1286–1295.

67. Douglas, J., Kelly, P., Evans, J. T., and Garcia, J. V. (1999) Efficient transduction of human lymphocytes and CD34+ cells via human immunodeficiency virus-based gene transfer vectors. *Hum. Gene Ther.* **10,** 935–945.

68. Douglas, J. L., Lin, W. Y., Panis, M. L., and Veres, G. (2001) Efficient human immunodeficiency virus-based vector transduction of unstimulated human mobilized peripheral blood CD34+ cells in the SCID-hu Thy/Liv model of human T cell lymphopoiesis. *Hum. Gene Ther.* **12,** 401–413.

69. Chen, W., Wu, X., Levasseur, D. N., et al. (2000) Lentiviral vector transduction of hematopoietic stem cells that mediate long-term reconstitution of lethally irradiated mice. *Stem Cells* **18,** 352–359.

70. Donahue, R. E., Sorrentino, B. P., Hawley, R. G., An, D.S., Chen, I. S., and Wersto, R. P. (2001) Fibronectin fragment CH-296 inhibits apoptosis and enhances ex vivo gene transfer by murine retrovirus and human lentivirus vectors independent of viral tropism in nonhuman primate CD34(+) cells. *Mol. Ther.* **3,** 359–367.

71. Haas, D. L., Case, S. S., Crooks, G. M., and Kohn, D. B. (2000) Critical factors influencing stable transduction of human CD34(+) cells with HIV-1-derived lentiviral vectors. *Mol. Ther.* **2,** 71–80.

72. Mikkola, H., Woods, N. B., Sjogren, M., et al. (2000) Lentivirus gene transfer in murine hematopoietic progenitor cells is compromised by a delay in proviral integration and results in transduction mosaicism and heterogeneous gene expression in progeny cells. *J. Virol.* **74,** 11911–11918.

73. Salmon, P., Kindler, V., Ducrey, O., Chapuis, B., Zubler, R. H., and Trono, D. (2000) High-level transgene expression in human hematopoietic progenitors and differentiated blood lineages after transduction with improved lentiviral vectors. *Blood* **96,** 3392–3398.

74. Yamada, K., Olsen, J. C., Patel, M., Rao, K. W., and Walsh, C. E. (2001) Functional correction of fanconi anemia group c hematopoietic cells by the use of a novel lentiviral vector. *Mol. Ther.* **3,** 485–490.

75. Saulnier, S. O., Steinhoff, D., Dinauer, M. C., et al. (2000) Lentivirus-mediated gene transfer of gp91phox corrects chronic granulomatous disease (CGD) phenotype in human X-CGD cells. *J. Gene Med.* **2,** 317–325.

76. May, C., Rivella, S., Callegari, J., et al. (2000) Therapeutic haemoglobin synthesis in beta-thalassaemic mice expressing lentivirus-encoded human beta-globin. *Nature* **406,** 82–86.

77. Chinnasamy, N., Chinnasamy, D., Toso, J. F., et al. (2000) Efficient gene transfer to human peripheral blood monocyte-derived dendritic cells using human immunodeficiency virus type 1-based lentiviral vectors. *Hum. Gene Ther.* **11,** 1901–1909.

78. Stripecke, R., Cardoso, A. A., Pepper, K. A., et al. (2000) Lentiviral vectors for efficient delivery of CD80 and granulocyte-macrophage- colony-stimulating factor in human acute lymphoblastic leukemia and acute myeloid leukemia cells to induce antileukemic immune responses. *Blood* **96,** 1317–1326.

79. Schroers, R., Sinha, I., Segall, H., et al. (2000) Transduction of human PBMC-derived dendritic cells and macrophages by an HIV-1-based lentiviral vector system. *Mol. Ther.* **1,** 171–179.

80. Dyall, J., Latouche, J. B., Schnell, S., and Sadelain, M. (2001) Lentivirus-transduced human monocyte-derived dendritic cells efficiently stimulate antigen-specific cytotoxic T lymphocytes. *Blood* **97,** 114–121.

81. Cremer, I., Vieillard, V., Sautes-Fridman, C., and De Maeyer, E. (2000) Inhibition of human immunodeficiency virus transmission to CD4+ T cells after gene transfer of constitutively expressed interferon beta to dendritic cells. *Hum. Gene Ther.* **11,** 1695–1703.

82. Gruber, A., Kan-Mitchell, J., Kuhen, K. L., Mukai, T., and Wong-Staal, F. (2000) Dendritic cells transduced by multiply deleted HIV-1 vectors exhibit normal phenotypes and functions and elicit an HIV-specific cytotoxic T-lymphocyte response in vitro. *Blood* **96,** 1327–1333.

83. Evans, J. T., Cravens, P., Lipsky, P. E., and Garcia, J. V. (2000) Differentiation and expansion of lentivirus vector-marked dendritic cells derived from human CD34(+) cells. *Hum. Gene Ther.* **11,** 2483–2492.

84. Neil, S., Martin, F., Ikeda, Y., and Collins, M. (2001) Postentry restriction to human immunodeficiency virus-based vector transduction in human monocytes. *J. Virol.* **75,** 5448–5456.

85. Miyoshi, H., Takahashi, M., Gage, F. H., and Verma, I. M. (1997) Stable and efficient gene transfer into the retina using an HIV-based lentiviral vector. *Proc. Natl. Acad. Sci. USA* **94,** 10319–10323.

86. Takahashi, M., Miyoshi, H., Verma, I. M., and Gage, F. H. (1999) Rescue from photoreceptor degeneration in the rd mouse by human immunodeficiency virus vector-mediated gene transfer. *J. Virol.* **73,** 7812–7816.

87. Wang, X., Appukuttan, B., Ott, S., et al. (2000) Efficient and sustained transgene expression in human corneal cells mediated by a lentiviral vector. *Gene Ther.* **7,** 196–200.

88. Galileo, D. S., Hunter, K., and Smith, S. B. (1999) Stable and efficient gene transfer into the mutant retinal pigment epithelial cells of the Mitf(vit) mouse using a lentiviral vector. *Curr. Eye Res.* **18,** 135–142.

89. Gallichan, W. S., Kafri, T., Krahl, T., Verma, I. M., and Sarvetnick, N. (1998) Lentivirus-mediated transduction of islet grafts with interleukin 4 results in sustained gene expression and protection from insulitis. *Hum. Gene Ther.* **9,** 2717–2726.

90. Leibowitz, G., Beattie, G. M., Kafri, T., et al. (1999) Gene transfer to human pancreatic endocrine cells using viral vectors. *Diabetes* **48,** 745–753.

91. Giannoukakis, N., Mi, Z., Gambotto, A., et al. (1999) Infection of intact human islets by a lentiviral vector. *Gene Ther.* **6,** 1545–1551.

92. Johnston, J. C., Gasmi, M., Lim, L. E., et al. (1999) Minimum requirements for efficient transduction of dividing and nondividing cells by feline immunodeficiency virus vectors. *J. Virol.* **73,** 4991–5000.

93. Hamaguchi, I., Woods, N. B., Panagopoulos, I., et al. (2000) Lentivirus vector gene expression during ES cell-derived hematopoietic development in vitro. *J. Virol.* **74,** 10778–10784.

94. Tarantal, A. F., O'Rourke, J. P., Case, S. S., et al. (2001) Rhesus monkey model for fetal gene transfer: studies with retroviral- based vector systems. *Mol. Ther.* **3,** 128–138.

95. Wolfgang, M. J., Eisele, S. G., Browne, M. A., et al. (2001) Rhesus monkey placental transgene expression after lentiviral gene transfer into preimplantation embryos. *Proc. Natl. Acad. Sci. USA* **98,** 10728–10732.

96. Park, F., Ohashi, K., Chiu, W., Naldini, L., and Kay, M. A. (2000) Efficient lentiviral transduction of liver requires cell cycling in vivo. *Nat. Genet.* **24,** 49–52.

97. Korin, Y. D. and Zack, J. A. (1998) Progression to the G_1b phase of the cell cycle is required for completion of human immunodeficiency virus type 1 reverse transcription in T cells. *J. Virol.* **72,** 3161–3168.

98. Sutton, R. E., Reitsma, M. J., Uchida, N., and Brown, P. O. (1999) Transduction of human progenitor hematopoietic stem cells by human immunodeficiency virus type 1-based vectors is cell cycle dependent. *J. Virol.* **73,** 3649–3660.

99. Schnell, T., Foley, P., Wirth, M., Munch, J., and Uberla, K. (2000) Development of a self-inactivating, minimal lentivirus vector based on simian immunodeficiency virus. *Hum. Gene Ther.* 11, 439–447.

100. Negre, D., Mangeot, P. E., Duisit, G., et al. (2000) Characterization of novel safe lentiviral vectors derived from simian immunodeficiency virus (SIVmac251) that efficiently transduce mature human dendritic cells. *Gene Ther.* **7,** 1613–1623.

101. Nakajima, T., Nakamaru, K., Ido, E., Terao, K., Hayami, M., and Hasegawa, M. (2000) Development of novel simian immunodeficiency virus vectors carrying a dual gene expression system. *Hum. Gene Ther.* **11,** 1863–1874.

102. Curran, M. A., Kaiser, S. M., Achacoso, P. L., and Nolan, G. P. (2000) Efficient transduction of nondividing cells by optimized feline immunodeficiency virus vectors. *Mol. Ther.* **1,** 31–38.

103. Wang, G., Slepushkin, V., Zabner, J., et al. (1999) Feline immunodeficiency virus vectors persistently transduce nondividing airway epithelia and correct the cystic fibrosis defect. *J. Clin. Invest.* **104,** R55–R62.

104. Olsen, J. C. (1998) Gene transfer vectors derived from equine infectious anemia virus. *Gene Ther.* **5,** 1481–1487.

105. Mitrophanous, K., Yoon, S., Rohll, J., et al. (1999) Stable gene transfer to the nervous system using a non-primate lentiviral vector. *Gene Ther.* **6,** 1808–1818.

106. Berkowitz, R., Ilves, H., Lin, W. Y., et al. (2001) Construction and molecular analysis of gene transfer systems derived from bovine immunodeficiency virus. *J. Virol.* **75,** 3371–3382.

107. Browning, M. T., Schmidt, R. D., Lew, K. A., and Rizvi, T. A. (2001) Primate and feline lentivirus vector rna packaging and propagation by heterologous lentivirus virions. *J. Virol.* **75,** 5129–5140.
108. Rizvi, T. A. and Panganiban, A. T. (1993) Simian immunodeficiency virus RNA is efficiently encapsidated by human immunodeficiency virus type 1 particles. *J. Virol.* **67,** 2681–2688.
109. Hofmann, W., Schubert, D., LaBonte, J., et al. (1999) Species-specific, postentry barriers to primate immunodeficiency virus infection. *J. Virol.* **73,** 10020–10028.

26

Lentiviral Vectors for the Delivery of DNA into Mammalian Cells

Roland Wolkowicz, Garry P. Nolan, and Michael A. Curran

1. Introduction

Vectors derived from oncoretroviruses, represented by the prototype Moloney murine leukemia virus (MMLV), are powerful tools for gene transfer into mammalian cells. Vectors derived from such viruses are able to carry an insert of up to 6.5 kb. Because Retroviridae and derived vectors insert their genome into the host chromosome, the transgene delivered by these viruses are stably expressed in the infected cells. From a safety standpoint, the vectors are designed to eliminate any need to carry viral genes or associated toxicities into the host cell. This also substantially reduces their potential immunogenicity. Finally, the titers achieved with these vectors can be very high, yielding efficient infection in a broad range of cell types.

The biology of oncoretroviridae-derived vectors renders them unable to transduce terminally differentiated or nondividing cells. This limitation is a significant reason for the development of lentiviral vectors. When it was discovered that HIV-1 (a member of the Lentiviral subclass of the Retroviridae) was able to infect nondividing cells such as macrophages, it was soon determined that vectors from Lentiviridae were capable of transducing many nondividing and primary cells. An added benefit was the fact that these vectors permit larger inserts than their oncoretroviral counterparts, up to 10 kb. The above properties render lentiviral vectors an attractive gene transfer tool.

From: *Methods in Molecular Biology, vol. 246:*
Gene Delivery to Mammalian Cells: Vol. 2: Viral Gene Transfer Techniques
Edited by: W. C. Heiser © Humana Press Inc., Totowa, NJ

The use of lentiviral vectors for gene transfer in general, and for gene therapy in human beings in particular, raises obvious safety concerns. In this respect, replication-competent recombinant (RCR) virus production is mostly avoided by separating *cis*-acting sequences (mainly for packaging, reverse transcription, and integration) and *trans*-acting sequences (viral proteins) onto different plasmids *(1)*. Furthermore, development of self-inactivating (SIN) vectors has been pursued, as will be discussed in the section concerning safety *(2–4)*.

An overview of the human immunodeficiency virus (HIV) and feline immunodeficiency virus (FIV) viral life cycles is useful to illustrate the features needed for the production of recombinant lentiviral vectors, and follows below.

1.1. Virus Life Cycle Overview: Relevance to Vectors

1.1.1. General

The life cycle of all retroviruses includes distinct steps characterized by binding and fusion with the target cell, entry, partial uncoating, reverse transcription, and integration into the host chromosome following nuclear import. The virus-production steps that lead to budding and maturation of infectious virions include transcription and transport of RNA into the cytoplasm, synthesis of the viral proteins, and their assembly into whole virions.

All Retroviridae bear in common three main genes, *gag, pol,* and *env*, which are responsible for the expression of the structural, enzymatic, and membrane proteins, respectively. Lentiviridae contain additional open reading frames (ORFs) encoding accessory and regulatory proteins engaged in transcriptional regulation, assembly of viral particles, and infectivity. The genome of HIV-1 is represented in **Fig. 1A** and that of FIV in **Fig. 1B**.

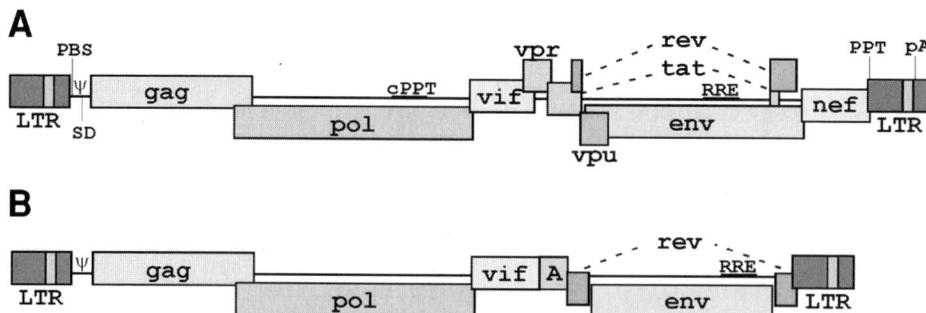

Fig. 1. Schematic representation of the viral genome. **(A)** HIV-1. **(B)** FIV. PBS, primer binding site; PPT, polypurine tract; pA, polyadenylation signal. Refer to the text for other abbreviations.

1.1.2. HIV-1

The 5' long terminal repeat (LTR) includes mainly binding recognition elements for various cellular transcription factors. Tat, Rev, and Nef are synthesized first from completely spliced viral mRNA. Tat ensures the efficient elongation of *de novo* synthesized mRNA via binding to the Tat/transactivator responsive element (TAR) *(5)*. Rev is needed for the transport of partially spliced and unspliced/full-length mRNA from the nucleus to the cytoplasm, a task it accomplishes by binding to the Rev-responsive element (RRE) present in the coding region of *env (6,7)*. Nef, although dispensable for in vitro virus production, plays a critical role in enhancing the in vivo pathogenesis of HIV through a combination of evasion and regulation of host-cell viability. Next, the *gag* and *pol* genes are translated and proteolytically processed into their component subunits. Gag carries the internal structural proteins of the virus: matrix (MA), capsid (CA), nucleoprotein (NC), and p6. The pol polyprotein yields the viral enzymatic proteins protease (PR), reverse transcriptase (RT), RNase H, and integrase (IN). In the endoplasmic reticulum (ER) *env* is translated into p160, subsequently cleaved into surface (SU) and transmembrane (TM) glycoproteins, and is ultimately incorporated onto the surface of the budding virion *(8)*.

The *cis* element, psi (ψ), present between the 5' LTR and the major splice donor site (SD), targets full-length RNA for encapsidation into nascent particles *(9)*, which also incorporate Vif, Vpr, and Nef. The role of the regulatory proteins of HIV-1 is beyond the scope of this introduction, but can be accessed through substantial literature on the subject, including some listed herein.

Mature viral particles bind to target cells via the interaction of SU and the primary cell-membrane receptor, CD4, in concert with coreceptors such as CCR5 for macrophages *(10)* and CXCR4 for T cells *(11)*. Following binding, conformational changes occur allowing fusion of the viral and cellular membranes, a prerequisite for virus entry *(8,12)*.

Once the virus enters the cell, uncoating occurs, revealing a ribonucleoprotein complex that includes reverse transcriptase. This complex ultimately synthesizes double-stranded DNA, the provirus, from the viral RNA template *(13,14)*. The process of reverse transcription involves multiple steps in which the RNA serves as a template for synthesis of the first (minus) strand of DNA, followed by degradation of the RNA template, and finally synthesis of the plus DNA strand *(15–18)*. During reverse transcription, the Vpr protein seems to interact with uracil DNA glycosylase (UDG), a cellular enzyme *(19)*, supposedly to minimize misincorporation of uracil into the cDNA in milieu poor in free dNTPs. Finally, the provirus is transported into the nucleus in a process mediated by Vpr *(20)*, IN *(21)*, and MA *(22)*, possibly by forming a complex in the central polypurine tract (cPPT), a sequence located in the center of the genome.

1.1.3. FIV

The life cycle of the feline immunodeficiency virus (FIV) is similar to that of HIV-1. Here we concentrate on the characteristics specific to the replication of this virus.

Unlike HIV-1, FIV contains not nine but three known accessory proteins encoded by *vif, ORFA/2,* and *rev.* The Vif protein seems to be important for efficient infectivity *(23)*. Tat-like, encoded by *ORFA/2,* is a weak transcriptional activator. FIV, unlike HIV-1, lacks an apparent TAR element *(24)*. Rev, as in HIV-1, binds to the RRE present in the coding region of Env, allowing for the nucleocytoplasmic transport of partially spliced and nonspliced mRNA *(25)*. Nonspliced RNA gives rise to the Gag and Gag-Pol polyproteins, the last generated by translation frameshifting *(26)*. Gag is also proteoliticaly cleaved into MA, CA, and NC, but Pol gives rise not only to PR, RT (including RNase H), and IN, but also to dUTPase *(27)*, circumventing the need for an "adaptor," like Vpr, which captures a cellular dUTPase in the case of HIV-1.

FIV is able to infect nondividing cells, possibly owing to the presence of karyophilic motifs in MA and IN, although these are less defined than are the nuclear-localization signals present in HIV-1 *(28,29)*. The other sequence involved in nuclear localization, namely the cPPT, in contrast to HIV-1, seems to differ substantially from its 3' counterpart *(30)*.

For binding, FIV utilizes mainly the CXCR4 receptor, and not CD4 as does HIV-1 *(31)*. CCR5 has also been implicated in FIV binding *(32)*. The main steps in the life cycle of FIV are largely similar to other lentiviruses, including reverse transcription of the genome, formation of a preintegration complex including MA and IN, and finally integration of the provirus into the host chromosome.

1.2. Lentiviral Vector Development

1.2.1. General

Lentiviral vectors are based on the MMLV prototype. The genetic information of the latter is divided into three independent vectors, the first carrying the *gag* and *pol* genes; the second carrying *env*; and the third, the transfer vector, carrying all of the necessary *cis*-acting sequences of the virus. Also, the transfer vector often carries a marker, alone or in addition to a gene of interest.

The presence of multiple regulatory genes in lentiviral vector systems differentiates them from MMLV-based vectors. These accessory genes for the most part have been deleted in recent generations of these systems. The genetic information in these systems is divided into at least three vectors, and in some cases even four (as an additional safety step). The virions can be pseudotyped with several envelopes, notably the amphotropic envelope of MMLV *(33,34)*. Nevertheless, the vesicular stomatitis virus envelope gene, *VSV-G*, is the most widely used. VSV-G utilizes a cell-membrane phospholipid as its receptor, granting it a very broad cell tropism *(35–37)*. Moreover, it is stable under strong centrifugal force, allowing concentration of the virus if needed.

1.2.2. HIV-1-Based Vectors

The first HIV-1-based vectors were intended for research purposes only. In those vectors, *env* was replaced either by a nonendogenous *env* gene to study the tropism of the virus, or by a transgene of interest to study expression *(38,39)*. The first HIV-based vector system for delivering genes was a three-vector system. The structural proteins of the virus were placed on two plasmids, whereas the *cis*-acting sequences and the neomycin gene expressed from the MMLV LTR were on a third plasmid *(40)*. This three-vector system is considered the first generation of vectors. The packaging vector included the entire genome but without the envelope, driven by the promoter of the human cytomegalovirus (pCMV) instead of the 5′ LTR, and the polyA tail of the insulin gene in place of the 3′ LTR. The second plasmid, the envelope plasmid, carried the env gene, generally *VSV-G (1)*. The third plasmid, the transfer vector, included all of the *cis*-acting sequences of the virus including the LTRs, psi (ψ), RRE, and the primer binding site, as well as sequences needed for transcription, packaging, and integration of the genome. This vector contained a transgene expressed from an internal promoter. (See **Fig. 2** for a representation of the packaging vectors and **Fig. 3** for the transfer vectors.)

In the next generation of vectors, the accessory genes of the virus were deleted in order to further diminish the appearance of RCRs. *Vif, Vpr, Vpu,* and *Nef* were deleted, with no detrimental effects on the infectivity of the virion produced *(41–43)*.

The third generation of vectors was constructed to circumvent the need for the Tat and Rev proteins. The need for Tat was overcome by substituting the U3 region of the 5′ LTR with pCMV or the Rous sarcoma virus (RSV) LTR as shown in **Fig. 2** *(43,44)*. The need for the Rev protein and its responsive element, RRE, was circumvented by a codon-optimized gag-pol gene that gives rise to a high level of gag-pol mRNA *(45)*.

Incorporation of *cis*-acting elements that may increase both the viral titer and levels of gene expression are under continuous investigation. These include the cPPT as we have seen, but also the matrix attachment region, MAR, from immunoglobulin-kappa (Igκ) *(46)*. More recent generations of vectors include the Woodchuck hepatitis virus posttranscriptional regulatory element (WPRE), which promotes efficient polyadenylation of nascent RNA and mRNA stabilization *(47)*. For many individuals, the current vector designs are adequate for most gene transfer purposes.

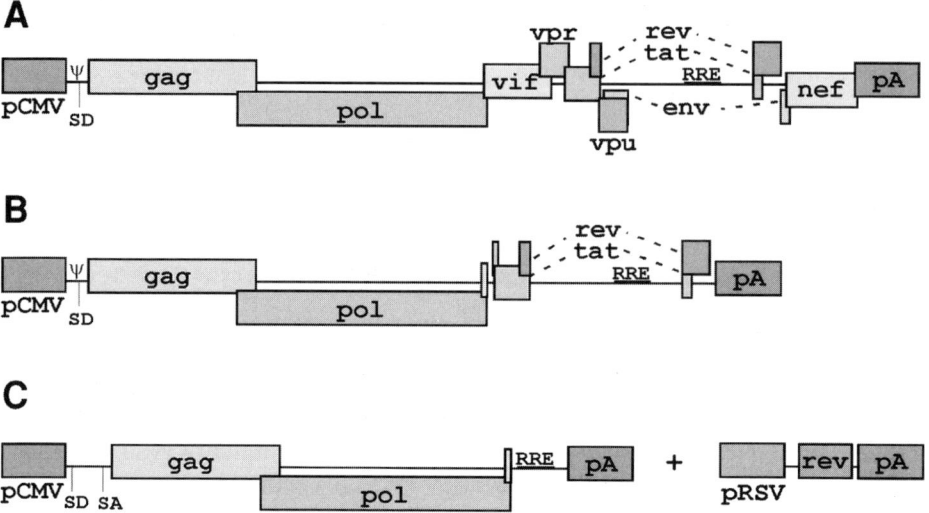

Fig. 2. Representation of the most widely used HIV-1-based packaging vectors.
(**A**) First generation. (**B**) Second generation. (**C**) Third generation. pA, polyadenylation
signal; SA, splice acceptor; pRSV, Rous sarcoma virus promoter. Refer to the text for
other abbreviations.

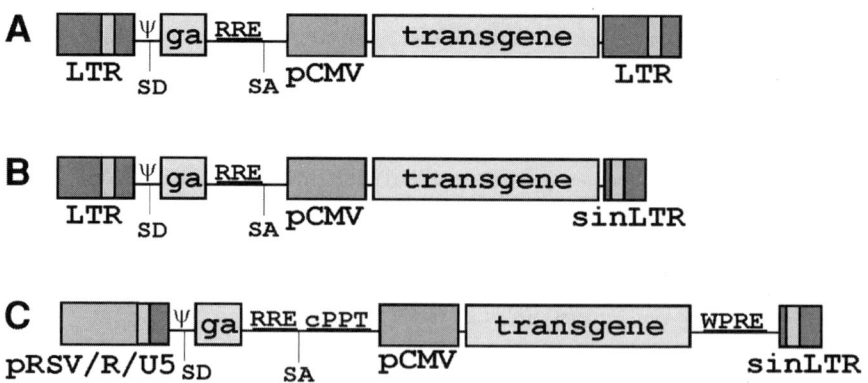

Fig. 3. Representation of the most widely used HIV-1-based transfer vectors. (**A**)
Second generation. (**B**) Third generation/self-inactivating vectors. (**C**) Last improved
version including cPPT and WPRE. ga, 5′ end of the gag gene; SA, splice acceptor;
sinLTR, self-inactivating LTR (i.e., most of the U3 region has been deleted); pRSV/
R/U5, a hybrid promoter with the U3 region of HIV-1 substituted with the Rous sarcoma virus promoter. Refer to the text for other abbreviations.

SIN vectors have been developed based on the third generation of plasmids. Deleting up to 400 bp of the U3 region of the 3′ LTR eliminates possible interference effects of the viral promoter in the host genome *(2,48)*. Moreover, this deletion facilitates the introduction of on/off activation systems or tissue-specific promoters.

1.2.3. FIV-Based Vectors

FIV was the first nonprimate lentivirus used to develop gene-transfer vectors systems *(26,28,49)*. The approach to developing vectors based on FIV has been similar to that for HIV-1 vectors. Again, there are three major plasmids: the packaging vector, the transfer vector, and the envelope vector.

Due to the very low transactional activity of the FIV LTR in human cells, the U3 region has been substituted with a pCMV. An internal promoter within the transfer vector is included to drive the expression of delivered genes. As with the HIV-1 based vectors, the packaging signal includes sequences from the 3′ end of the 5′ LTR and the 5′ end of *gag*. Again, the second and third generations of the vectors remove *env* and the regulatory genes of the virus, *vif, ORFA/2*, and *rev*. The deletion of Rev is overcome by the introduction of a constitutive transport element (CTE) in the packaging and transfer vectors. Additionally, *VSV-G* is supplied in *trans*. (*See* **Fig. 4** for a representation of the FIV-based vector systems.)

The latest version of the vectors is a SIN vector that appears to be more efficient than its wild-type 3′ LTR counterpart (M. Curran, unpublished data). The efficiency of infection with current FIV-based vectors is about 1–5 × 10⁶ infectious units (IU)/mL, a value approaching the efficiency of the HIV-1 based vectors. Additional vectors that include WPRE sequence and cPPT or cPPT-like sequences are under investigation (M. Curran, in preparation).

1.3. Safety

In order to diminish the possibility of RCR generation, the genetic information of these viruses has been separated onto more than one vector. In general, the sequences acting *in trans*, mainly the genes coding for the viral proteins, are placed in two separate vectors (*gag-pol* on one vector and *env* on a second). Other sequences, which are present in *cis* for virus encapsidation, reverse transcription, and integration, are placed on yet a third vector. The rationale behind this approach is to diminish as much as possible the chance of productive recombination events between sequences. Minimizing homologous sequences between different vectors also reduces the risk of generating RCRs, with some laboratories taking steps to even recode the codons that encode the entire gag-pol gene.

Because of safety concerns over the possibility of RCR production with HIV-based vectors, SIN vectors have also been developed *(2,4,48,50,51)*. These vec-

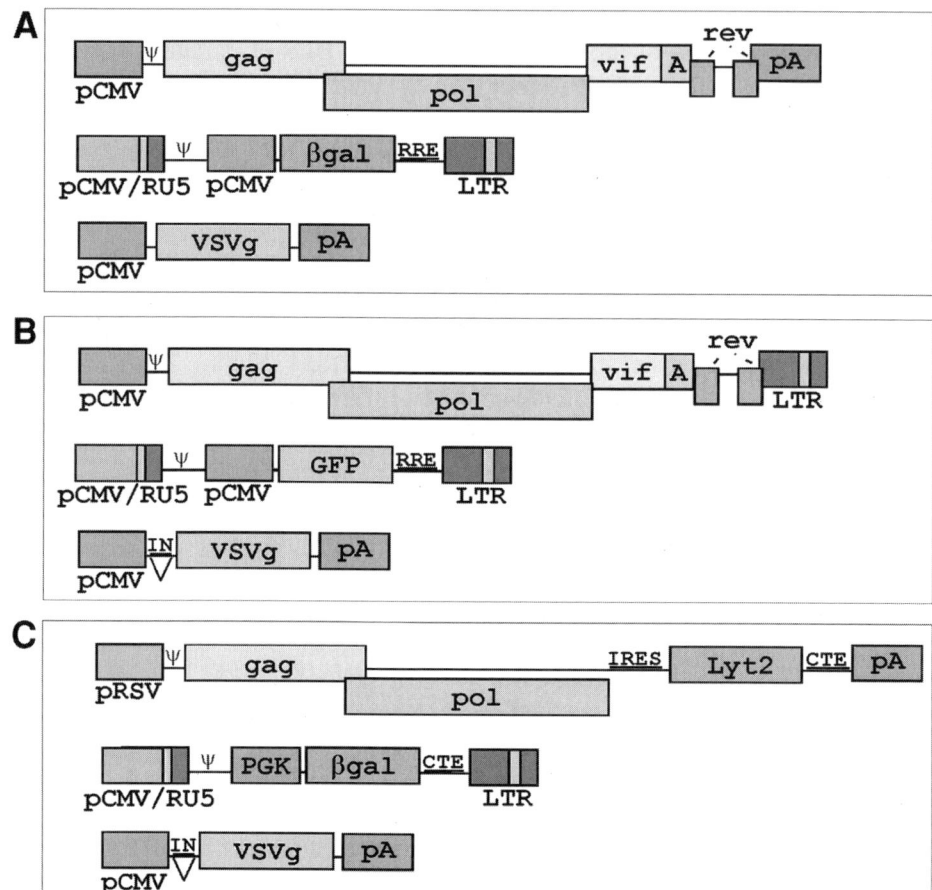

Fig. 4. Representation of the most widely used FIV-based vector systems, including the packaging, transfer and envelope vectors. (**A**) Second generation. This is the version used by Johnston et al. *(28)*. In Poeschla et al. *(26)*, the RRE is upstream of the pCMV. (**B**) Second generation based on Curran et al. *(49)*. (**C**) Third generation. The packaging vector may or not include an internal ribosome entry site (IRES) and a marker gene like Lyt2. pA, polyadenylation sequence, pCMV/RU5, a hybrid promoter with the U3 region of HIV-1 substituted with the cytomegalovirus immediate early promoter, β-gal, *E. coli* β-galactosidase gene; GFP, green fluorescent protein gene; IN, intron sequence for higher gene expression; PGK, human phosphoglycerate kinase promoter; VSVg, the vesicular stomatitis virus envelope gene. Refer to the text for other abbreviations.

tors have been divested of most of the 3' LTR U3 region bearing the major transcriptional functions of the HIV genome. The 3' LTR becomes the 5' LTR during the process of reverse transcription in the target cells. Consequently, the viral DNA sequence inserted into the target genome is promoter-less. Thus, even a full recombinant would be very unlikely to transcribe full-length viral RNA.

Still, some in vivo concerns remain regarding the use of HIV-based vectors in humans. A theoretical, if remote, possibility for mobilization of these vectors by the wild-type virus may exist, especially in patients with AIDS. Nonprimate lentiviral vectors like those based on FIV do not have the negative connotation of their HIV-based counterparts, as patients may be reluctant to be treated with HIV-based gene therapy. Vectors derived from viruses that are nonpathogenic in humans may be more compelling for clinical use for that reason, but some have made the argument that such vectors might potentially introduce nonhuman pathogen sequences into humans. Although FIV has been shown to use at least one receptor existing in humans, the CXCR4 molecule, it does not have any observable pathogenic effect. Therefore, the use of FIV-based vectors for human gene therapy may have a promising future, but the area requires further consideration and discussion until a consensus can be reached.

The user is strongly cautioned to consider the nature of the virions being produced. This includes full compliance with all relevant institutional safety rules, as well as local and governmental regulatory agencies and guidelines. Of particular concern would be expression of known oncogenes or toxins, as the worker might potentially come in contact with such infectious materials. Appropriate containment guidelines should be followed strictly.

2. Materials
2.1. Cell Line and Culture Medium

1. Human 293T cells. 293T are a human embryonic kidney cell line originally transformed with sheared type 5 Adenovirus DNA and subsequently with SV40 large T-antigen mutant (tsA 1609). Transfected with the *Escherichia coli* neo^r gene, the cells were selected for resistance to neomycin (ATCC, Manassas, VA, Cat. no. CCRL-1573).
2. Six-cm diameter tissue-culture plates (other sizes can be used).
3. Phosphate-buffered saline (PBS): 1.0% NaCl, 0.025% KCl, 0.14% Na_2HPO_4, 0.025% KH_2PO_4 (all w/v) at pH 7.2.
4. Dulbecco's modified Eagle's medium (DMEM), 0.1-µm filtered, with 4500 mg/L D-glucose, 2 mM L-glutamine, 110 mg/L sodium pyruvate and pyridoxine hydrochloride (Invitrogen/GIBCO, Carlsbad, CA, Cat. no. 11995-065).
5. Fetal bovine serum (FBS).

6. Antibiotics: penicillin (10,000 U/mL) and streptomycin (10,000 µg/mL).
7. L-Glutamine (200 m*M*) (Invitrogen/GIBCO, Cat. no.12381-018).
8. Trypsin-EDTA: 0.05% Trypsin, 0.53 m*M* EDTA · 4Na.

2.2. Transfection Reagents

2.2.1. Calcium Phosphate

1. Chloroquine (50 m*M* in distilled water) (Sigma, St. Louis, MO, Cat. no. C-6628).
2. CaCl$_2$ (2 *M* in distilled water, sterilized by filtration through a 0.2-µm filter) (Mallinckrodt, St. Louis, MO, Cat. no. 4160).
3. HEPES sodium salt (Sigma, Cat. no. H-7006).
4. Na$_2$HPO$_4$ stock solution: 5.25g dissolved in 500 mL of distilled water, approx 74 m*M* (Fisher, Pittsburgh, PA, Cat. no. S374-500).
5. 2X HBS: 8 g of NaCl, 6.5 g of HEPES, 10 mL of Na$_2$HPO$_4$ stock solution; bring volume to 500 mL. The pH should be exactly 7.0, buffered with NaOH or HCl.

2.2.2. FuGENE 6

1. FuGENE 6 (Roche Corporation, Indianapolis, IN, Cat. no. 1 814 443).
2. DMEM or Opti-MEM (Invitrogen/GIBCO, Cat. no. 31985-070).

2.3. DNA Vectors

Purify DNA plasmids (packaging vector, transfer vector, and envelope vector) in large scale in order to minimize endotoxin contamination (*see* **Notes 1** and **2**).

2.4. Collection of Virus Supernatant

1. Three- or 5-mL syringes.
2. 0.45-µm low-protein-binding syringe filters (Pall Corp., East Hills, NY, Cat. no. 4184).
3. Fifteen-mL polypropylene tubes.

2.5. Concentration of Virus

1. Fifty-mL Oak Ridge polypropylene copolymer tubes (Nalgene Nunc, Rochester, NY, Cat. no. 3139-0050).
2. Tris-NaCl-EDTA buffer (TNE): 50 m*M* Tris-HCl, pH 7.8, 130 m*M* NaCl, and 1 m*M* EDTA. Autoclave and store at 4°C.
3. 0.45-µm low-protein-binding 115-mL filters (Nalgene Nunc, Cat. no. 245-0045).

2.6. Infection Reagents

1. Polybrene (hexadimethrine bromide) (1000X stock: 5-mg/mL in PBS) (Sigma, Cat. no. H-9268).
2. RetroNectin (fibronectin) (PanVera, Madison, WI, Cat. no. TAK T100) or RetroNectin-coated plates (PanVera, Cat. no. TAK T110).

3. Methods

3.1. Preparation of the Virus Producing Cells

1. Plate $1.5–2.0 \times 10^6$ 293T cells in 3 mL of DMEM media supplemented with 10% FBS and 1% penicillin/streptomycin per 6-cm plate for calcium phosphate transfection, or $1.0–1.5 \times 10^6$ 293T cells in 3 mL of media per 6-cm plate for transfection with FuGENE 6. If needed, other sizes of plates can be used, but keep the number of cells proportional to the surface area of the plate (*see* **Note 3**).
2. Incubate the cells for 24 h at 37°C in a humidified incubator containing 5% CO_2.

3.2. Transfection

3.2.1. DNA

The amount of DNA used for transfection is subject to the experimental goals, especially regarding the ratio between the different vectors. Here we give only the amounts that have worked best in our hands. Those amounts are always based on a 6-cm diameter plate. When preparing the DNA for calcium phosphate precipitation (**Subheading 3.2.2.**), mix the plasmids in a 15-mL polypropylene tube. When preparing the plasmids for transfection using FuGENE 6 (**Subheading 3.2.3.**), mix the plasmids in a microfuge tube.

1. For the HIV-1-based system, for each transfection, combine 2 µg of packaging vector, 3 µg of transfer vector, and 3 µg of envelope vector.
2. For the FIV-based system, for each transfection, combine 5 µg of packaging vector, 3µg of transfer vector, and 2 µg of envelope vector (*see* **Note 4**).

3.2.2. Transfection by Calcium Phosphate

Cells should be 40–60% confluent at the time of transfection.

1. Add 2 µL of 50 m*M* chloroquine stock to the cells plated in **Subheading 3.1.** (see **Note 5**).
2. To the plasmid mixture prepared in **Subheading 3.2.1.**, add 61 µL of 2 *M* $CaCl_2$.

3. Add sterile distilled water at room temperature to bring the volume to 500 µL. Briefly centrifuge the tube to bring all of the liquid to the bottom.
4. Add 500 µL of 2X HBS and immediately mix with an automatic pipettor, bubbling vigorously for 2–10 s (*see* **Note 6**).
5. Add the DNA-HBS mixture dropwise to the cells; swirl the plate gently to ensure homogenous mixing (*see* **Note 7**).
6. Transfer the plate to a 37°C incubator with a humidified atmosphere containing 5% CO_2 for 8 h.
7. Change the medium. Add 2-3 mL of DMEM supplemented with 10% FBS.
8. Change the medium again after 16–24 h and incubate at 37°C in a humidified incubator containing 5% CO_2 for 24 h.
9. Transfer the plate to a 32°C incubator with a humidified atmosphere containing 5% CO_2 for 8 h prior to harvesting the virus (around 40 h post-transfection) (*see* **Note 8**).

3.2.3. Transfection by FuGENE 6

Cells should be 40–60% confluent at the time of transfection.

1. Add 266 µL of DMEM (serum free) to a microfuge tube.
2. Add 10 µL of FuGENE 6 to the tube. The reagent should be at room temperature (*see* **Note 9**).
3. To the plasmid mixture prepared in **Subheading 3.2.1.**, add the FuGENE 6-containing medium in a dropwise manner and mix carefully.
4. Incubate for 15 min at room temperature.
5. Add the mixture dropwise to the cells. Swirl the plate gently to ensure homogenous mixing.
6. Incubate the cells at 37°C for 24 h.
7. Change the media. Add 2–3 mL of fresh DMEM supplemented with 10% FBS (*see* **Note 10**).
8. Transfer the plate to a 32°C incubator with a humidified atmosphere containing 5% CO_2 for 8 h prior to harvesting the virus (around 40 h post-transfection) (*see* **Note 8**).

3.3. Collection of Virus Supernatant

1. Collect the medium containing the virus from the cells transfected in **Subheading 3.2.2.** or **3.2.3.** 48 h posttransfection. If desired, more medium can be added to the plate for repetitive virus supernatant collection. This can be done at least twice within the next 24-h time period.
2. Transfer the supernatant into a 15-mL tube and spin at 300–500g for 5 min to remove the debris (*see* **Note 11**).
3. Carefully collect the supernatant avoiding the pellet. This is the nonconcentrated viral supernatant.

4. Filter the supernatant through a 0.45-μm low-protein-binding syringe filter. This virus stock can be used for infection. If it will be used within 3 d, store it at 4°C. Otherwise, store the stock at −80°C (*see* **Note 12**).

3.4. Large-Scale Virus Production

This procedure may be used when very high titers are needed. Nonconcentrated viral supernatant may not be efficient for infecting some cell lines that are difficult to infect.

1. Inoculate 12.5×10^6 293T cells per 15-cm dish in 18 mL of DMEM supplemented with 5% FBS (*see* **Note 13**). Prepare at least two plates per sample. Incubate the cells overnight at 37°C in a humidified incubator containing 5% CO_2.
2. Transfect cells using FuGENE 6 as described in **Subheading 3.2.3.**, **steps 1–5**, increasing all volumes, including plasmid DNA, six- to seven-fold. Incubate cells at 37°C in a humidified incubator containing 5% CO_2.
3. Forty-eight h after transfection, collect the supernatants from both plates into one sterile 50-mL polypropylene tube (*see* **Note 14**).
4. Transfer the filtered supernatant into an autoclaved 50-mL polycarbonate Oak Ridge tube and close the tube securely.
5. Centrifuge at 50,000*g* for 1.5 h.
6. After centrifugation a small white/translucent pellet should be visible. Decant supernatant carefully.
7. Resuspend the pellet in 300–1000 μL of TNE. Wash the bottom walls of the tube to recover the entire pellet (*see* **Notes 15** and **16**).
8. Store the virus stock at 4°C for no more than 3 d or aliquot and freeze in cryotubes at −80°C.

3.5. Infection of Target Cells

A target cell is any cell that might eventually be infected by the viral particle. Because the systems presented here, both HIV-1-based and FIV-based, are pseudotyped with the VSV-G envelope, the tropism of the virions are extremely broad. Because of the nature of these vectors, the marker carried by these vectors will be integrated into nondividing cells. Different cell types, of course, will be infected with different efficiencies.

3.5.1. Basic Infection of Adherent Cells

1. Plate the target cells so that they will be 40–70% confluent at the time of infection (*see* **Note 17**).
2. Carefully aspirate the medium without disturbing the attached cells.
3. Add the desired amount of virus stock, generally 400–1000 μL (*see* **Note 18**).

4. Add medium used to grow the target cells, up to a volume of 3 mL for a 6-cm plate and 10 mL for a 10-cm plate (*see* **Note 19**).
5. Add polybrene to a final concentration of 5 µg/mL.
6. Incubate the cells at 32°C in a humidified atmosphere containing 5% CO_2 for 8–16 h.
7. Change the medium and transfer the plates to a 37°C humidified incubator containing 5% CO_2 for at least 24–36 h (*see* **Note 20**).
8. At this point the cells should be infected and actively expressing the transgene. If a reporter gene was present on the transfer vector, assay its expression by fluorescence-activated cell sorting (FACS) or another suitable method to determine the percentage of cells successfully infected, and/or to separate the infected and noninfected cells.

3.5.2. Basic Infection of Nonadherent Cells

1. Spin down desired number of cells at 300–500g for 5 min (*see* **Notes 17** and **21**).
2. Aspirate the medium and add the desired volume of viral supernatant, generally 400–1000 µL (*see* **Note 18**).
3. Proceed from **step 4** as for adherent cells (*see* **Subheading 3.5.1.**).

3.5.3. Spin Infection

Spin infection is used to infect nonadherent cells. Centrifugation seems to increase contact between viral particles and cells. As a consequence, the infection efficiency increases. This method also increases infection efficiency of adherent cells, although perhaps to a lesser degree than suspension cells.

1. Plate cells in multiwell plates (not in individual dishes) in order to spin them easily. Choose the appropriate multiwell plate for the experiment (e.g., 6-well to 96-well plates). The plating density of the cells will vary depending on the cell size and virus concentration (i.e., the more concentrated the virus solution, the more dense the cells can be). It is generally best to plate the cells densely enough that they are viable, but not so densely that they aggregate and reduce the cell surface area available for infection. When infecting adherent cells, inoculate the cells, allow the cells to attach overnight, then aspirate the medium and proceed with the experiment.
2. Infect the plates as described in **Subheading 3.5.1.**, **steps 3–5**: add virus and polybrene. Seal the plates with parafilm or tape (*see* **Note 22**).
3. Place the plates on plate carriers in a swinging bucket rotor centrifuge and centrifuge at 1500g for 90 min at 32°C.

4. Remove the seal and transfer the plates to a 32°C humidified incubator containing 5% CO_2 for 8–16 h (*see* **Note 20**).
5. Change the medium and transfer the plates to a 37°C incubator with a humidified atmosphere containing 5% CO_2 for at least 24–36 h.
6. At this point the cells should be infected and actively expressing the transgene. If a reporter gene was present on the transfer vector, assay its expression by FACS or another suitable method to determine the percentage of cells successfully infected, and/or to separate the infected and noninfected cells.

3.5.4. Infection With RetroNectin

This method may be used for cells that are difficult to infect or need special treatment for growth. Moreover, some cells are very sensitive to polycationic reagents and RetroNectin (fibronectin) will overcome the need for polybrene.

1. Coat flat-bottom, multiwell, nontissue-culture treated plates with Retro-Nectin at 30 µg/mL (*see* **Note 23**).
2. Incubate the plates at room temperature for 2 h.
3. Remove the RetroNectin solution from the plates (it can be reused up to five times). Add medium for target cells and incubate for 30 min at 37°C.
4. Add cells and viral supernatant (*see* **Notes 17, 19** and **22**).
5. Add polybrene to a final concentration of 5 µg/mL (*see* **Note 24**).
6. Spin at 1500g for 90 min (*see* **Subheading 3.5.3., step 3**) (*see* **Note 25**).
7. Incubate the cells at 32°C in a humidified incubator containing 5% CO_2 for 8–16 h.
8. Change the medium and transfer the cells to 37°C in a humidified incubator containing 5% CO_2 for at least 24–36 h (*see* **Note 20**).
9. At this point the cells should be infected and actively expressing the transgene. If a reporter gene was present on the transfer vector, assay its expression by FACS or another suitable method to determine the percentage of cells successfully infected, and/or to separate the infected and noninfected cells.

4. Notes

1. DNA could be phenol extracted for further purification but there is no obligation to do so. Maxi-prep kits for DNA extraction generally produce good yields and sufficient purification. Try to avoid DNA minipreps for small experiments since the DNA preparation may be less pure and the virus production may be inefficient.

2. For the HIV-1-based system, a plasmid carrying regulatory proteins like Vpr or Rev might be included for specific research goals or needs. Depending on the cell type, some regulatory proteins may influence viral infection efficiency.

3. The quality of the cells is critical for the production of virus. The cells should look healthy and be 40–60% confluent at the time of transfection. Lower, and especially higher, concentrations of cells will drastically decrease the virus production. Plating cells in a small volume until they attach can minimize convective currents that might give rise to uneven distribution across the plate. Additional medium can be added carefully after the cells have attached (usually 3–5 h).

4. If other vectors are added such as plasmids carrying regulatory proteins, the amount should be investigated. For *vpr* or *rev* plasmids, for example, we recommend 1.5 μg of each.

5. With the addition of 1 mL of DNA mixture, the final concentration of chloroquine will be 25 μ*M*. Chloroquine is toxic to the cells; do not leave the cells in the chloroquine containing medium for more than 8 h. Alternatively, add less chloroquine and incubate the cells longer.

6. Bubbling time is absolutely critical and will be different for every HBS batch. This parameter should be optimized for every new batch.

7. A very fine and delicate granular black precipitate should appear on the top of the cells. This precipitate will be more visible after several hours. A coarse precipitate with large particles is a sign of inefficient transfection.

8. The temperature of incubation is actually a compromise between the best temperature for the cells (37°C) and the best temperature for the stability of the recombinant virus (32°C). Incubation at 32°C rather than 37°C is subsequently optional.

9. We have found that 1 μL of FuGENE 6 per μg of DNA is generally optimal. Add the reagent directly to the medium without touching the walls of the tube.

10. FuGENE 6 is not toxic to the cells. Nevertheless, it is advisable to change the medium 24 h after transfection.

11. This centrifugation step removes cellular debris as well as producer cells (293T cells) that could be carried over during viral infection. If it is crucial to avoid possible cellular contamination when infecting target cells, the viral supernatant should be filtered.

12. Freezing the virus stock decreases the infectivity by 50%. Avoid multiple rounds of freeze-thaw cycles by aliquoting the virus. At 4°C, infectivity decreases by 50% every 36–48 h.

13. In order to decrease the amount of serum proteins pelleted during virus precipitation, it is recommended to grow virus-producing cells in medium containing less FBS than normal growth medium.

14. Filter the supernatant through a 115 mL, 0.45-μm, low-protein-binding filter. Filtration will avoid possible toxic effects of VSV-G fragments on the cells, but will decrease virus titer. This step is optional.

15. The volume of TNE will determine the concentration of the virus. Some cells do not tolerate the toxic effect of VSV-G owing to its fusogenic nature. The final TNE volume should be determined experimentally.

16. This step may be repeated in order to increase the viral yield. The second resuspension can be done in medium used to grow the cells rather than in TNE.

17. The number of cells as well as the volume of viral supernatant should be determined experimentally for each cell type and virus stock. In general, infecting highly confluent cells will decrease the efficiency of infection. As a starting point, for a 6 cm plate, inoculate $0.5–1.0 \times 10^6$ adherent cells 12–24 h prior to infection. For nonadherent cells, inoculate 1.0×10^6 cells just before infection.

18. The amount of viral supernatant required will depend on the virus titer, the quality of the virus stock, and target cell being infected. If the virus has been concentrated, a lower volume will be needed.

19. Adding medium is a compromise between giving to the cells sufficient medium to grow in and diluting the virus, which decreases infection efficiency.

20. It is important to change the medium because the polycationic molecule polybrene, although known to enhance retroviral transduction, is toxic to cells. Some cells are more sensitive to polybrene than are others, therefore, the optimal time to incubate the cells with polybrene should be empirically determined for each cell line.

21. Nonadherent cells can be infected by the basic procedure described here, however, they are more efficiently infected by spin infection (**Subheading 3.5.3.**). It is also possible to spin infect adherent cells.

22. The final volume within each well should be sufficient to cover the cells (so that they do not dry out) and small enough to avoid spilling over during centrifugation. We recommend 50–60% of normal volume.

23. **Steps 1–3** can be skipped if you use commercially available RetroNectin-coated plates.

24. There is no experimental evidence of the additive effect of polybrene with recombinant fibronectin (RetroNectin) on virus infection efficiency. This

step is optional and it is not recommended with cells sensitive to polycationic reagents.

25. There is no experimental evidence for the additive effect of centrifugation with recombinant RetroNectin on virus-infection efficiency. Skip this step if cells are sensitive to centrifugation.

References

1. Naldini, L., Blomer, U., Gallay, P., Ory, D., Mulligan, R., Gage, F. H., Verma, I.M., and Trono, D. (1996) In vivo gene delivery and stable transduction of nondividing cells by a lentiviral vector. *Science* **272,** 263–267.

2. Miyoshi, H., Blomer, U., Takahashi, M., Gage, F. H., and Verma, I. M. (1998) Development of a self-inactivating lentivirus vector. *J. Virol.* **72,** 8150–8157.

3. Wu, X., Wakefield, J. K., Liu, H., Xiao, H., Kralovics, R., Prchal, J. T., and Kappes, J. C. (2000) Development of a novel trans-lentiviral vector that affords predictable safety. *Mol. Ther.* **2,** 47–55.

4. Mangeot, P. E., Negre, D., Dubois, B., Winter, A. J., Leissner, P., Mehtali, M., et al. (2000) Development of minimal lentivirus vectors derived from simian immunodeficiency virus (SIVmac251) and their use for gene transfer into human dendritic cells. *J. Virol.* **74,** 8307–8315.

5. Emerman, M. and Malim, M. H. (1998) HIV-1 regulatory / accessory genes: keys to unraveling viral and host cell biology. *Science* **280,** 1880–1884.

6. Neville, M., Stutz, F., Lee, L., Davis, L. I., and Rosbash, M. (1997) The importin-beta family member Crm1p bridges the interaction between Rev and the nuclear pore complex during nuclear export. *Curr. Biol.* **7,** 767–775.

7. Malim, M. H., Hauber, J., Le, S. Y., Maizel, J. V., and Cullen, B. R. (1989) The HIV-1 rev trans-activator acts through a structured target sequence to activate nuclear export of unspliced viral mRNA. *Nature* **338,** 254–257.

8. Wyatt, R. and Sodroski, J. (1998) The HIV-1 envelope glycoproteins: fusogens, antigens, and immunogens. *Science* **280,** 1884–1888.

9. Clever, J. L. and Parslow, T. G. (1997) Mutant human immunodeficiency virus type 1 genomes with defects in RNA dimerization or encapsidation. *J. Virol.* **71,** 3407–3414.

10. Deng, H., Liu, R., Ellmeier, W., Choe, S., Unutmaz, D., Burkhart, M., et al. (1996) Identification of a major co-receptor for primary isolates of HIV-1. *Nature* **381,** 661–666.

11. Feng, Y., Broder, C. C., Kennedy, P. E., and Berger, E. A. (1996) HIV-1 entry cofactor: functional cDNA cloning of a seven-transmembrane, G protein-coupled receptor. *Science* **272,** 872–877.

12. Binley, J. and Moore, J. P. (1997) HIV-cell fusion. The viral mousetrap. *Nature* **387,** 346–348.

13. Baltimore, D. (1970) RNA-dependent DNA polymerase in virions of RNA tumour viruses. *Nature* **226,** 1209–1211.

14. Temin, H. (1970) RNA-dependent DNA polymerase in virions of Rous sarcoma virus. *Nature* **226,** 1211–1213.

15. Rhim, H., Park, J., and Morrow, C. D. (1991) Deletions in the tRNA(Lys) primer-binding site of human immunodeficiency virus type 1 identify essential regions for reverse transcription. *J. Virol.* **65,** 4555–4564.

16. Mizrahi, V. (1989) Analysis of the ribonuclease H activity of HIV-1 reverse transcriptase using RNA.DNA hybrid substrates derived from the gag region of HIV-1. *Biochemistry* **28,** 9088–9094.

17. Charneau, P., Mirambeau, G., Roux, P., Paulous, S., Buc, H., and Clavel, F. (1994) HIV-1 reverse transcription. A termination step at the center of the genome. *J. Mol. Biol.* **241,** 651–662.

18. Zennou, V., Petit, C., Guetard, D., Nerhbass, U., Montagnier, L., and Charneau, P. (2000) HIV-1 genome nuclear import is mediated by a central DNA flap. *Cell* **101,** 173–185.

19. Bouhamdan, M., Benichou, S., Rey, F., Navarro, J. M., Agostini, I., Spire, B., et al. (1996) Human immunodeficiency virus type 1 Vpr protein binds to the uracil DNA glycosylase DNA repair enzyme. *J. Virol.* **70,** 697–704.

20. Popov, S., Rexach, M., Ratner, L., Blobel, G., and Bukrinsky, M. (1998) Viral protein R regulates docking of the HIV-1 preintegration complex to the nuclear pore complex. *J. Biol. Chem.* **273,** 13347–13352.

21. Tsurutani, N., Kubo, M., Maeda, Y., Ohashi, T., Yamamoto, N., Kannagi, M., and Masuda, T. (2000) Identification of critical amino acid residues in human immunodeficiency virus type 1 IN required for efficient proviral DNA formation at steps prior to integration in dividing and nondividing cells. *J. Virol.* **74,** 4795–4806.

22. Haffar, O. K., Popov, S., Dubrovsky, L., Agostini, H., Tang, T., Pushkarsky, I., et al. (2000) Two nuclear localization signals in the HIV-1 matrix protein regulate nuclear import of the HIV-1 pre-integration complex. *J. Mol. Biol.* **299,** 359–368.

23. Tomonaga, K., Norimine, J., Shin, Y. S., Fukasawa, M., Miyazawa, T., Adachi, A., et al. (1992) Identification of a feline immunodeficiency virus gene which is essential for cell-free virus infectivity. *J. Virol.* **66,** 6181–6185.

24. de Parseval, A. and Elder, J. H. (1999) Demonstration that orf2 encodes the feline immunodeficiency virus transactivating (Tat) protein and characterization of a unique gene product with partial rev activity. *J. Virol.* **73,** 608–617.

25. Phillips, T. R., Lamont, C., Konings, D. A., Shacklett, B. L., Hamson, C. A., Luciw, P. A., and Elder, J. H. (1992) Identification of the Rev transactivation and Rev-responsive elements of feline immunodeficiency virus. *J. Virol.* **66,** 5964–5971.

26. Poeschla, E. M., I. F. Wong-Staa, and D. J. Looney. (1998) Efficient transduction of nondividing human cells by feline immunodeficiency virus lentiviral vectors. Nat. Med. **4,** 354–357.

27. Elder, J. H., Lerner, D. L., Hasselkus-Light, C. S., Fontenot, D. J., Hunter, E., Luciw, P. A., et al. (1992) Distinct subsets of retroviruses encode dUTPase. *J. Virol.* **66,** 1791–1794.

28. Johnston, J. C., Gasmi, M., Lim, L. E., Elder, J. H., Yee, J. K., Jolly, D. J., et al. (1999) Minimum requirements for efficient transduction of dividing and nondividing cells by feline immunodeficiency virus vectors. *J. Virol.* **73,** 4991–5000.

29. Mitrophanous, K., Yoon, S., Rohll, J., Patil, D., Wilkes, F., Kim, V., et al (1999) Stable gene transfer to the nervous system using a non-primate lentiviral vector. *Gene Ther.* **6,** 1808–1818.

30. Whitwam, T., Peretz, M., and Poeschla, E. (2001) Identification of a central DNA flap in feline immunodeficiency virus. *J. Virol.* **75,** 9407–9414.

31. Poeschla, E. M. and Looney, D. J. (1998) CXCR4 is required by a nonprimate lentivirus: heterologous expression of feline immunodeficiency virus in human, rodent, and feline cells. *J. Virol.* **72,** 6858–6866.

32. Kovacs, E. M., Baxter, G. D., and Robinson, W. F. (1999) Feline peripheral blood mononuclear cells express message for both CXC and CC type chemokine receptors. *Arch. Virol.* **144,** 273–285.

33. Chesebro, B., Wehrly, K., and Maury, W. (1990) Differential expression in human and mouse cells of human immunodeficiency virus pseudotyped by murine retroviruses. *J. Virol.* **64,** 4553–4557.

34. Stitz, J., Buchholz, C. J., Engelstadter, M., Uckert, W., Bloemer, U., Schmitt, I., and Cichutek, K. (2000) Lentiviral vectors pseudotyped with envelope glycoproteins derived from gibbon ape leukemia virus and murine leukemia virus 10A1. *Virology* **273,** 16–20.

35. Akkina, R. K., Walton, R. M., Chen, M. L., Li, Q. X., Planelles, V., and S. Chen, I. (1996) High-efficiency gene transfer into CD34+ cells with a human immunodeficiency virus type 1-based retroviral vector pseudotyped with vesicular stomatitis virus envelope glycoprotein G. *J. Virol.* **70,** 2581–2585.

36. Poeschla, E., Corbeau, P., and Wong-Staal, F. (1996) Development of HIV vectors for anti-HIV gene therapy. Proc. Natl. Acad. Sci. USA. **93,** 11395–11399.

37. Reiser, J. (2000) Production and concentration of pseudotyped HIV-1-based gene transfer vectors. *Gene Ther.* **7,** 910–913.

38. Page, K. A., Landau, N. R., and Littman, D. R. (1990) Construction and use of a human immunodeficiency virus vector for analysis of virus infectivity. *J. Virol.* **64,** 5270–5276.

39. Landau, N. R., Page, K. A., and Littman, D. R. (1991) Pseudotyping with human T-cell leukemia virus type I broadens the human immunodeficiency virus host range. *J. Virol.* **65,** 162–169.

40. Parolin, C., Dorfman, T., Palu, G., Gottlinger, H., and Sodroski, J. (1994) Analysis in human immunodeficiency virus type 1 vectors of cis-acting sequences that affect gene transfer into human lymphocytes. *J. Virol.* **68,** 3888–3895.

41. Zufferey, R., Nagy, D., Mandel, R. J., Naldini, L., and Trono, D. (1997) Multiply attenuated lentiviral vector achieves efficient gene delivery in vivo. *Nat. Biotechnol.* **15,** 871–875.

42. Kafri, T., Blomer, U., Peterson, D. A., Gage, F. H., and Verma, I. M. 1997 Sustained expression of genes delivered directly into liver and muscle by lentiviral vectors. *Nat. Genet.* **17,** 314–317.

43. Kim, V. N., Mitrophanous, K., Kingsman, S. M., and Kingsman, A. J. (1998) Minimal requirement for a lentivirus vector based on human immunodeficiency virus type 1. *J. Virol.* **72,** 811–816.

44. Dull, T., Zufferey, R., Kelly, M., Mandel, R. J., Nguyen, M., Trono, D., and Naldini, L. (1998) A third-generation lentivirus vector with a conditional packaging system. *J. Virol.* **72,** 8463–8471.

45. Kotsopoulou, E., Kim, V. N., Kingsman, A. J., Kingsman, S. M., and Mitrophanous, K. A. (2000) A Rev-independent human immunodeficiency virus type 1 (HIV-1)-based vector that exploits a codon-optimized HIV-1 gag-pol gene. *J. Virol.* **74,** 4839–4852.

46. Park, F. and Kay, M. A. (2001) Modified HIV-1 based lentiviral vectors have an effect on viral transduction efficiency and gene expression in vitro and in vivo. *Mol. Ther.* **4,** 164–173.

47. Zufferey, R., Donello, J. E., Trono, D., and Hope, T. J. (1999) Woodchuck hepatitis virus posttranscriptional regulatory element enhances expression of transgenes delivered by retroviral vectors. *J. Virol.* **73,** 2886–2892.

48. Zufferey, R., Dull, T., Mandel, R. J., Bukovsky, A., Quiroz, D., Naldini, L., and Trono, D. (1998) Self-inactivating lentivirus vector for safe and efficient in vivo gene delivery. *J. Virol.* **72,** 9873–9880.

49. Curran, M. A., Kaiser, S. M., Ahacoso, P. L., and Nolan, G. P. (2000) Efficient transduction of nondividing cells by optimized feline immunodeficiency virus vectors. *Mol. Ther.* **1,** 31–38.

50. Deglon, N., Tseng, J. L., Bensadoun, J. C., Zurn, A. D., Arsenijevic, Y., Pereira de Almeida, L., et al. (2000) Self-inactivating lentiviral vectors with enhanced transgene expression as potential gene transfer system in Parkinson's disease. *Hum. Gene Ther.* **1,** 179–190.

51. Xu, K., Ma, H., McCown, T. J., Verma, I. M., and Kafri, T. (2001) Generation of a stable cell line producing high-titer self-inactivating lentiviral vectors. *Mol. Ther.* **3,** 97–104.

27

Stable Gene Delivery to CNS Cells Using Lentiviral Vectors

Deborah J. Watson, Brian A. Karolewski, and John H. Wolfe

1. Introduction

Recombinant viral vectors have been used to study a variety of fundamental issues in developmental neurobiology, as well as pathogenesis and treatments for various neurodegenerative diseases. Lentiviral vectors are valuable tools for neurobiology research owing to their ability to transduce nondividing cells, such as neurons, and to introduce therapeutic or reporter genes into central nervous system (CNS) cells in vivo and in vitro.

1.1. Lentiviral Gene Delivery to CNS Cells

Lentiviruses such as HIV-1 are plus-stranded RNA retroviruses. Lentivirus preintegration complexes interact with the nuclear pore and undergo active transport into the nucleus of nondividing cells (reviewed by Fouchier and Malim [1]), where the proviral DNA is integrated into the genomic DNA of the host cell. This feature is the primary reason why lentiviruses are being developed as gene-transfer vectors for postmitotic cells in the CNS (2,3). Other lentiviral vectors have been developed using HIV-2 (4), FIV (5,6), EIAV (7,8), and combinations of SIV and HIV-1 (9).

Lentiviral vectors have a number of additional advantages for transduction of CNS cells. These vectors may be able to mediate long-term expression of therapeutic genes because the recombinant provirus is integrated into the host genome and can persist for the lifetime of the cell. The methods to produce replication-

From: *Methods in Molecular Biology, vol. 246:*
Gene Delivery to Mammalian Cells: Vol. 2: Viral Gene Transfer Techniques
Edited by: W. C. Heiser © Humana Press Inc., Totowa, NJ

defective vectors with recombinant genomes (*see* **Subheading 3.**) theoretically eliminate the possibility of transferring any HIV-1 viral genes to the target cell. Self-inactivating (SIN) vectors reduce the probability of oncogenesis by promoter insertion. In SIN vectors, viral promoter activity is deleted from the integrated provirus by deletions in the U3 region of the 3′ long terminal repeat (LTR) that are copied during reverse transcription to the 5′ LTR. Removal of viral promoter activity eliminates competition with the internal promoter in the expression cassette. In addition, in the integrated provirus, transcription is initiated downstream of the encapsidation (ψ) sequence, reducing the chance that a vector genome can be encapsidated *(10)*. Lentiviral vectors may have the capacity to carry as much as 8 kb of genetic material into the target cell (approx twice the capacity of adeno-associated virus [AAV] vectors; *[11]*), facilitating the transfer of larger expression cassettes. This capacity should allow for the generation of a wide variety of useful experimental vectors, with multiple expression cassettes and promoters, bicistronic or tricistronic cassettes linked by internal ribosomal entry sites, and regulatable or tissue-specific promoters. To increase gene delivery in brain, newer generations of lentiviral vectors incorporate the central polypurine tract (cPPT), an approx 180 bp region derived from the *gag* region, which increases nuclear import of the proviral DNA *(12,13)* and transduction efficiency in the brain *(14)*.

Lentiviral vectors are also proving to be tremendously useful for transduction of various neural cell types in vitro. Examples of transducible primary CNS cells include motor neurons *(15)*, cerebellar granule neurons *(16–18)*, cortical neurons *(9)*, hippocampal neurons *(8)*, astrocytes *(19)*, and macrophages and microglia *(20,21)*. Perhaps even more intriguing is the opportunity to transduce cultures of primary and immortalized neural progenitor cells *(22–26)*. Using viral vectors for ex vivo transduction can aid in the selection of neural stem and progenitor cells and provide a way to mark them for lineage tracing. In addition, transduced neural stem or progenitor cells can be retransplanted into the developing, mature, or injured brain in order to track their migration, phenotypic differentiation, and cell fate.

1.2. Targeted Gene Delivery in the CNS Using Pseudotyped Lentiviral Vectors

Another feature of the lentiviral vector system is that the virions can carry a surface protein that bypasses the usual HIV receptors and co-receptors, thus changing or expanding the range of cell types that the vector can bind to and enter. This is done by pseudotyping, which involves replacing the HIV-1 envelope glycoprotein with an envelope glycoprotein from another virus, such as the vesicular stomatitis virus glycoprotein (VSV-G). A major cellular receptor for VSV-G is thought to be phosphatidylserine, a lipid component of plasma membranes *(27)*. Pseudotyping with VSV-G allows transduction of a wide range of cell types from various species, including the mouse, that HIV-1 does not normally infect *(28)*. With these conditions, the major factor governing which cell types express the transgene is the vector-encoded internal promoter. Importantly, for in vivo experiments in the CNS in which small volumes of vector are injected into brain structures, the stability of VSV-G also allows concentration of the vector by ultracentrifugation. One major disadvantage to the use of VSV-G, however, is the difficulty of generating stable packaging cell lines expressing VSV-G at high levels. A number of inducible systems are currently being developed to overcome this problem *(29–35)*.

Envelope glycoproteins from other viruses have also been used to pseudotype HIV vectors and target their entry to certain cell populations. Pseudotyped lentiviral and retroviral vectors have been generated with surface glycoproteins from a variety of different enveloped viruses, including Ebola virus (Zaire and Reston subtypes), Marburg virus, rabies virus, lymphocytic choriomeningitis virus (LCMV), Mokola virus, human foamy virus, gibbon ape leukemia virus (GaLV), murine leukemia virus (amphotropic envelope), influenza virus (HA), avian leukosis-sarcoma virus (ALSV-A), and respiratory syncitial virus (F and G envelope proteins) *(8,16,36–43)*, although not every viral envelope glycoprotein is suitable for this purpose *(44)*. For example, we recently showed that pseudotyping lentiviral vectors with the Mokola or MuLV amphotropic envelope leads to dramatically different transduction patterns within the hippocampus *(18)*.

1.3. Lentiviral Vector Production

The development of stable packaging cell lines that produce high titers of lentiviral vectors has been hindered by the toxicity of constitutive VSV-G expression (*see 29,30,33*). For this reason, many groups continue to use transient triple transfection to generate their vector stocks. Although this requires large amounts of high-quality plasmid DNA, the method is reliable in producing 10^6 or higher transducing units per mL of culture supernatant. At least three plasmids are required: one encoding the envelope glycoprotein (most commonly VSV-G), one encoding the packaging proteins (minimally including *gag* and *pol*, often including *tat* and *rev* on the same plasmid, and in some cases the accessory genes *vif, vpr, vpu,* and *nef*), and one encoding the genome (including the intact or self-inactivating LTR's, the packaging signal (ψ), the promoter and cDNA of interest and, in some cases, posttranslational regulatory elements and the central polypurine tract (cPPT), which increase titer and expression *(12,32,45)*.

2. Materials
2.1. Transfection

1. A highly transfectable cell line such as human embryonic kidney 293T (available from ATCC, Rockville, MD, Cat. no. CRL 1573).

2. Growth media for 293T cells: Dulbecco's modified Eagle's medium (DMEM) with glutamine, supplemented with 100 U/mL penicillin, 100 µg/mL streptomycin, and 10% fetal bovine serum (PBS).

3. Poly-D-lysine (Sigma, St. Louis, MO, Cat. no. P0899).

4. Borate buffer (dissolve 2.38 g of boric acid and 1.27 g of borax in 500 mL of water; filter-sterilize; store at 4°C up to 6 mo).

5. Reagents for calcium phosphate transfection: 2 M $CaCl_2$, 2X HeBS (16.4 g of NaCl, 11.9 g of HEPES acid, 0.21 g of Na_2HPO_4, adjust the pH to 7.05–7.12, adjust the volume to 1 L of water and filter-sterilize) (*see* **Note 1**). The CalPhos Transfection Kit from Clontech (Palo Alto, CA, Cat. no. K2051-1) also works well.

6. High-quality plasmid DNA: transfer plasmid, packaging plasmid, envelope plasmid purified through cesium chloride/ethidium bromide equilibrium centrifugation or an anion-exchange matrix (e.g., Qiagen® plasmid purification kit, Qiagen, Valencia, CA). For 10-cm plates, use 20 µg of transfer plasmid, 15 µg of packaging plasmid, and 5 µg of envelope plasmid *(46)*. For 15-cm plates, use 45 µg of transfer plasmid, 33.75 µg of packaging plasmid, and 11.25 µg of envelope plasmid.

7. Polybrene (hexadimethrine bromide; Sigma, Cat. no. H9268), 800 µg/mL stock solution in PBS. Store in aliquots at −20°C.

8. Large polyallomer centrifuge tubes (such as Beckman Coulter, Fullerton, CA, Cat. no. 326823) for concentrating the vector in a Beckman SW28 ultracentrifuge rotor.

2.2. Stereotactic Surgery

1. Anaesthesia: scale to weigh the mice, ketamine (Ketaset, Fort Dodge Animal Health, Fort Dodge, IA) and Xylazine (Phoenix Pharmaceutical Inc, St. Joseph, MO), 0.9% sterile saline, disposable sterile syringes with 27 G needles (*see* **Note 2**).

2. Surgical Preparation: A small electric shaver, Betadine Surgical Scrub (Henry Schein Veterinary and Medical Supply Catalog, Melville, NY, Cat. no. 690-2680), a scalpel, small surgical clips (such as Bulldog-type serrefines, Fine Science Tools, Foster City, CA, Cat. no. 18050-28), stereotaxic frame (such as Kopf Small Animal Stereotaxic Frame Model 900, David Kopf Instruments, Tujunga, CA), low-temperature fine tip cautery (Henry Schein, Cat. no. 100-2222), Artifical Tears Ointment (Phoenix Pharmaceutical Inc, Cat. no. XC 50120)

3. Drilling and injection: Drill and drill bits (such as Foredom FM3545 control with MH145 handpiece, available from Kopf Instruments), 10-µL Hamilton syringe with a blunt 30 G needle for vector injections; pump (such as a Stoelting 310; Stoelting Co, Wood Dale, IL), a two-arm halogen light.

4. Postoperative care: absorbable suture thread (such as coated Vicryl with C3 needle, Ethicon, Somerville, NJ), heating pad.
5. Perfusion: Dissecting tray or styrofoam block covered with foil and placed in a glass dish, sterile PBS (4°C), fixative (such as 4% paraformaldehyde in PBS, pH 7.4, 4°C), scissors for cutting the skin, sharp, fine scissors, large blunt forceps, hemostat or needles, butterfly needle (23 × 3/4 in infusion set, Abbott Laboratories, Chicago, IL, Cat. no. 4565), three-way stopcock (Henry Schein, Cat. no. 507-5840), infusion pump and tubing.
6. A manual for stereotaxic surgery such as *Stereotaxic Surgery in the Rat: A Photographic Series (47)*. (ajkirbyco@pobox.com). Other reviews have been published recently *(48,49)*.
7. A mouse brain atlas such as *The Mouse Brain in Stereotaxic Coordinates (50)*. Additional guides are available on-line at http://www.nervenet.org/mbl/mbl.html.

2.3. Titering β-Galactosidase Vectors by X-Gal Histochemistry

1. Glutaraldehyde, 25% aqueous solution (Sigma, Cat. no. G-6257), diluted to 0.5% in PBS immediately before use.
2. PBS containing 1 mM magnesium chloride.
3. N,N,-Dimethylformamide (Fisher, Pittsburgh, PA, Cat. no. P119-500), 5-bromo-4-chloro-3-indolyl-β-D-galactoside (X-gal, Sigma, Cat. no. B4252). To prepare a 40X stock (40 mg/mL), add 0.2 g of X-gal to 5 mL of dimethylformamide and store at −20°C in a foil-wrapped tube.
4. Potassium ferricyanide $K_3Fe(CN)_6$ (Fisher, Cat. no. P232-500), potassium ferrocyanide $K_4Fe(CN)_6$•$3H_2O$ (Fisher, Cat. no. P236-500), PBS. To prepare a 20X stock of potassium cyanide solution (20X KCN), add 1.64 g of $K_3Fe(CN)_6$ and 2.10 g of $K_4Fe(CN)_6$•$3 H_2O$ to 50 mL of 1X PBS and store in a foil-wrapped tube at room temperature for up to a month.
5. 1 M magnesium choride.
6. To prepare the X-gal substrate solution, mix 15 µL of 1 M $MgCl_2$ with 14 mL of PBS. Add 750 µL of 20X KCN and 375 µL of 40X X-gal. Filter through a 0.2-micron filter. This produces ~ 15 mL of substrate solution, enough for one 24-well plate.

3. Methods
3.1. Transfection (see Note 3)

1. Coat 10-cm or 15-cm diameter tissue-culture plates 4 h to overnight at room temperature with 1 mg/mL of poly-D-lysine in borate buffer. Rinse the plates well with sterile water before plating the cells (*see* **Note 4**). If you plan to concentrate the vector in an SW28 rotor, we recommend starting

Table 1
Solution Volumes Required for Cell Transfection*

Component	10-cm plate	15-cm plate
Growth medium	9.0 mL	20.0 mL
Plasmid DNA/H_2O	438 μL	985.5 μL
2 M $CaCl_2$	62 μL	139.5 μL
2X HeBS	500 μL	1.125 mL

*See **Subheading 3.1.** for details.

with three 15-cm plates so that the approx 33 mL of viral supernatant will completely fill one centrifuge tube in the SW28 rotor (**Subheading 3.2., step 3**). The rotor holds a total of six tubes.

2. Trypsinize 293T cells and inoculate 1.5×10^6 cells per 10-cm plate or 2.1×10^7 cells per 15-cm plate (*see* **Note 5**).
3. The following day, gently change the medium to 9 mL (10-cm plate) or 20 mL (15-cm plate) 2–4 h prior to transfection (*see* **Table 1**).
4. For each plate to be transfected, combine the three plasmid DNAs and add water to bring the volume to 438 μL (for each 10-cm plate) or 985.5 μL (for each 15-cm plate). Add 62 μL (10-cm plate) or 139.5 μL (15-cm plate) of 2 M $CaCl_2$ (*see* **Note 6**).
5. While mixing thoroughly, add the DNA/CaCl2 mixture dropwise to 500 μL (10-cm plate) or 1.125 mL (15-cm plate) of 2X HeBS. Let the mixture stand at room temperature for 20 min to allow formation of the precipitate, then add dropwise to the plate of cells (*see* **Note 7**).
6. Eight to twelve hours after transfection, aspirate the transfection mixture and gently rinse the cells.
7. Replace the medium with 5 mL (10-cm) or 11 mL (15-cm) of culture medium (*see* **Note 8**).
8. Collect and replace the medium every 24 h for 3 d. For concentration on the day of collection, keep the viral supernatant on ice until ready to process as described in **Subheading 3.2.** Otherwise, freeze the viral supernatant at −80°C.

3.2. Collection and Concentration of the Viral Supernatant

The highest viral titers are obtained by concentrating each batch of viral supernatant on the day it is collected. For convenience, it is possible to freeze each batch of viral supernatant after filtration and then thaw all the batches together on ice and concentrate them together. However, because of some loss during the freeze-thaw procedure and because the titer normally drops after the first day of collection, the final titer may be considerably lower. See **Notes 9–12** on biosafety and laboratory handling of lentiviral vectors.

1. Centrifuge the viral supernatant for 5 min at 200g, 4°C, to remove cellular debris.
2. Filter the viral supernatant through a 0.45-μm filter (*see* **Note 13**).
3. Centrifuge the medium at 50,000g for 2 h at 4°C to concentrate the viral supernatant (*see* **Note 14**).
4. Carefully aspirate or drain the supernatant from the tube. The pellet can be resuspended in the remaining supernatant, generally 200–300 μL. Keep the tube on ice while resuspending the pellet with a pipetman. Store aliquots of the vector at −80°C (*see* **Note 15**).

3.3. Titering Lentiviral Vectors

For reporter genes, we find that a functional titer (i.e., a value in transducing units/mL) is the most useful measure of the amount of vector produced (*see* **Note 16**).

1. For vectors encoding β-gal, plate 5 × 10^4 293T cells per well in a poly-L-lysine-coated 24-well plate with 0.5 mL of media (*see* **Subheading 2.1.2.**).
2. One day later, replace the media with viral supernatant diluted in media containing 8 μg/mL polybrene for a final volume of 0.5 mL. The dilutions should be duplicate 10-fold dilutions from 1:10 to 1:100,000.
3. Forty-eight hours after transduction, assay the cells by X-gal histochemistry. Remove and add all solutions very gently. Aspirate the medium from the cells and rinse with PBS. Add 0.5 mL of 0.5% glutaraldehyde in PBS per well in a fume hood for 10 min at room temperature. Discard the glutaraldehyde into an approved waste container in the fume hood and rinse the cells twice for 10 min each with PBS containing 1 mM MgCl$_2$. Add 0.5 mL of X-gal substrate solution to each well. Incubate 2 h to overnight at 37°C. Stop the reaction by removing the substrate solution and adding 0.5 mL of PBS to each well.
4. Count the number of blue colonies in each well and multiply by the dilution factor to obtain the titer. Expected titers are in the range of 10^5–10^6 transducing units per mL.

3.4. Assays for Replication-Competent Recombinants

1. Several assays for replication competent recombinants are detailed in another chapter in this volume (*see* Chapter 25).

3.5. Anaesthesia

1. A combination of ketamine and xylazine can be used to induce anaesthesia. Weigh the mouse and inject the anaesthesia intraperitoneally to a dose of 100 mg/kg (ketamine) and 5–10 mg/kg (xylazine) for adult mice. A fresh dilution of the mixture should be prepared for each group of mice (*see* **Note 17**).

3.6. Surgical Preparation of the Animals

1. Shave the head of the mouse with an electric razor. Place a drop of Artificial Tears Ointment on each eye. Swab the scalp with betadine, then make an incision along the midline to expose the skull.
2. Mount the mouse in a stereotaxic frame. Make sure the animal's head is level and fixed in position by the earbars and the toothbar. Attach small surgical clips to the scalp and let them hang to the side of the head to hold the scalp out of the way. Mount a low-temperature cauterizer or pen into the needle guide and align it to bregma, the intersection of the saggital and coronal sutures. Based on the coordinates in a mouse brain atlas, adjust the anterior-posterior and medial-lateral coordinates so that the cauterizer or pen lies over the structure of interest. Gently make a small mark on the skull to guide the drilling, then replace the cauterizer or pen with the drill in the arm of the stereotaxic frame. Carefully drill a hole through the skull, constantly raising the drill to check the progress.
3. In the mouse brain, commonly used coordinates include (in mm, relative to bregma): [AP −2, DV +2, ML +2] for hippocampus; [AP 0, DV +2.5, ML +2] for striatum; [AP 0, DV +1, ML +2] for cortex; and [AP −2, DV +4, ML +1.5] for thalamus. It is possible to inject multiple structures at one injection site by raising the needle after the first injection; for example, striatum and cortex, or thalamus and hippocampus can be injected together.

3.7. Injection of Virus

1. On the stereotaxic frame, replace the drill with the Hamilton syringe attached to the pump. Swing the arm of the needle-holder to the side in order to load the needle with vector.
2. Thaw the vector on ice and keep it on ice at all times. Flick the tube or pipet up and down a few times to mix the vector before loading it into the needle. If the needle clogs repeatedly, briefly centrifuge the vector before use. One–two μL can be injected into each site in the adult mouse brain, with a maximum of 10 μL per brain.
3. Beginning at bregma, reposition the needle over the drill hole according to the coordinates of the target region. Lower the needle until the tip just passes the skull, stopping before the dura and cortex are penetrated. Note

the reading of the scale and slowly lower the needle the appropriate depth into the brain. Be sure that the needle has penetrated the dura; if the drilling was performed carefully, the dura may remain intact and will not be easily penetrated by a fine, blunt needle. Once the needle is at the target depth, lower the needle an additional 0.5 mm, then raise it again before starting the injection.

4. Inject the vector: various investigators use from 1–5 μL per site, with injection rates ranging from 0.1–2 μL/min. We recommend 0.2 μL/min and no more than 1–2 μL per site or a total of 10 μL of volume injected per mouse brain.

5. Leave the needle in place at least 3 min before raising the needle. Raise the needle very slowly. The goal is to allow the tissue to seal behind the needle, avoiding backflow of the injected vector out of the hole.

3.8. Postoperative Animal Care

1. Release the mouse from the stereotaxic frame and wet the scalp gently with a cotton swab.
2. Suture the scalp. Cover the floor of a cage with paper towels and put it on a heating pad. Wet a few food pellets and place them on the floor of the cage.
3. When the animals have fully recovered from the anaesthesia (usually 1–3 h) remove the cage from the heating pad, remove the paper towels, and replace it in the ventilated rack.
4. Check the mice daily to make sure the scalp wound is healing. Mice will often chew out the sutures. When surgery is performed with appropriately sterilized equipment, it is usually not necessary to use antibiotic ointment on the wound or to include antibiotics in the drinking water.

3.9. Perfusion

1. Time points for sacrifice are determined by the experimental goal but should be no sooner than 3 d after transduction in order for expression to reach detectable levels. Initial experiments may be performed a week after transduction. Depending on the promoter, expression may be observed for a year or more.
2. Deeply anaesthetize the mouse using 145 mg/kg ketamine and 30 mg/kg xylazine (*see* **Subheading 3.5.** above).
3. For perfusion, set up a pump with the inflow tube in a bottle of fixative on ice and the outflow tube connected via the three-way valve to both the butterfly needle infusion set and a 50-mL syringe containing cold PBS. Before beginning the perfusion, run the pump so that fixative flows through the tubing and out the needle. Turn off the pump, close the valve to the fixative, and use the syringe to flush the tubing and needle with PBS; this

should allow PBS from the syringe to be flushed through the animal first. After closing the three-way valve to the syringe port and turning on the pump, fixative should then flow through with no interruptions in the flow of fluid into the animal. A flow rate of 5 mL/min is desirable.

4. Place the mouse belly-up with the tail towards you on a dissecting tray or a styrofoam block covered with foil in a flat glass dish. Confirm that the animal is deeply anesthetized. If necessary, use needles to pin the mouse in place. Make a midline incision through the skin to expose the abdominal wall. Starting from the lower abdomen, hold the muscle layer up with forceps and slice open the abdominal wall with sharp scissors, being careful to not puncture any underlying organs. When the ribcage is reached, hold the base of the ribcage up with the forceps and slice through the diaphragm to the left in order to open the chest cavity and avoid the heart. Slice to the left, then to the right to expose the heart. Flip the rib cage up towards the animal's nose and pin it but do not cut it off. Make an additional lateral slice on one side of the chest wall so blood and fixative can drain out of the chest cavity into the dissecting tray.

5. The heart should still be beating. Locate the right atrium, the small, dark red chamber in the upper left rear area of the heart. Using very sharp scissors, make a small snip as far away from the rest of the heart as possible. Blood will begin to flow into the chest cavity. Holding the heart with large blunt forceps, quickly pierce the left ventricle with a butterfly needle and begin perfusing with PBS. Do not insert the needle too deeply. If fluid begins to flow from the nose or mouth of the animal, the needle penetrated into the right ventricle. In this case, withdraw and reposition the needle. After a few minutes of perfusion, the needle can be clamped with a hemostat to hold it in place, or fixed in place with needles. We commonly use about 10 mL of PBS for an adult mouse. Watch for the liver to lighten in color as an indicator. After the blood has been flushed out of the mouse, switch to fixative and continue the perfusion. Use about the same volume (in milliliters) of fixative as the mouse weighs (in grams). Fixative should be infused at a very slow flow rate in order to avoid damaging the tissue. The tail will often curl or the paws may twitch in response to the fixative.

6. Carefully open the scalp and skull and remove the brain. Some investigators prefer to postfix the brain in the same fixative overnight at 4°C before embedding the tissue.

4. Notes

1. The pH of the 2X HeBS is critical for high transfection efficiency and each batch should be carefully tested.

2. Ketamine is a DEA non-narcotic Schedule 3 drug and must be purchased by an investigator with a DEA license.
3. Many methods of transfection are available. We have continued to use the calcium phosphate method of transfection because it has proved efficient, reproducible, and inexpensive for large transfections. However, the composition of the medium should be compatible with the transfection method (for example, serum-free medium is recommended for use with some lipid transfection reagents).
4. HEK293T cells are extremely fragile and media should always be changed with great care. Coating the culture plates with poly-D-lysine helps the cells to adhere during transfection. Cells should be split at a ratio of approx 1:6 every 4 d.
5. After extensive passage, the transfection efficiency of a culture of 293T cells will decline. It is advisable to freeze several vials of low-passage cells so a liquid nitrogen stock is available when this happens.
6. If multiple plates are to be transfected, the DNA/CaCl2/HeBS solutions can be prepared in bulk and divided between the plates.
7. The ratio of transfection mixture to culture medium should not exceed 1:10.
8. Make sure the plates sit flat in the incubator so the medium is evenly distributed. If necessary, place a weight on top of the lid to minimize evaporation.
9. All waste material should be collected in a biohazard bag inside the tissue culture hood, and the bag should be sealed before removing it from the hood. All contaminated material should be autoclaved before disposal.
10. Use of glass and sharps should be minimized.
11. A virucide, such as Conflikt detergent and disinfectant (Fisher, Cat. no. 04-355-34), should be used to wipe down the hoods and pipetmen after use.
12. All viral production should be restricted to one hood if possible and producer cultures should be kept in a separate incubator if possible.
13. It is absolutely essential to filter the vector before transferring it to target cells. Contamination of a target cell culture with the producer cells can lead to uninterpretable data. We find that gravity flow preserves the viability of the vector to a greater extent than forcing the viral supernatant through the filter with the syringe plunger or vacuum.
14. It is best to use a swinging bucket rotor so that the pellet will be collected at the bottom of the tube rather than on the side. After concentration, the viral pellet can be very sticky and require scraping and pipetting to remove it from the tube.
15. Vector frozen in serum-containing medium survives freezing better than vector frozen in serum-free medium.

16. Alternative approaches to determining virus titer include *in situ* hybridization or immunohistochemistry to detect expression of the transgene in the titering cells or FACS analysis. However, some proteins are difficult to detect, and in this case the use of a p24 gag ELISA (Alliance HIV-1 p24 ELISA kit, Perkin Elmer Life Sciences, Boston, MA, cat. no. NEK050001KT) or a reverse transcriptase assay may be useful. The issue of positive controls, which includes positive control samples, and sensitivity of these assays should be carefully considered. Investigators using lentiviral vectors as experimental tools may not have access to a Biosafety level 3 facility in which to work with wild type HIV or, to minimize the risk of cross-contamination, may not wish to work with wild type HIV in their laboratories. In some cases, media from the transfected producer cells can serve as positive controls. Another alternative is to construct vectors with an internal ribosomal entry site and a marker gene downstream of the cDNA of interest, although the lower abundance of the downstream protein requires a sensitive detection method. Some investigators produce a vector encoding a reporter gene such as β-gal in parallel with the vector of interest and use it to estimate the titer of the second vector, with the caveat that the packaging efficiencies may not be identical.

17. Monitor the mouse's breathing and pinch response throughout the procedure and adjust the anaesthesia if necessary.

Acknowledgments

This work was supported by NIH grants DK42707, DK46637, and NS38690 (JHW), and an institutional NRSA grants in neurovirology (NS07180) (DJW) and Comparative medicine (RR07063) (BAK), an individual NRSA fellowship (NS11024) (DJW) and the Alavi-Dabiri Family Postdoctoral Award (DJW).

References

1. Fouchier, R. A. and Malim, M. H. (1999) Nuclear import of human immunodeficiency virus type-1 preintegration complexes. *Adv. Virus Res.* **52,** 275–299.
2. Naldini, L., Blomer, U., Gage, F. H., Trono, D., and Verma, I. M. (1996) Efficient transfer, integration, and sustained long-term expression of the transgene in adult rat brains injected with a lentiviral vector. *Proc. Natl. Acad. Sci. USA* **93,** 11382–11388.
3. Costantini, L. C., Bakowska, J. C., Breakefield, X. O., and Isacson, O. (2000) Gene therapy in the CNS. *Gene Ther.* **7,** 93–109.
4. Arya, S. K., Zamani, M., and Kundra, P. (1998) Human immunodeficiency virus type 2 lentivirus vectors for gene transfer: expression and potential for helper virus-free packaging. *Hum. Gene Ther.* **9,** 1371–1380.
5. Poeschla, E. M., Wong-Staal, F., and Looney, D. J. (1998) Efficient transduction of nondividing human cells by feline immunodeficiency virus lentiviral vectors. *Nat. Med.* **4,** 354–357.

6. Johnston, J. C., Gasmi, M., Lim, L. E., Elder, J. H., Yee, J. K., Jolly, D. J., et al. (1999) Minimum requirements for efficient transduction of dividing and nondividing cells by feline immunodeficiency virus vectors. *J. Virol.* **73,** 4991–5000.

7. Olsen, J. C. (1998) Gene transfer vectors derived from equine infectious anemia virus. *Gene Ther.* 5: 1481–1487.

8. Mitrophanous, K., Yoon, S., Rohll, J., Patil, D., Wilkes, F., Kim, V., et al. (1999) Stable gene transfer to the nervous system using a non-primate lentiviral vector. *Gene Ther.* **6,** 1808–1818.

9. White, S. M., Renda, M., Nam, N. Y., Klimatcheva, E., Zhu, Y., Fisk, J., et al. (1999) Lentivirus vectors using human and simian immunodeficiency virus elements. *J. Virol.* **73,** 2832–2840.

10. Bukovsky, A. A., Song, J. P., and Naldini, L. (1999) Interaction of human immunodeficiency virus-derived vectors with wild-type virus in transduced cells. *J. Virol.* **73,** 7087–7092.

11. Cui, Y., Iwakuma, T., and Chang, L. J. (1999) Contributions of viral splice sites and cis-regulatory elements to lentivirus vector function. *J. Virol.* **73,** 6171–6176.

12. Follenzi, A., Ailles, L. E., Bakovic, S., Geuna, M., and Naldini, L. (2000) Gene transfer by lentiviral vectors is limited by nuclear translocation and rescued by HIV-1 pol sequences. *Nat. Genet.* **25,** 217–222.

13. Zennou, V., Petit, C., Guetard, D., Nerhbass, U., Montagnier, L., and Charneau, P. (2000) HIV-1 genome nuclear import is mediated by a central DNA flap. *Cell* **101,** 173–185.

14. Zennou, V., Serguera, C., Sarkis, C., Colin, P., Perret, E., Mallet, J., and Charneau, P. (2001) The HIV-1 DNA flap stimulates HIV vector-mediated cell transduction in the brain. *Nat. Biotechnol.* **19,** 446–450.

15. Cisterni, C., Henderson, C. E., Aebischer, P., Pettmann, B., and Deglon, N. (2000) Efficient gene transfer and expression of biologically active glial cell line-derived neurotrophic factor in rat motoneurons transduced with lentiviral vectors. *J. Neurochem.* **74,** 1820–1828.

16. Mochizuki, H., Schwartz, J. P., Tanaka, K., Brady, R. O., and Reiser, J. (1998) High-titer human immunodeficiency virus type 1-based vector systems for gene delivery into nondividing cells. *J. Virol.* **72,** 8873–8883.

17. Nomura, T., Yabe, T., Mochizuki, H., Reiser, J., Becerra, S. P., and Schwartz, J. P. (2001) Survival effects of pigment epithelium-derived factor expressed by a lentiviral vector in rat cerebellar granule cells. *Dev. Neurosci.* **23,** 145–152.

18. Watson, D. J., Kobinger, G. P., Passini, M. A., Wilson, J. M., and Wolfe, J. H. (2002) Targeted transduction patterns in the mouse brain by lentivirus vectors pseudotyped with VSV, Ebola, Mokola, LCMV or MuLV envelope proteins. *Mol. Ther.* **5,** 528–537.

19. Ericson, C., Wictorin, K., and Lundberg, C. (2002) Ex vivo and in vitro studies of transgene expression in rat astrocytes transduced with lentiviral vectors. *Exp. Neurol.* **173,** 22–30.

20. Hein, A., Martin, J. P., Koehren, F., Bingen, A., and Dorries, R. (2000) In vivo infection of ramified microglia from adult cat central nervous system by feline immunodeficiency virus. *Virology* **268**, 420–429.
21. Mordelet, E., Kissa, K., Calvo, C. F., Lebastard, M., Milon, G., van der Werf, S., et al. (2002) Brain engraftment of autologous macrophages transduced with a lentiviral flap vector: an approach to complement brain dysfunctions. *Gene Ther.* **9**, 46–52.
22. Englund, U., Ericson, C., Rosenblad, C., Mandel, R. J., Trono, D., Wictorin, K., and Lundberg, C. (2000) The use of a recombinant lentiviral vector for ex vivo gene transfer into the rat CNS. *Neuroreport* **11**, 3973–3977.
23. Arsenijevic, Y., Villemure, J. G., Brunet, J. F., Bloch, J. J., Deglon, N., Kostic, C., et al. (2001) Isolation of multipotent neural precursors residing in the cortex of the adult human brain. *Exp. Neurol.* **170**, 48–62.
24. Englund, U., Fricker-Gates, R. A., Lundberg, C., Bjorklund, A., and Wictorin, K. (2002) Transplantation of human neural progenitor cells into the neonatal rat brain: extensive migration and differentiation with long-distance axonal projections. *Exp. Neurol.* **173**, 1–21.
25. Hughes, S. M., Moussavi-Harami, F., Sauter, S. L., and Davidson, B. L. (2002) Viral-mediated gene transfer to mouse primary neural progenitor cells. *Mol. Ther.* **5**, 16–24.
26. Watson, D., Longhi, L., Lee, E., Fulp, C., Fujimoto, S., Royo, N., et al. (2003) Genetically modified NT2N human neuronal cells mediate long-term gene expression as CNS grafts in vivo and improve functional cognitive outcome following experimental traumatic brain injury. *J. Neuropathol. Exp. Neurol.* **62**, 368–380.
27. Schlegel, R., Tralka, T. S., Willingham, M. C., and Pastan, I. (1983) Inhibition of VSV binding and infectivity by phosphatidylserine: is phosphatidylserine a VSV-binding site? *Cell* **32**, 639–646.
28. Burns, J. C., Friedmann, T., Driever, W., Burrascano, M., and Yee, J. K. (1993) Vesicular stomatitis virus G glycoprotein pseudotyped retroviral vectors: concentration to very high titer and efficient gene transfer into mammalian and nonmammalian cells. *Proc. Natl. Acad. Sci. USA* **90**, 8033–8037.
29. Kafri, T., van Praag, H., Ouyang, L., Gage, F. H., and Verma, I. M. (1999) A packaging cell line for lentivirus vectors. *J. Virol.* **73**, 576–584.
30. Klages, N., Zufferey, R., and Trono, D. (2000) A stable system for the high-titer production of multiply attenuated lentiviral vectors. *Mol. Ther.* **2**, 170–176.
31. Kaul, M., Yu, H., Ron, Y., and Dougherty, J. P. (1998) Regulated lentiviral packaging cell line devoid of most viral cis-acting sequences. *Virology* **249**, 167–174.
32. Sparacio, S., Pfeiffer, T., Schaal, H., and Bosch, V. (2001) Generation of a flexible cell line with regulatable, high-level expression of HIV Gag/Pol particles capable of packaging HIV-derived vectors. *Mol. Ther.* **3**, 602–612.
33. Pacchia, A. L., Adelson, M. E., Kaul, M., Ron, Y., and Dougherty, J. P. (2001) An inducible packaging cell system for safe, efficient lentiviral vector production in the absence of HIV-1 accessory proteins. *Virology* **282**, 77–86.

34. Farson, D., Witt, R., McGuinness, R., Dull, T., Kelly, M., Song, J., et al. (2001) A new-generation stable inducible packaging cell line for lentiviral vectors. *Hum. Gene Ther.* **12,** 981–997.
35. Zhang, B., Xia, H. Q., Cleghorn, G., Gobe, G., West, M., and Wei, M. Q. (2001) A highly efficient and consistent method for harvesting large volumes of high-titre lentiviral vectors. *Gene Ther.* **8,** 1745–1751.
36. Reiser, J., Harmison, G., Kluepfel-Stahl, S., Brady, R. O., Karlsson, S., and Schubert, M. (1996) Transduction of nondividing cells using pseudotyped defective high-titer HIV type 1 particles. *Proc. Natl. Acad. Sci. USA* **93,** 15266–15271.
37. Wool-Lewis, R. J. and Bates, P. (1998) Characterization of Ebola virus entry by using pseudotyped viruses: identification of receptor-deficient cell lines. *J. Virol.* **72,** 3155–3160.
38. Chan, S. Y., Speck, R. F., Ma, M. C., and Goldsmith, M. A. (2000) Distinct mechanisms of entry by envelope glycoproteins of Marburg and Ebola (Zaire) viruses. *J. Virol.* **74,** 4933–4937.
39. Stitz, J., Buchholz, C. J., Engelstadter, M., Uckert, W., Bloemer, U., Schmitt, I., and Cichutek, K. (2000) Lentiviral vectors pseudotyped with envelope glycoproteins derived from gibbon ape leukemia virus and murine leukemia virus 10A1. *Virology* **273,** 16–20.
40. Lewis, B. C., Chinnasamy, N., Morgan, R. A., and Varmus, H. E. (2001) Development of an avian leukosis-sarcoma virus subgroup A pseudotyped lentiviral vector. *J. Virol.* **75,** 9339–9344.
41. Kobinger, G. P., Weiner, D. J., Yu, Q. C., and Wilson, J. M. (2001) Filovirus-pseudotyped lentiviral vector can efficiently and stably transduce airway epithelia in vivo. *Nat. Biotechnol.* **19,** 225–230.
42. Desmaris, N., Bosch, A., Salaün, C., Petit, C., Prévost, M.-C., Tordo, N., et al. (2001) Production and neurotropism of lentivirus vectors pseudotyped with lyssavirus envelope glycoproteins. *Mol. Ther.* **4,** 149–156.
43. Beyer, W. R., Westphal, M., Ostertag, W., and von Laer, D. (2002) Oncoretrovirus and lentivirus vectors pseudotyped with lymphocytic choriomeningitis virus glycoprotein: Generation, concentration, and broad host range. *J. Virol.* **76,** 1488–1495.
44. Christodoulopoulos, I. and Cannon, P. M. (2001) Sequences in the cytoplasmic tail of the gibbon ape leukemia virus envelope protein that prevent its incorporation into lentivirus vectors. *J. Virol.* **75,** 4129–4138.
45. Zufferey, R., Donello, J. E., Trono, D., and Hope, T. J. (1999) Woodchuck hepatitis virus posttranscriptional regulatory element enhances expression of transgenes delivered by retroviral vectors. *J. Virol.* **73,** 2886–2892.
46. Naldini, L., Blomer, U., Gallay, P., Ory, D., Mulligan, R., Gage, F. H., et al. (1996) In vivo gene delivery and stable transduction of nondividing cells by a lentiviral vector. *Science* 272, 263–267.
47. Cooley, R. K. and Vanderwold, C. H. (1990) *Stereotaxic Surgery in the Rat: A Photographic Series.* A. J. Kirby Co., London, Ontario.

48. Brooks, A. I., Halterman, M., Chadwick, C., Davidson, B., Haak-Frendscho, M., Radel, C., et al. (1998) Reproducible and efficient murine CNS gene delivery using a microprocessor-controlled injector. *J. Neurosci. Meth.* **80,** 137–147.
49. Messier, C., Émond, S., and Ethier, K. (1999) New techniques in stereotaxic surgery and anesthesia in the mouse. *Pharm. Chem. Behav.* **63,** 313–318.
50. Franklin, K. B. J. and Paxinos, G. (1997) *The Mouse Brain in Stereotaxic Coordinates*. Academic Press, San Diego, CA.
51. Kafri, T., Gene delivery by lentiviral vectors, in *Gene Delivery to Mammalian Cells in Culture: Methods and Protocols*, (Heiser, W. C. ed.), Humana Press, Totowa, NJ.

28

Gene Delivery to Hematopoietic Stem Cells Using Lentiviral Vectors

Hiroyuki Miyoshi

1. Introduction

Hematopoietic stem cells (HSCs) are clonogenic cells capable of both self-renewal and multilineage differentiation. An efficient method for gene transfer into HSCs is required for exploring HSC biology as well as for gene therapy of hematopoietic disorders. Retroviral vectors have been the most widely used vectors for gene transfer to HSCs. However, retroviral vectors require cell division for integration, limiting their use for gene transfer into HSCs that are exclusively quiescent. Although prestimulation of HSCs with cytokines can enhance gene-transfer efficiency *(1–7)*, exposure to cytokines also stimulates HSCs to differentiate, resulting in the reduction of long-term repopulating capacity *(8–15)*. In contrast, lentiviral vectors based on the human immunodeficiency virus type 1 (HIV-1) can efficiently transduce human CD34+ cells without cytokine prestimulation and long-term multilineage expression of the transgene is detected in nonobese diabetic/severe combined immunodeficient (NOD/SCID) mice after transplantation *(16–22)*. Murine HSCs can also be easily transduced with lentiviral vectors without cytokine prestimulation *(23–26)*.

In this chapter, the protocols for transduction of human CD34+ cells that contain HSCs are described. The procedures include the isolation of human CD34+ cells, preparation of lentiviral vectors, transduction of human CD34+ cells, and in vivo analysis of transduced HSCs using NOD/SCID mice.

From: *Methods in Molecular Biology, vol. 246:*
Gene Delivery to Mammalian Cells: Vol. 2: Viral Gene Transfer Techniques
Edited by: W. C. Heiser © Humana Press Inc., Totowa, NJ

2. Materials

2.1. Reagents and Equipment

1. Ficoll-Paque PLUS (Amersham Biosciences, Uppsala, Sweden).
2. Magnetic cell separator MidiMACS, VarioMACS, or SuperMACS II (Miltenyi Biotec Inc., Auburn, CA).
3. MACS CD34 MicroBeads and FcR blocking reagent provided by Direct CD34 progenitor cell isolation kit (Miltenyi Biotec Inc.).
4. MACS separation column MS or LS (Miltenyi Biotec Inc.).
5. Ultracentrifuge with swinging-bucket rotors (Beckman SW28 and SW55 rotors or equivalent).

2.2. Solutions and Culture Medium

1. Phosphate-buffered saline (PBS) containing 2 mM EDTA.
2. MACS buffer: PBS containing 2 mM EDTA and 0.5% bovine serum albumin (BSA).
3. 2.5 M CaCl$_2$: Sterilize the solution by passing it through a pre-wetted 0.22-μm filter. Store the solution in aliquots at −20°C.
4. 2 × BBS: 50 mM BES, 280 mM NaCl, 1.5 mM Na$_2$HPO$_4$. Adjust the pH to 6.95 with NaOH. Sterilize the solution by passing it through a 0.22-μm filter. Store the solution in aliquots at −20°C.
5. Hank's balanced salt solution (HBSS) (Invitrogen/GIBCO, Carlsbad, CA).
6. Serum-free culture medium (SFM): Iscove's modified Dulbecco's medium (IMDM) (Invitrogen/GIBCO) supplemented with 10% StemSpan BIT 9500 Serum Substitute (Stem Cell Technologies Inc., Vancouver, BC).
7. Complete DMEM medium: Dulbecco's modified Eagle's medium (DMEM) (Invitrogen/GIBCO) supplemented with 10% fetal bovine serum (FBS), 2 mM L-glutamine, 100 U/mL penicillin, and 100 μg/mL streptomycin sulfate.

3. Methods

3.1. Isolation of Human CD34⁺ Hematopoietic Progenitor Cells

Human HSCs are enriched in the CD34⁺ fraction. Mononuclear cells from cord blood (CB), bone marrow (BM), or granulocyte colony-stimulating factor (G-CSF)-mobilized peripheral blood (PB) are obtained by Ficoll-Paque density gradient centrifugation. CD34⁺ cells are then magnetically labeled using MACS CD34 MicroBeads and enriched on positive selection columns in the magnetic field of the MACS separator.

1. Collect fresh human CB, BM, or G-CSF-mobilized PB samples and treat with an anticoagulant (e.g., heparin) (*see* **Note 1**).

2. Dilute the cells with 4 volumes of PBS containing 2 mM EDTA.
3. Carefully layer 35 mL of diluted cell suspension over 15 mL of Ficoll-Paque PLUS in a 50-mL conical centrifuge tube.
4. Centrifuge at 500g for 45 min at 20°C in a swinging-bucket rotor with no brake.
5. Aspirate the upper layer leaving the mononuclear cell layer undisturbed at the interphase.
6. Using a sterile pipet, carefully transfer the mononuclear cell (MNC) layer to a new 50-mL conical centrifuge tube.
7. Wash the cells by adding excess PBS containing 2 mM EDTA and centrifuge at 300g for 10 min at 20°C.
8. Resuspend the cell pellet in a final volume of 300 μL of MACS buffer per 10^8 total cells. For less than 10^8 total cells, use 300 μL.
9. Add 100 μL of FcR blocking reagent per 10^8 total cells to the cell suspension and incubate for 10 min at 4°C.
10. Add 100 μL of MACS CD34 MicroBeads per 10^8 total cells, mix well, and incubate for 30 min at 4°C.
11. Wash the cells by adding excess PBS containing 2 mM EDTA and centrifuge at 300g for 10 min at 4°C.
12. Resuspend the cell pellet in a final volume of 500 μL of MACS buffer per 10^8 total cells.
13. Choose a positive selection MACS column type (MS or LS) according to the number of total unseparated cells and place it with column adaptor in the magnetic field of the MACS separator. Fill and rinse with buffer (MS: 500 μL; LS: 3 mL).
14. Pass the cells through 30-μm nylon mesh to remove clumps. Wet filter with buffer before use.
15. Apply the cells to the column, allow the cells to pass through the column and wash with buffer (MS: 2 mL; LS: 12 mL).
16. Remove column from separator, place column on a suitable tube, pipet MACS buffer on top of column (MS: 1 mL; LS: 5 mL) and elute retained cells using the plunger supplied with the column. Repeat this step twice for better cell yield.
17. For greater than 90% CD34$^+$ cell purity, repeat the magnetic separation (**steps 13**, **15**, and **16**). Apply the eluted cells to a fresh MS column, wash, and elute retained cells in MACS buffer.
18. Wash the cells by adding excess SFM and centrifuge at 300g for 10 min at 4°C.
19. Resuspend CD34$^+$ cells in SFM at a concentration of 5 × 10^6 cells/mL (*see* **Note 2**). CD34$^+$ cells can be stored in SFM with 10% dimethylsulfoxide (DMSO) in a liquid nitrogen freezer.

3.2. Preparation of Lentiviral Vectors

HIV-1-based lentiviral vectors pseudotyped with the vesicular stomatitis virus G glycoprotein (VSV-G) are generated by transient transfection of four plasmids, the packaging construct (pMDLg/pRRE) *(27)*, the VSV-G-expressing construct (pMD.G) *(28)*, the Rev-expressing construct (pRSV-Rev) *(27)*, and the self-inactivating (SIN) lentiviral vector construct *(29)*, into 293T cells. The packaging construct, pMDLg/pRRE, in which all accessory genes (*vif, vpr, vpu,* and *nef*) and regulatory genes (*tat* and *rev*) have been deleted, can be used to produce lentiviral vectors for transduction of CD34$^+$ hematopoietic progenitor cells *(18,19,21,22)*.

1. Harvest exponentially growing 293T cells by trypsinization and seed 5 × 10^6 cells in poly-lysine coated 10-cm dish in 10 mL of complete DMEM medium (*see* **Note 3**).
2. Incubate the cells for 24 h at 37°C in a humidified incubator with an atmosphere of 10% CO_2. The cells should be about 75% confluent at the time of transfection.
3. Transfection: Prepare a total volume of 450 μL of plasmid DNA solution containing 17 μg of the SIN vector plasmid (*see* **Note 4**), 12 μg of the packaging plasmid (pMDLg/pRRE), 5 μg of the Rev-expressing plasmid (pRSV-Rev), and 5 μg of the VSV-G-expressing plasmid (pMD.G) in a 5-mL polystyrene tube (Falcon 2058) or a 1.5-mL microfuge tube. Add 50 μL of 2.5 *M* CaCl$_2$ and then 500 μL of 2 × BBS with gentle mixing. Incubate the mixture for 10–20 min at room temperature. Mix gently by pipetting and transfer the calcium phosphate-DNA solution dropwise to the dish. Rock the dish gently to mix the medium.
4. Incubate the cells for 12–16 h at 37°C in a humidified incubator with an atmosphere of 3% CO_2.
5. Remove the medium by aspiration and add 7.5 mL of prewarmed (37°C) complete DMEM medium.
6. Incubate for 48 h at 37°C in a humidified incubator with an atmosphere of 10% CO_2.
7. Remove the vector-containing medium and filter through a 0.45-μm filter (*see* **Note 5**).
8. Transfer the filtered medium to a conical ultracentrifuge tube that fits a Beckman SW28 rotor (or equivalent) and centrifuge at 50,000*g* (19,400 rpm in a Beckman SW28 rotor) for 2 h at 20°C.
9. Discard the supernatant and resuspend the pellet in an appropriate volume (up to 50 μL per dish) of HBSS (or IMDM) by pipetting. Try to avoid introducing bubbles. When dealing with small-scale preparations of vector, the following step can be omitted; store the vector suspension at −80°C.

10. Transfer the vector suspension to an ultracentrifuge tube that fits a Beckman SW55 rotor (or equivalent) and fill the tube with HBSS. Centrifuge at 50,000g (24,000 rpm in a Beckman SW55 rotor) for 2 h at 20°C.
11. Discard the supernatant and resuspend the pellet in an appropriate volume (up to 10 μL per dish) of HBSS (or IMDM) by pipetting. Try to avoid introducing bubbles.
12. Transfer the vector suspension to a tightly capped tube. Vortex the tube until no solid material is visible. If any debris is still visible, remove it by brief centrifugation and discard. Store the vector suspension in aliquots at −80°C (*see* **Note 6**).

3.3. Transduction of CD34+ Cells by Lentiviral Vectors

1. Incubate CD34+ cells in SFM for 12–24 h before transduction at 37°C in a humidified incubator with an atmosphere of 5% CO_2. (If CD34+ cells are frozen, thaw the cells, wash, and resuspend in SFM.)
2. Count the number of viable cells using the Trypan Blue exclusion method (*see* **Note 7**).
3. Wash the cells by adding excess SFM and centrifuging at 300g for 10 min at 4°C.
4. Resuspend the cells in SFM at a concentration of 5×10^6 cells/mL and transfer them to a 5-mL polypropylene round-bottom tube or a 96-well round-bottom cell culture dish (*see* **Note 8**).
5. Thaw the lentiviral vector stock and mix well by pipetting. Add the vector suspension to the cells at a multiplicity of infection (MOI) of 50–300 (*see* **Note 9**).
6. Incubate the vector-cell mixture for 5–12 h at 37°C in a humidified incubator with an atmosphere of 5% CO_2 (*see* **Note 10**).
7. Wash the cells with HBSS and centrifuge at 300g for 10 min at 4°C.
8. Resuspend the cells in an appropriate volume of HBSS. (The resuspension medium and volume should be determined based on subsequent assays.)

3.4. Analysis of Transduced Human CD34+ Cells in NOD/SCID Mice

In vitro assays for primitive hematopoietic cells, such as colony-forming cell (CFC) assays and long-term culture-initiating cells (LTC-IC) assays, can be used for analysis of transduced CD34+ cells. However, HSCs can only be assayed by their ability to repopulate conditioned recipients. NOD/SCID mice are the most popular transplant recipients for human HSCs *(30,31)*.

1. Irradiate NOD/SCID mice (8–10 wk old) at a sublethal dose (about 300 centigrays) 2–12 h before transplantation.

2. Intravenously inject $2–4 \times 10^5$ transduced CD34$^+$ cells via the tail vein using 1-mL syringe with a 27 G needle.
3. At 6 wk or more after transplantation, collect blood samples from PB, spleen, and BM.
4. Analyze blood samples by flow cytometry with fluorescent antibodies against human specific cell-surface antigens or transgene products (**Fig. 1**).

4. Notes

1. HSCs are enriched in CB compared with BM or mobilized PB *(32)*.
2. The purity of CD34$^+$ cells is usually greater than 95% as determined by flow cytometry.
3. For large-scale preparation, 15-cm dishes can be used. Scale up all materials in proportion to the area of the plates.
4. The latest SIN lentiviral vectors contain the central polypurine tract (cPPT) sequence, which improves the transduction efficiency *(17,19,21)*, and the woodchuck hepatitis virus posttranscriptional regulatory element (WPRE), which enhances the expression of the transgene *(33)*.
5. It is possible to add fresh medium and to collect vector-containing medium a second time from the cells. Add 7.5 mL of prewarmed (37°C) complete DMEM medium to the dish immediately after withdrawing the medium and continue incubation for another 48 h. Vector titers will be about 75% lower.
6. The titer of vectors is determined by measuring the amount of HIV-1 p24 gag antigen using an ELISA kit. (Commercial kits are available from PerkinElmer, Inc. (Boston, MA) and Advanced Biotechnologies, Inc., Columbia, MD). One ng of p24 would represent 5,000–10,000 infectious units (IU) on 293T or HeLa CD4$^+$ cells. In the case of vectors containing marker genes, the titer of vectors can be determined by infection of 293T or HeLa CD4$^+$ cells with serial dilutions of the vector stocks. Most vector stocks after two rounds of ultracentrifugation have titers ranging between 5×10^8 and 2×10^9 IU/mL. The titer of the vector stocks remains stable for at least 6 mo when they are stored in closed tubes at $-80°C$. In general, the vector stocks can be frozen and thawed several times with gradually loss of infectivity. If the yields of the vector are low, the number of infectious viral particles should be determined in samples taken at various stages during the purification to determine where losses are occurring.
7. To assay viable cells by Trypan Blue exclusion, mix 50 µL of cells with 50 µL of 0.4% Trypan Blue for 5 min. Transfer the cells to a hemocytometer and score the fraction of clear cells (alive) and dark blue cells (dead).
8. Using a smaller volume of SFM during transduction tends to increase the transduction efficiency *(34)*.

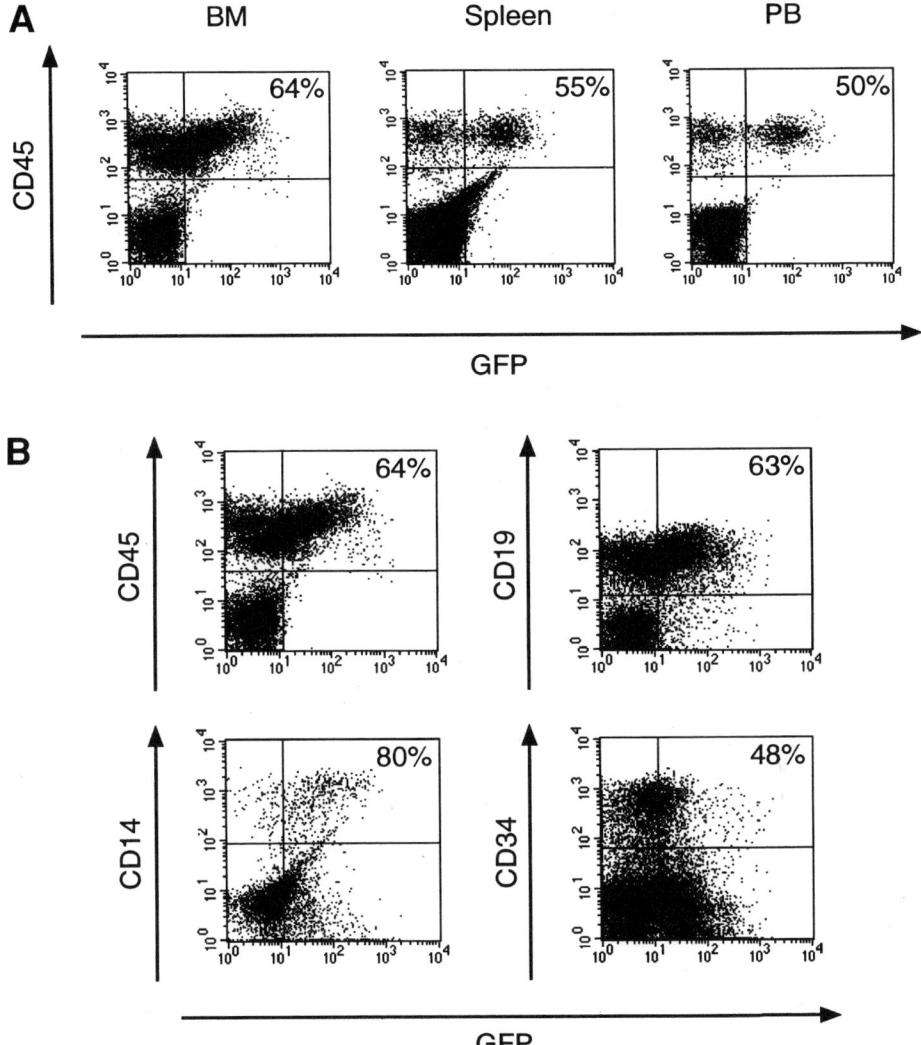

Fig. 1. Flow cytometry analysis of the transgene expression in NOD/SCID mice transplanted with lentiviral vector-transduced human CD34+ cells. Human CB CD34+ cells were transduced with an HIV-1-based lentiviral vector containing the green fluorescence protein (GFP) gene under the control of the cytomegalovirus (CMV) promoter and transplanted into sublethally irradiated NOD/SCID mice. At 15 wk posttransplantation, mononuclear cells were isolated from BM, spleen, and PB and analyzed by flow cytometry using FACSCalibur (Becton-Dickinson, Franklin Lakes, NJ) with CEL-LQuest software. **(A)** BM, spleen, and PB cells were stained with phycoerythrin (PE)-conjugated anti-human CD45. **(B)** BM cells were stained with phycoerythrin (PE)-conjugated anti-human CD14 (myeloid cells), CD19 (B cells), or CD34 (progenitor cells). Percentages of GFP+ cells in engrafted human cell subpopulation are indicated.

9. Increases in MOI, addition of polybrene, and RetroNectin (recombinant human fibronectin fragment CH-296) (Takara Bio Inc., Otsu, Japan) have little effect on the transduction efficiency.
10. Incubation for 5 h is enough for transduction but longer incubation may improve the transduction efficiency.

References

1. Cheng, L., Du, C., Lavau, C., Chen, S., Tong, J., Chen, B. P., et al. (1998) Sustained gene expression in retrovirally transduced, engrafting human hematopoietic stem cells and their lympho-myeloid progeny. *Blood* **92,** 83–92.
2. Conneally, E., Eaves, C. J., and Humphries, R. K. (1998) Efficient retroviral-mediated gene transfer to human cord blood stem cells with in vivo repopulating potential. *Blood* **91,** 3487–3493.
3. Marandin, A., Dubart, A., Pflumio, F., Cosset, F. L., Cordette, V., Chapel-Fernandes, S., et al. (1998) Retrovirus-mediated gene transfer into human CD34+38low primitive cells capable of reconstituting long-term cultures in vitro and nonobese diabetic-severe combined immunodeficiency mice in vivo. *Hum. Gene Ther.* **9,** 1497–1511.
4. Schilz, A. J., Brouns, G., Knoss, H., Ottmann, O. G., Hoelzer, D., Fauser, A. A., et al. (1998) High efficiency gene transfer to human hematopoietic SCID-repopulating cells under serum-free conditions. *Blood* **92,** 3163–3171.
5. van Hennik, P. B., Verstegen, M. M., Bierhuizen, M. F., Limon, A., Wognum, A. W., Cancelas, J. A., et al. (1998) Highly efficient transduction of the green fluorescent protein gene in human umbilical cord blood stem cells capable of cobblestone formation in long-term cultures and multilineage engraftment of immunodeficient mice. *Blood* **92,** 4013–4022.
6. Cavazzana-Calvo, M., Hacein-Bey, S., de Saint Basile, G., Gross, F., Yvon, E., Nusbaum, P., et al. (2000) Gene therapy of human severe combined immunodeficiency (SCID)-X1 disease. *Science* **288,** 669–672.
7. Dorrell, C., Gan, O. I., Pereira, D. S., Hawley, R. G., and Dick, J. E. (2000) Expansion of human cord blood CD34(+) CD38(−) cells in ex vivo culture during retroviral transduction without a corresponding increase in SCID repopulating cell (SRC) frequency: dissociation of SRC phenotype and function. *Blood* **95,** 102–110.
8. van Beusechem, V. W., Bart-Baumeister, J. A., Hoogerbrugge, P. M., and Valerio, D. (1995) Influence of interleukin-3, interleukin-6, and stem cell factor on retroviral transduction of rhesus monkey CD34+ hematopoietic progenitor cells measured in vitro and in vivo. *Gene Ther.* **2,** 245–255.
9. Peters, S. O., Kittler, E. L., Ramshaw, H. S., and Quesenberry, P. J. (1996) Ex vivo expansion of murine marrow cells with interleukin-3 (IL-3), IL-6, IL-11, and stem cell factor leads to impaired engraftment in irradiated hosts. *Blood* **87,** 30–37.
10. Yonemura, Y., Ku, H., Hirayama, F., Souza, L. M., and Ogawa, M. (1996) Interleukin 3 or interleukin 1 abrogates the reconstituting ability of hematopoietic stem cells. *Proc. Natl. Acad. Sci. USA* **93,** 4040–4044.

11. Yonemura, Y., Ku, H., Lyman, S. D., and Ogawa, M. (1997) In vitro expansion of hematopoietic progenitors and maintenance of stem cells: comparison between FLT3/FLK-2 ligand and KIT ligand. *Blood* **89,** 1915–1921.
12. Bhatia, M., Bonnet, D., Kapp, U., Wang, J. C., Murdoch, B., and Dick, J. E. (1997) Quantitative analysis reveals expansion of human hematopoietic repopulating cells after short-term ex vivo culture. *J. Exp. Med.* **186,** 619–624.
13. Gothot, A., van der Loo, J. C., Clapp, D. W., and Srour, E. F. (1998) Cell cycle-related changes in repopulating capacity of human mobilized peripheral blood CD34(+) cells in non-obese diabetic/severe combined immune-deficient mice. *Blood* **92,** 2641–2649.
14. Rebel, V. I., Tanaka, M., Lee, J. S., Hartnett, S., Pulsipher, M., Nathan, D. G., et al. (1999) One-day ex vivo culture allows effective gene transfer into human nonobese diabetic/severe combined immune-deficient repopulating cells using high-titer vesicular stomatitis virus G protein pseudotyped retrovirus. *Blood* **93,** 2217–2224.
15. Szilvassy, S. J., Meyerrose, T. E., and Grimes, B. (2000) Effects of cell cycle activation on the short-term engraftment properties of ex vivo expanded murine hematopoietic cells. *Blood* **95,** 2829–2837.
16. Miyoshi, H., Smith, K. A., Mosier, D. E., Verma, I. M., and Torbett, B. E. (1999) Transduction of human CD34+ cells that mediate long-term engraftment of NOD/SCID mice by HIV vectors. *Science* **283,** 682–686.
17. Follenzi, A., Ailles, L. E., Bakovic, S., Geuna, M., and Naldini, L. (2000) Gene transfer by lentiviral vectors is limited by nuclear translocation and rescued by HIV-1 pol sequences. *Nat. Genet.* **25,** 217–222.
18. Guenechea, G., Gan, O. I., Inamitsu, T., Dorrell, C., Pereira, D. S., Kelly, M., et al. (2000) Transduction of human CD34+ CD38-bone marrow and cord blood-derived SCID-repopulating cells with third-generation lentiviral vectors. *Mol. Ther.* **1,** 566–573.
19. Sirven, A., Pflumio, F., Zennou, V., Titeux, M., Vainchenker, W., Coulombel, L., et al. (2000) The human immunodeficiency virus type-1 central DNA flap is a crucial determinant for lentiviral vector nuclear import and gene transduction of human hematopoietic stem cells. *Blood* **96,** 4103–4110.
20. Woods, N. B., Fahlman, C., Mikkola, H., Hamaguchi, I., Olsson, K., Zufferey, R., et al. (2000) Lentiviral gene transfer into primary and secondary NOD/SCID repopulating cells. *Blood* **96,** 3725–3733.
21. Sirven, A., Ravet, E., Charneau, P., Zennou, V., Coulombel, L., Guetard, D., et al. (2001) Enhanced transgene expression in cord blood CD34(+)-derived hematopoietic cells, including developing T cells and NOD/SCID mouse repopulating cells, following transduction with modified trip lentiviral vectors. *Mol. Ther.* **3,** 438–448.
22. Gao, Z., Golob, J., Tanavde, V. M., Civin, C. I., Hawley, R. G., and Cheng, L. (2001) High levels of transgene expression following transduction of long-term nod/scid-repopulating human cells with a modified lentiviral vector. *Stem Cells* **19,** 247–259.
23. Barrette, S., Douglas, J. L., Seidel, N. E., and Bodine, D. M. (2000) Lentivirus-based vectors transduce mouse hematopoietic stem cells with similar efficiency to moloney murine leukemia virus-based vectors. *Blood* **96,** 3385–3391.

24. Chen, W., Wu, X., Levasseur, D. N., Liu, H., Lai, L., Kappes, J. C., and Townes, T. M. (2000) Lentiviral vector transduction of hematopoietic stem cells that mediate long-term reconstitution of lethally irradiated mice. *Stem Cells* **18**, 352–359.

25. Mikkola, H., Woods, N. B., Sjogren, M., Helgadottir, H., Hamaguchi, I., Jacobsen, S. E., et al. (2000) Lentivirus gene transfer in murine hematopoietic progenitor cells is compromised by a delay in proviral integration and results in transduction mosaicism and heterogeneous gene expression in progeny cells. *J. Virol.* **74**, 11911–11918.

26. Tahara-Hanaoka, S., Sudo, K., Ema, H., Miyoshi, H., and Nakauchi, H. (2002) Lentiviral vector-mediated transduction of murine CD34($-$) hematopoietic stem cells. *Exp. Hematol.* **30**, 11–17.

27. Dull, T., Zufferey, R., Kelly, M., Mandel, R. J., Nguyen, M., Trono, D., and Naldini, L. (1998) A third-generation lentivirus vector with a conditional packaging system. *J. Virol.* **72**, 8463–8471.

28. Naldini, L., Blomer, U., Gallay, P., Ory, D., Mulligan, R., Gage, F. H., et al. (1996) In vivo gene delivery and stable transduction of nondividing cells by a lentiviral vector. *Science* **272**, 263–267.

29. Miyoshi, H., Blomer, U., Takahashi, M., Gage, F. H., and Verma, I. M. (1998) Development of a self-inactivating lentivirus vector. *J. Virol.* **72**, 8150–8157.

30. Shultz, L. D., Schweitzer, P. A., Christianson, S. W., Gott, B., Schweitzer, I. B., Tennent, B., et al. (1995) Multiple defects in innate and adaptive immunologic function in NOD/LtSz-scid mice. *J. Immunol.* **154**, 180–191.

31. Dick, J. E. (1996) Normal and leukemic human stem cells assayed in SCID mice. *Semin. Immunol.* **8**, 197–206.

32. Wang, J. C., Doedens, M., and Dick, J. E. (1997) Primitive human hematopoietic cells are enriched in cord blood compared with adult bone marrow or mobilized peripheral blood as measured by the quantitative in vivo SCID-repopulating cell assay. *Blood* **89**, 3919–3924.

33. Zufferey, R., Donello, J. E., Trono, D., and Hope, T. J. (1999) Woodchuck hepatitis virus posttranscriptional regulatory element enhances expression of transgenes delivered by retroviral vectors. *J. Virol.* **73**, 2886–2892.

34. Haas, D. L., Case, S. S., Crooks, G. M., and Kohn, D. B. (2000) Critical factors influencing stable transduction of human CD34($+$) cells with HIV-1-derived lentiviral vectors. *Mol. Ther.* **2**, 71–80.

29

Delivery of Genes to the Eye Using Lentiviral Vectors

Masayo Takahashi

1. Introduction

The primary aim of gene transfer into the retinal cells has been to investigate the developmental mechanisms of the retinal cells or to reverse retinal diseases. Retroviruses have been used to investigate the differentiation of retinal cells, to study the embryonic retina in vivo or explant organ culture, and to trace the fate of the cells that were dividing at the time of gene transfer *(1–5)*.

Using adenovirus, Bennett et al. showed the possibility of using gene therapy to correct degenerative diseases of the central nervous system (CNS) *(6)*. However, owing to the short duration of the gene expression, adenovirus is not suitable for correcting chronic diseases. Currently, lentivirus *(7–9)* and adeno-associated virus vectors *(10–14)* are being used for studying and correcting gene therapy of retinal degenerative diseases.

Using an HIV vector carrying the green fluorescent protein (GFP) gene expressed from the cytomegalovirus (CMV) promoter, we showed that efficient and long-lasting gene expression could be obtained in the retina *(7,8)* (**Fig. 1**). Moreover, gene expression was restricted to the photoreceptor cells and was more efficient with the rhodopsin promoter. Similar results were reported using adeno-associated virus (AAV) vector *(10)*. Using a lentivirus vector carrying the phosphodiesterase beta subunit (PDEβ) gene, the mutation of which causes retinal degeneration called retinitis pigmentosa in *rd* mice, photoreceptor cells were rescued from degeneration in *rd* mice for at least 6 mo by PDEβ transduction using HIV-based lentivirus vector *(9)*.

From: *Methods in Molecular Biology, vol. 246:*
Gene Delivery to Mammalian Cells: Vol. 2: Viral Gene Transfer Techniques
Edited by: W. C. Heiser © Humana Press Inc., Totowa, NJ

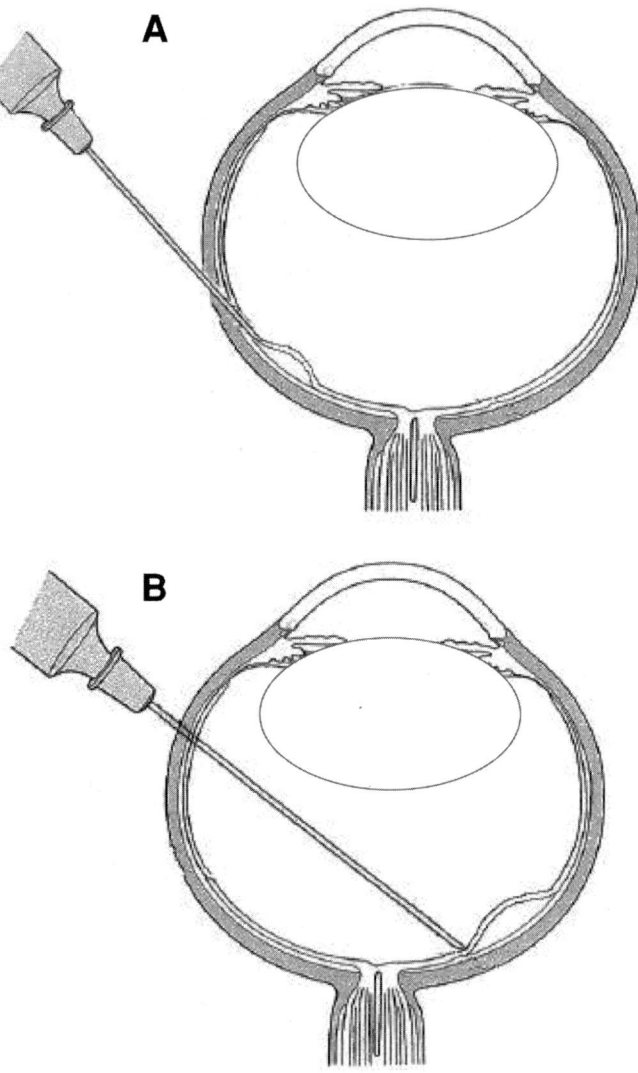

Fig. 1. Subretinal injection. (**A**) scleral approach. (**B**) vitreous approach.

The following procedure describes methods for delivering genes to the retina. The method differs depending on the size of the eye, ranging from postnatal d 1 to the human eye.

2. Surgical Instruments

2.1. Injection for the Small Eye

1. Fine forceps.
2. Spring handle scissors.
3. Tapered glass pipet needle (*see* **Note 1**) or 27 G needle.
4. Hamilton syringe (10 µL).
5. Concaved contact lens.
6. Stereomicroscope for surgery.

2.2. Vitrectomy for the Large Eye

In addition to the instruments in **Subheading 2.1.**, add the following:

1. Vitrectomy equipment (e.g., VT-5000 (Nidek, Gamagori, Japan) including 23 G vitreous cutter, 2.5- or 4-mm infusion tube, irrigation tube, illumination fiber).
2. Speculum, V-lance knife or microscalpel.
3. Concaved contact lens and ring for vitrectomy surgery.
4. 500 mL of balanced salt solution (BSS).
5. Subretinal injection cannula (32 G needle) with 1-mL syringe.
6. 5-0 and 8-0 polyglactin.

3. Methods

3.1. Gene Delivery to the Subretinal Space of Rats and Mice

Gene delivery to the subretinal space may be made with the scleral approach or the vitreous approach (**Fig. 1**). The scleral approach is better for small eyes such as mouse eyes. The vitreous approach is applicable for the bigger eyes than adult rat eyes. For both approaches, the first few steps are the same. (See ref. *15* for anatomy of the eyeball.)

1. Anesthetize rats or mice with an intraperitoneal injection of pentobarbital (40 mg/kg). Anesthetize the cornea with eye drops containing 2% xylocaine.
2. Obtain full mydriasis using eye drops containing 0.5% Tropicamide (*see* **Note 2**).
3. Cut or tear the conjunctiva at the limbus to expose the quadrant of sclera (*see* **Note 3**).
4. Push the eyelids with forceps to dislocate the eyeball slightly.

5. Load a Hamilton syringe with the lentivirus vector (at a titer of 3×10^8 infectious units (IU)/mL or higher) to be injected. The volume of the vector depends on the size of the eyeball. For a newborn mouse, inject 1.5 µL; for an adult rat, inject 3 µL.

3.1.1. Scleral Approach (Fig. 1A)

1. Rotate the eyeball by pinching the conjunctiva at the limbus with fine forceps.
2. Make a scleral perforation (no deeper than the choroid) near the equator with a 27 G needle or microscalpel.
3. Insert a tapered glass pipet needle connected to the Hamilton syringe (containing the lentivirus vector from **Subheading 3.1.**, **step 5**) with a tube (**Fig. 2**, *see* **Note 4**) through the incision toward the subretinal space (parallel to the spherical surface).

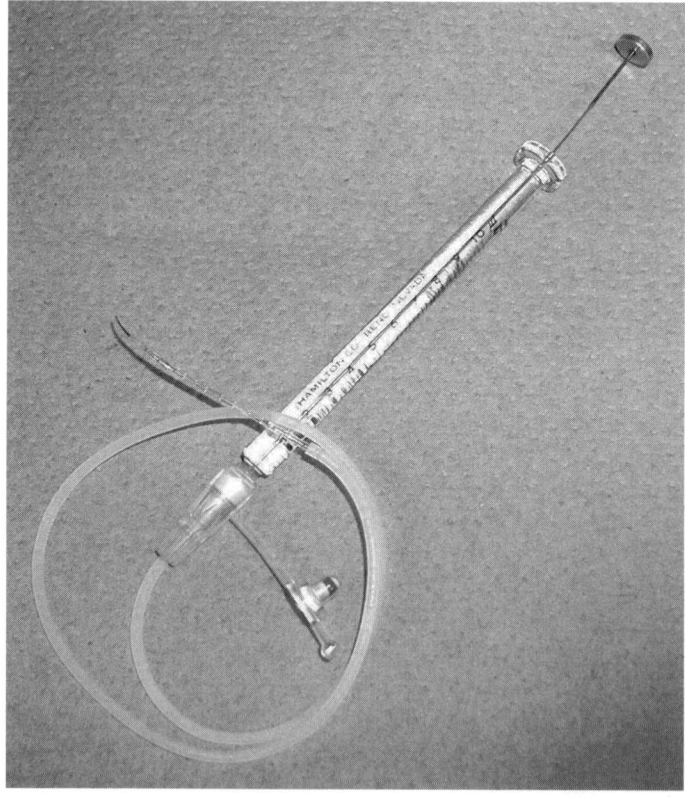

Fig. 2. Hamilton syringe with tapered glass pipet needle connected by tube.

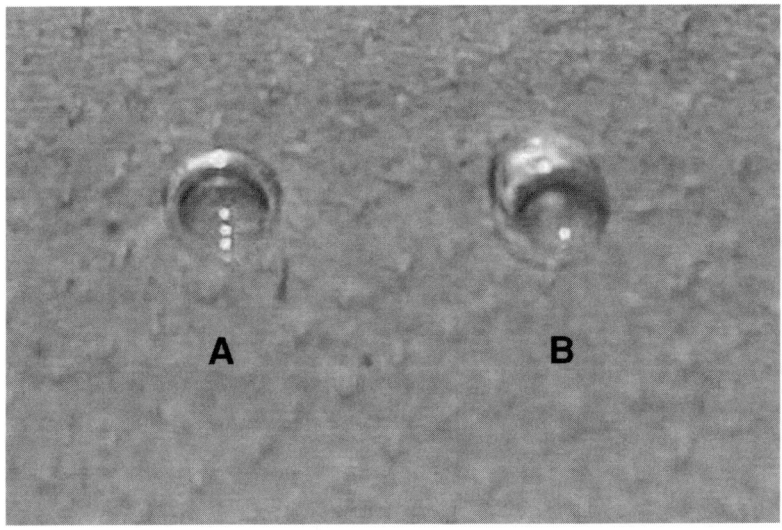

Fig. 3. Concaved contact lenses (for rats). (**A**) flat lens for observation of posterior (central) retina. (**B**) 30° angled lens for observation of peripheral retina

4. Inject the virus suspension into the subretinal space.
5. Confirm gene delivery into the subretinal space by the appearance of a focal retinal detachment. This may be observed under a surgical microscope and will appear as a circle around the retina when viewed through the concaved contact lens (**Fig. 3**) or through a glass compression slide.

3.1.2. Vitreous Approach (Fig. 1B)

1. Observe the retina through a concaved contact lens.
2. Perforate the sclera at the equator into the vitreous cavity with a 24 G needle.
3. Insert a tapered glass pipet needle connected to a Hamilton syringe (containing the lentivirus vector from **Subheading 3.1., step 5**) (*see* **Note 4**) through the incision into the subretinal space by perforating the retina.
4. Inject the virus suspension into subretinal space.

3.2. Gene Delivery to the Retina of Monkey (Fig. 4, see Note 4)

1. Anesthetize the monkey with an intramuscular injection of ketamine (10 mg/kg) and an intraperitoneal injection of pentbarbital (15 mg/kg).
2. Make lateral tarsorrhaphies (cut the lateral canthus for 2–3 mm with scissors) to open the eyelids as wide as possible.
3. Open the eyelids with a speculum.
4. Obtain full mydriasis using eye drops containing 0.5% Tropicamide.

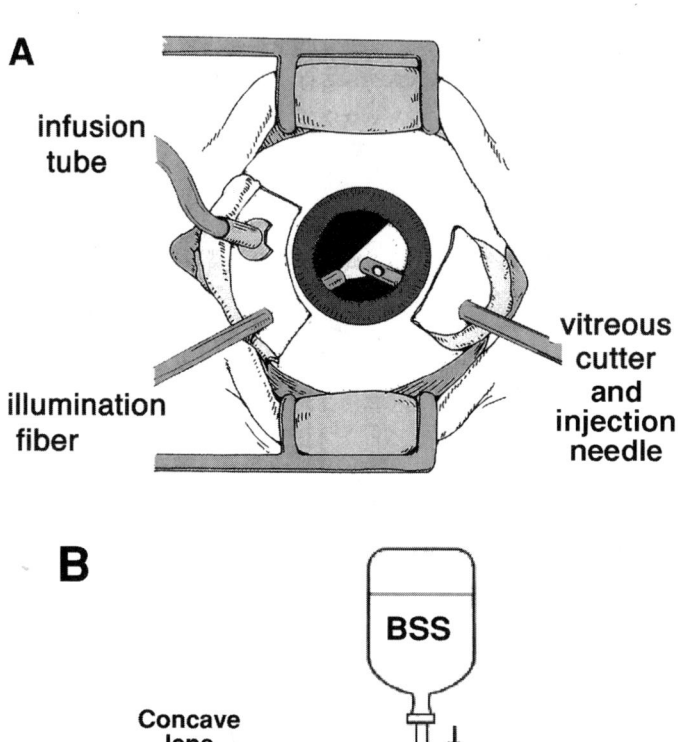

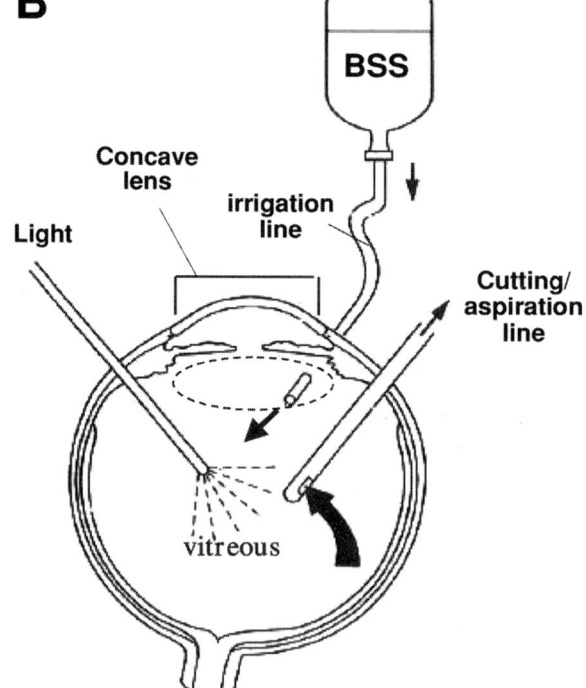

Fig. 4. Illustration of the three-port vitrectomy surgery. (**A**) birds-eye illustration. (B) cross-section illustration.

5. Perforate the sclera into the vitreous space with a V-lance knife 2.5 mm posterior to the limbus at 2, 4, and 10 o'clock (three-port method; port = perforation).
6. Suture an infusion tube containing BBS at the 4 o'clock port using 5-0 polyglactin. The infusion tube should be devoid of air bubbles.
7. Flow the sterile BBS from the infusion tube. Remove the air from the tube before inserting it into the vitreous space.
8. Suture the contact lens ring with 8-0 polyglactin at the limbus and put the concaved contact lens in the contact lens ring for vitrectomy.
9. Put the illumination fiber at 2 o'clock and the vitreous cutter at 10 o'clock. Remove the vitreous completely using the vitreous cutter.
10. Fill a Hamilton syringe with vector and connect it to a 32 G needle bent at a 120° angle. Insert the needle (subretinal injection cannula) at the 10 o'-clock port. Perforate the retina gently. The virus should have a titer of 3×10^8 infectious units or greater. For a monkey, inject a volume of about 50–100 µL *(12)*.
11. Inject the viral suspension into the subretinal space and make a focal retinal detachment.
12. Remove the infusion tube and suture the three ports with 8-0 polyglactin.
13. Wash the ocular surface with antibiotics (for example, penicillin) and put antibiotic ointment (such as Ecolicin) in the eye.

3.3. Gene Delivery to the Vitreous Space

This procedure can be used for any kind of animal. However, this method is much less efficient at delivering genes to the retina than is delivery into subretinal space via the scleral approach or the vitreous approach, especially in the large eye.

1. Anesthetize the animal as described above.
2. Perforate the sclera and choroid (and retina) with a 27 G needle into the vitreous space. Make the perforation from the equator toward the posterior pole to avoid perforating the lens capsule in small animals such as rats and mice. Make the perforation at the pars plana (where the retina changes into ciliary epithelium, several millimeters from the limbus) in larger animals.
3. Inject the viral suspension into the vitreous space. The volume depends on the capacity of the vitreous space of the animal. For mice weighing about 15 g, inject approx 4 µL; for rats weighing about 250 g, inject approx 8 µL.

3.4. Transfection Efficiency

1. Anesthetize animal to euthanasia.
2. Perfuse the animal with 4% paraformaldehyde transcardially as follows. Cut the skin and the ribs at the midline of the chest with scissors and open

the thorax cavity. Put the perfusion tube into the left ventricle of the heart. Make a small cut in the right auricle to allow the blood to flow out. Perfuse the animal with saline containing 5 IU/mL heparin until all of the blood is replaced by saline, then continue the perfusion with 4% paraformaldehyde in 0.1 M phosphate buffer (PB) at 20 (mL/min).

3. Enucleate the eyeball (*see* **Note 7**) and excise the anterior segments (*see* **Note 8**) of the eyes to make eyecups (remove the cornea, iris and lens).
4. Fix the eyecup in 4% paraformaldehyde in 0.1 M PB at 4°C for 1–2 h.
5. Immerse the eyecup in 30% sucrose at 4°C overnight.
6. Freeze the eyecup in cryopreservation compound (OCT compound) with dry ice.
7. Section the eyecup in a cryostat into 10 µm thick sections.
8. Pick up the sections onto coated glass slides.
9. Mount the cover glass with glycerin or another anti-quenching agent of fluorescence.
10. If the GFP gene was delivered, observe expression of the protein using a fluorescence microscope (*see* **Note 5**). Use immunohistochemical methods for observing expression of other genes.

4. Notes

1. Heat the thin portion of Pasteur pipette until the glass is red and soft. Pull the tip of the pipet with forceps to make them long and tapered. Cut the tapered portion to make a needle. Grind the tip of the glass needle with sandpaper to make it smooth.
2. Full mydriasis can be obtained in 15 min. Put the tropicamide in the eye and continue with the procedure.
3. Leave the small flap of conjunctiva at the limbus so that you can hold and rotate the eyeball by pinching the conjunctival flap.
4. Fill the connecting tube with oil to transmit the pressure. A 27 G needle connected to a Hamilton syringe also can be used for adult rat and mouse eyes.
5. For simple animal experiments, you can omit the infusion port and removal of vitreous. In that case the equipment of vitrectomy are not necessary. Follow the methods of vitreous approach for small animals. However, in large animals, retinal tear and retinal detachment may occur by pulling the vitreous with the needle if you do not remove the vitreous.
6. These procedures have been applied to the study of reversal of photoreceptor degeneration in the retinitis pigmentosa model mice (*rd* mice). We have evaluated the efficacy of gene transfer into photoreceptor cells (**Fig. 5**).

   In eyes injected with an HIV vector expressing the GFP gene from the CMV promoter, the GFP gene was intensely expressed in a high proportion

HIV-CMV　　　HIV-Rhodopsin　　　MLV-CMV

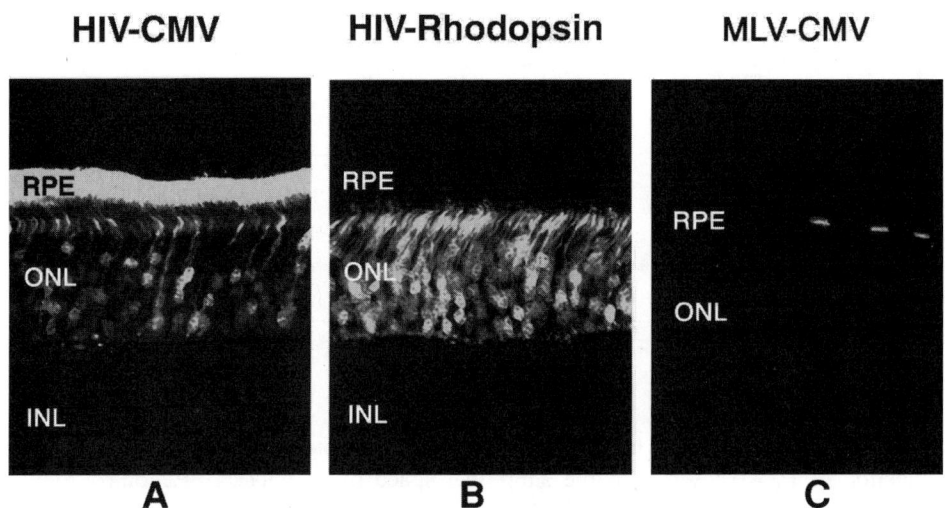

Fig. 5. Expression of GFP in the retina of rat pups 12 wk after injection. (**A**) HIV vector containing the GFP gene expressed from the CMV promoter. (**B**) HIV vector containing the GFP gene expressed from the rhodopsin promoter. Note that only photoreceptor cells are GFP positive. (**C**) MLV vector containing the GFP gene expressed from the CMV promoter. Adapted with permission from ref. (**8**). RPE, retinal pigment epithelium; ONL, outer nuclear layer; INL, inner nuclear layer.

of retinal pigment epithelial (RPE) cells as well as in a large number of photoreceptor cells. In contrast, we observed only a few GFP positive photoreceptor cells when the GFP gene was delivered using an MMLV vector (retrovirus vector). Injection of the HIV vector expressing the GFP gene from the rhodopsin promoter resulted in GFP expression only in photoreceptor cells. GFP expression in RPE cells was completely quenched by using the rhodopsin promoter.

Because of these results, we used an HIV vector with a rhodopsin promoter to rescue photoreceptor cells in *rd* mice transferring the correct PDEβ gene, the mutation of which causes the retinitis pigmentosa. We confirmed that the rhodopsin positive photoreceptor cells survived 6 wk after the injection of rhodopsin-PDEβ vector (**Fig. 6**). On the other hand, photoreceptor cells injected with a rhodopsin-GFP vector degenerated and disappeared completely after 6 wk. These results confirm the feasibility of gene therapy using an HIV vector for recessive forms of retinal degenerative diseases.

7. For small animals, cut the conjunctiva around the cornea with spring handle scissors. Dislocate the eyeball by pushing the eyelids with forceps. In-

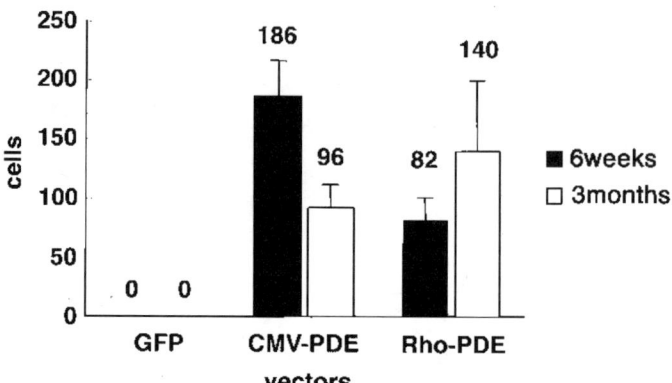

Fig. 6. The number of rhodopsin-positive photoreceptor cells 6 wk and 3 mo after injection of HIV vectors into the subretinal space of *rd* mice. GFP, control vector. Adapted with permission from ref. *(9)*.

sert a scissors behind the eyeball and cut the optic nerve, muscles and connective tissues simultaneously.

For large animals such as monkey, incise the conjunctiva around the cornea and expose the rectus muscles that attach to the eyeball. Cut the four rectus muscles at the insertion and make the eyeball freely movable. Insert a scissors behind the eyeball and cut the optic nerve.

8. Perforate the eyeball at the equator or 2-mm posterior from the limbus with microscalpel under a surgical microscope. Insert a blade of the scissors into the vitreous space and cut the eyeball (sclera, choroid, and retina) parallel to the limbus. Cut the eyeball into two pieces (posterior eyecup and anterior segment).

References

1. Turner, D. L. and Cepko, C. L. (1987) A common progenitor for neurons and glia persists in rat retina late in development. *Nature* **328**, 131–136.
2. Fields-Berry, S. C., Hallidaym A. L., and Cepko, C. L. (1992) A recombinant retrovirus encoding alkaline phosphatase confirms clonal boundary assignment in lineage analysis of murine retina. *Proc. Natl. Acad. Sci. USA* **89**, 693–697.
3. Lillien, L. (1995) Changes in retinal cell fate induced by overexpression of EGF receptor. *Nature* **377**, 158–162.
4. Tomita, K., Ishibashi, M., Nakahara, K., Ang, S. L., Nakanishi, S., Guillemot, F., and Kageyama, R. (1996) Mammalian hairy and Enhancer of split homolog 1 regulates differentiation of retinal neurons and is essential for eye morphogenesis. *Neuron* **16**, 723–734.

5. Bao, Z. Z. and Cepko C. L. (1997) The expression and function of Notch pathway genes in the developing rat eye. *J. Neurosci.* **17**, 1425–1434.

6. Bennett, J., Tanabe, T., Sun, D., Zeng, Y., Kjeldbye, H., Gouras, P., and Maguire, A. M. (1996) Photoreceptor cell rescue in retinal degeneration (rd) mice by in vivo gene therapy. *Nat. Med.* **2**, 649–654.

7. Naldini, L., Blomer, U., Gallay, P., Ory, D., Mulligan, R., Gage, F. H., et al. (1996) In vivo gene delivery and stable transduction of nondividing cells by a lentiviral vector. *Science* **272**, 263–267.

8. Miyoshi, H., Takahashi, M., Gage, F. H., and Verma, I.M. (1996) Stable and efficient gene transfer into the retina using an HIV-based lentiviral vector. *Proc. Natl. Acad. Sci. USA* **94**, 10319–10323.

9. Takahashi, M., Miyoshi, H., Verma, I. M., and Gage, F. H. (1999) Rescue from photoreceptor degeneration in the rd mouse by human immunodeficiency virus vector-mediated gene transfer. *J. Virol.* **73**, 7812–7816.

10. Flannery, J. G., Zolotukhin, S., Vaquero, M. I., LaVail, M. M., Muzyczka, N., and Hauswirth, W. W. (1997) Efficient photoreceptor-targeted gene expression in vivo by recombinant adeno-associated virus. *Proc. Natl. Acad. Sci. USA* **94**, 6916–6921.

11. Ali, R. R., Sarra, G. M., Stephens, C., Alwis, M. D., Bainbridge, J. W., Munro, P. M., et al. (2000) Restoration of photoreceptor ultrastructure and function in retinal degeneration slow mice by gene therapy. *Nat. Genet.* **25**, 306–310.

12. Bennett, J., Maguire, A. M., Cideciyan, A. V., Schnell, M., Glover, E., Anand, V., et al. (1999) Stable transgene expression in rod photoreceptors after recombinant adeno-associated virus-mediated gene transfer to monkey retina. *Proc. Natl. Acad. Sci. USA* **96**, 9920–9925.

13. Flannery, J. G., Zolotukhin, S., Vaquero, M. I., LaVail, M. M., Muzyczka, N., and Hauswirth, W. W. (1997) Efficient photoreceptor-targeted gene expression in vivo by recombinant adeno-associated virus. *Proc. Natl. Acad. Sci. USA* **94**, 6916–6921.

14. Jomary, C., Vincent, K. A., Grist, J., Neal, M. J., and Jones, S. E. (1997) Rescue of photoreceptor function by AAV-mediated gene transfer in a mouse model of inherited retinal degeneration. *Gene Ther.* **4**, 683–690.

15. Forrester, J., Dick, A., McMenamin P., Lee, W. (eds.) (1996) Anatomy of the eye. Structure of the eye. Structure of the eye. Saunders, NY, pp. 13–50.

30

Lentiviral Transduction of Human Dendritic Cells

Roland Schroers and Si-Yi Chen

1. Introduction

Dendritic cells (DCs) are potent antigen-presenting cells (APCs) that play a pivotal role in stimulating antigen-specific T cells in vivo *(1,2)*. The cardinal properties of DCs are: (1) the ability to take up, process, and present antigens; (2) the ability to migrate through different tissues into lymphoid organs; and (3) the ability to interact with and stimulate T cells *(3)*. Because of their unique capability of generating primary CD4+ and CD8+ T-cell responses, DCs are of particular interest for immunotherapeutic approaches to infectious disease and cancer.

There are many ways to load DCs with antigen and to subsequently induce specific immune responses in vivo and in vitro. One strategy relies on genetically modified peripheral blood mononuclear cell (PBMC)-derived DCs to establish expression of a gene that encodes a specific antigen. In contrast to pulsing DCs with antigenic peptides or proteins, genetic engineering of DCs has the advantages of providing multiple epitopes for major histocompatibility class (MHC) class I—and MHC class II-restricted immune responses as well as a continuous supply of antigen for presentation alone or together with immuno-modulatory proteins (cytokines, for example).

Although numerous gene delivery systems, including virus-based and non-virus-based methods are available, most of them show low rates of gene transfer to DCs and, thus, are not optimal for DC transduction. Nonviral gene-transfer

From: *Methods in Molecular Biology, vol. 246:*
Gene Delivery to Mammalian Cells: Vol. 2: Viral Gene Transfer Techniques
Edited by: W. C. Heiser © Humana Press Inc., Totowa, NJ

methods, including calcium phosphate precipitation, electroporation, and liposomal transfection, have been demonstrated to be inefficient in transducing human DCs *(4)*. Adenoviral vectors by contrast, are able to achieve high gene transfer rates in human PBMC-derived DCs and are widely used to transduce these cells. However, their utility may be hampered by the high multiplicity of infection (MOI) required for efficient transduction and by their own inherent immunogenicity. Additionally, adenoviruses cannot integrate their genetic information into the genomes of target cells so that the expression of the transferred genes is transient *(5)*.

Oncoretroviral vectors based on murine leukemia viruses have been successfully used to genetically modify CD34$^+$ DC progenitors from bone marrow *(6)*, cord blood *(7)*, and peripheral blood *(8)*. Because these vectors cannot transduce nondividing cells, their gene-transfer efficacy is restricted to DC progenitors that proliferate before terminal differentiation. Thus, using conventional oncoretroviral vectors to transfer genes to postmitotic PBMC-derived DCs is technically daunting, if not impossible.

HIV-1 and other lentivirus-based vector systems can efficiently transduce many nondividing and terminally differentiated cell types, both in vitro and in vivo *(9,10*; *see also* Chapter 25). Lentivectors stably integrate their provirus into the genomes of transduced cells, resulting in sustained expression of transgenes. These characteristics qualify lentivectors as a potential tool for DC transduction and analysis of in vitro and in vivo immune responses to candidate antigens. Analysis of lentiviral gene transfer to PBMC-derived DCs by different investigators *(11–14)* has provided independent evidence that lentivectors can efficiently transduce PBMC-derived DCs which then can elicit antigen-specific T-cell responses *(13,14)*.

In this chapter, we describe the procedures for capturing PBMC-derived DCs, for producing and titrating of VSV-G pseudotyped lentivectors, and for transducing DCs with lentiviruses (*see* **Note 1**).

2. Materials

2.1. Sources of Cells, Culture Media, Cytokines, and Reagents

1. Dulbecco's modified Eagle's medium (DMEM), RPMI 1640, fetal bovine serum (FBS), penicillin, streptomycin sulfate, glutamine, Lymphoprep™, and distilled H$_2$O are available from Invitrogen (Carlsbad, CA).
2. Serum-free CellGro® DC medium can be purchased from Cellgenix (Freiburg im Breisgau, Germany).
3. Recombinant human interleukin-4 (rhIL-4; specific activity, 1.0×10^4 IU/μg), recombinant human granulocyte-macrophage colony-stimulating

factor (rhGM-CSF; specific activity, 1.0×10^5 IU/μg), recombinant human interleukin-1β (rhIL-1 β; specific activity, 1.1×10^5 IU/μg), recombinant human interleukin-6 (rhIL-6; specific activity, 1.0×10^5 IU/μg), and recombinant human tumor necrosis factor-α (rhTNF-α; specific activity, 4.0×10^4 IU/μg) are available from R&D Systems (Minneapolis, MN).

4. Prostaglandin E_2 (PGE$_2$) and chloroquine diphosphate salt are both available from Sigma (St. Louis, MO).
5. Chemically competent *Escherichia coli* bacterial cells, strain HB101, can be purchased from Promega (Madison, WI).
6. 293T human embryonic kidney (HEK 293T) cells are available from ATCC (Rockville, MD) with permission of Dr. G. P. Nolan (Stanford University, CA).
7. Ion-exchange chromatography (IEC) columns for endotoxin-free DNA purification from bacterial cultures are available from Qiagen Inc. (Chatsworth, CA)

2.2. Plasmids

1. Subclone the transgene into the lentiviral transfer plasmid vector (e.g., pTRIP(U3-CMV; **ref. 15**) according to standard methods **(16)**.
2. For cloning purposes and large-scale plasmid preparation, transform the lentiviral transfer plasmid and the packaging plasmid (e.g., pCMVΔ8.91; **ref. 17**) into *Escherichia coli*, strain HB101. The envelope plasmid (e.g., pMD.G; **ref. 9**) can be transformed into *E. coli*, strain DH5α.
3. Amplify and extract plasmid DNA from bacterial cells and purify endotoxin-free DNA by IEC according to the manufacturer's recommendations (e.g., Qiagen).

2.3. Solutions and Cell Culture Medium

1. Culture medium for HEK 293T cells: DMEM with 10% FBS, 100 IU/mL penicillin, 100 μg/mL streptomycin sulfate, and 2 mM glutamine.
2. Culture medium for DC: CellGro® serum-free DC medium supplemented with 1000 IU/mL rhGM-CSF and 1000 IU/mL rhIL-4. For in vitro maturation of DCs, also add 10 ng/mL rhTNF-α, 10 ng/mL rhIL-1β, 1000 IU/mL rhIL-6, and 1 μg/mL PGE$_2$ *(18)*.
3. HeBS buffer for calcium phosphate precipitation: 280 mM NaCl, 50 mM HEPES, 1.5 mM Na$_2$HPO$_4$, 10 mM KCl, 12 mM dextrose. Adjust pH to 7.05–7.12 at room temperature. Filter-sterilize (0.22-μm membrane), divide into aliquots, and store at -20°C (*see* **Note 2**).
4. Calcium chloride solution for calcium phosphate precipitation: 2.5 M CaCl$_2$. Filter-sterilize (0.22-μm membrane) and store at room temperature.

5. Chloroquine solution (100X): 10 m*M* chloroquine. Filter-sterilize through a 0.22-μm membrane, divide into aliquots, and store at $-20°C$ (solution is light-sensitive).

3. Methods

3.1. Culture of Human PBMC-Derived DCs

1. Draw 20 mL of peripheral blood into heparinized blood collection tubes.
2. Dilute peripheral blood 1:1 with RPMI 1640 cell-culture medium (25°C) in a 50-mL sterile plastic tube inside a tissue-culture hood.
3. Pipette 15 mL of Lymphoprep™ reagent (25°C) into two 50-mL sterile centrifuge tubes, and carefully overlay each with 20 mL of RPMI-diluted blood.
4. Centrifuge the tubes at 400*g* for 40 min at 25°C (with centrifuge brake off).
5. Carefully pipet the interphase fraction containing the PBMCs into a fresh 50-mL sterile plastic tube, add 20–30 mL of RPMI 1640, and centrifuge at 450*g* for 10 min at 25°C.
6. Resuspend the cell pellet by finger-flicking, add 30 mL of RPMI 1640, and centrifuge at 400*g* for 5 min at 25°C; repeat this step once.
7. Resuspend the cell pellet in 3–5 mL of warm (37°C) CellGro® DC medium (without cytokines). Mix a small sample of the cells with an equal volume of trypan blue (0.4%) and count the cells in a hemocytometer; $3–5 \times 10^7$ viable PBMCs are routinely obtained with this procedure.
8. Adjust PBMC density to 5×10^6 /mL with CellGro® DC medium (without cytokines), and transfer the cell suspension to a sterile tissue-culture flask (25 cm^2 for 5–7 mL, 75 cm^2 for 8–12 mL).
9. Incubate for 2 h at 37°C and 5% humidified CO_2.
10. Carefully remove by pipetting the culture medium that contains nonadherent PBMCs and discard.
11. Rinse off nonadherent PBMCs by adding 8–15 mL of warm phosphate-buffered saline (PBS), carefully move flask in cross-shape, and remove PBS; repeat this step three times, discarding the wash each time.
12. Add warm (37°C) CellGro® DC medium containing rhIL-4 and rhGM-CSF (e.g., 10 mL for a 75 cm^2 flask).
13. Two days later (DC culture d 2), add fresh cytokines (1000 IU/mL rhIL-4 and 1000 IU/mL rhGM-CSF) directly into the culture flask.
14. On DC culture d 4, add 1000 IU/mL rhIL-4 and 1000 IU/mL rhGM-CSF. To facilitate in vitro DC maturation, add 10 ng/mL rhTNF-α, 10 ng/mL rhIL-1β, 1000 IU/mL rhIL-6, and 1 μg/mL PGE_2.
15. On d 7 or 8 of culture, the majority of the cells are nonadherent, and the large adherent cells (20–40% of all cells) are mature DCs by morphologic and immunophenotypic criteria.

3.2. Lentiviral Vector Production and Titration

1. The day before transfection, seed $0.8–1.5 \times 10^7$ HEK 293T cells per 15-cm tissue-culture plate into 15 mL of DMEM, 10% FBS. One d later, the cells should be 70–80% confluent. It is recommeded starting with five 15-cm tissue-culture plates.

2. Mix 1,000 µL of chloroquine solution (100X) with 11.5 mL of warm DMEM medium and add 2.5 mL of the solution carefully to a single 15-cm tissue-culture plate just prior to transfection.

3. In a 50-mL sterile plastic tube, mix 350 µg of lentiviral transfer plasmid, 150 µg of packaging plasmid, and 150 µg of envelope plasmid. Adjust to 5,625 µL with distilled H_2O.

4. Add 625 µL of 2.5 M $CaCl_2$ solution (25°C).

5. While bubbling the mixture with a pipet-aid, slowly add (one drop every other second) 6250 µL of HeBS buffer solution (25°C). Vortex the precipitation mixture for 5 s at maximum speed.

6. Wait for 1 min (not longer!), and then add 2.5 mL of the calcium phosphate-DNA precipitate mixture dropwise onto each tissue-culture plate. Distribute the precipitate by carefully moving the dish in cross-shape. Incubate at 37°C and 5% humidified CO_2.

7. Six to eight hours later, a very fine precipitate should be visible all over the plate by phase-contrast microscopy (*see* **Note 2**). Remove the media containing chloroquine as well as the co-precipitated calcium phosphate and vector DNA. Slowly add 20 mL of warm (37°C) DMEM medium. Incubate plates at 37°C and 5% humidified CO_2.

8. After 60–72 h, harvest the tissue-culture medium and centrifuge at 3000–4000g to separate the supernatant containing lentivector particles from cellular debris. At this point, the supernatant should contain between 10^5 and 10^6 vector particles per mL.

9. To concentrate the VSV-G pseudotyped lentiviruses (100–1000-fold), centrifuge the supernatant at 100,000g for 2 h at 4°C in an ultracentrifuge (e.g., Sorvall Surespin™ 630 rotor). Each ultracentrifuge tube has a volume of 36 mL.

10. Discard the supernatant and resuspend the pellet in 3–5 mL of CellGro® medium by up-and-down pipetting and end-over-end rotation at 4°C for 2 h. Pool the resuspended vector pellets from different tubes, then centrifuge at 3000–4000g to remove protein precipitates, and repeat **step 9**.

11. Resuspend in 100–200 µL of CellGro® medium per ultracentrifuge tube, divide into aliquots, and freeze concentrated virus at −80°C. Depending on the starting volume and initial vector titer, a final concentration of 10^8 to 10^9 transducing vector particles per mL is expected. Starting with five tissue-

culture plates a total amount of 5×10^7 transducing vector particles can be expected.

3.3. Titering Lentivectors Expressing Green Fluorescent Protein

1. Seed 1×10^5 HEK 293T cells per well (1–2 mL) into 6-well tissue-culture plates 24 h before lentiviral transduction.
2. Prepare serial 10-fold dilutions of the lentiviral vector concentrate in 1000 µL of DMEM.
3. Remove cell culture medium from the 6-well plates, and carefully add vector dilutions to adherent HEK 293T cells.
4. After 3 d, determine the frequency of reporter gene expressing 293T cells by fluorescence-activated cell sorting (FACS) analysis.
5. Estimate the lentivector titer according to the following formula: titer = F $\times 2 \times C_0/V \times D$ (D = virus dilution factor; V = volume of inoculum; F = frequency of reporter-positive 293T cells; C_0 = number of target cells at the time of seeding). Because roughly one cell division occurs between the time of seeding and the time of transduction (24 h), the total number of 293T cells at the time of transduction can be estimated as twice the number at the time of seeding.

3.4. Lentiviral Transduction of DCs

1. Transfer DCs on culture d 3 or later (from **Subheading 3.1.** following **step 13**) to a 24-well tissue culture plate at a density of $1–5 \times 10^5$ DCs per well (count only the large cells).
2. Add concentrated lentivirus to achieve vector concentrations of 1×10^9 to 1×10^7 in a transduction volume of 200 µL.
3. Incubate at 37°C, 5% CO_2 for 8–12 h.
4. Adjust the culture volume to 1000 µL per well using CellGro® DC medium. Depending on the lentiviral vector (transcription control by LTR vs internal promoter; e.g., CMV vs EF1α), transgene expression can be detected 24–48 h after transduction.

4. Notes

1. The following procedures were used to transduce human PBMC-derived DCs with a VSV-G pseudotyped lentiviral vector encoding hrGFP under the transcriptional control of an internal CMV promoter (TRIPΔU3-CMV hrGFP). DCs (5×10^5) were transduced on d 3 of culture in CellGro® medium at vector concentrations of 5×10^8, 5×10^7, and 5×10^6 "infectious" units (IU)/mL. One d after transduction, in vitro maturation of DCs were induced by addition of rhTNF-α (10 ng/mL), rhIL-1β (10 ng/mL), rhIL-6 (1000 IU/mL), and PGE_2 (1 µg/mL). Three d after transduction,

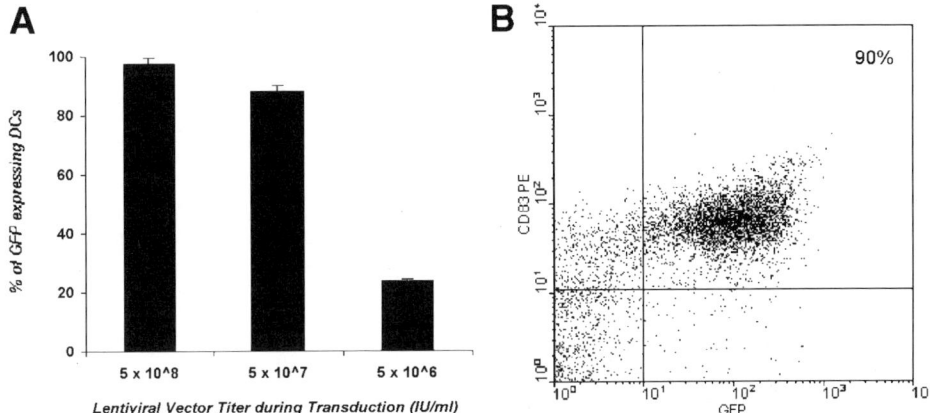

Fig. 1. Lentiviral transduction of DC. DC were transduced at d 3 of culture at various concentrations of the lentivector TRIPΔU3-CMV hrGFP. In vitro maturation of DC was facilitated by addition of a TNF-α, IL-1β, IL-6, PGE$_2$. Three d after transduction, cellular expression of the reporter protein hrGFP and of the DC cell-surface marker CD83 were analyzed by FACs. As depicted in **Fig. 1A**, transduction efficiencies of approx 98% at a lentiviral titer of 5 $\times$ 10^8 IU/mL and 25% at a titer of 5 $\times$ 10^6 IU/mL were observed. The majority of GFP expressing DC were found to be positive for CD83 (**Fig. 1B**)

FACS analyses were performed to determine the percentages of hrGFP-expressing DCs in the cell population with high FSC (forward scatter) and SSC (side scatter) profiles characteristic of mature DCs (**Fig. 1A**). In addition, the immunophenotypic features of the transduced DCs were analyzed. The majority of transduced (green fluorescent protein [GFP]+) DCs expressed high levels of DC-typical markers: CD80, CD86, HLA-ABC, HLA-DR, and CD83 (**Fig. 1B**). These results demonstrate that human PBMC-derived DCs can be efficiently transduced by lentiviral vectors.

2. Efficient lentivector production requires optimization of transient transfection of HEK 293T cells by DNA-calcium phosphate coprecipitation. The main factors influencing the precipitation reaction are the concentrations of calcium and phosphate as well as the pH of the HeBS buffer. Other parameters such as temperature, DNA concentration, and reaction time determine the quality of the DNA-calcium phosphate precipitate *(19)*. As a guideline, transfection efficiencies of 60–80% (e.g., determined by a GFP-encoding transfer lentivector) should be achieved; otherwise the conditions for the precipitate reaction should be carefully tested and optimized.

References

1. Steinman, R. M. (1991) The dendritic cell system and its role in immunogenicity. *Ann. Rev. Immunol.* **9,** 271–296.
2. Banchereau, J. and Steinman, R. M. (1998) Dendritic cells and the control of immunity. *Nature* **392,** 245–252.
3. Hart, D. N. J. (1997) Dendritic cells: unique leukocyte populations which control the primary immune response. *Blood* **90,** 3245–3287.
4. Arthur, J. F., Butterfield, L. H., Roth, M. D., Bui, L. A., Kiertscher, S. M., Lau, R., et al. (1997) A comparison of gene transfer methods in human dendritic cells. *Cancer Gene Ther.* **4,** 17–25.
5. Diao, J., Smythe, J. A., Smyth, C., Rowe, P. B., and Alexander, I. E. (1999) Human PBMC-derived dendritic cells transduced with an adenovirus vectorinduce cytotoxic T-lymphocyte responses against a vector-encoded antigen in vitro. *Gene Ther.* **6,** 845–853.
6. Szabolcs, P., Gallardo, H. F., Ciocon, D. H., Sadelain, M., and Young, J. W. (1997) Retrovirally transduced human dendritic cells express a normal phenotype and potent T-cell stimulatory capacity. *Blood* **90,** 2160–2167.
7. Bello-Fernandez, C., Matyash, M., Strobl, H., Pickl, W. F., Majdic, O., Lyman, S. D., and Knapp, W. (1997) Efficient retrovirus-mediated gene transfer of dendritic cells generated from CD34+ cord blood cells under serum-free conditions. *Hum. Gen. Ther.* **8,** 1651–1658.
8. Chischportich, C., Bagnis, C., Galindo, R., Mannoni, P. (1999) Expression of the nisLacZ gene in dendritic cells derived from retrovirally transduced peripheral blood CD34+ cells. *Haematologica* **84,** 195–203.
9. Naldini, L., Blomer, U., Gallay, P., Ory, D., Mulligan, R., Gage, F. H., et al. (1996) In vivo gene delivery and stable transduction of nondiving cells by a lentiviral vector. *Science* **272,** 263–267.
10. Miyoshi, H., Smith, K. A., Mosier, D. E., Verma, I. M., and Torbett, B. E. (1999) Transduction of human CD34+ cells that mediate long-term engraftment of NOD/SCID mice by HIV vectors. *Science* **283,** 682–686.
11. Chinnasamy, N., Chinnasamy, D., Toso, J. F., Lapointe, R., Candotti, F., Morgan, R. A., and Hwu, P. (2000) Efficient gene transfer to human peripheral blood monocyte-derived dendritic cells using human immunodeficiency virus type 1-based lentiviral vectors. *Hum. Gene Ther.* **11,** 1901–1909.
12. Schroers, R., Sinha, I., Segall, H., Schmidt-Wolf, I. G., Rooney, C. M., Brenner, M. K., et al. (2000) Transduction of human PBMC-derived dendritic cells and macrophages by an HIV-1-based lentiviral vector system. *Mol. Ther.* **1,** 171–179.
13. Dyall, J., Latouche, J. B., Schnell, S., and Sadelain, M. (2001) Lentivirus-transduced human monocyte-derived dendritic cells efficiently stimulate antigen-specific cytotoxic T lymphocytes. *Blood* **97,** 114–121.
14. Gruber, A., Kan-Mitchell, J., Kuhen, K. L., Mukai, T., and Wong-Staal, F. (2000) Dendritic cells transduced by multiply deleted HIV-1 vectors exhibit normal phenotypes and functions and elicit an HIV-specific cytotoxic T-lymphocyte response in vitro. *Blood* **96,** 1327–1333.

15. Dardalhon, V., Herpers, B., Noraz, N., Pflumio, F., Guetard, D., Leveau, C., et al. (2001) Lentivirus-mediated gene transfer in primary T cells is enhanced by a central DNA flap. *Gene Ther.* **8,** 190–198.

16. Sambrook, J., Fritsch, E. F., and Maniatis, T. (eds.) (1989) *Molecular Cloning: A Laboratory Manual,* 2nd ed. Cold Spring Harbor Laboratory Press, Cold Spring Harbor, NY.

17. Zufferey, R., Nagy, D., Mandel, R. J., Naldini, L., and Trono, D. (1997) Multiply attenuated lentiviral vector achieves efficient gene delivery in vivo. *Nat. Biotechnol.* **15,** 871–875.

18. Jonuleit, H., Kuhn, U., Muller, G., Steinbrink, K., Paragnik, L., Schmitt, E., et al. (1997) Pro-inflammatory cytokines and prostaglandins induce maturation of potent immunostimulatory dendritic cells under fetal calf serum-free conditions. *Eur. J. Immunol.* **27,** 3135–3142.

19. Jordan, M., Schallhorn, A., and Wurm, F. M. (1996) Transfecting mammalian cells: optimization of critical parameters affecting calcium-phosphate precipitate formation. *Nucleic Acids Res.* **24,** 596–601.

VI

DELIVERY USING RETROVIRUSES

31

Gene Transfer by Retroviral Vectors

An Overview

Nikunj Somia

1. Introduction

Viruses have evolved to deliver their genetic cargo to cells and, due to the pathogenicity of some viruses, this process has been the subject of a great deal of study. In this respect, retroviruses came to the fore in the early 1900s with the demonstration by Ellermann and Bang (1908) and by Rous (1911) that chicken leukosis was caused by a virus, now referred to as avian sarcoma/leukosis virus (ASLV). This began a body of work that led to the identification of virus-induced tumors in mammalian species and retroviruses (as they are now called) were identified as the causative agents for a number of pathologies from tumors to acquired immunodefiencey syndrome (AIDS) (elegantly reviewed in refs. *[1,2]*). Retroviruses were characterized as RNA viruses that replicate through a DNA intermediate. The study of retroviruses contributed and led to the elucidation of a number of diverse biological phenomena as the oncogenes carried by these viruses began to be identified as receptors, kinases, and transcription factors *(3)*. The idea that these viruses could be used to ferry a gene of choice, rather than the viral genome, springs from the study of Rous sarcoma virus (RSV).

In pioneering studies looking at the infection of single cells by single viral particles, it was observed that a single viral particle of RSV was sufficient to transform a cell. However, the transformed cell could not produce infectious

From: *Methods in Molecular Biology, vol. 246:*
Gene Delivery to Mammalian Cells: Vol. 2: Viral Gene Transfer Techniques
Edited by: W. C. Heiser © Humana Press Inc., Totowa, NJ

transforming particles *(4,5)*. For this to occur, a second, superinfection, of this cell with a replication competent, but nontransforming retrovirus is required. This led to the idea of a replication defective retrovirus that carried the transforming gene but lacked viral genes needed for replication. Furthermore the products of these missing genes could be provided *in trans* if the genome of a replication competent virus was present in the same cell *(6,7)*.

The classification of retroviruses has gone through some change as the field has evolved. Initially, the viruses were grouped based on virion morpholgy as viewed under the electron microscope (types B, C, and D). Subsequently, nucleotide sequencing of the viral genomes led to the present classification of the retroviruses into seven genera. These can be further studied at http://www.ncbi. nlm.nih.gov/ICTVdb/Ictv/fr-index.htm. Today, with regard to vectors, reference is also made as a pedestrian classification to "simple" and "complex" retroviruses. The former refers to "oncoviruses" such as Murine Leukemia Virus (MLV) and Avian Leukosis Virus (ALV) currently in the Alpharetrovirus and Gammaretrovirus group (formally in the C-type; see **Fig.1**). These viruses typically contain *gag*, *pol*, and *env* genes. The "complex" retroviruses comprise viruses from groups such as the lentivirus and spumaviruses. These contain additional accessory genes along with the basic *gag*, *pol*, and *env* genes *(8)*. Viral vectors based on both simple and complex viruses have been developed. This overview will only consider the viral vectors based on MLV and ALV, because lentiviral vectors are covered by Tal Kafri (Chapter 25) and other reviews cover foamy virus vectors (from the spumavirus group) *(9)*.

2. The Retrovirus Life Cycle

2.1. Envelopes and Receptors

The retroviral life cycle (*see* **Fig. 2**) begins with the binding of the retrovirus envelope to a receptor on the target cell. The envelope protein is synthesized from the viral genome as a polyprotein with a signal sequence and so enters the secretion pathway. The polyprotein is cleaved in this pathway to form the surface (SU) and transmembrane (TM) portions of the envelope, linked by di-sulphide bonds *(10)*. Retroviral receptors and entry co-factors determine the tropism of the retrovirus. Hence retroviruses are also referred to by the nature of the target cells they infect *(11)*. For example the MLV class of viruses are called ecotropic (infecting only murine cells), amphotropic (infecting most mammalian cells), or xenotropic (infecting most mammalian cells except some murine cells). This tropism is mainly determined by the sequence variation in the otherwise highly related envelope proteins that determine binding to different receptors. The receptors for this particular family of retroviruses are a basic amino acid transporter (ecotropic) *(12)*, phosphate transporters (amphotropic)

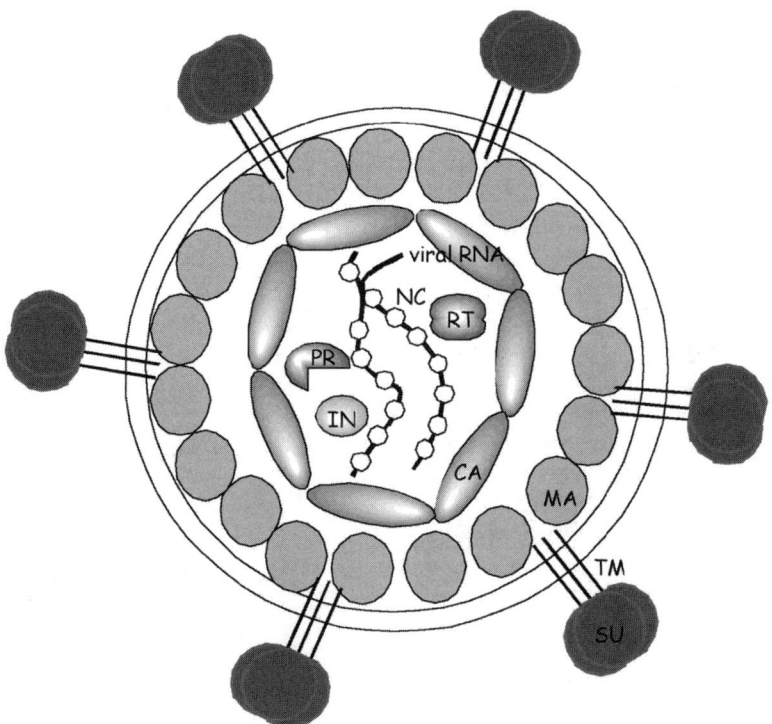

Fig. 1. Schematic structure of retroviruses. An 80–100 nm particle that contains the viral RNA genome encapsulated by protein and lipid molecules. A typical retrovirus consists of double-stranded RNA encapsulated by a protein core surrounded by a lipid bilayer. The protein core is comprised of matrix (MA), capsid (CA), and nucleocapsid (NC). These proteins are the proteolytic cleavage products of a polyprotein which is the product of the *gag* gene. Also contained in the core are three enzymes, protease (PR), reverse transcriptase (RT), and integrase (IN), which are derived by proteolysis of a polyprotein product of the *pol* gene. The lipid bilayer is studded with the product of the envelope (*env*) gene of the retrovirus, a polyprotein that is cleaved into surface (SU) and a transmembrame (TM) portions.

(13,14), and a putative G-protein coupled protein of unknown function (xeno-tropic) *(15–17)*. The pathology caused by retroviral infection is reflected in part by the spectrum of cells that are infected, and this in turn is a function of enve-lope-receptor interaction.

To date, crystal structures for the SU portion *(18)* and separately the TM *(19)* portions of the ecotropic envelope and the HIV envelope *(19)* have been eluci-dated. These show structural homology to defined structures of the influenza

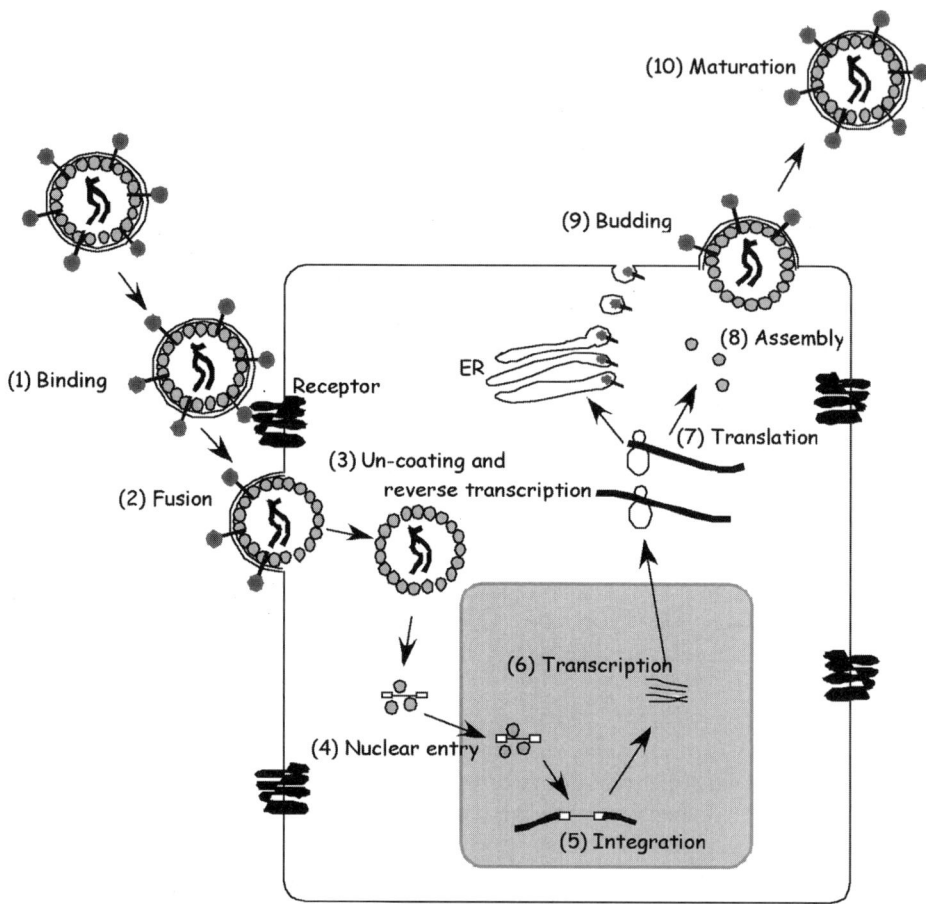

Fig. 2. Retroviral life cycle. This is an illustration of the typical life cycle of most retroviruses. (1) Binding of the retroviral envelope protein to the receptor on the target cell. (2) Binding is followed by a fusion event between the lipid membrane of the virus and the cellular membrane. (3) The nucleoprotein complex uncoats and the RNA genome is copied into DNA. (4) Nuclear entry requires cell division for MLV and ALV but is an active process for HIV. (5) The viral DNA is integrated into the host chromosome. (6) The viral DNA is transcribed into full length and spliced messages that are (7) translated. (8) The full-length RNA and viral proteins are assembled and (9) bud from the cell and the action of the protease results in (10) maturation of the virus, which repeats the life cycle. Adapted from Freed E.O. (*Virology* **251**, 1–15, 1998).

virus surface protein HA. The biology of this latter protein has been studied in some detail, and so it is presumed that events following envelope binding to the receptor follow a similar pathway *(20)*. Specifically, binding induces a conformational change that exposes a fusogenic peptide sequence present on the N-terminus (the ectodomain) of TM. This mediates, for most retroviruses, a fusion between the viral membrane and the cell membrane. In the case of some, i.e., ecotropic envelope, binding to the receptor initiates endocytosis of the viral particle. The acidic pH in the endosome is thought to mediate another conformational change that fully licenses the fusogenic peptide and leads to a fusion between the membrane of the endosome and the viral membrane. In either case, the end result is the entry of the viral core into the cytoplasm of the target cell.

2.2. Reverse Transcription and Viral Replication

On entering the cell the retroviral particle uncoats, a process that is still not clearly understood. The result is a nucleoprotein complex (consisting of the structural and catalytic components of the retrovirus, as well as the two copies of the RNA genome). There is some evidence that limited DNA synthesis occurs before entry of the virus into the target cell *(21)*; however, the bulk of the reverse transcription and DNA synthesis occurs after viral entry. The block to synthesis and the environment making the template permissive for synthesis are not yet defined, though genetic evidence (point mutants) suggests that the structural proteins of the retrovirus are critical in this process *(22)*. Furthermore the role of host proteins in this process needs to be elucidated, as exemplified in the case of HIV, where cyclophilins have been shown to bind to gag and are needed for early entry events *(23)*.

The main actor in the process of DNA synthesis is reverse transcriptase, a DNA polymerase that can utilize both RNA and DNA as a template. This enzyme also has a nuclease activity (RNase H) that hydrolyses RNA in an RNA:DNA duplex *(24)*. The strategy of replication is better understood by illustrating the DNA stage of the virus (the proviral stage) and the how the virus copes with the ends for replication. **Figure 3** shows a schematic of the proviral genome, defined by the beginning and end of a long terminal repeat (LTR) *(25,26)*.

These repeats are further divided into three regions: U3, R, and U5. The machinery of transcription is directed by sequences in the U3 region, which constitutes the promoter/enhancer region of the retrovirus. The R region is delineated by the start of transcription in the 5′ LTR and the polyadenlyation site at the 3′ LTR. This results in the transcription of a genomic RNA molecule that misses the U3 at one end and the U5 at the other. The replication of the RNA molecule illustrated in **Fig. 4** counters this problem, by utilizing the existing U3 and U5 twice as templates for DNA synthesis. This feature of the replication cycle is used in the development of SIN and double copy vectors, as outlined later *(27,28)*.

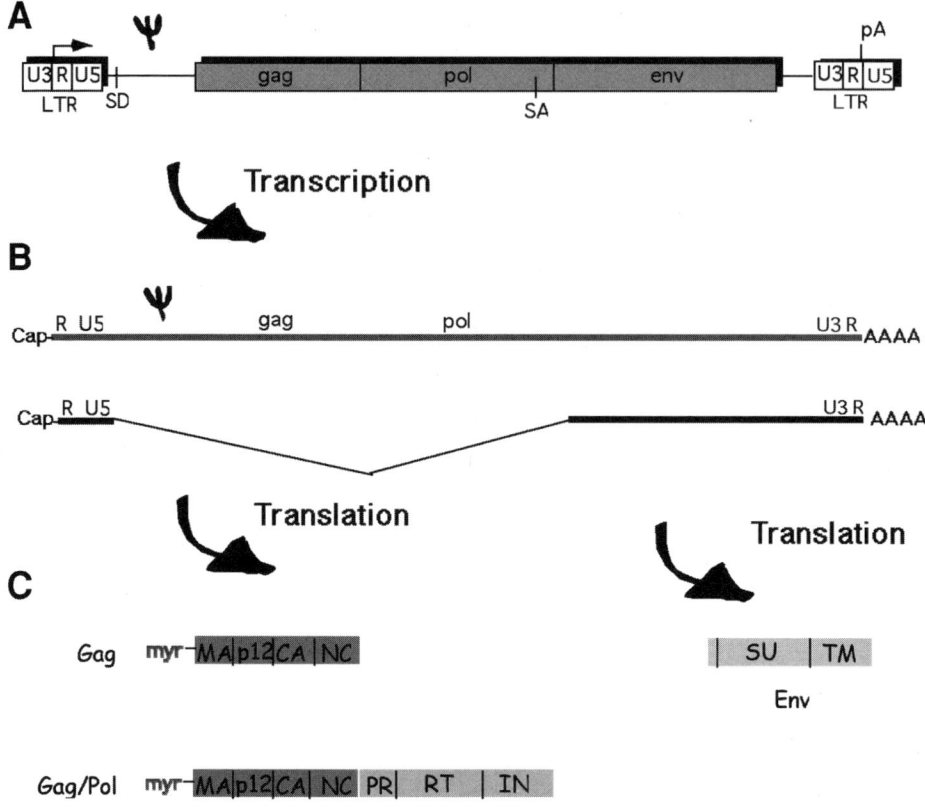

Fig. 3. The genomic structure of **(A)** an integrated provirus and its **(B)** transcription and **(C)** translation products. (A) The viral genome consisting of the gag, pol, and env genes with the long terminal repeats (LTRs) subdivided into the U3, R, and U5 regions. SD and SA are splice donor and acceptor sites. ψ defines the packaging sequence. (B) This integrated DNA is transcribed into both full length and spliced transcript. (C) A Gag and Gag-Pol polypeptide are translated from the full genomic transcript, while the envelope (Env) protein is translated from the subgenomic transcript. The Gag polyprotein has a myristate modification on its N-terminus, and encodes the matrix (MA), p12, capsid (CA), and nucleocapsid (NC), which are liberated by proteolysis. Readthrough suppression or ribosome frameshifting results in the production of a larger Gag-Pol polyprotein. This also encodes the protease (PR), reverse transcriptase (RT), and integrase (IN) proteins.

Fig. 4. Replication of the retrovirus genome. DNA synthesis is initiated by reverse transcriptase from a primer tRNA molecule that binds a primer binding site (pbs). This reaction proceeds to the end of the 5′ LTR, and the DNA:RNA dependent RNAse H of RT removes the hybridized RNA. The cDNA hybridizes to the R region within the

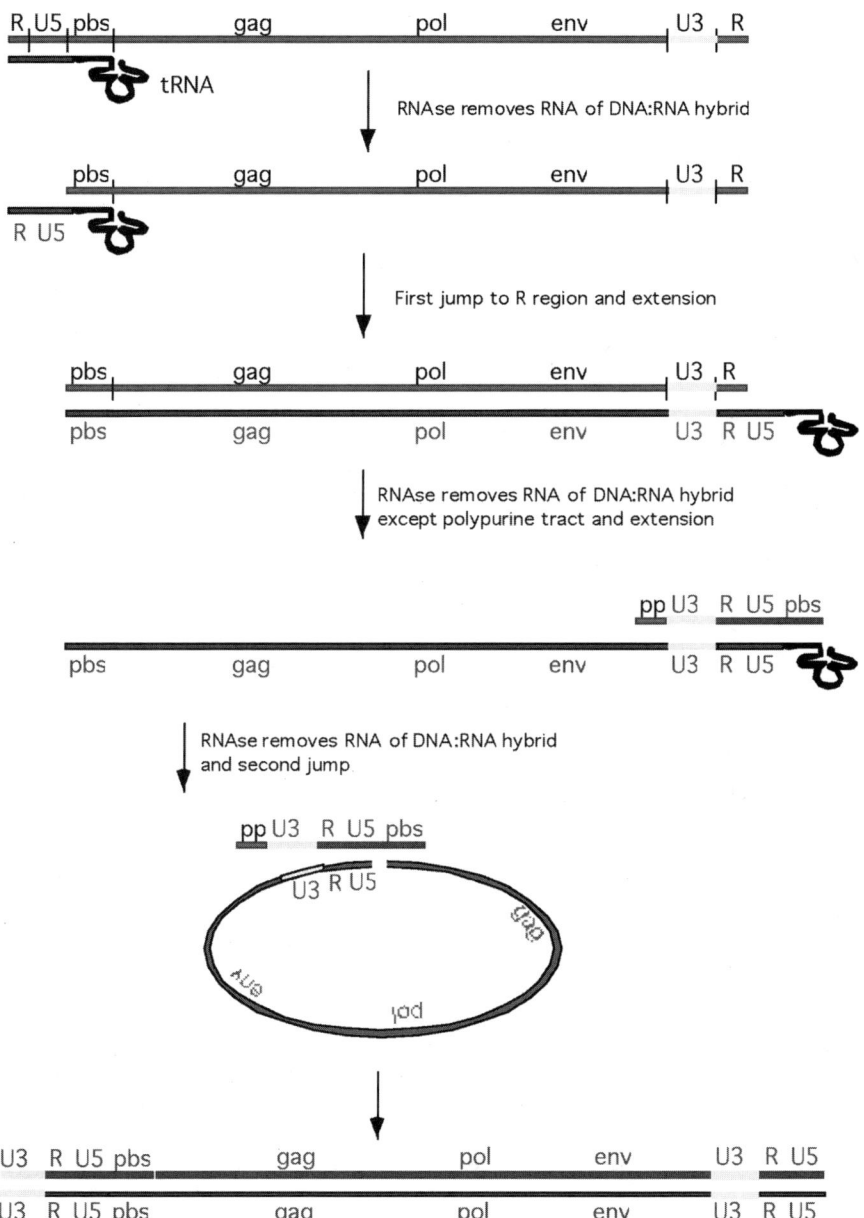

3′-LTR (first jump), then the DNA is extended the length of the genome. RNase then removes most of the hybrid RNA except a small polypurine (PP) tract. RNase removes the tRNA and the short DNA strand hybridizes to the pbs sequence at the 3′-LTR of the first DNA strand (second jump). Reverse transcriptase then completes copying the second DNA strand. It is this double-stranded DNA that integrates into the host chromosome as the provirus.

2.3. Integration

Following replication, the viral DNA is integrated into the host chromosomal DNA. The reverse transcription reaction results in a nucleoprotein complex comprising viral and cellular host proteins. The exact nature of this pre-integration complex for various viruses is still uncertain, owing partly to the small fractions of the complexes that form a true pre-integration complex. However, viral integrase is the defining member of the complex, as well as a linear viral DNA that is the substrate for the integrase. Other proteins include the capsid protein for Moloney murine leukemia virus (MMLV) and the matrix protein for HIV. This pre-integration complex has to be positioned next to chromosomal DNA, and hence has to cross the nuclear membrane. As discussed by Kafri in this volume, this can be achieved by lentiviruses by virtue of nuclear localization signals in various proteins of HIV. For MLV, this juxtapositioning requires the breakdown of the nuclear membrane at cell division. Once in the nucleus, the linear DNA product has a number of fates. Three nonproductive paths are: (1) intermolecular ligation of the linear molecule, giving rise to a 2 LTR circle; (2) an intermolecular recombination between the two LTRs in the linear molecule, giving rise to a one LTR circle; and (3) an autointegration product that results in circles with varying fragments of the viral DNA. The productive path is the integration of the linear viral DNA into the chromosomal DNA. The integrase enzyme acts on the blunt-ended viral DNA and removes the terminal two bases from the 3′ ends, and these are used in the integration reaction, where the 3′ hydroxyl groups of the viral DNA attack the phosphodiester bonds of the chromosomal DNA. The two ends of the viral genome are integrated into the host chromosome separated by 4–6 bases depending on the viral integrase *(29,30)*. These single-stranded bases are then replicated by a polymerase reaction in the nucleus, resulting in base duplication at the integration site. The integration site is defined with respect to the viral DNA, is somewhat random with respect to the chromosomal DNA, but is presumably constrained by the access to DNA bound in nucleosomes. Indeed, recent studies on HIV suggest that these integration events occur preferentially into transcriptionally active regions of the host chromosome *(31)*. Large-scale mutagenesis efforts utilizing retroviral integration have also identified hotspots for integration *(32)*.

2.4. Gene Expression

Once integrated into the genome, the retrovirus DNA, the provirus, behaves like a cellular RNA polymerase II-directed gene. The U3 region of the retroviral RNA contains various enhancers that dictate binding of cellular transcription factors. This region also contains a TATA box that recruits the basal transcription machinery to the promoter. A striking feature of the U3 region is the large number of transcription factor binding sites, some of which, in any given cell type, may be redundant *(33,34)*. Thus, this dense packaging of binding sites allows the virus to replicate in a variety of target cells. The composition of the viral LTRs will dictate the efficiency of transcription for any given cell type and also any pathology that is caused by the virus *(35)*. For example, transcription of MMLV is seen in a number of cell types from fibroblasts to lymphoid cells, but is repressed in embryonal carcinoma and embryonal stem cells. This restriction is due to the presence of negative regulatory elements in the proviral DNA *(36)*. The mouse mammary tumor virus (MMTV) has steroid responsive regulatory elements that dictates the expression of the viral RNA *(37)*. Hence, the expression of viral RNAs can be both repressed and enhanced by various signal-transduction pathways.

Transcription initiation occurs at the start of the R region in 5' the LTR, and the RNA is processed by polyadenylation at the end of the R region in the 3' LTR *(38)*, the result of a polyadenylation signal in the U3 region (in MMTV and ASLV) or at the R/U5 boundary region (in MMLV and HIV). The presence of a polyadenylation signal at R/U5 boundary would result in a very short transcript at the 5' LTR because transcription initiation occurs at the beginning of the R region (*see* **Fig. 3**). This problem is overcome by negative regulation of the 5'-polyadenylation signal due to the proximity of the capsite and/or the splice sites *(39)*. The RNA is also processed by splicing. For the C-type retroviruses, two transcripts and possibly three *(40)* are produced (*see* **Fig. 3**): a genomic LTR-LTR message, and a subgenomic message that codes for the envelope protein. This simple choice is probably mediated by a complex set of signals from suppression of splicing to transport elements that ferry the genomic message out from the nucleus *(41)*. The result is the appearance of a constant steady-state ratio of genomic and subgenomic messages.

The genomic message serves as the template for translation of two polyproteins, Gag and Gag-Pol. The translation of Gag is controlled because there are some ATG codons ahead of the *bona fide* initiation methionine. A number of mechanisms have been proposed to explain this, and the complex secondary structure that also serves as the packaging sequence may in part be responsible. For MLV, this region is proposed to act as an internal ribosome entry site *(42)*. The Gag protein is co-translationally modified by addition of a fatty acid to its N-terminus, typically myristate. This modification is necessary but not sufficient for binding of the Gag and Gag-Pol polyproteins to membranes, where the retrovirus is assembled *(43,44)*.

The translation of the gag polyprotein occurs from the initiating methionine to the stop codon. However the Gag-Pol polyprotein requires a bypass of this stop codon. This is achieved by either readthrough suppression (e.g., for MLV) *(45)* or frameshift suppression (e.g., for ALV) *(46)*. For readthrough suppression, the stop codon for Gag (UAG in MLV) is misread as a glutamine (CAG codon). This readthrough is dictated by *cis*-acting sequences and results in one Gag-Pol product for every 20 Gag proteins. For frameshift suppression, *cis*-acting sequences dictate a slippage of the ribososme by −1, leading again to about 5% Gag-Pol product. The translation of envelope protein occurs from the subgenomic message and its signal sequence directs its co-translational insertion into the endoplasmic reticulum (ER) where the protein is processed in a complex manner. Envelope is heavily glycosylated, cleaved into a surface (SU) and a transmembrane (TM) component by ER resident protease systems and oligomerizes into trimers and higher structures *(10)*. Once synthesized, the Gag, Gag-Pol, and Env proteins are all directed to a site on the membrane for assembly. The choice and mechanism by which these sites are chosen are as yet not known. The Gag protein is the driving force for this assembly; indeed, the 2000 or so molecules of Gag found in each cell are sufficient to generate a core particle encapsidating viral RNA *(47)*.

2.5. Packaging of the RNA

It became clear from the study of transforming retroviruses that sequences 5′ of the *gag* gene contained sequences that direct the preferential packaging of full length viral RNA into particles. This region, referred to as Psi (ψ) or the encapsidation sequence (E), was experimentally shown to be important for directing packaging by deletion analysis *(48,49)*. Similar analysis established that sequences within the gag gene contained an extended packaging sequence that can improve the efficiency of packaging, though this may depend on the cell type or levels of expression of the transcript *(50)*. These loss-of-function analysis were complemented with gain-of-function assays that demonstrated that these sequences could also direct packaging of heterologous transcripts. Furthermore, comparison of this region among retroviruses shows no sequence homology, and hence it may be the secondary structure of this region that directs the preferential packaging (some 200-fold over a cellular RNA) of this transcript *(51)*. The RNA is packaged as a dimer and the lack of packaging by mutants in the nucleocapsid (NC) of Gag implicates this zinc-finger protein in the packaging process *(52,53)*. There is also the incorporation of a specific tRNA that acts as the primer for RT, and this tRNA is highly enriched over other tRNA molecules in the viral particle. Mutants in RT do not show this enrichment, implicating this protein in recruiting the tRNA into the virion *(54,55)*.

2.6. Budding

The assembled components at the membrane bud as an immature particle from the membrane. The controlling mechanisms for this process are unclear. Electron microscopy of wild-type and mutant viruses has revealed that on budding the virion particles mature into infectious particles. The processing of the Gag and Gag-Pol polyproteins by the viral protease liberates the individual proteins that lead to the change in morphology of the particle. It is believed that this processing also changes the function of the Gag-derived proteins, allowing them to play a role in the infection process *(56,57)*. The binding of this virus through its envelope to its cognate receptor on the surface of the host cell will initiate the life cycle once more.

3. Packaging Cell Lines

3.1. Elements from the Life Cycle Needed for Packaging

Retroviral vectors are made by subverting the life cycle of a retrovirus to effect efficient gene transfer. There are basically two elements that comprise the system: a vector and a helper. The latter provides all of the proteins *in trans* to generate a virus particle. However, the RNA produced lacks the packaging sequence and so cannot be packaged into a viral particle. The vector produces a transcript that contains all of the control elements for reverse transcription and integration (provided by components of the LTR), the packaging sequence, and the gene of interest (*see* **Fig. 5**).

This transcript is packaged into the viral particle and buds off the cell. A retrovirus transduction is then simply achieved by taking the culture media from this producer cell and incubating it on target cells. The envelope then binds the receptor (providing it is present) and recapitulates the first part of the viral life cycle. Upon integration the gene of interest is expressed. No more viruses can be produced because the system now lacks any viral proteins.

3.2. Early Cells Lines and Refinements

The earliest naturally occurring packaging cells were those that were used to produce virus particles that caused transformation. These cells produced both a virus that ferried the oncogene and a wild-type virus. Using this approach, genes other than oncogenes were used *(58)*, but the utility of such a system was limited as a result of the presence of replication-competent virus that also infected the target cell. Cell lines that carried a virus that had a defective packaging signal, and a vector that carried a gene and the packaging signal between the LTRs, gave this system of gene transfer greater utility. These first-generation packaging cells, e.g., ψ-2 (based on MMLV) *(49)* and C3A2 (based on avian REV-A) *(59)*, produced vectors with titers of 10^7/mL. This is a measure of the efficiency of gene transfer, that is, 10^7 cells were transduced with a marker gene using 1 mL of the supernatant from the producer cells. In early 1980 (and to an extent today), this viral mode of gene transfer was orders of magnitude better than the physical methods for gene transfer. The envelope proteins on these viral vectors

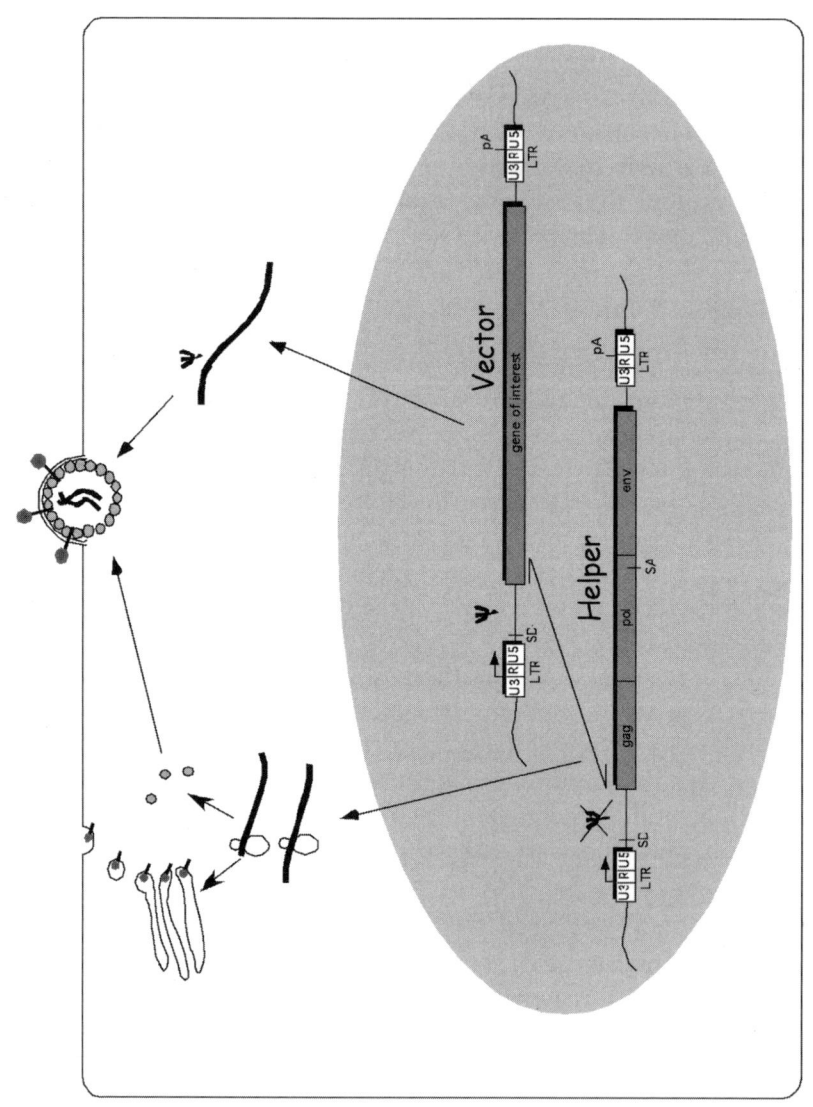

(determined by the parent virus used to make the packaging cell lines) had a limited tropism for murine, avian, and some mammalian cells. The development of ψ-AM *(60)* presented a cell line with an amphotropic envelope that was permissive for a number of different mammalian species. This greatly increased the utility of retroviral vectors.

3.3. The Production of Replication-Competent Helper Virus

The problem with the first generation of packaging cell lines was that it required one recombination event (*see* **Fig. 5**) between the vector and helper sequences to generate a replication-competent retrovirus (RCR). These viruses can be detected in a number of assays from the polymerase chain reaction (PCR; e.g., to check for continued transfer of gag-pol genes) *(61)* to biological assays *(62)*. The most sensitive of these is the marker rescue assay, in which a cell that harbors an integrated, selectable, viral vector is incubated with supernatant from a packaging cell line. The presence of RCR will result in the mobilization of the selectable viral vector, which can be detected by infection and selection on virgin cells.

In order to eliminate the production of RCR, which could have oncogenic potential if injected into an animal, further refinements were made to the design of packaging cell lines. PA317 *(63,64)* exemplifies one cell line that was carefully constructed to retain the ability to produce high-titer virus (still in the 10^7/mL range). In this line, the helper virus has multiple deletions; in addition to the packaging sequence, the splice donor site and the 3' LTR are deleted. These deletions require two recombination events to generate RCR. Furthermore, a reduction in the homology in the vector construct that drives the homologous recombination further reduces the occurrence of RCR. This is exemplified by the changes made in the wobble position of the coding sequence of gag by Morgenstern and Lund *(65)*.

The final refinement in packaging cell lines was to split the helper genome;

Fig. 5. This figure illustrates the two components, helper and vector, needed to make a packaging cell line. In its simplest form, the helper virus codes for all the proteins (Gag, Pol, and Env) that are needed for virus production, but it contains a deletion in the packaging sequence ψ. The vector contains the gene of interest between the two LTRs of the virus, which are necessary for viral replication and integration. The transcript produced from this vector contains the ψ sequence and this RNA is packaged into the virions that are produced. Note that a recombination event between the helper and vector sequences, shown above by a line between the two, would result in the generation of a replication-competent virus.

these are the lines of choice to generate vectors. In these cells, exemplified by ψCRE, ψCRIP *(66)*, GP+E-86 *(67)*, and GP+*env*AM12 *(68)*, the gag-pol coding region is separated from the envelope coding region. These modifications require three separate recombination events to generate RCR. The genes can also be controlled using hetrologous control elements (other than the LTR and polyadenylation sequence) further reducing the homology between the helper and vector sequence.

A consideration should also be given to the species from which the packaging cell is derived. Early evidence demonstrated that vectors from early packaging cell lines were inactivated by the human blood complement system. This was demonstrated to be due to Gal(alpha 1–3)Gal terminal carbohydrate modification of surface proteins *(69)*. This can be overcome by generating the vectors in human cells (as opposed to the early lines that were produced in murine cells or avian cells) *(70)*. The human cell lines (such as FLYA13 and FLYRD18 *[71]*) also lack VL30 RNA. This RNA is endogenous to murine cells, and is an ancient retrovirus that is mutant in the coding region for gag-pol and env. However, it contains a packaging sequence that is recognized by the MLV proteins and competes with vector RNA for packaging. In the target cell, it is converted to DNA and integrated into the host chromosome *(72)*. This also highlights a potential source of contamination in viral gene-transfer methods compared to physical methods. Until and unless we have a probe for a biological agent that we can monitor, we have some uncertainty as to what is being transferred from the packing cell line to the target cell, apart from the vector.

3.4. Stable/Transient Production of Virus

The previous discussion focused on the establishment of stable packaging cell lines that are defined by the stable integration into the chromosomal DNA of the helper and vector components. Early cell clones with stable helper functions were tested by transfecting the vector construct and testing the supernatant 48 h later for vector production to isolate the optimal clone for further development *(49)*. However, vector produced in this manner was of low titer, because only a fraction of the cells were transfected. The optimal clone was then transfected with the vector, and stable integrants were selected, hence all cells contained helper and vector. This procedure is laborious and time-consuming. In order to generate high-titer vector, Landau and Littman *(73)* used COS-7 cells that allow replication of plasmids containing the SV40 origin of replication (ori). They also made helper and vector plasmids with the SV40 ori; transient transfection of these plasmids into COS-7 cells gave a high-titer virus after 48 h. However, the overall transfection efficiency in this system was still low. More recently the human embryonic kidney cell line, 293 (and its derivative 293T), has gained popularity in generating high-titer virus transiently. This cell line can be transfected with very high efficiencies (up to 100%) and so generates high titer virus on transfection with helper and vector components *(74–76)*. This has accelerated the use of retroviral vectors because high-titer virus can be generated quickly, with little effort, and has led to the development of technologies such as cDNA libraries in retroviral vectors *(77,78)*.

3.5. Pseudotyping

Cells that can be transduced with a retroviral vector are limited in the first instance by the presence of a receptor for the envelope protein. A phenomenon that was observed early in the study of viruses, pseudotyping, can help expand or restrict the host range of the vector. If a cell is simultaneously infected with two different viruses, the envelope proteins can mix and be cross-packaged *(79,80)*. This observation has been extended to viral vectors, in that with split genomes for the helper, the envelope protein from any number of enveloped viruses can be incorporated into the vector particle. An early example of this was the incorporation of the Gibbon ape Leukemia virus (GALV) envelope into viral particles composed of MLV gag-pol proteins *(81)*. This technique has the advantage in that it can increase or enable the infection of other types of cells, if these cells express the receptor for the envelope. For example, the receptor for the GALV envelope seems to be a more abundant on a range of mammalian cells that are not as readily infected with the amphotropic envelope *(82)*. An envelope that has gained tremendous popularity is the G protein from the vesicular stomatitis virus (VSV-G). The receptor for this envelope has been proposed to be a lipid component of the membrane, and this envelope has a very broad tropism, enabling transduction to mammals, birds, and fish. It has the added advantage that it can withstand the shearing force of ultracentrifugation, and so can be quantitatively concentrated for applications in vivo *(83)*. A number of different envelopes have now been pseudotyped into MMLV, including Ebola *(84)*. Why and how psuedotyping is possible remains to be determined *(85)*. There are some signals on the C-terminus portions of envelope proteins that can inhibit pseudotyping, and removal of these can facilitate incorporation of the envelope into the virus particle. This has been observed for the HIV-1 envelope in MMLV *(86)*. Finally the geometry of infection can be reversed, and the receptor of the envelope can be pseudotyped into the viral vector, and this will only infect cells that express the viral envelope *(87–89)*. Hence, the choice of packaging cell line or construct used is determined by the ability of the envelope to mediate binding and fusion with the target cell.

4. Vector Design

Along with a packaging cell line, a vector, containing the gene of interest, is the second component of the retroviral delivery system. The elements that are cardinal to the vector are the LTRs, the packaging sequence (ψ), and the poly-purine tract (PPT) (*see* **Fig. 4**). Around these sequences, investigators have placed a number of genes and control elements. The gene of interest is cloned between the LTRs and a choice has to be made as to the promoter elements that will dictate the transcription of the gene. This promoter must obviously be ac-tive in the target cell. The polyadenlyation signal for the transcript is usually de-rived from the 3′ LTR, because an internal signal would truncate the transcript, and this would eliminate the 3′ LTR, making the vector uninfectious. **Figure 6** illustrates the various popular designs for vectors. The choice of vector is de-termined by the efficiency of transduction of the target cell. For example, if the transduction efficiency is low, it may be prudent to include a selectable marker in the vector that allows selection for transduced cell.

4.1. Promoter Elements in Vectors

The transgenes ferried on vectors can contain a combination of promoter el-ements that dictate transcription of the gene. The efficiency of a particular pro-moter in the target cell is obviously a prime consideration in the choice of vec-tor. One of the simplest vectors is illustrated in **Fig. 6A**. In this design, the gene is located between the two LTRs, and on infection the enhancer elements in the U3 dictate transcription of the cDNA. The splice donor and acceptor sites can be present or absent in these vector types; the retention of these signals can in-crease the viability of RCR (if it occurs) as well as add to the efficiency of gene expression, as seen in vectors such as MFG (mutational frameshift in gag) *(90,91)*. A major consideration in using these vectors is the ability of the tran-scription elements in the U3 to direct transcription in the target cell. For exam-ple, in embryonic carcinoma and stem cells these vectors are effectively si-lenced. However, vectors have been developed that can overcome inhibition to expression by selection for mutations in and around the U3 region that allow expression in these cell types *(36)*.

In some cases there is a need to select cells that have been transduced with a vector and this requires the expression of two genes, one coding for a selectable marker and the other for the desired transgene. The expression of two genes can be achieved using a number of strategies. **Figure 6B** illustrates one that utilizes the differential splicing of the retroviral transcript. The first protein is synthe-sized from the full-length LTR-LTR transcript, whereas a second protein is syn-thesized from the spliced message *(92)*, although this can be problematic if a cryptic splice acceptor is present in the cDNA contained in the intron *(93)*. **Fig-ure 6C** illustrates an approach where an internal promoter and the LTR dictate the transcription of two transcripts. Note that both promoters use the same polyadenylation signal in the LTR of the vector *(94)*. Another approach to the expression of two proteins is the use of an internal ribosome entry site that al-

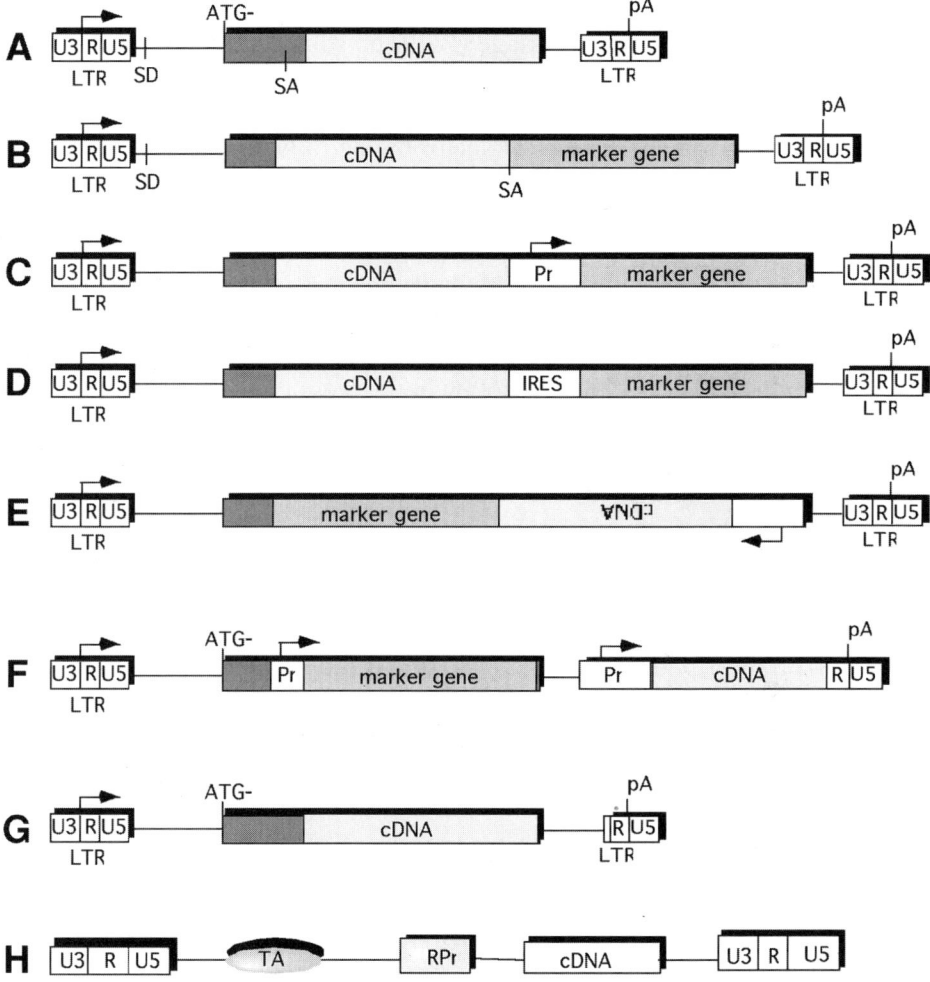

Fig. 6. Illustrates the design of a number of vectors. The simplest form is the gene of interest between the LTRs (**A**). Designs that allow expression of multiple proteins (**B–F**), of expression from adjacent cellular promoters (**G**), and regulated expression (**H**). Pr, promoter element; IRES, internal ribosome entry site; RPr, regulated promoter; TA, transactivator for the regulated promoter.

lows termination and downstream reinitiating of translation. These bi-cistronic messages exploit internal ribosome entry sites (IRES) that allow cap-independent translation of messages *(95)*. Introns from cellular genes can mediate control of transcription of that gene. In such cases this intron needs to be included in the vector. However this intron may be spliced out in the packaging cell. Also, in cases where a cryptic polyA or splice signal is present in the cDNA, this would truncate the LTR-LTR message (96). The solution to these problems is to place the cDNA (introns) in the opposing direction to the vector transcript (**Fig. 6E**) (97). This requires promoter and polyA signals to be added to the vector, and may result in lower titers of vector owing to two opposing promoters, that can reduce the concentration of the LTR-LTR message. Another approach to this problem is to generate the viral RNA in the cytoplasm using alphavirus vectors that replicate in the cytoplasm (98). The expression of a selectable marker allows the isolation of cells that have been transduced. Genes that confer resistance to drug selection are widely used (e.g., neomycin phosphotransferase) (94). Fluorescence activated cell sorting (FACS) has also been applied to the isolation of transduced cells following expression of enzymes that generate a fluorescent substrate (e.g., β-galactosidase) or the expression of fluorescent proteins (e.g., green fluorescent protein, GFP) (99,100). Finally the ectopic expression of a cell surface protein has been used to sort transduced cells using FACS (101) or with conjugated magnetic beads (102).

Protein fusions have also been effectively used to transduce two functions into a cell from one message (103). A number of systems are available for the regulated expression of transgenes (**Fig. 6H**). Common to all these is the production of a transcriptional activator (TA) that is responsive to a cell permeable small molecule. These TAs act on a responsive promoter (RPr) to direct the transcription of the transgene. Retroviral vectors have been designed to deliver both the TA and the RPr-transgene to cells. For example, using the popular tetracycline-inducible system, the two components of this system have been delivered on one or two vectors (104).

Innovative designs in vectors utilize the understanding of the reverse-transcription reaction where the U3 of the LTR is duplicated in the target cell (**Fig. 4**). Double-copy (DC) vectors have a transcription unit placed into the U3 region at the 3'-LTR (**Fig. 6F**). On completion of the RT reaction, this is copied to the U3 in the 5'-LTR and two copies of the gene are integrated into the target cell (105). Self-inactivating (SIN) vectors (**Fig. 6G**) have deletions in the U3 region at the 5'-LTR such that on transduction these vectors are no longer competent to direct transcription (106). On integration, expression of the transgene has to be directed by nearby cellular enhancer elements or internal promoters. The former allow the use of these vectors in enhancer trap experiments (*see* below), and the latter allow better tissue-specific expression, because internal tissue-specific promoters will have no interference from U3 transcription enhancer elements. A more recent application is the design of self-excising vectors, where a Cre recombinase target sequence (LoxP) is added into the 3' U3 region and the gene for Cre recombinase is added in the body of the vector. On integration, the LoxP sites are duplicated and flank the Cre recombinase coding sequence. The expression of Cre recombinase results in recombination between the two LoxP sites, causing the body of the vector to be excised from the host genome *(107)*. This system is designed for the pulsed expression of the Cre recombinase in target cells, but any other gene introduced into the body of the vector would have a pulsed expression in the target cell.

4.2. Uses of Vectors

Retroviral vectors have been developed for gene delivery in several diverse areas of molecular biology. The major limitation is that the cells must divide (but this can be overcome by the use of lentiviral vectors). The simplest application is in the production of cell lines that express a transgene introduced on a retroviral vector. As outlined earlier, there is a choice of vectors to choose from, and transgenic clones can be easily isolated by co-expression of a selectable marker. Furthermore, these genes can be regulated if this is required. This observation has not been translated to whole animals because MLV vectors are silenced during embryonic development. Transgenic animals are possible with lentiviral vectors *(108,109)*.

Second, in the field of development, because retroviruses integrate into the host chromosome at a single position, the location of the virus can be used to great effect to follow the lineage of a cell and its progeny *(110)*.

A third application for gene delivery using retroviral vectors is in gene therapy *(111)*. For MLV-based vectors, this has been restricted to the delivery of therapeutic genes to dividing cells. In particular, gene delivery to stem cells shows great promise *(112)*, though a wide range of tissues can be transduced. Fourth, MLV vectors have been useful in the delivery of toxic genes to cancer cells, which are actively dividing, in contrast to neighboring normal cells *(113)*.

A fifth area in which retroviral vectors have been useful is in gene discovery. The integrating nature of the virus can reveal function by insertional inactivation of a gene into the host chromosome. This has been achieved in cell culture and, more recently, efforts have been directed to inactivation of genes in embryonic stem (ES) cells and the generation of knockout mice from these cells. These efforts are directed towards understanding the function of genes in whole animals *(114)*. The insertion of a SIN vector with a reporter gene makes gene expression dependent on flanking cellular promoters. This enhancer trap technique has led to the identification of spatially and temporally controlled genes in cells and whole animals *(115)*. Insertion of the vector upstream of a gene can also result in the activation of genes directed by the U3 enhancer elements of the LTR. This has led to the identification of a number of transforming oncogenes *(116)*.

Finally, retroviral vectors have been developed for constructing cDNA libraries. Retroviral vectors allow the transfer and identification of gene function by complementation of mutants or by ectopic expression of the gene *(77,78)*.

5. Future Developments

Preliminary studies suggest that these vectors will still be developed further. In the near future, retroviral vectors will be developed that drive the expression of short interfering RNAs (siRNA) in target cells, adding "knockdown" to the "knockout" functions of vectors *(117)*. There is a need for robust insulating sequences that will enable positional independent expression from these vectors. In some applications, there is a need to insulate cellular genes from the effects of enhancer sequences in the vector *(118)*. A further improvement for these applications would be to direct the site-specific integration of virus into the genome *(119)*. Tissue-specific promoters in vectors will be aided by the analysis of the human genome sequence and the data from expression analysis from expressed sequence tag (EST) databases. Finally, there are early studies aimed at targeting the infection of vectors to particular cell types by alterations of the envelope proteins of the virus *(120)*.

References

1. Vogt, P. K. (1996) Peyton Rous: homage and appraisal. *Faseb J.* **10(13)**, 1559–1562.
2. Vogt, P. K. (1997) Historical Introduction to the General Properties of Retroviruses. in *Retroviruses* (Coffin, J. M., Hughes, S. H., and Varmus, H. E., eds) Cold Spring Harbor Laboratory Press, Cold Spring Harbor, New York, pp. 1–26.
3. Varmus, H. (1988) Retroviruses. *Science* **240(4858)**, 1427–1435.
4. Hanafusa, H., Hanafusa, T., and Rubin, H. (1963) The defectiveness of Rous sarcoma virus. *Proc. Natl. Acad. Sci. USA* **49**, 572–580.
5. Temin, H. M. (1963) Separation of morphological conversions and virus production in Rous Sarcoma virus infection. *Cold Spring Harbor Symp. Quant. Biol.* **27**, 407–414.
6. Hanafusa, H. (1965) Analysis of the defectiveness of Rous sarcoma virus. III. Determining influence of a new helper virus on the host range and susceptibility to interference of RSV. *Virology* **25**, 248–255.
7. Hanafusa, H. (1970) Virus production by Rous saroma cells. *Curr. Top. Microbiol. Immunol.* **51**, 114–123.
8. Vogt, V. M. (1997) in *Retroviruses* (Coffin, J. M., Hughes, S. H., and Varmus, H. E., eds.), Cold Spring Harbor Laboratory Press, Cold Spring Harbor, New York, pp. 27–70.
9. Pandya, S., Klimatcheva, E., and Planelles, V. (2001) Lentivirus and foamy virus vectors: novel gene therapy *Expert Opin. Biol. Ther.* **1(1)**, 17–40.
10. Hunter, E. and Swanstrom, R. (1990) Retrovirus envelope glycoproteins. *Curr. Top. Microbiol. Immunol.* **157**, 187–253.
11. Overbaugh, J., Miller, A. D., and Eiden, M. V. (2001) Receptors and entry cofactors for retroviruses include single and multiple transmembrane-spanning proteins

as well as newly described glycophosphatidylinositol-anchored and secreted proteins. *Microbiol. Mol. Biol. Rev.* **65(3),** 371–389.

12. Albritton, L. M., Tseng, L., Scadden, D., and Cunningham, J. M. (1989) A putative murine ecotropic retrovirus receptor gene encodes a multiple membrane-spanning protein and confers susceptibility to virus infection. *Cell* **57(4),** 659–666.

13. Miller, D. G., Edwards. R. H., and Miller, A. D. (1994) Cloning of the cellular receptor for amphotropic murine retroviruses reveals homology to that for gibbon ape leukemia virus. *Proc. Natl. Acad. Sci. USA* **91(1),** 78–82.

14. van Zeijl, M., Johann, S. V., Closs, E., Cunningham, J., Eddy, R., Shows, T. B., and O'Hara, B. (1994) A human amphotropic retrovirus receptor is a second member of the gibbon ape leukemia virus receptor family. *Proc. Natl. Acad. Sci. USA* **91(3),** 1168–1172.

15. Tailor, C. S., Nouri, A., Lee, C. G., Kozak, C., and Kabat, D. (1999) Cloning and characterization of a cell surface receptor for xenotropic and polytropic murine leukemia viruses. *Proc. Natl. Acad. Sci. USA* **96(3),** 927–932.

16. Yang, Y. L., Guo, L., Xu, S., Holland, C. A., Kitamura, T., Hunter, K., and Cunningham, J. M. (1999) Receptors for polytropic and xenotropic mouse leukaemia viruses encoded by a single gene at Rmc1. *Nat. Genet.* **21(2),** 216–219.

17. Battini, J. L., Rasko, J. E., and Miller, A. D. (1999) A human cell-surface receptor for xenotropic and polytropic murine leukemia viruses: possible role in G protein-coupled signal transduction. *Proc. Natl. Acad. Sci. USA* **96(4),** 1385–1390.

18. Fass, D., Davey, R. A., Hamson, C. A., Kim, P. S., Cunningham, J. M., and Berger, J. M. (1997) Structure of a murine leukemia virus receptor-binding glycoprotein at 2.0 angstrom resolution. *Science* **277(5332),** 1662–1666.

19. Fass, D., Harrison, S. C., and Kim, P. S. (1996) Retrovirus envelope domain at 1.7 angstrom resolution. *Nat. Struct. Biol.* **3(5),** 465–469.

20. Eckert, D. M. and Kim, P. S. (2001) Mechanisms of viral membrane fusion and its inhibition. *Annu. Rev. Biochem.* **70,** 777–810.

21. Trono, D. (1992) Partial reverse transcripts in virions from human immunodeficiency and murine leukemia viruses. *J. Virol.* **66(8),** 4893–4900.

22. Goff, S. P. (2001) Intracellular trafficking of retroviral genomes during the early phase of infection: viral exploitation of cellular pathways. *J. Gene Med.* **3(6),** 517–528.

23. Braaten, D., Franke, E. K., and Luban, J. (1996) Cyclophilin A is required for an early step in the life cycle of human immunodeficiency virus type 1 before the initiation of reverse transcription. *J. Virol.* **70(6),** 3551–3560.

24. Tanese, N. and Goff, S. P. (1988) Domain structure of the Moloney murine leukemia virus reverse transcriptase: mutational analysis and separate expression of the DNA polymerase and RNase H activities. *Proc. Natl. Acad. Sci USA* **85(6),** 1777–1781.

25. Coffin, J. M., Hageman, T. C., Maxam, A. M., and Haseltine, W. A. (1978) Structure of the genome of Moloney murine leukemia virus: a terminally redundant sequence. *Cell* **13(4),** 761–773.

26. Coffin, J. M. (1979) Structure, replication, and recombination of retrovirus genomes: some unifying hypotheses. *J. Gen. Virol.* **42(1),** 1–26.
27. Telesnitsky, A. and Goff, S. P. (1997) Reverse transcriptase and the generation of Retroviral DNA. in *Retroviruses* (Coffin, J. M., Hughes, S. H., and Varmus, H. E., eds.), Cold Spring Harbor Laboratory Press, Cold Spring Harbor, New York, pp. 121–160.
28. Wilhelm, M., Wilhelm, F. X. (2001) Reverse transcription of retroviruses and LTR retrotransposons. *Cell Mol. Life Sci.* **58(9),** 1246–1262.
29. Hindmarsh, P., and Leis, J. (1999) Retroviral DNA integration. *Microbiol. Mol. Biol. Rev.* **63(4),** 836–843.
30. Brown, P. O. (1997) Integration. in *Retroviruses* (Coffin, J. M., Hughes, S. H., and Varmus, E., eds.), Cold Spring Harbor Laboratory Press, Cold Spring Harbor, New York, pp. 161–204.
31. Schroder, A. R., Shinn, P., Chen, H., Berry, C., Ecker, J. R., and Bushman, F. (2002) HIV-1 integration in the human genome favors active genes and local hotspots. *Cell* **110(4),** 521–529.
32. Stryke, D., Kawamoto, M., Huang, C. C., Johns, S. J., King, L. A., Harper, C. A., et al. (2003) BayGenomics: a resource of insertional mutations in mouse embryonic stem cells. *Nucleic Acids Res.* **31(1),** 278–281.
33. Ruddell, A. (1995) Transcription regulatory elements of the avian retroviral long terminal repeat. *Virology* **206(1),** 1–7.
34. Graves, B. J., Johnson, P. F., and McKnight, S. L. (1986) Homologous recognition of a promoter domain common to MSV LTR and the HSV tk gene *Cell* **44(4),** 565–576.
35. Lewis, A. F., Stacy, T., Green, W. R., Taddesse-Heath, L., Hartley, J. W., and Speck, N. A. (1999) Core-binding factor influences the disease specificity of Moloney murine leukemia virus. *J. Virol.* **73(7),** 5535–5547.
36. Pannell, D., and Ellis, J. (2001) Silencing of gene expression: implications for design of retrovirus vectors. *Rev. Med. Virol.* **11(4),** 205–217.
37. Ostrowski, M. C., Huang, A. L., Kessel, M., Wolford, R. G., and Hager, G., L. (1984) Modulation of enhancer activity by the hormone responsive regulatory element from mouse mammary tumor virus. *EMBO J.* **3(8),** 1891–1899.
38. Boris-Lawrie, K., Roberts, T. M., and Hull, S. (2001) Retroviral RNA elements integrate components of post-transcriptional gene expression. *Life Sci.* **69(23),** 2697–2709.
39. Furger, A., Monks, J., and Proudfoot, N. J. (2001) The retroviruses human immunodeficiency virus type 1 and Moloney murine leukemia virus adopt radically different strategies to regulate promoter-proximal polyadenylation. *J. Virol.* **75(23),** 11735–11746.
40. Dejardin, J., Bompard-Marechal, G., Audit, M., Hope, T. J., Sitbon, M., and Mougel, M. (2000) A novel subgenomic murine leukemia virus RNA transcript results from alternative splicing. *J. Virol.* **74(8),** 3709–3714.

41. Rabson, A. B. and Graves, B. J. (1997) Synthesis and processing of Viral DNA. in *Retroviruses* (Coffin, J. M., Hughes, S. H., and Varmus, H. E., eds.), Cold Spring Harbor Laboratory Press, Cold Spring Harbor, New York, pp. 205–262.

42. Berlioz, C. and Darlix, J. L. (1995) An internal ribosomal entry mechanism promotes translation of murine leukemia virus gag polyprotein precursors. *J. Virol.* **69(4)**, 2214–2222.

43. Schultz, A. M., Henderson, L. E., and Oroszlan, S. (1988) Fatty acylation of proteins. *Annu. Rev. Cell Biol.* **4**, 611–647.

44. Scheifele, L. Z., Rhoads, J. D., and Parent, L. J. (2003) Specificity of plasma membrane targeting by the rous sarcoma virus gag protein. *J. Virol.* **77(1)**, 470–480.

45. Yoshinaka, Y., Katoh, I., Copeland, T. D., and Oroszlan, S. (1985) Murine leukemia virus protease is encoded by the gag-pol gene and is synthesized through suppression of an amber termination codon. *Proc. Natl. Acad. Sci. USA* **82(6)**, 1618–1622.

46. Jacks, T. and Varmus, H. E. (1985) Expression of the Rous sarcoma virus pol gene by ribosomal frameshifting. *Science* **230(4731)**, 1237–1242.

47. Vogt, V. M. (1996) Proteolytic processing and particle maturation. *Curr. Top. Microbiol. Immunol.* **214**, 95–131.

48. Watanabe, S. and Temin, H. M. (1982) Encapsidation sequences for spleen necrosis virus, an avian retrovirus, are between the 5′ long terminal repeat and the start of the gag gene. *Proc. Natl. Acad. Sci. USA* **79(19)**, 5986–5990.

49. Mann, R., Mulligan, R. C., and Baltimore, D. (1983) Construction of a retrovirus packaging mutant and its use to produce helper-free defective retrovirus. *Cell* **33(1)**, 153–159.

50. Yu, S. S., Kim, J. M., and Kim, S. (2000) High efficiency retroviral vectors that contain no viral coding sequences. *Gene Ther.* **7(9)**, 797–804.

51. Berkowitz, R., Fisher, J., and Goff, S. P. (1996) RNA packaging. *Curr. Top. Microbiol. Immunol.* **214**, 177–218.

52. Meric, C. and Spahr, P. F. (1996) Rous sarcoma virus nucleic acid-binding protein p12 is necessary for viral 70S RNA dimer formation and packaging. *J. Virol.* **60(2)**, 450–459.

53. Gorelick, R. J., Henderson, L. E., Hanser, J. P., and Rein, A. (1988) Point mutants of Moloney murine leukemia virus that fail to package viral RNA: evidence for specific RNA recognition by a "zinc finger-like" protein sequence. *Proc. Natl. Acad. Sci. USA* **85(22)**, 8420–8424.

54. Sawyer, R. C. and Hanafusa, H. (1979) Comparison of the small RNAs of polymerase-deficient and polymerase-positive Rous sarcoma virus and another species of avian retrovirus. *J. Virol.* **29(3)**, 863–871.

55. Levin. J. G. and Seidman, J. G. (1981) Effect of polymerase mutations on packaging of primer tRNAPro during murine leukemia virus assembly. *J. Virol.* **38(1)**, 403–408.

56. Crawford, S. and Goff, S. P. (1985) A deletion mutation in the 5′ part of the pol gene of Moloney murine leukemia virus blocks proteolytic processing of the gag and pol polyproteins. *J. Virol.* **53(3)**, 899–907.

57. Katoh, I., Yoshinaka, Y., Rein, A., Shibuya, M., Odaka, T., and Oroszlan, S. (1985) Murine leukemia virus maturation: protease region required for conversion from "immature" to "mature" core form and for virus infectivity. *Virology* **145(2)**, 280–292.

58. Miller, A. D., Jolly, D. J., Friedmann, T., and Verma, I. M. (1983) A transmissible retrovirus expressing human hypoxanthine phosphoribosyltransferase (HPRT): gene transfer into cells obtained from humans deficient in HPRT. *Proc. Natl. Acad. Sci. USA* **80(15)**, 4709–4713.

59. Watanabe, S. and Temin, H. M. (1983) Construction of a helper cell line for avian reticuloendotheliosis virus cloning vectors. *Mol. Cell Biol.* **3(12)**, 2241–2249.

60. Cone, R. D. and Mulligan, R. C. (1984) High-efficiency gene transfer into mammalian cells: generation of helper-free recombinant retrovirus with broad mammalian host range. *Proc. Natl. Acad. Sci. USA* **81(20)**, 6349–6353.

61. Martineau, D., Klump, W. M., McCormack, J. E., DePolo, N. J., Kamantigue, E., Petrowski, M., et al. (1997) Evaluation of PCR and ELISA assays for screening clinical trial subjects for replication-competent retrovirus. *Hum. Gene Ther.* **8(10)**, 1231–1241.

62. Peebles, P. T. (1975) An in vitro focus-induction assay for xenotropic murine leukemia virus, feline leukemia virus C, and the feline—primate viruses RD-114/CCC/M-7. *Virology* **67(1)**, 288–291.

63. Miller, A. D. and Buttimore, C. (1986) Redesign of retrovirus packaging cell lines to avoid recombination leading to helper virus production. *Mol. Cell Biol.* **6(8)**, 2895–2902.

64. Miller, A. D. (2002) PA 317 retrovirus packaging cells. *Mol. Ther.* **6(5)**, 572–575.

65. Morgenstern, J. P. and Land, H. (1990) Advanced mammalian gene transfer: high titre retroviral vectors with multiple drug selection markers and a complementary helper-free packaging cell line. *Nucleic Acids Res.* **18(12)**, 3587–3596.

66. Danos, O. Mulligan, R. C. (1988) Safe and efficient generation of recombinant retroviruses with amphotropic and ecotropic host ranges. *Proc. Natl. Acad. Sci. USA* **85(17)**, 6460–6464.

67. Markowitz, D., Goff, S., and Bank, A. (1988) A safe packaging line for gene transfer: separating viral genes on two different plasmids. *J. Virol.* **62(4)**, 1120–1124.

68. Markowitz, D., Goff, S., and Bank, A. (1988) Construction and use of a safe and efficient amphotropic packaging cell line. *Virology* **167(2)**, 400–406.

69. Takeuchi, Y., Porter, C. D., Strahan, K. M., Preece, A. F., Gustafsson, K., Cosset, F. L., et al. (1996) Sensitization of cells and retroviruses to human serum by (alpha 1-3) galactosyltransferase. *Nature* **379(6560)**, 85–88.

70. DePolo, N. J., Harkleroad, C. E., Bodner, M., Watt, A. T., Anderson, C. G., Greengard, J. S., et al. (1999) The resistance of retroviral vectors produced from human cells to serum inactivation in vivo and in vitro is primate species dependent. *J. Virol.* **73(8)**, 6708–6714.

71. Cosset, F. L., Takeuchi, Y., Battini, J. L., Weiss, R. A., and Collins, M. K. (1995) High-titer packaging cells producing recombinant retroviruses resistant to human serum. *J. Virol.* **69(12),** 7430–7436.
72. Patience, C., Takeuchi, Y., Cosset, F. L., and Weiss, R. A. (1998) Packaging of endogenous retroviral sequences in retroviral vectors produced by murine and human packaging cells. *J. Virol.* **72(4),** 2671–2676.
73. Landau, N. R. and Littman, D. R. (1992) Packaging system for rapid production of murine leukemia virus vectors with variable tropism. *J. Virol.* **66(8),** 5110–5113.
74. Finer, M. H., Dull, T. J., Qin, L., Farson, D., and Roberts, M. R. (1994) kat: a high-efficiency retroviral transduction system for primary human T lymphocytes. *Blood* **83(1),** 43–50.
75. Pear, W. S., Nolan, G. P., Scott, M. L., and Baltimore, D. (1993) Production of high-titer helper-free retroviruses by transient transfection. *Proc. Natl. Acad. Sci. USA* **90(18),** 8392–8396.
76. Soneoka, Y., Cannon, P. M., Ramsdale, E. E., Griffiths, J. C., Romano, G., Kingsman, S. M., and Kingsman, A. J. (1995) A transient three-plasmid expression system for the production of high titer retroviral vectors. *Nucleic Acids Res.* **23(4),** 628–633.
77. Kitamura, T., Onishi, M., Kinoshita, S., Shibuya, A., Miyajima, A., and Nolan, G. P. (1995) Efficient screening of retroviral cDNA expression libraries. *Proc. Natl. Acad. Sci. USA* **92(20),** 9146–9150.
78. Somia, N. V., Schmitt, M. J., Vetter, D. E., Van Antwerp, D., Heinemann, S. F., and Verma, I. M. (1999) LFG: an anti-apoptotic gene that provides protection from Fas-mediated cell death. *Proc. Natl. Acad. Sci. USA* **96(22),** 12667–12672.
79. Zavada, J. (1972) Pseudotypes of vesicular stomatitis virus with the coat of murine leukaemia and of avian myeloblastosis viruses. *J. Gen. Virol.* **15(3),** 183–191.
80. Huang, A. S., Besmer, P., Chu, L., and Baltimore, D. (1973) Growth of pseudotypes of vesicular stomatitis virus with N-tropic murine leukemia virus coats in cells resistant to N-tropic viruses. *J. Virol.* **12(3),** 659–662.
81. Miller, A. D., Garcia, J. V., von Suhr, N., Lynch, C. M., Wilson. C., and Eiden, M. V. (1991) Construction and properties of retrovirus packaging cells based on gibbon ape leukemia virus. *J. Virol.* **65(5),** 2220–2224.
82. Barrette, S., Douglas, J., Orlic, D., Anderson, S. M., Seidel, N. E., Miller, A. D., and Bodine, D. M. (2000) Superior transduction of mouse hematopoietic stem cells with 10A1 and VSV-G pseudotyped retrovirus vectors. *Mol. Ther.* **1(4),** 330–338.
83. Yee, J. K., Friedmann, T., and Burns, J. C. (1994) Generation of high-titer pseudotyped retroviral vectors with very broad host range. *Methods Cell Biol.* **43 Pt A,** 99–112.
84. Yang, Z., Delgado, R., Xu, L., Todd, R. F., Nabel., E. G., Sanchez, A., and Nabel, G. J. (1998) Distinct cellular interactions of secreted and transmembrane Ebola virus glycoproteins. *Science* **279(5353),** 1034–1037.

85. Zavada, J. (1982) The pseudotypic paradox. *J. Gen. Virol.* **63(Pt 1)**, 15–24.
86. Hohne, M., Thaler, S., Dudda, J. C., Groner, B., and Schnierle, B. S. (1999) Truncation of the human immunodeficiency virus-type-2 envelope glycoprotein allows efficient pseudotyping of murine leukemia virus retroviral vector particles. *Virology* **261(1)**, 70–78.
87. Mebatsion, T., Finke, S., Weiland, F., and Conzelmann, K. K. (1997) A CXCR4/CD4 pseudotype rhabdovirus that selectively infects HIV-1 envelope protein-expressing cells. *Cell* **90(5)**, 841–847.
88. Balliet, J. W. and Bates, P. (1998) Efficient infection mediated by viral receptors incorporated into retroviral particles. *J. Virol.* **72(1)**, 671–676.
89. Somia, N. V., Miyoshi, H., Schmitt, M. J., and Verma, I. M. (2000) Retroviral vector targeting to human immunodeficiency virus type 1-infected cells by receptor pseudotyping. *J. Virol.* **74(9)**, 4420–4424.
90. Riviere, I., Brose, K., and Mulligan, R. C. (1995) Effects of retroviral vector design on expression of human adenosine deaminase in murine bone marrow transplant recipients engrafted with genetically modified cells. *Proc. Natl. Acad. Sci. USA* **92(15)**, 6733–6737.
91. Krall, W. J., Skelton, D. C., Yu, X. J., Riviere, I., Lehn, P., Mulligan, R. C., and Kohn, D. B. (1996) Increased levels of spliced RNA account for augmented expression from the MFG retroviral vector in hematopoietic cells. *Gene Ther.* **3(1)**, 37–48.
92. Cepko, C. L., Roberts, B. E., and Mulligan, R. C. (1984) Construction and applications of a highly transmissible murine retrovirus shuttle vector. *Cell* **37(3)**, 1053–1062.
93. Korman, A. J, Frantz, J. D., Strominger, J. L., and Mulligan, R. C. (1987) Expression of human class II major histocompatibility complex antigens using retrovirus vectors. *Proc. Natl. Acad. Sci USA* **84(8)**, 2150–2154.
94. Miller, A. D. and Rosman, G. J. (1989) Improved retroviral vectors for gene transfer and expression. Biotechniques **7(9)**, 980–990.
95. Adam, M. A., Ramesh, N., Miller, A. D., and Osborne, W. R. (1991) *J. Virol.* **65(9)**, 4985–4990.
96. Li, Q., Emery, D. W., Fernandez, M., Han, H., and Stamatoyannopoulos, G. (1999) Development of viral vectors for gene therapy of beta-chain hemoglobinopathies: optimization of a gamma-globin gene expression cassette. *Blood* **93(7)**, 2208–2216.
97. Jonsson, J. J., Habel, D. E., and McIvor, R. S. (1995) Retrovirus-mediated transduction of an engineered intron-containing purine nucleoside phosphorylase gene. *Hum. Gene Ther.* **6(5)**, 611–623.
98. Wahlfors, J. J. and Morgan, R. A. (1999) Production of minigene-containing retroviral vectors using an alphavirus/retrovirus hybrid vector system. *Hum. Gene Ther.* **10(7)**, 1197–206.
99. Fiering, S. N., Roederer, M., Nolan, G. P., Micklem, D. R., Parks, D. R., and Herzenberg, L. A. (1991) Improved FACS-Gal: flow cytometric analysis and sorting of viable eukaryotic cells expressing reporter gene constructs. *Cytometry* **12(4)**, 291–301.

100. Persons, D. A., Allay, J. A., Allay, E. R., Smeyne, R. J., Ashmun, R. A., Sorrentino, B. P., and Nienhuis, A. W. (1997) Retroviral-mediated transfer of the green fluorescent protein gene into murine hematopoietic cells facilitates scoring and selection of transduced progenitors in vitro and identification of genetically modified cells in vivo. *Blood* **90(5),** 1777–1786.
101. Planelles, V., Haislip, A., Withers-Ward, E. S., Stewart, S. A., Xie, Y., Shah, N. P., and Chen, I. S. (1995) A new reporter system for detection of retroviral infection. *Gene Ther.* **2(6),** 369–376.
102. Zhang, J. and Sapp, C. M. (2001) A novel retroviral vector that allows the magnetic selection of infected cells. *J. Virol. Methods* **94(1–2),** 1–6.
103. Lupton, S. D., Brunton, L. L., Kalberg, V. A., and Overell, R. W. (1991) Dominant positive and negative selection using a hygromycin phosphotransferase-thymidine kinase fusion gene. *Mol. Ther. Cell Biol.* **11(6),** 3374–3378.
104. Hofmann, A., Nolan, G. P., and Blau, H. M. (1996) Rapid retroviral delivery of tetracycline-inducible genes in a single autoregulatory cassette. *Proc. Natl. Acad. Sci. USA* **93(11),** 5185–5190.
105. Hantzopoulos, P. A., Sullenger, B. A., Ungers, G., and Gilboa, E. (1989) Improved gene expression upon transfer of the adenosine deaminase minigene outside the transcriptional unit of a retroviral vector. *Proc. Natl. Acad. Sci. USA* **86(10),** 3519–3523.
106. Yu, S. F., von Ruden, T., Kantoff, P. W., Garber, C., Seiberg, M., Ruther, U., et al. (1986) Self-inactivating retroviral vectors designed for transfer of whole genes into mammalian cells. *Proc. Natl. Acad. Sci. USA* **83(10),** 3194–3198.
107. Silver, D. P. and Livingston, D. M. (2001) Self-excising retroviral vectors encoding the Cre recombinase overcome Cre-mediated cellular toxicity. *Mol. Cell* **8(1),** 233–243.
108. Pfeifer, A., Ikawa, M., Dayn, Y., and Verma, I. M. (2002) Transgenesis by lentiviral vectors: lack of gene silencing in mammalian embryonic stem cells and preimplantation embryos. *Proc. Natl. Acad. Sci. USA* **99(4),** 2140–2145.
109. Lois, C., Hong, E. J., Pease, S., Brown, E. J., and Baltimore, D. (2002) Germline transmission and tissue-specific expression of transgenes delivered by lentiviral vectors. *Science* **295(5556),** 868–872.
110. Lemischka, I. R. (1993) Retroviral lineage studies: some principals and applications. *Curr. Opin. Genet. Dev.* **3(1),** 115–118.
111. Hu, W. S. and Pathak, V. K. (2000) Design of retroviral vectors and helper cells for gene therapy. *Pharmacol. Rev.* **52(4),** 493–511.
112. Hawley, R. G. (2001) Progress toward vector design for hematopoietic stem cell gene therapy. *Curr. Gene Ther.* **1(1),** 1–17.
113. Solly, S. K., Trajcevski. S., Frisen, C., Holzer, G. W., Nelson. E., Clerc, B., et al. (2003) Replicative retroviral vectors for cancer gene therapy. *Cancer Gene Ther.* **10(1),** 30–39.
114. Stanford, W. L., Cohn, J. B., and Cordes, S. P. (2001) RET: a poly A-trap retrovirus vector for reversible disruption and expression monitoring of genes in living cells. Gene-trap mutagenesis: past, present and beyond. *Nat. Rev. Genet.* **2(10),** 756–768.

115. Ishida, Y. and Leder, P. (1999) RET: a poly A-trap retrovirus vector for reversible disruption and expression monitoring of genes in living cells. *Nucleic Acids Res.* **27(24),** e35.
116. Suzuki, T., Shen, H., Akagi, K., Morse, H. C., Malley, J. D., Naiman, D. Q., et al. (2002) New genes involved in cancer identified by retroviral tagging. *Nat. Genet.* **32(1),** 166–174.
117. Barton, G. M. and Medzhitov, R. (2002) Retroviral delivery of small interfering RNA into primary cells. *Proc. Natl. Acad. Sci. USA* **99(23),** 14943–14945.
118. Ramezani, A., Hawley, T. S., and Hawley, R. G. (2003) *Blood* **13,** 13.
119. Bushmanl, F. (2002) Targeting retroviral integration? *Mol. Ther.* **6(5),** 570–571.
120. Buchholz, C. J., Stitz, J., and Cichutek, K. (1999) Retroviral cell targeting vectors. *Curr. Opin. Mol. Ther.* **1(5),** 613–621.

32

Gene Delivery to Cells in Culture Using Retroviruses

Nikunj Somia

1. Introduction

Moloney leukemia virus-based vectors can be generated in cells that express the products of three retroviral genes, *gag*, *pol*, and *env*. There are a number of cell lines such as PG13 *(1)* and FLYA13 *(2)*, known as packaging cells, that have been established that stably express these genes. When these cells are transfected with vector DNA, they will generate retroviral transducing particles in the supernatant of the cells. The transducing particles can be produced by transient transfection of the packaging cells or from cells that have stably integrated the vector DNA into the packaging cell line. This is particularly useful if the vector will be required in large amounts for an extended period of time. The titer that can be achieved from transient transfection of DNA will be proportional to the transfection efficiency. In this respect, cells derived from the human embryonic kidney cell line 293 *(3)* are particularly useful because they can be transfected very efficiently (typically 90–99%) *(4–7)*. Here I outline a method to generate vector transiently. Methods to generate stable packaging cell lines can be found elsewhere *(8)*.

The products of the *gag* and *pol* gene are obligatory to the retroviral life cycle and to vector production. These can be provided on helper plasmid vectors such as pGP or pVPack-GP. The *env* gene will determine the tropism of the retroviral vector. Murine ecotropic envelope restricts the infection to rodent cells, whereas amphotropic envelopes mediate infection of most mammalian cell types. These genes are encoded on plasmids such as pVPack Env, pE-eco, and pE-ampho. In addition, the envelope protein can be replaced with the entry mol-

From: *Methods in Molecular Biology, vol. 246:*
Gene Delivery to Mammalian Cells: Vol. 2: Viral Gene Transfer Techniques
Edited by: W. C. Heiser © Humana Press Inc., Totowa, NJ

ecules of other enveloped viruses. Envelope proteins may be used that can either restrict the cell tropism or greatly expand it. A common molecule used to pseudotype retroviral vectors is the entry molecule of the vesicular stomatitis virus, the G protein (VSV-G). This molecule imparts a very broad tropism and is encoded by plasmids pVSV-G, and one in the pVPackEnv series. Finally the pCL vector series (7) encodes the gag, pol, and env genes on the same plasmid. Transfection of these expression plasmids and the vector plasmid enables production of retroviral vector particles.

The vector DNA consists of the LTRs of a retrovirus and the packaging sequence of the virus that enables the retroviral RNA (or in this case the vector RNA) to be recognized from other RNA molecules in the cytoplasm and to be packaged into a viral particle. The LTRs are critical because they contain all of the signals for proper conversion of the RNA into DNA and the necessary signal for the integration of the DNA into the host genome. The packaging sequence is vital not only for its obvious role in packaging but because it also contains the binding site for a tRNA molecule that comes with the retroviral RNA and is the priming site of reverse transcriptase. Apart from these two features, vector design can be varied to include internal promoters, selectable markers, reporter genes, internal ribosome entry site sequences, and so on. These points are fully covered in the accompanying overview chapter to this section of the book (see Chapter 31).

2. Materials

2.1. Packaging Cell Lines and Vector Plasmids

Table 1 outlines some sources for packaging cell lines and vector plasmids. Most of the vectors include a selectable marker (see **Note 1**). Those marked with * are designed to provide regulated expression of the transgene. Cells based on 293 cells that stably express Gag and Gag/Pol proteins are GP2-293 and ANJOU 65. Cells that additionally express an ecotropic envelope function are EcoPack and BOSC 23. Amphopack, Bing, and ProPak-A express an amphotropic envelope. Finally, ProPak X expresses a xenotropic envelope (i.e., it can infect most mammalian cells, except some murine cells). This is not an exhaustive list of cell lines available, but represents cells in which gag/pol and env functions are separated, and also cell lines that are readily available. The 293T/17 cell line (from ATCC) can also be used and must be transfected with the vector plasmid along with helper plasmids encoding Gag/Pol and an envelope.

2.2. Solutions

1. 2X BBS: 50 mM BES (N,N-bis(2-hydroxyethyl)-2-aminoethane-sulfonic acid), 280 mM NaCl, and 1.5 mM Na$_2$HPO$_4$ (adjust the pH to 6.95 with NaOH).
2. 2.5 M CaCl$_2$.
3. 0.01% Crystal violet in 70% methanol.

Table 1
Common Sources of Cell Lines, Vectors, and Helper Plasmids

Source	Packaging cell line	Vector	Helper plasmid
Nature Technology Corporation 4701 Innovation Drive Lincoln, NE 68521 402-472-6530	HTam HTgp	VLVectors *(9)*	
Retrogen Inc. 6645 Nancy Ridge Drive San Diego, CA 92121 858-455-8411		PRT-X PRT-XNeo	pPkg pEco pAmpho
Imgenex Corporation 11175 Flintkote Ave., Ste. E San Diego, CA 92121 858-642-0978		pCL series *(7)*	pCL-Eco pCLAmpho pCL-10A1 *(7)*
BD Biosciences Clontech 1020 East Meadow Circle Palo Alto, CA 94303 877-232-8995	AmphoPack EcoPack2 GP2-293 RetroPack™ PT67 *(10)*	LRCX and MSCV series RevTet-On/ Off vectors*	pVSV-G Vector
ARIAD Pharmaceuticals Inc. 26 Lansdowne St., Cambridge, MA 02139 617-494-0400		ARGENT Regulated transcription Retrovirus*	
Stratagene 11011 N. Torrey Pines Road La Jolla, CA 92037		pFB vector	pVPack-GP pVPack Env
Takara Distributed by: PanVera Corporation 501 Charmany Drive, Madison, WI 53719 800-791-1400		pDON-A1	pGP pE-eco pE-ampho
ATCC American Type Culture Collection P.O. Box 1549 Manassas, VA 20108 703-365-2700	ANJOU 65 BOSC 23 *(5)* Bing PG13 *(1)* ProPak-A *(11)* ProPak-X *(12)* PT67 *(10)*		

2.3. Plasmids

The DNA can be made using a variety of methods. CsCl or Qiagen-based (Qiagen, Valencia, CA) purification of the plasmids generally produce plasmids that transfect cells well. If a large number of constructs are to be tested, miniprep DNA can be prepared by alkaline lysis, however, the transfection efficiency, and hence the vector titer, will be lower.

3. Methods

This transfection procedure is one essentially developed by Chen and Okayama *(13)*. Other procedures have been reported to work equally well.

3.1. Generation of the Viral Vector

The number of plates required will be determined by the amount of vector needed. Generally, one plate is sufficient for producing enough virus for initial screening and analysis. The vector titer is generally around 10^6 colony forming (CFU)/ mL (*see* **Note 1**).

1. Day 1. Seed 293T cells into 10 cm plates with a 1:5 split from a near confluent 10 cm plate. The cell density should be about 5×10^6 cells/plate (*see* **Note 2**). Prepare as many plates as needed and one extra for a "mock" transfection (no DNA) as a negative control.
2. Day 2. Transfect the vector and envelope plasmid. The volumes given are for one plate but these may be scaled up proportionally when multiple plates are to be transfected.
   a. In a 1.5-mL microfuge tube, mix 10 µg of the helper expression construct (e.g., pCLAmpho) with 15 µg of vector DNA (e.g., pCLNCX); bring the volume to 450 µL with sterile distilled water;
   b. Add 50 µL of 2.5 M $CaCl_2$ and vortex.
   c. Add 500 µL of 2X BBS dropwise. Mix by inverting the tube a few times. Incubate for 5–10 min at room temperature.
3. Add the 1 mL DNA mixture evenly on top of the 293T cells. Swirl the supernatant and DNA mix to get complete mixing.
4. Incubate the cells overnight in a humidified incubator with 3% CO_2 at 37°C (*see* **Note 3**).
5. Day 3. Change the media and add 5–10 mL of fresh media. Incubate for 48 h in a humidified incubator with 10% CO_2 (*see* **Note 4**).
6. Day 5. Harvest the media and filter through a 0.45-µm filter (*see* **Note 5**). The media can now be stored (*see* **Note 6**) or used immediately. If the vector contains a selectable marker it can be titered (*see* **Subheading 3.3.**) prior to using it to infect target cells. However, if the vector contains a cDNA or nonselectable gene, it is preferable to perform the infection directly.

3.2. Infection of Target Cells

Infection of the target cells and suceptibility to infection will depend on the envelope used (*see* **Note 7**). For example, using the pCLAmpho helper will generate virus that can infect most mammalian cells.

1. Day 1 (*see* **Note 8**). Inoculate the cells to be infected so that they are approx 15–20% confluent the following day (*see* **Note 9**). The size of dish will be determined by the total number of cells to be infected.
2. Day 2. Aspirate the media and add supernatant containing the viral vector (*see* **Note 10**). If desired, add polybrene or Retronectin to increase transduction efficiency (*see* **Note 11**).
3. Day 3. Remove the viral supernatant and add fresh media. For cell types that are poorly infected (i.e., some primary cell cultures) or to ensure that 100% of the cells are infected, the cells maybe further infected by changing the retroviral supernatant and adding fresh retroviral supernatant. This infection can be repeated numerous times (generally four to five) or until a toxic effect is apparent on the infected cells. Though entry of the vector into cells is rapid, the infection should be allowed to proceed for 12–24 h (*see* **Note 1**).
4. Day 4. Apply the appropriate selection to the cells either with a drug, if the vector contained a selectable marker, or any other selection that may be pertinent to a genetic screen (*see* **Note 12**). Assay for a reporter gene should be done about 48 h after infection to allow time for expression and establishment of steady-state levels of protein expression.

3.3. Measuring Virus Titer

1. Day 1. Inoculate 4 mL (approx 5×10^5) test cells (e.g., 3T3 or HeLa) into eight 6-cm dishes. Add polybrene to the medium if the titer must be determined in its presence (*see* **Note 8**).
2. Day 2. Add 1 mL, 100 μL, or 10 μL of viral supernatant to the plates in duplicate. Also include a negative control (i.e., uninfected cells).
3. Day 3. Trypsinze the cells and split them 1/20 into 10-cm dishes with the appropriate selection (*see* **Note 11**). Plate the cells in duplicate.
4. Day 5–8. The cells will form colonies on the plates; when these are visible to the eye, stain and count the cells. Aspirate the media and rinse in PBS. Cover the surface area of the tissue dish with 0.01% crystal violet. Stain the cells for 20 min, then wash the plates gently with water. The titer is expressed as CFU/mL and is calculated by dividing by the volume of supernatant used for the infection (1.0, 0.1, or 0.01 mL) and multiplying by 20. Ideally, the one set of plates should have 50–100 colonies and a mean should be taken between.the duplicate plates (*see* **Note 12**).

4. Notes

1. Optimizing viral production. A retroviral vector containing a reporter that can easily be followed (e.g., β-galactosidase or green fluorescent protein [GFP]) greatly simplifies the optimization of vector production. In this case use pCLNC-GFP for the vector, and visualize the transfection efficiency by viewing the 293 cells under a fluorescent microscope. The transfected cells will appear green, and almost all the cells should be transfected. If not, double-check on the transfection mixtures. Also the virus will transduce both G418 resistance and the GFP gene. Hence, infection can be monitored using the CFU assay by counting the number of GFP positive cells in a FACS scanner 48 h after infection.

2. A proportion of 293 cells will detach from the dishes during the transfection procedure. To produce supernatant with the highest titer of virus, the plates should be coated with gelatin or poly-L-lysine to reduce detachment of the cells. These plates may be prepared by covering tissue culture dishes with 0.1% gelatin in PBS or 0.002% poly-L-lysine in PBS. Incubate the plates for 30 min to overnight at room temperature. Aspirate the reagent and rinse the plates with sterile, distilled water. Plate cells on the coated surface as usual.

3. Though the Chen and Okayama procedure recommends a 3% CO_2 incubator, this can be a problem where an extra incubator has to be set aside for transfections. We have used a 5% CO_2 incubator for this step and it will also suffice. After overnight incubation the media will be pink and a fine precipitate all over the plate can be observed. The cells will also look rounded; this is normal and not a cause for concern.

4. This media will ultimately contain the retroviral vector. Hence, if a higher titer per mL of medium is desired, it can be replaced with a lower volume of medium. Five mL is the lowest volume we have used on a 10-cm plate without having to worry about the medium evaporating and the dish drying out. Also at this stage, the medium can be changed for another medium that the **target** cells prefer. Usually this is not a problem, though some medium (like serum-free medium) will result in a slightly lower titer.

5. For maximal recovery of virus particles, a second harvest can be done by replacing the medium after the first collection and collecting the second harvest of the virus 72 h posttransfection. This is possible because the maximal expression from transient transfection of plasmids is 48–72 h.

6. The media that contains the transducing retrovirus vector can be used immediately to infect target cells or it may be stored in aliquots frozen at −70°C for later use. At this temperature the virus is stable for many months. Repeated freeze/thawing should be avoided.

7. Some hamster cells are very poorly infected by vectors carrying envelopes from mouse retroviruses. This block can be overcome by treating the cells with the glycosylation inhibitor tunicamycin *(14)*. We have also found that vectors pseudotyped with the VSV-G envelope do not exhibit this block.

8. If coordination of vector production and production and target cell infection is desired, note that d 4 of the vector production protocol is d 1 of the infection protocol.

9. Because Moloney murine leukemia virus-based vectors can only infect dividing cells, the split should be such that the cells will go through some divisions before the dish becomes confluent. Hence, the number of cells seeded will depend on the cell type. For 3T3 cells, a 1/10 split from a confluent 10-cm dish (at a density of about 10^6 cells/dish) is optimal. Split the cells so that they will not become cofluent in a 24-h period. Infection of such cells will reduce the effective titer.

10. The amount of supernatant used will depend on the desired end. To select for stable clones the multiplicity of infection (MOI) should be 0.1–1, whereas for the analysis of expression from a mixed population of cells, the MOI can be as high as 10.

11. On some cell types, addition of polybrene can enhance the titer by two- to fivefold. However, polybrene can be toxic to some cells, so this should be determined before it is used. Typically, 4 µg/mL of polybrene is added to the supernatant. A nontoxic protein, RetroNectin (Takara Shuzo Co, Shiga, Japan), a fibronectin fragment, has also been reported to increase transduction rates.

12. For pCLNCX derived vectors this is G418. The final concentration in the medium must be determined for each cell type with a kill curve and ranges up to 750 µg/mL (active concentration) for 3T3 cells and up to 500 µg/mL for HeLa cells. A dose-response curve should be established for the selection drug on the cell type being tested.

13. Note that the titer is a term given to the transduction efficiency of the vector on a particular cell type. The efficiency on another cell type may be lower. For example, the virus titer determined in HeLa cells and a B-cell line may well differ by two orders of magnitude.

References

1. Miller, A. D., Garcia. J. V., von Suhr, N., Lynch, C. M., Wilson, C., and Eiden, M. V. (1991) Construction and properties of retrovirus packaging cells based on gibbon ape leukemia virus. *J. Virol.* **65(5)**, 2220–2224.

2. Cosset, F. L., Takeuchi, Y., Battini, J. L., Weiss, R. A., and Collins, M. K. (1995) High-titer packaging cells producing recombinant retroviruses resistant to human serum. *J. Virol.* **69(12)**, 7430–7436.

3. Graham, F. L. and van der Eb, A. J. (1973) A new technique for the assay of infectivity of human adenovirus 5 DNA. *Virology* **52(2),** 456–467.

4. Finer, M. H., Dull, T. J., Qin, L., Farson, D., and Roberts, M. R. (1994) Kat: a high-efficiency retroviral transduction system for primary human T lymphocytes. *Blood* **83(1),** 43–50.

5. Pear, W. S., Nolan, G. P., Scott, M. L., and Baltimore, D. (1993) Production of high-titer helper free retroviruses by transient trasfection. *Proc. Natl. Acad. Sci. USA* **90(18),** 8392–8396.

6. Soneoka, Y., Cannon, P. M., Ramsdale, E. E., Griffiths, J. C., Romano, G., Kingsman, S. M., and Kingsman, A. J. (1995) A transient three-plasmid expression system for the production of high-titer retroviral vectors. *Nucleic Acids Res.* **23(4),** 628–633.

7. Naviaux, R. K., Costanzi, E., Haas, M., and Verma, I. M. (1996) The pCL vector system: rapid production of helper-free, high-titer, recombinant retroviruses. *J. Virol.* **70(8),** 5701–5705.

8. Miller, A. D., Miller, D. G., Garcia, J. V., and Lynch, C. M. (1993) Use of retroviral vectors for gene transfer and expression. *Methods Enzymol.* **217,** 581–599.

9. Chakraborty, A. K., Zink, M. A., and Hodgson, C. P. (1995) Expression of VL30 vectors in human cells that are targets for gene therapy. *Biochem. Biophys. Res. Commun.* **209(2),** 677–683.

10. Miller, A. D. and Chen, F. (1996) Retrovirus packaging cells based on 10A1 murine leukemia virus for production of vectors that use multiple receptors for cell entry. *J. Virol.* **70(8),** 5564–5571.

11. Rigg, R. J., Chen, J., Dando, J. S., Forestell, S. P., Plavec, I., and Bohnlein, E. (1996) A novel human amphotropic packaging cell line: high titer, complement resistance, and improved safety *Virology* **218(1),** 290–295.

12. Forestell, S. P., Dando, J. S., Chen, J., de Vries, P., Bohnlein, E., and Rigg, R. J. (1997) Novel retroviral packaging cell lines: Complementary tropisms and improve vector production for efficient gene transfer. *Gene Ther.* **4(6),** 600–610.

13. Chen, C. and Okayama, H. (1987) High efficiency transformation of mammalian cells by plasmid DNA. *Mol. Cell Biol.* **7(8),** 2745–2752.

14. Miller, D. G. and Miller, A. D. (1992) Tunicamycin treatment of CHO cells abrogates multiple blocks to retrovirus infection, one of which is due to a secreted inhibitor. *J. Virol.* **66(1),** 78–84.

33

Retrovirus-Mediated Gene Transfer to Tumors

Utilizing the Replicative Power of Viruses to Achieve Highly Efficient Tumor Transduction In Vivo

Christopher R. Logg and Noriyuki Kasahara

1. Introduction

Vectors derived from retroviruses have been widely studied as tools for gene transfer into mammalian tissue in vivo. One application for which retroviral vectors have received particular attention is gene transfer into tumor cells for treatment of cancer. Simple retroviruses, such as murine leukemia virus (MLV), and the vectors derived from them, require cell division for infection and thus possess a degree of inherent specificity for the rapidly dividing cells of neoplastic tissue. This unique property and the ease with which retroviral vectors are manipulated and produced have provided much of the impetus for their use in experimental and clinical cancer gene-therapy studies.

1.1. Use of Conventional Replication-Defective Retroviral Vectors for Cancer Gene Therapy

The vast majority of work carried out with mammalian retroviral vectors has utilized replication-defective forms, which cannot replicate beyond a single transduction event. These vectors are most commonly produced using one of two approaches. One employs "packaging cells," which stably express the *gag, pol,* and *env* viral proteins necessary for production of infectious virus particles. Introduction into these cells of a vector construct, which consists of a viral

From: *Methods in Molecular Biology, vol. 246:*
Gene Delivery to Mammalian Cells: Vol. 2: Viral Gene Transfer Techniques
Edited by: W. C. Heiser © Humana Press Inc., Totowa, NJ

genome in which the viral protein-coding sequences have been replaced by a transgene of interest, results in the subsequent production of virus particles carrying the vector genome. The other approach uses simultaneous transfection of cells with one or two plasmids for the expression of *gag, pol,* and *env* and another plasmid encoding the transfer vector in order to produce vector particles transiently. A typical set of plasmids used for transient production of defective retrovirus vectors is diagrammed in **Fig 1A**.

The initial results of animal studies using defective retroviral vectors to deliver genes into tumors were encouraging. Introduction of vector-producing cells (VPCs) producing a vector containing the herpes simplex virus thymidine kinase gene into murine gliomas, followed by administration of the nucleoside analog ganciclovir, resulted in regression or elimination of the tumors *(4)*. In several of the early studies, tumor models were generated by injection of either pre-transduced tumor cells *(8)* or equal numbers of VPCs and tumor cells *(19)*, resulting in artificially high levels of transduction, which facilitated effective vector-mediated anti-tumor responses. However, stringent tests of the ability of retroviral vectors to transduce tumor cells in vivo, utilizing direct injection of

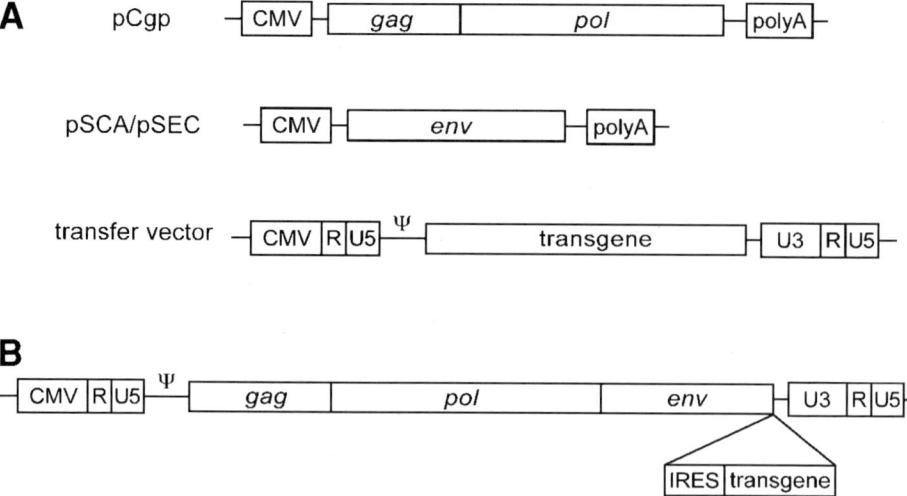

Fig. 1. Structure of plasmids used to generate replication-defective and replication-competent retroviral vectors by transient transfection. (**A**) Generation of defective vectors requires separate plasmids for expression of viral proteins and for production of the transfer vector RNA. Because the *gag-pol* and *env* expression constructs lack the packaging signal (ψ), only the vector RNA is packaged into virions and transferred into target cells. (**B**) RCR vectors are generated from a single plasmid consisting of a complete MLV genome containing an IRES-transgene cassette inserted immediately downstream of the *env* gene.

vector-containing supernatant or VPCs into preestablished tumors, followed by quantitation of transduction efficiency, yielded disappointing results. A large number of such studies have shown that, typically, fewer than 10% of cells in a given tumor mass can be transduced following repeated injections of defective retrovirus vectors or packaging cells directly into preestablished tumors.

Clinical trials using these vectors have reported even more discouraging results. In the largest clinical study of cancer gene therapy to date, VPCs were injected at a large number of sites within brain tumors that had been subjected to surgical debulking *(18)*. Quantitation of the gene-transfer level resulting from this treatment demonstrated that no more than 0.002% of the tumor cells in any examined patient had been transduced. Other clinical trials have reported similar results. Although 100% efficiency of gene transfer is not required for an effective anti-tumor response, owing to the bystander effect of suicide genes, vectors with higher efficiency are greatly needed.

1.2. Use of Replication-Competent Retrovirus (RCR) Vectors for Cancer Gene Therapy

1.2.1. Advantages vs Risks of Using RCR Vectors

With the aim of improving in vivo transduction efficiency by retroviral vectors, we have developed vectors derived from MLV that retain all of the native viral sequences and thus possess the ability to replicate *(13,14)*. The structure of these vectors, depicted in **Fig. 1B**, consists of an intact MLV genome containing an internal ribosome entry site (IRES)-transgene cassette inserted between the *env* gene and 3' untranslated region. As these vectors can replicate beyond the initial infection event, each transduced cell becomes, in effect, a VPC and greatly improved in vivo transduction efficiencies are possible. In animal tumor models, we have observed transduction efficiencies of >90% using replicating vectors.

As uncontrolled spread of replication-competent virus can result in insertional mutagenesis and carcinogenesis, the possibility of generating replication-competent retrovirus (RCR) during production of retroviral vectors has been a foremost concern of investigators in the field. This concern was heightened following the development of fatal lymphomas in 3 out of 10 rhesus macaques injected with bone marrow cells contaminated with RCR *(6)*. However, the macaques that developed malignancies were severely immunocompromised, and similar experiments using less severe immunosuppression resulted in no evidence of RCR pathology *(2,3)*.

Furthermore, the initial rationale for use of retroviral vectors in cancer gene therapy would still hold true for RCR vectors, i.e., unlike the newer HIV-based lentiviral vectors, MLV-based vectors can only transduce cells that are actively dividing *(15)* and because the majority of normal cells in many tissues are quiescent, transduction would be selective for tumor cells. In addition, incorporation of a suicide gene into the RCR vector would itself constitute a built-in safeguard, as even noncancerous cells infected by the virus would eventually be eliminated by treatment, and the spread of the vector would be inherently self-limited.

1.2.2. Design of RCR Vectors

To construct MLV-based RCR vectors, we insert an IRES-transgene cassette precisely at the junction of the *env* gene and the 3' untranslated region (UTR) (**Fig. 1B**). This configuration allows highly efficient replication and complete transduction of cell cultures at extremely low multiplicities of infection, and intratumoral transduction levels reaching 75–100% in nude mouse tumor xenograft models after injection of even 10^4 transducing units of viral supernatant, with no detectable emergence of revertants that had deleted the transgene *(14)*. The efficiency of transgene expression is dependent on the strength of the IRES sequence used. We have found good transgene expression with use of the encephalomyocarditis virus (EMCV) IRES in most cell lines we have tested. The total length and specific sequence of the IRES-transgene insert may also be of importance: for example, there appears to be a limitation of approx 1.3 kb that can be stably tolerated as an insertion at the *env*-3' UTR junction, and larger insertions result in decreased replication kinetics and rapid deletion of the inserted sequences *(13)*. Furthermore, certain sequences appear to impair the replicative ability of the virus even when the total insert length is well within the tolerated range. Of course, other configurations such as insertion of an internal promoter-transgene cassette or alternative splice acceptor-transgene cassette are also possible, either at the same insert site or at other sites, such as the U3 region of the 3' LTR. However, these alternative configurations may also result in complications such as promoter interference and increased genomic instability.

Further modifications to the vector genome may include incorporation of additional mechanisms for selective targeting of tumor cells. This may be achieved at the level of physical targeting via modification of the viral envelope *(10)*, or by transcriptional targeting via replacement of the viral enhancer/promoter elements with tissue-specific regulatory sequences *(5,12)*. At the level of physical targeting, a variety of naturally occurring retroviral envelope sequences (e.g., ecotropic, amphotropic, GALV, etc.) can be used as the *env* gene within the RCR genome, or alternatively, specific targeting ligand sequences can be incorporated into the *env* gene, although this approach has encountered significant difficulties in achieving efficient levels of targeted transduction. For transcriptional targeting, a variety of tissue-specific promoters have been used in conventional replication-defective vectors, and we have recently demonstrated that MLV replication can be stringently restricted to prostate-derived cells by incorporation of the prostate-specific probasin promoter into the U3 region of the LTR *(12)*.

1.3. Cell-Culture and Animal Model Experiments Using RCR Vectors

1.3.1. Testing RCR Vectors on Cancer Cells in Culture

In general, RCR vectors have proven to be an exceedingly useful and forgiving tool for gene delivery to cancer cells, both in culture as well as in vivo. Even if low-titer vector preparations are used, the virus will eventually spread throughout the culture or tumor, as long as the cells are permissive for MLV replication.

Permissiveness can vary significantly among different tumor cell lines. In addition to the mitotic rate, the infectivity of MLV-based RCR vectors in a given cell line depends on a variety of other factors, including the expression levels of cellular proteins required for viral entry, integration, transcription of the provirus, and assembly of virus particles. Generally speaking, RCR vectors containing the amphotropic envelope easily infect the large majority of mammalian tumor cell lines we have tested, and ecotropic variants readily infect a similarly broad proportion of tested tumor cell lines of mouse or rat origin. Replication kinetics, however, can vary widely between cell lines. In the case of human tumor cell lines, the presence or absence of the appropriate cell-surface receptor (e.g., the Pit-2 phosphate transporter for RCR vectors encoding the amphotropic envelope) and the cell proliferative rate are likely to be among the most important factors determining the permissiveness to infection. It is therefore advisable to confirm the susceptibility of a cell line for a specific RCR vector prior to use in an animal tumor model involving the cell line. In the event that low infectivity is observed, it is possible to vary the envelope gene in order to obtain more efficient viral entry. If propagation kinetics appears to be impaired owing to post-entry blocks to viral replication, it is sometimes possible to isolate a variant of the vector that is better adapted to its particular cancer cell host by molecular evolution through prolonged passage in culture.

In permissive cell lines, IRES-transgene insertions of up to 1.3 kb in length are stably retained for more than seven serial cell-free passages of the RCR vector, under conditions in which 1:100 dilutions of conditioned medium from each passage of infected cells are used to infect each subsequent culture, with cultures of 10^6 cells being completely infected at each passage. As each of the other 99 aliquots from the conditioned media of each passage are similarly capable of transducing 10^6 cells, this suggests that on the order of $100^7 \times 10^6 = 10^{20}$ cells can be transduced starting from an initial 100–200 µL of RCR vector stock without evidence of genomic instability.

1.3.2. Testing RCR Vectors in Tumor Models In Vivo

Once efficient viral replication has been confirmed in culture, one can progress to studies involving in vivo transduction of solid tumors derived from the same cell type. Again, as RCR vector spread is highly efficient, even a single dose of low titer viral supernatant can be sufficient to ultimately result in transduction of the entire tumor mass. The kinetics of transduction appears to be largely dependent on the initial size of the tumor mass, the mitotic index of the tumor cells, the dosage of the viral inoculum, and the susceptibility of the cell type to infection. These parameters, and how they affect the overall transduction efficiency of the replicating vector, should also be assessed empirically for different tumor models tested.

We have found it convenient to employ RCR vectors expressing the green fluorescent protein (GFP) marker gene to first determine the kinetics of vector spread in different tumor models. After dissection of tumor samples at different time points, the cells are enzymatically disaggregated and analyzed by flow cytometry to determine the percentage of GFP-expressing cells. Alternatively, newer methods for direct in vivo imaging using sensitive CCD cameras (to detect GFP) or microPET (to detect conversion of fluoridated prodrug analog compounds) might also be employed if such facilities are readily available.

Understanding the kinetics of RCR vector spread in the particular tumor model to be used will greatly facilitate the design of subsequent experiments in which therapeutic strategies such as RCR-mediated delivery of suicide genes are to be tested. The timing of prodrug administration after inoculation of the RCR vector is a particularly important consideration in this regard. Adequate time for vector spread should be allowed prior to administration of the prodrug, in order to maximize the percentage of infected cells within the tumor, which are thus rendered susceptible to the prodrug. Naturally, as this percentage approaches 100%, the therapeutic efficacy achieved is correspondingly greater. Once the infected cells are killed, RCR vector spread is terminated. Thus, administration of the prodrug too early (i.e., prior to adequate levels of vector spread within the tumor) will not only result in insufficient levels of cell killing to effectively inhibit tumor growth, but will also prematurely abort any further vector spread. However, it should also be kept in mind that most suicide gene/prodrug converting enzyme systems can achieve some degree of bystander effect by killing untransduced cells through local diffusion of toxic pro-drug metabolites via gap junctions. This will not only serve to enhance the therapeutic effectiveness of suicide gene therapy, but will also provide a means of terminating RCR vector spread even in cases where some percentage of the vector population might have deleted or mutated the transgene.

2. Materials

2.1. Cells and Cell Culture

1. 293T human embryonic kidney *(7)*, and NMU rat mammary carcinoma *(2)* cell lines. Both cell lines can be obtained from the ATCC (Manassas,

VA). All cells are maintained at 37°C in a humidified 5% CO_2 atmosphere.

2. Dulbecco's modified Eagle's medium (DMEM, 4.5 g/L glucose) with 10% fetal bovine serum (FBS), 100 U/mL of penicillin, and 100 μg/mL of streptomycin. Used as growth media for 293T and NMU cells.
3. Trypsin 0.05%, with 0.53 mM ethylenediaminetetraacetic acid (EDTA) in Hank's balanced salt solution (HBSS).

2.2. Transient Production of Replication-Defective and Replication-Competent Retrovirus Vectors

1. DMEM with 10% FBS. Growth media without antibiotics.
2. Serum-free DMEM (SF DMEM).
3. Plasmids for generating replication-defective vectors: packaging plasmids for expression of *gag*, *pol*, and *env*, e.g., pCgp for *gag-pol* expression and pSCA for amphotropic *env* expression, or similar constructs; transfer vector plasmid encoding GFP, e.g., pMX-GFP or similar vector (*see* **Note 1**).
4. Plasmids for generation of replication-competent vectors encoding GFP (e.g., pACE-GFP) or cytosine deaminase (e.g., pACE-CD) (*see* **Notes 2** and **3**).
5. Lipofectamine 2000 (LF2K, Invitrogen, Carlsbad, CA).
6. Sterile 5-mL syringes.
7. Sterile 0.45-μm syringe filters, such as Millex-GV (Millipore, Bedford, MA) or Acrodisc Supor (Pall Gelman Sciences, Ann Arbor, MI) filters.
8. Polybrene (hexadimethrine bromide; Sigma, St. Louis, MO): 5 mg/mL (1000X) in water. Sterilize with a 0.2-μm syringe filter and store aliquots at −20°C.

2.3. Establishment of Subcutaneous Tumor Model

1. Homozygous nude (nu/nu) BALB/c mice, 6–8 wk old.
2. Phosphate-buffered saline (PBS), pH 7.4 (Invitrogen).
3. Sterile 1-mL tuberculin syringes.
4. Sterile 27 G hypodermic needles.
5. Matrigel basement membrane matrix solution (BD Biosciences, San Jose, CA).

2.4. FACS Analysis of Vector Spread in Tumors

All reagents, plasticware, and scalpels should be sterile.

1. Tumor dissociation media: DMEM containing 10% FBS, 200 U/mL collagenase type I, 5 μM azidothymidine (AZT), 100 U/mL penicillin, and 100 μg/mL streptomycin.

For type I collagenase, CLS-1 from Worthington Biochemical (Lakewood, NJ) or 1A from Sigma are used. AZT stock solution (Sigma) is 10 mg/mL and stable for several months when stored at 4°C in the dark.

2. HBSS (Invitrogen, Carlsbad, CA).
3. Growth media: DMEM with 10% FBS, 100 U/mL penicillin and 100 μg/mL streptomycin.
4. 9-cm Petri dishes.
5. Scalpels.
6. 6-cm Tissue-culture dishes.
7. 50-mL polypropylene centrifuge tubes.
8. Trypsin 0.05%, with 0.53 mM EDTA in HBSS.
9. Nylon mesh cell strainers, 40–70-μm pore size, sterile.
10. Round-bottom, 12 × 75 mm polystyrene FACS tubes.
11. Propidium iodide (PI), 500 μg/mL (100X) in PBS. Store at 4°C in dark.
12. Flow cytometer equipped with a 488-nm argon laser and detectors for red and green fluorescence.

2.5. Immunohistochemical Analysis of Transduction in Tumors

1. Necropsy tools.
2. Aluminum or plastic tissue molds.
3. OCT tissue embedding compound (Sakura Finetek, Torrance, CA).
4. Isopentane (2-methylbutane).
5. Liquid nitrogen.
6. Stainless-steel beaker.
7. Cryostat.
8. Superfrost Plus microscope slides (Fisher Scientific, Pittsburgh, PA).
9. PBS, pH 7.4.
10. Acetone.
11. H_2O_2/PBS solution: 0.3% H_2O_2 in PBS.
12. Polyclonal rabbit anti-GFP antibody. Either antibody sc-8334 (Santa Cruz Biotechnology, Santa Cruz, CA) or ab290 (AbCam, Cambridge, UK) work well.
13. Avidin-biotin complex (ABC)-horseradish peroxidase (HRP) immunohistochemistry kit. Suitable kits are available from several suppliers, including Pharmingen (San Diego, CA) and Zymed (South San Francisco, CA). If a nonrabbit primary antibody is used, make sure to use a kit specific for the species in which the primary antibody was raised. The kit should contain the following: blocking reagent, biotinylated anti-rabbit secondary antibody, and avidin- or streptavidin-conjugated HRP (avidin-HRP).
14. Diaminobenzidine (DAB) substrate (included with some ABC kits).

15. Mayer's hematoxylin solution.
16. Crystal/Mount aqueous mounting media (Biomeda, Foster City, CA).

2.6. PCR Detection of Intratumoral and Extratumoral Spread of Vector

1. Tissue lysis solution: 50 mM Tris-HCl, pH 8.0, 10 mM EDTA, 0.5% SDS.
2. Proteinase K: 10 mg/mL in 50% 10 mM Tris-HCl and 50% glycerol. Store aliquots at −20°C.
3. DNase-free RNase A, 5–50 mg/mL.
4. 5 M Sodium acetate.
5. Isopropanol.
6. 70% ethanol.
7. Tris-EDTA: 10 mM Tris-HCl, 1 mM EDTA, pH 8.0.
8. Primers for detection of GFP, 10 μM in TE: Upstream: 5'-ATGGTGAG-CAAGGGCGAGGA-3'; Downstream: 5'-GCTACTTGTACAGCTCGTC-CATGC-3'.
9. Internal control primers (for mouse β-casein), 10 μM in TE: Upstream: 5'-GATGTGCTCCAGGCTAAAGTT-3'; Downstream: 5'-AGAAACGGA ATGTTGTGGAGT-3'.
10. Platinum PCR SuperMix (Invitrogen).

2.7. Tumor Growth Inhibition Following RCR-Mediated Transduction of Suicide Gene

1. PBS, pH 7.4.
2. 5-Fluorocytosine (5-FC; Sigma, St. Louis, MO): 10 mg/mL in PBS. This solution is stable at 4°C in the dark for several months.
3. Sterile 1-cc tuberculin syringes.
4. Sterile 27 G hypodermic needles.

3. Methods

3.1. Transient Production of Replication-Defective and Replication-Competent Retrovirus Vectors

Production and use of retrovirus vectors should be performed at Biosafety Level 2. Work should be carried out in a certified laminar flow hood with HEPA filters. Treat all virus-containing liquid waste with 2–5% bleach before discarding. Discard used plastic labware used for cell culture into biohazard containers.

1. One d prior to transfection, plate 293T cells onto a 6-cm dish so that at the start of the transfection procedure they are 90–95% confluent. Ensure that the cells do not clump and are evenly distributed over the surface of the dish.

Transfection at cell densities under 90% will result in lower titers (*see* **Note 4**).

2. Two h before transfection, completely remove media on cells and replace with 3 mL of antibiotic-free growth media (*see* **Note 5**).
3. In a 5-mL polystyrene tube, dilute the plasmid DNA into 300 µL of SF DMEM (*see* **Note 6**).

   a. For defective vectors: use 2 µg of vector plasmid, 1.5 µg of *gag-pol* plasmid, and 1.5 µg of envelope plasmid.
   b. For RCR vectors: use 5 µg of single RCR vector-encoding plasmid.

4. In another polystyrene tube, dilute 18 µL of LF2K into 300 µL of SF DMEM and incubate for 5 min at room temperature. Longer incubation may result in reduced transfection efficiency.
5. Combine the diluted DNA and the diluted LF2K and incubate for 20–30 min at room temperature.
6. Distribute the DNA-LF2K mixture over the cells. Gently rock plate to mix. Incubate cells overnight at 37°C in CO_2 incubator.
7. Eighteen to 24 h after adding the LF2K-DNA complexes to the cells, replace the media with 3 mL of fresh antibiotic-free growth media. Avoid dislodging cells by gently pipetting media onto side of dish.
8. Approximately 20 h after the change of media on d 3, collect the cell supernatant and replace with 3 mL of fresh antibiotic-free growth media. Filter the supernatant through a 0.45-µm syringe filter. If it will be used within 5 d, store the supernatant at 4°C until use. For longer-term storage, keep supernatant at −70°C (*see* **Note 7**).
9. Repeat collection of supernatant approx 20 h after the previous day's harvest, filter supernatant, and store at 4°C or −70°C as required (*see* **Note 8**).

3.2. Establishment and Transduction of Subcutaneous Mouse Tumor Model (see Note 9)

All procedures employing animals should adhere to institutional guidelines and be approved by the local Institutional Animal Care and Use Committee.

3.2.1. Preparation of Tumor Cells for Subcutaneous Inoculation

1. Grow tumor cell cultures (e.g., NMU cells) to provide sufficient cells to inoculate the number of mice to be used. Grow an extra 10–20% cells to allow for cell loss during preparation for injection (*see* **Notes 10** and **11**).
2. Remove media and wash cells with PBS.
3. Harvest cells using trypsin-EDTA, then add 4–5 volumes of serum-free DMEM (*see* **Note 12**).

4. Transfer cells to 50-mL centrifuge tubes and pellet by centrifugation at 200–300*g* for 5 min.
5. Remove supernatant and resuspend cells in serum-free DMEM to a density of 1–2 × 10⁷ cells/mL (*see* **Note 13**).

3.2.2. Subcutaneous Tumor Inoculation Protocol (see Note 14)

1. Using an alcohol swab, wipe the skin of nude mice at the flanks where the tumor cells are to be injected.
2. Just before injection, gently pipet the cell suspension up and down to mix cells evenly.
3. Draw the cell suspension into a 1-cc syringe and attach a 27 G needle.
4. Insert the needle 5–10 mm through the skin into the sc space, retract the needle slightly, and, while gently pinching skin at site of needle entry, slowly inject 100 µL of cells. Lack of resistance during injection indicates that the needle is properly positioned.

3.2.3. Injection of Vector into Tumors

Using RCR vectors, pre-established tumors can be very efficiently transduced by injection with vector supernatant even after reaching a size of 100–500 mm³ (corresponding to a spherical volume 0.5–1 cm in diameter). NMU tumors usually take 2–4 wk to reach this size, although the growth rate of individual tumors can vary significantly. In contrast, conventional replication-defective retroviral vectors generally show much poorer transduction efficiency in pre-established tumors of this size, with distribution limited to the area immediately surrounding the injection site.

1. Using an alcohol swab, wipe skin of mice at the flanks where vector preparations are to be injected.
2. Draw the vector prepared in **Subheading 3.1.** into a 1-cc syringe and attach a 30 G needle.
3. Insert the needle through the skin, aiming for the center of the tumor mass, and very slowly inject 100 µL of vector. Varying the angle of the needle during the process of infusion often results in better dispersal of the vector inoculum within the tumor. However, avoid piercing the tumor at more than one site, as this can cause leakage of the vector from the tumor.

3.3. Harvest of Tumor and Extratumoral Tissues for Analysis of Vector Spread

For RCR vectors, tumor and extratumoral tissues are removed from euthanized mice at two or more time points after injection of the vector to assess vector spread. For example, with NMU tumors of 1-cm diameter injected with 10^{4-5} transducing units of ACE-GFP vector supernatant in nude mice, we have found that in most tumors, less than 20% of the tumor mass is transduced after 3 wk, whereas the majority of viable cells are transduced by 6 wk after vector injection (*see* **Note 15**). For conventional replication-defective vectors, the transduction level using a comparable dosage is generally less than 2%, and does not increase with time.

3.3.1. Euthanasia of Mice

This procedure should be carried out in an exhaust hood. Wear gloves to prevent exposure to halothane.

1. Place 10–12 cotton balls at the bottom of a 1000-mL glass beaker.
2. Pour 4–5 mL of halothane onto cotton balls.
3. Place a circular piece of aluminum foil, cut to the size of the beaker bottom, on top of the soaked cotton balls.
4. Place mouse on aluminum foil circle and completely seal top of beaker with second piece of foil. Two mice can be euthanized in one beaker.
5. Wait 10–15 min, until mouse displays no sign of movement.
6. Remove mouse and surgically open thoracic cavity to ensure death.

3.3.2. Dissection of Mice

The same mice sacrificed at a given time point can be used for harvest of tumor tissues to assess gene-transfer efficiency by both RCR vectors and conventional replication-defective vectors, as well as for harvest of various nontumor tissues to assess systemic extratumoral spread of RCR vectors. Furthermore, the same tumor removed from an individual mouse can be divided into portions to be processed in parallel for flow cytometric analysis, immunohistochemistry, and extraction of genomic DNA for PCR. Note, however, that some tumors may not contain enough viable, non-necrotic tissue for all three assays. Thus far, we have not been able to detect RCR spread to any extratumoral tissues even with highly sensitive genomic PCR methods; however, because monitoring the potential systemic spread of RCR vectors is an important biosafety consideration, we always process extratumoral tissues for PCR.

1. Remove tumors on both flanks of mouse (*see* **Note 16**).
2. Place each tumor into a petri dish and trim away necrotic, connective, or fatty tissue with a scalpel.
3. Divide tumors into three parts as follows: a 500–800 mg portion to be processed for flow cytometric analysis, approx 20 mg for extraction of genomic DNA, and the remaining tissue for immunohistochemistry. Portions of tumor to be used for immunohistochemistry need not be thicker than 5

mm. The center of sizable tumors usually contains necrotic tissue made up of dead and dying cells. Ensure that each portion of the tumor contains enough viable tissue for analysis, although this may not be possible in experiments where suicide gene therapy is used effectively to inhibit tumor growth.

4. Process each portion of tumor as described in **Subheading 3.4.1.** for flow cytometric analysis, **Subheading 3.5.1.** for histological sectioning, and **Subheading 3.6.1.** for extraction of DNA. Samples to be used for DNA extraction can be stored at $-70°C$ for later processing.

5. Remove tissue samples from the following solid organs for each animal: brain, lung, spleen, liver, small intestine, large intestine, and skin. Be careful not to contaminate any tissue with fragments from any other tissue, and do not re-use scalpels and dishes that have previously been used for dissection of tumor samples.

6. Wash each tissue with PBS to remove excess blood; place approx 20 mg of each tissue into a microfuge tube, and proceed to **step 10**.

7. Harvest bone marrow from the mice as follows: remove the femurs and dissect away excess muscle and ligament tissue. Cut off the distal tip of each femur with scissors. Using a 5-cc syringe with a 23 G needle forcefully inserted through the proximal tip, wash 1–2 mL of PBS through the hollow of each bone into a 15-mL centrifuge tube.

8. Centrifuge the tube at $500g$ for 5 min to pellet the bone marrow cells.

9. Remove supernatant and transfer bone marrow cells to a microfuge tube.

10. For each sample of extratumoral tissue harvested from **steps 6** and **9**, begin DNA extraction procedure (**Subheading 3.6.1.**) or place cells in storage at $-70°C$ for later processing.

3.4. Flow Cytometric Analysis of Vector Transduction in Tumors

Flow cytometry is used to quantitate transduction levels in tumors injected with GFP-encoding vectors (**Fig. 2**). Flow cytometric quantitation of transduction levels using other transgenes is also possible, assuming the transgene encodes another fluorescent protein or a cell-surface protein which can be labeled by an antibody tagged with a fluorescent marker.

3.4.1. Tumor Dissociation Procedure

In this procedure, tumor tissue is disaggregated by overnight incubation in collagenase-containing medium. AZT is included in the media to prevent further replication of the vector. This procedure is carried out under sterile conditions.

1. Transfer tumor tissue to petri dish and remove necrotic, fibrous, and fatty tissue with scalpel.

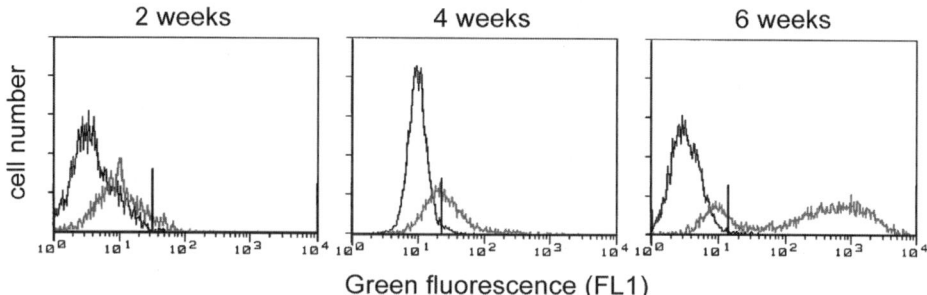

Fig. 2. Flow cytometric analysis of RCR vector spread within a sc tumor model at 2, 4, and 6 wk after intratumoral injection of ZAPd-GFP ecotropic RCR vector encoding GFP. Tumors were removed from mice at the indicated time points following injection of vector, dissociated into single-cell suspensions, and analyzed for GFP expression by flow cytometry. Gray histograms represent tumors injected with ZAPd-GFP, and black histograms represent control tumors not injected with vector.

2. Moisten tissue with just enough HBSS to keep moist and mince into small pieces (1–2 mm across) using two scalpels.
3. Transfer minced tissue to a 50-mL centrifuge tube containing 10 mL of HBSS.
4. Swirl tube to gently agitate tissue pieces.
5. Allow pieces to briefly settle and remove supernatant.
6. Repeat wash with another 10 mL of HBSS.
7. Transfer approx 1 g of tissue pieces to a 6-cm culture dish containing 5 mL of tumor dissociation media (*see* **Note 17**).
8. Incubate at 37°C overnight.
9. Gently disperse loose tissue clumps by pipetting with a 10-mL pipet and transfer dissociated tissue and supernatant to a 50-mL centrifuge tube. If cells adhere to the dish, add 2 mL of trypsin-EDTA, incubate until the cells have detached, and combine with the cells in centrifuge tube.
10. Centrifuge at 200–300g for 4 min.
11. Remove the supernatant and resuspend cell pellet in 2 mL of HBSS.
12. Remove remaining tissue clumps by passing the resuspended cells through a 40-µm cell strainer into a FACS tube. Add additional HBSS if suspension is too dense for straining.

3.4.2. Flow Cytometric Analysis of GFP Expression in Dissociated Tumors

1. To stain dead cells and cell fragments, add PI solution to the suspension to a final concentration of 5 µg/mL and mix by gentle pipetting.

2. Allow the staining to proceed for 5 min at room temperature before analyzing cells.
3. Analyze at least 5,000 events on the flow cytometer, recording side scatter, forward scatter, and red (>650 nm) and green (500–550 nm) fluorescence (*see* **Note 18**).
4. Exclude dead cells from analysis by gating out the cell population labeled by PI and hence exhibiting high levels of red fluorescence, either during the run ("live gate") or during subsequent offline analysis.
5. Determine the percentage of cells that exhibit green fluorescence but are negative for PI staining. This percentage corresponds to the level of transduction in the tumor sample.

3.5. Immunohistochemical Analysis of Vector Transduction in Tumors

Immunohistochemical staining is another sensitive method for demonstrating the presence of specific vector-encoded proteins in transduced tissues, and furthermore provides additional information regarding the physical distribution of the vector and cell types transduced within the tumor. This section describes in detail a method optimized for immunohistochemical detection of GFP in tumors injected with RCR or replication-defective vectors encoding this marker. **Figure 3** shows representative results from transduced tumor tissues examined by immunohistochemistry using a GFP-specific antibody. Immunohistochemistry also has the advantage of being flexible and suitable for use with vectors that do not contain a marker gene detectable by flow cytometric methods. Hence, with use of an appropriate primary antibody, the protocol here may be adapted to detect any specific protein encoded by transgenes carried by replicative or nonreplicative vectors, as well as the *gag, pol,* and *env* genes of the RCR vector.

3.5.1. Preparation of Tumor Samples for Frozen Sectioning

1. Label one tissue mold for each tumor sample to be analyzed and fill each halfway with OCT. Tumor tissue should be processed promptly after removal from animal, as described in **Subheading 3.3.2.**
2. Place tumor samples in OCT within pre-labeled molds.
3. Add enough additional OCT to cover samples.
4. Fill a stainless steel beaker two-thirds full with isopentane and float the beaker in liquid nitrogen. Keep the isopentane from freezing by stirring regularly with forceps.
5. Place tissue mold containing sample in the chilled isopentane and allow it to submerge.
6. After the OCT completely solidifies (approx 1 min), remove mold and store at −70°C until sectioning is carried out (*see* **Note 19**).

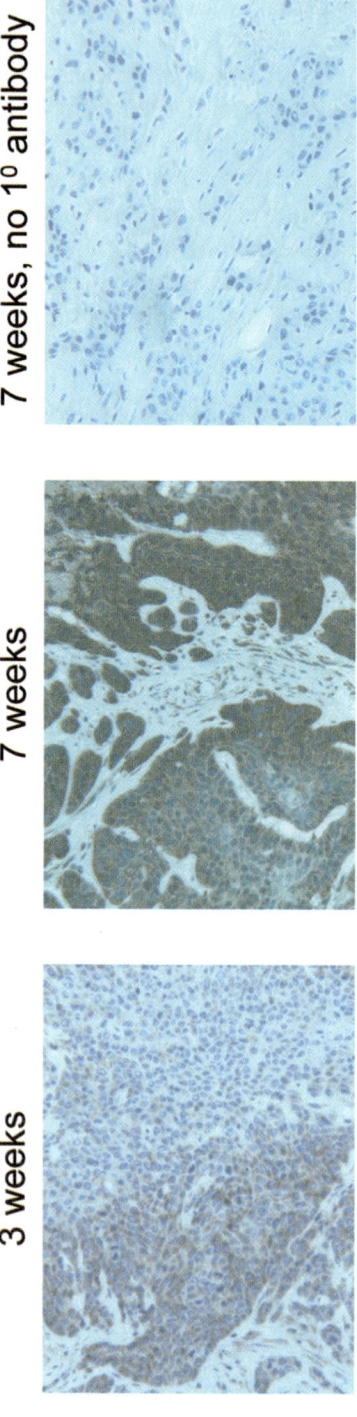

Fig. 3. Immunohistochemical staining of NMU tumors injected with ecotropic RCR vector encoding GFP. Tumors were removed from mice at 3 or 7 wk following intratumoral injection of RCR vector ZAPd-GFP and were stained using a primary antibody to GFP and the ABC–horseradish peroxidase method. Staining reveals increased transduction levels over time (left and middle sections) and staining in the absence of primary antibody (right section) demonstrates the specificity of the signal.

3.5.2. Making Frozen Tissue Sections

1. Equilibrate the tissue blocks to the temperature of the cryostat (−20 to −22°C).
2. Label microscope slides using solvent resistant pen.
3. Cut sections 4–6 μm thick on cryostat.
4. Allow sections to melt flatly onto pre-labeled microscope slides.
5. Allow sections to air dry for 2 h to overnight.

If staining will not be carried out within 24 h of sectioning, unfixed, air-dried sections can be temporarily stored on slides at −70°C in a sealed container. If staining will be carried out the next day, air-dry the slides overnight.

3.5.3. Staining of Tissue Sections

Each step below is to be carried out at room temperature unless otherwise specified. To prevent evaporation of reagents and drying of sections during the staining procedure, all incubation steps following fixation should be carried out in a humidity chamber (*see* **Note 20**).

1. Immediately before staining, fix sections by immersing in chilled acetone for 5 min.
2. Wash slides by incubating twice in PBS for 5 min.
3. Incubate slides in H_2O_2/PBS solution for 5 min to quench endogenous peroxidase activity.
4. Wash slides in PBS for 5 min. Repeat once.
5. To prevent nonspecific staining, cover each section with 2–4 drops of blocking reagent, prepared according to the manufacturer's recommendations, for 15 min. If blocking reagent is not included in the kit, use normal (non-immune) serum from same species in which the secondary antibody was produced, diluted 1:20 in PBS.
6. Remove excess blocking reagent by tapping and blotting sides of slides with absorbent paper.
7. Immediately add primary antibody, diluted 500-fold in PBS, to sections and incubate for 1–2 h (*see* **Note 21**).
8. Wash slides in PBS for 5 min. Repeat once.
9. Incubate in biotinylated secondary antibody at manufacturer's recommended dilution for 30 min.
10. Wash slides in PBS for 5 min. Repeat once.
11. Incubate in avidin-HRP (or streptavidin-HRP) solution, prepared according to manufacturer's instructions, for 20 min.
12. Add a few drops of DAB substrate, enough to cover the sample, and incubate for 5–10 min, until the desired level of staining has been achieved. The formation of brown precipitate can be monitored under a microscope.

13. Wash slides in distilled water for 5 min.
14. Apply enough hematoxylin solution to cover each section and incubate for 1–4 min, until sufficient blue counterstaining is achieved (*see* **Note 22**).
15. Wash slides gently under running tap water for 5 min.
16. Apply 3 drops of aqueous mounting media and rotate each slide to spread media over the entire section, covering an area about the size of a quarter. Do not apply a coverslip.
17. Incubate slides in oven at 37°C for 1–2 h or 50°C for 30–40 min.
18. Remove slides from oven and allow to equilibrate to room temperature. The sections are then ready for microscopic examination.

3.6. PCR Analysis of Vector Spread and Transduction in Tumors

The polymerase chain reaction (PCR) is used to detect transduction by retroviral vectors within tumors and, in particular, to assess the potential spread of RCR vectors to extratumoral tissues (**Fig. 4**). Owing to the sensitivity of polymerase chain reaction (PCR), care must be taken during extraction of genomic DNA and setting up of the PCR reactions to avoid cross-contamination of samples.

3.6.1. Extraction of Genomic DNA from Tumor and Extratumoral Tissues

1. Mince each tissue sample (from **Subheading 3.3.2.**) inside a microfuge tube using a scalpel.
2. Promptly add 600 µL of tissue lysis solution.
3. Add 8 µL of proteinase K solution and mix by inverting tube several times.
4. Incubate at 55°C for 3 h to overnight, or until tissue has dissolved. Once the tissue dissolves, samples can be stored for several days at room temperature.
5. Add a volume of RNase A solution equivalent to 50 µg of RNase A.
6. Mix the sample by inverting tube several times and incubate for 30 min at 37°C.
7. Allow the sample to cool to room temperature.
8. Add 210 µL of 5 *M* sodium acetate and mix thoroughly by inverting several times.
9. Centrifuge the tube at ≥13,000*g* for 4 min to pellet precipitated proteins.
10. If a compact pellet is not visible at the bottom of the tube, incubate on ice for 5 min and repeat the centrifugation.
11. Pipet the supernatant into a clean microfuge tube.
12. Add 650 µL of isopropanol to the sample and mix by inverting tube 20 times until white threadlike DNA precipitate is visible.
13. Pellet DNA by centrifuging tube at ≥13,000*g* for 2 min.
14. Pour out the supernatant without displacing pellet and add 1 mL of 70% ethanol to wash the DNA of residual salt.

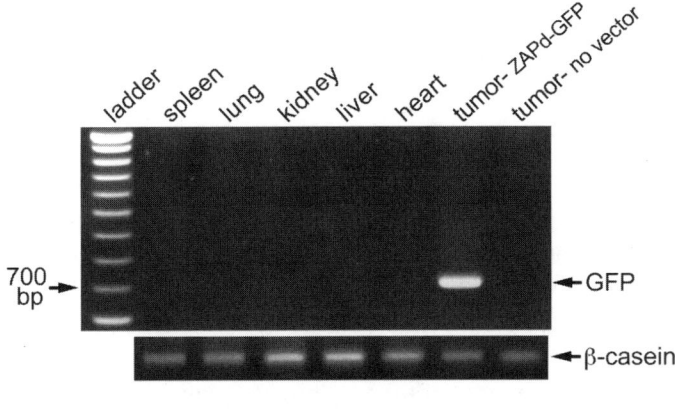

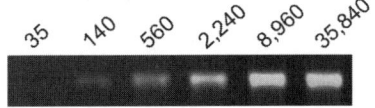

Fig. 4. Analysis of extratumoral spread of ZAPd-GFP ecotropic RCR vector by PCR amplification of the GFP transgene from genomic DNA. Genomic DNA from tumors and various other tissues, isolated 7 wk after intratumoral injection of vector, was used in PCR amplification of the GFP transgene (top). A 520-bp region of the β-casein gene was amplified as an internal control (middle). The sensitivity of the assay was determined by amplification using a series of serial dilutions of pZAPd-GFP plasmid DNA as template in the presence of untransduced genomic DNA (bottom), and was found to be capable of detecting down to between 35 and 140 copies of the provirus in a background of 100,000 genome equivalents. The results suggest that the RCR vector is well-confined to the tumor

15. Invert the tube several times and centrifuge at ≥13,000*g* for 1 min.
16. Pour off the ethanol without disturbing the pellet and leave the tube open for 15–20 min to allow residual ethanol to evaporate.
17. Add 100 µL of TE and allow DNA to rehydrate by incubating overnight at room temperature or for 1–2 h at 55°C. Periodically tapping the tube will help dissolve the DNA.
18. After the DNA is completely dissolved, determine the concentration of the sample by standard spectrophotometry (i.e., optical absorbance at 260 nm); the expected concentration should be on the order of 0.2–1.0 µg/µL. The size of the DNA fragments should also be verified by electrophoresis on a 0.7% agarose gel; most of the DNA should be larger than 30 kb in size, otherwise PCR results may be unreliable.

3.6.2. PCR Protocol

This procedure details the conditions for amplification of the GFP transgene of ACE-GFP within genomic DNA of mouse tissues. Using the appropriate primers and cycling parameters, this procedure can be adapted for detection of any transgene inserted into a retroviral vector.

1. For each DNA sample to be tested, add the following to an autoclaved, thin-walled PCR tube: 45 µL Platinum PCR SuperMix, 1 µL upstream GFP primer, 1 µL downstream GFP primer, 1 µL upstream β-casein primer, 1 µL downstream β-casein primer, 0.5 µg genomic DNA (volume can be 0.5–5 µL).
2. Thoroughly mix contents of tube by pipetting, and keep on ice until cycling is performed.
3. If the thermal cycler is not equipped with a "hot top," overlay each reaction with one or two drops of mineral oil to prevent evaporation.
4. Load tubes into thermal cycler preheated to 94°C and incubate for 2 min.
5. Perform 30 cycles of PCR as follows:

    a. Denaturation at 94°C for 40 s.
    b. Annealing at 58°C for 1 min.
    c. Elongation at 72°C for 1 min.

6. Run 5 µL of each reaction and 0.5 µg of 100-bp DNA ladder on a 1% agarose gel. The PCR-amplified GFP and β-casein products are approx 720 and 520 bp in size, respectively (**Fig. 4**).

3.7. Tumor Growth Inhibition Following RCR-Mediated Transduction of Suicide Gene

The kinetics of RCR-mediated gene transfer, as assessed by serial time-point analysis of the spread of GFP marker gene transduction within tumors using the protocols described in **Sections 3.4 and 3.5**, can be used to determine the optimal timing of pro-drug administration when similar vectors carrying a suicide gene are employed in the same tumor model. In general, better results will be obtained if the vector is allowed a suitable interval of time to spread throughout a majority of the tumor tissue prior to pro-drug administration, as cell killing by the suicide gene effect will also effectively terminate further vector spread.

3.7.1. Prodrug Administration Schedule

1. For an NMU tumor of a 500 mm³ volume injected with RCR vectors expressing cytosine deaminase, allow 2–4 wk for vector spread to occur prior to administration of prodrug. Keep in mind that the tumor volume increases roughly with the cube of the diameter. Therefore, a 1-cm diameter tumor

has a volume approx eightfold larger than that of a 0.5 cm diameter tumor and may require a significantly longer period to achieve a similar level of transduction. In this case, the therapeutic effect of suicide gene transfer will be more effective if a significantly longer period of time is allotted for vector spread to occur (e.g., 5–6 wk), as long as the animals can tolerate the large tumor burden.

2. Initiate prodrug administration by daily intraperitoneal (ip) injection of 5-FC solution at a dosage of 500 mg/kg/d.
3. Continue pro-drug administration for at least 7 consecutive days.
4. Repeat pro-drug administration cycles as necessary if tumor regrowth occurs.

3.7.2. Measurement of Tumor Volume

1. Tumor size should be measured with calipers in two dimensions every 3–4 d, and the volume calculated as length $\times$ width^2 $\times$ 0.5.
2. Statistical analysis (e.g., Student's t-test) should be performed to calculate the significance of differences in the mean $\pm$ standard deviation of tumor volumes between the treatment and control groups (*see* **Note 23**).

4. Notes

1. Replication-defective vector constructs: plasmid pCgp is used for expression of the *gag-pol* proteins and plasmid pSCA is used for expression of the amphotropic *env* protein *(9)*. If only mouse or rat cells are to be transduced, pSCA can be substituted with the ecotropic *env* expression plasmid pSEC *(9)*. Retroviral vectors may also be pseudotyped with the vesicular stomatitis virus glycoprotein (VSV-G), which allows transduction of a broad range of cell types, including mouse, rat, and human *(1)*. To pseudotype the vectors with VSV-G, pSCA is substituted with a VSV-G expression plasmid, such as pMD.G *(17)*. Plasmid pMX-GFP *(11)* is used as the transfer vector, although any MLV-based vector containing a GFP transgene may be used. Other packaging and transfer vector constructs developed for transient production of defective retrovirus vectors may also be used, including those of the pHIT *(20)* and pCL *(16)* systems.
2. Replication-competent retroviral vector constructs: Plasmids pACE-GFP *(12)* or pACE-CD are used to generate vectors for transducing cells with the GFP or CD transgenes, respectively. Both of these vectors contain the 4070A amphotropic envelope. pACE-CD was generated by amplifying the yeast CD cDNA by PCR from plasmid pCR-Blunt-CD (kindly provided by Dr. P. Roy-Burman, University of Southern California) and inserting it into pACE-GFP in place of the GFP transgene. Plasmid pZAPd-GFP is a vari-

ant of pACE-GFP containing an ecotropic envelope *(14)* and can be used for transduction of mouse or rat cells.

3. When designing an RCR vector with a novel insert, several points should be kept in mind. Inserts should not contain a transcription termination signal. Sequences in the 3′ UTR and 3′ LTR, which are downstream of the transgene insertion site, are required for retroviral replication and termination of transcription before these sequences will prevent replication of the vector. Avoid using inserts with an overall size larger than 1.3 kb. When using the 550-bp EMCV IRES, the upper limit for transgene size is therefore approx 750 bp. Inserts larger than this tend to be rapidly deleted during vector replication. Also avoid using inserts that contain direct or inverted repeats. These sequences may promote inter-repeat deletion during transfection and vector propagation. Such deletion mutants usually have a replicative advantage over full-length vector and will eventually overtake the replicating vector population. Do not include introns in the transgene. The splicing pattern of the resulting vector will be difficult to predict and may produce unintended splicing forms which interfere with vector replication. Furthermore, because retroviruses replicate through an RNA stage, inserted introns generally will not persist within the vector genome beyond a single replication cycle.

4. It is important that the 293T cells used for transfection be as healthy as possible at the time of transfection. While cells are being maintained prior to transfection, they should not be allowed to grow to overconfluence and the passage number should be kept as low as possible by preparing adequate numbers of frozen aliquots. When the cells are split for transfection, it is important that they are evenly dispersed upon reattachment to the plate. If the cells reattach in clumps, the cultures will not reach confluence during the transfection procedure and the resulting titers will be lower. 293T cells adhere weakly to tissue-culture plastic and care must be taken when media is added or removed from cultures to avoid disturbing the cells. Polylysine-coated tissue culture dishes can be used in place of standard dishes to improve cell adherence and transfection efficiency.

5. As lipofection reagents increase the permeability of the cell membrane, the use of media containing antibiotics during transfection can result in the buildup of high intracellular levels of these drugs. The resulting cytotoxicity will impair the production of vector particles, resulting in lower titers.

6. It is important that the media used in **steps 3** and **4** not contain serum because serum proteins can interfere with the formation of the lipid-DNA complexes required for efficient transfection. After the complexes have formed in the absence of serum, subsequent exposure to serum, as in **step 6**, will not degrade their activity.

7. Titers of MLV-based vectors containing an amphotropic envelope can drop by 50% or more upon freezing. Storage of the vectors at 4°C results in relatively small (<10%) losses in titer after 2–3 d. Thus, although highest titers are guaranteed by using vector immediately after production, if short-term storage is necessary, maintain vector at 4°C rather than freeze. For longer-term storage, vector should be kept at −70ºC. Inclusion of cryoprotectants, such as dimethyl sulfoxide (DMSO) or glycerol, in vector preparations does not prevent loss in titer during freezing.

8. For transfections involving GFP-encoding vectors, transfection efficiency can be conveniently assessed by analysis of the transfected cells by flow cytometry for GFP fluorescence. Briefly, wash the transfected cells in PBS after harvest of the virus-containing conditioned medium and trypsinize in 2 mL of trypsin-EDTA solution. After the cells have detached from the dish, add 4 mL of PBS and gently pipet the cells up and down to produce a single-cell suspension, then analyze by flow cytometry as described in **Subheading 3.4.2.** The percentage of GFP-positive cells directly corresponds to the efficiency of the transfection. The same procedure can be used to determine the infectious titer of retroviral vector preparations after transduction of cultured tumor cells.

9. Prior to using tumor cell lines other than NMU to establish tumors for in vivo transduction, as noted in **Subheading 1.3.1.**, it is advisable that the permissiveness of each cell line be determined in vitro to ensure that the RCR vectors can replicate efficiently in the cell line of interest. In order to quickly assess the permissiveness of a particular cell line or cell type for RCR vector, we generally first test the efficiency of replication in cultured cells using a vector encoding GFP, which allows easy and rapid assessment of infectivity by fluorescence microscopy or flow cytometry

10. Make sure that a sufficient number of cells is grown to account for the number of animals in the vector + pro-drug treatment group, as well as all control groups, which ideally should include a completely untreated group, a group treated with vector alone, and a group treated with pro-drug alone. Each sc tumor requires injection of $1–2 \times 10^6$ cells, and each mouse can be used to grow two tumors, one on each flank. At 100% confluency, NMU cells reach a density of approx 70,000 cells/cm², and thus one confluent 10-cm dish of NMU cells contains roughly 5.5×10^6 cells.

11. NMU cells should be growing in log phase and between 70–90% confluence at the time of harvest for injection. Cultures that are overconfluent may result in a reduced tumor-establishment rate. To minimize host immune response to injection of tumor-cell suspension, make sure to wash cells well in PBS before trypsinizing and to not include FBS in the DMEM used for final resuspension of cells.

12. Clumping of cells after trypsinization can be minimized by first adding one volume of DMEM, mixing the cell suspension by gentle pipetting, then adding the remaining 3–4 volumes of DMEM. The cell suspension can be kept at room temperature prior to injections but should be used as soon as possible to maintain maximum viability.

13. Using the described procedure, we have found that approx 85% of injections with NMU cells result in development of palpable tumors. The use of other cell lines, however, may result in a different rate of tumor formation. To improve the frequency of tumor formation using a cell line with a low take rate, Matrigel basement membrane matrix (BD Biosciences) can be included in the cell suspension and is in some cases necessary for efficient tumor establishment. Matrigel, which is liquid at 2–8°C and gels at room temperature, is used at a 50:50 dilution with DMEM to resuspend pelleted cells. Note that Matrigel is quite viscous, and will require thorough yet gentle pipetting to disperse cells evenly within the suspension.

14. Although not required for this simple injection procedure, use of general anesthesia may allow tumor inoculation to be performed with less stress and more consistency. For this purpose, an intramuscular or intraperitoneal injection of a ketamine/xylazine mixture, at a dose of 90 mg/kg ketamine and 10 mg/kg xylazine, is appropriate. The resulting anesthesia lasts for 20–30 min. Hypothermia is a common cause of mortality in anesthetized mice. To prevent heat loss in mice, keep the animals on a blanket or other insulating material during anesthesia.

15. The rate of tumor growth may vary significantly among mice and over time. Tumor-bearing mice must therefore be observed with sufficient frequency (at least 3 times per week) to ensure that they can be euthanized before tumors ulcerate or reach a size that interferes with normal activity.

16. The dissections should be performed as quickly as possible to preserve cellular viability. In particular, do not allow the tissue to dry out or the cells will become unrecoverable. If necessary, tissue samples may be immersed in PBS for a short period of time prior to processing. Tissues to be used for immunohistochemical analysis should also be promptly frozen as described in **Subheading 3.5.1.** Tissue samples to be used for extraction of DNA can be frozen and stored in microfuge tubes at −70°C indefinitely if DNA extraction will not be carried out immediately.

17. Tumors derived from other cell lines may require an enzyme preparation other than type I collagenase for optimal dissociation. We have found that treatment with type I collagenase is too harsh for certain tumor types and results in poor recovery of intact cells. The appropriate method of dissociation for other tumor types must be determined empirically.

18. GFP can be detected using the FL1 channel and PI can be detected using the FL3 channel. The cytometer uses the 488-nm argon laser line for both GFP and PI.

19. If liquid nitrogen is not available, freezing may be carried out in the same manner using isopentane cooled on dry ice or in a −70°C freezer. Once tissue is embedded in OCT, it should not be allowed to thaw prior to sectioning. Thawing and refreezing of tissue will result in significant loss of morphological integrity and reduced antigenicity. Frozen, OCT-embedded tissues can be stored for 2–3 mo at −70°C without significant loss of morphological integrity.

20. To prevent artifactual staining, reagents should be applied evenly over the surface of the section. False-positive staining is often observed around the edges of a section owing to uneven application of reagents and drying of section during staining.

21. To validate the specificity of any observed signal, a control procedure in which PBS is used in place of the primary antibody should also be carried out in parallel on each section. Only the sections from vector-transduced tumors that were incubated with primary antibody should generate a signal.

22. Counterstaining should be relatively weak compared to the immunostaining, and care should be taken to not allow the hematoxylin to overstain the sections. If the counterstaining is too intense, it may obscure the visibility of immunoreactive regions in the sections. Ideally, the intensity of hematoxylin staining should be just enough to visualize overall cell structure.

23. The effect of RCR-mediated suicide gene transfer is monitored by the measurement of tumor growth over time or by evaluation of survival of tumor-bearing animals. It should be noted that analysis of tumor volume measurements is often complicated by the fact that tumor sizes become progressively more heterogeneous as they become bigger, resulting in larger standard deviations and difficulty in determining the significance of the differences. Also, it should be kept in mind that, particularly in the case of large preestablished tumors that have been successfully treated, tumor volumes may not show evidence of regression due to the persistence of necrotized or scar tissue. Conversely, in some cases we have observed some degree of tumor growth inhibition even after injection of wild-type retrovirus lacking any functional transgene; however, this did not appear to affect long-term outcome, suggesting that outgrowth of resistant cells readily occurs.

Survival analysis using the Kaplan-Meyer method is a useful approach for analyzing therapeutic efficacy *(21)*. In this case, careful consideration is required to ensure that adequate statistical power is employed. Our own statistical power analysis suggests that $n = 14$ animals per group would be required to achieve a significant result at $p < 0.05$ using Fisher's exact test, assuming the reference group has 0% survival and the treatment of interest has 50% surviving at the same time of follow-up. To assess any potential effects from the RCR vector alone or the prodrug alone, we always employ additional reference groups corresponding to these controls.

References

1. Burns, J. C., Friedmann, T., Driever, W., Burrascano, M., and Yee, J. K. (1993) Vesicular stomatitis virus G glycoprotein pseudotyped retroviral vectors: concentration to very high titer and efficient gene transfer into mammalian and nonmammalian cells. *Proc. Natl. Acad. Sci. USA* **90,** 8033–8037.
2. Cohen, L. A. (1982) Isolation and characterization of a serially cultivated, neoplastic, epithelial cell line from the N-nitrosomethylurea induced rat mammary adenocarcinoma. *In Vitro* **18,**565–575.
3. Cornetta, K., Moen, R. C., Culver, K., Morgan, R. A., McLachlin, J. R., Sturm, S., et al. (1990) Amphotropic murine leukemia retrovirus is not an acute pathogen for primates. *Hum. Gene Ther.* **1,** 15–30.
4. Culver, K. W., Ram, Z., Wallbridge, S., Ishii, H., Oldfield, E. H., and Blaese, R. M. (1992) In vivo gene transfer with retroviral vector-producer cells for treatment of experimental brain tumors. *Science* **256,** 1550–1552.
5. Diaz, R. M., Eisen, T., Hart, I. R., and Vile, R. G. (1998) Exchange of viral promoter/enhancer elements with heterologous regulatory sequences generates targeted hybrid long terminal repeat vectors for gene therapy of melanoma. *J. Virol.* **72,** 789–795.
6. Donahue, R. E., Kessler, S. W., Bodine, D., McDonagh, K., Dunbar, C., Goodman, S., et al. (1992) Helper virus induced T cell lymphoma in nonhuman primates after retroviral mediated gene transfer. *J. Exp. Med.* **176,**1125–1135.
7. DuBridge, R. B., Tang, P., Hsia, H. C., Leong, P. M., Miller, J. H., and Calos, M. P. (1987) Analysis of mutation in human cells by using an Epstein-Barr virus shuttle system. *Mol. Cell Biol.* **7,** 379–387.
8. Ezzeddine, Z. D., Martuza, R. L., Platika, D., Short, M. P., Malick, A., Choi, B.,and Breakefield, X. O. (1991) Selective killing of glioma cells in culture and in vivo by retrovirus transfer of the herpes simplex virus thymidine kinase gene. *New Biol.* **3,** 608–614.
9. Han, J. Y., Zhao, Y., Anderson, W. F., and Cannon, P. M. (1998) Role of variable regions A and B in receptor binding domain of amphotropic murine leukemia virus envelope protein. *J. Virol.* **72,** 9101–9108.
10. Han, X., Kasahara, N., and Kan, Y. W. (1995) Ligand-directed retroviral targeting of human breast cancer cells. *Proc. Natl. Acad. Sci. USA* **92,** 9747–9751.
11. Liu, X., Constantinescu, S. N., Sun, Y., Bogan, J. S., Hirsch, D., Weinberg, R. A., and Lodish, H. F. (2000) Generation of mammalian cells stably expressing multiple genes at predetermined levels. *Anal. Biochem.* **280,** 20–28.

12. Logg, C. R., Logg, A., Matusik, R. J., Bochner, B. H., and Kasahara, N. (2002) Tissue-specific transcriptional targeting of a replication-competent retroviral vector. *J. Virol.* **76,** 12783–12791.
13. Logg, C. R., Logg, A., Tai, C. K., Cannon, P. M., and Kasahara, N. (2001) Genomic stability of murine leukemia viruses containing insertions at the Env-3′ untranslated region boundary. *J. Virol.* **75,** 6989–6998.
14. Logg, C. R., Tai, C. K., Logg, A., Anderson, W. F., and Kasahara, N. (2001) A uniquely stable replication-competent retrovirus vector achieves efficient gene delivery in vitro and in solid tumors. *Hum. Gene Ther.* **12,** 921–932.
15. Miller, D. G., Adam, M. A., and Miller, A. D. (1990) Gene transfer by retrovirus vectors occurs only in cells that are actively replicating at the time of infection [published erratum appears in Mol Cell Biol 1992 Jan;12(1):433]. *Mol. Cell Biol.* **10,** 4239–4242.
16. Naviaux, R. K., Costanzi, E., Haas, M., and Verma, I. M. (1996) The pCL vector system: rapid production of helper-free, high-titer, recombinant retroviruses. *J. Virol.* **70,** 5701–5705.
17. Ory, D. S., Neugeboren, B. A., and Mulligan, R. C. (1996) A stable human-derived packaging cell line for production of high titer retrovirus/vesicular stomatitis virus G pseudotypes. *Proc. Natl. Acad. Sci. USA* **93,** 11400–11406.
18. Rainov, N. G. (2000) A phase III clinical evaluation of herpes simplex virus type 1 thymidine kinase and ganciclovir gene therapy as an adjuvant to surgical resection and radiation in adults with previously untreated glioblastoma multiforme. *Hum. Gene Ther.* **11,** 2389–2401.
19. Ram, Z., Walbridge, S., Shawker, T., Culver, K. W., Blaese, R. M., and Oldfield, E. H. (1994) The effect of thymidine kinase transduction and ganciclovir therapy on tumor vasculature and growth of 9L gliomas in rats. *J. Neurosurg.* **81,** 256–60.
20. Soneoka, Y., Cannon, P. M., Ramsdale, E. E., Griffiths, J. C., Romano, G., Kingsman, S. M., and Kingsman, A. J. (1995) A transient three-plasmid expression system for the production of high titer retroviral vectors. *Nucleic Acids Res.* **23,** 628–633.
21. Parmar, M. K. B. and Machin, D. (1995) *Survival Analysis: A Practical Approach.* John Wiley and Sons, Chichester, UK.

34

Delivery of Genes to Hematopoietic Stem Cells

Masafumi Onodera

1. Introduction

Bone marrow hematopoiesis is maintained by hematopoietic stem cells (HSC) *(1)*. Because of their unique features to self-renew and differentiate along all lineages of hematopoietic cells, even a single HSC can completely reconstitute bone marrow hematopoiesis of irradiated recipients *(2)*. Therefore, HSCs are considered to be the ideal target cell population in gene-therapy fields for genetic disorders that are susceptible to bone marrow transplantation *(3)*. However, because most HSCs are quiescent, it is difficult to transduce them using retroviral vectors *(4)*. Furthermore, retroviral vectors, especially Moloney murine leukemia virus (MMLV)-based retroviral vectors that have been commonly used in gene-therapy clinical trials, are very susceptible to *de novo* methylation in immature cells such as embryonic stem (ES) cells, embryonal carcinoma cells (EC), and HSCs, resulting in shut off/silencing of the transgene expression in vivo *(5)*. This is another obstacle for successful gene delivery into HSCs.

At least two avenues exist for improving transduction efficiency of HSCs and for obtaining sustained expression of a gene of interest in vivo. First, the components of the gene-transfer system, including retroviral vectors and packaging cell lines, can be improved. Second, culture conditions have been established for expanding HSCs as well as hematopoietic progenitor cells while maintaining HSC activities. The method of gene delivery into HSCs described here focuses on three points:

From: *Methods in Molecular Biology, vol. 246:*
Gene Delivery to Mammalian Cells: Vol. 2: Viral Gene Transfer Techniques
Edited by: W. C. Heiser © Humana Press Inc., Totowa, NJ

1. Vector constructs and packaging cell lines suitable for gene delivery into HSCs;
2. Culture conditions for maintaining stem cell activities of HSCs during transduction; and
3. Assays for evaluating the successful gene delivery into HSCs.

Although CD34+ cells obtained from cord blood have often been used as human hematopoietic progenitors in recent gene-transfer experiments, this chapter mainly describes methods of gene delivery to mouse HSCs. A brief summary of gene-transfer into human CD34+ cells is described in **Note 1**.

2. Materials

2.1. Retroviral Vectors and Expression Vectors for Viral Structural Proteins

Details of the retroviral structure and life cycle have been described in Chapter 31 on gene delivery by retroviral vectors. However, it should be noted that several unique retroviral vectors have recently been constructed and used for stem cell gene therapy (**Fig. 1**).

1. GCsap *(6)*: This vector is a simplified retroviral vector constructed by removing a dominant selectable marker such as neomycin phosphotransferase II (Neo) and inserting splice donor and acceptor sequences to generate subgenomic RNA that can be translated with high efficiency. When transduced into hematopoietic cells, it allows higher expression of the transgene per copy of virus in the transduced cells than do conventional vectors with dominant selectable markers.
2. MND *(7)*: This retroviral vector for stem cell gene therapy has been constructed to have resistance to *de novo* methylation by replacing the original LTR and primer binding site derived from MMLV with the myeloproliferative sarcoma virus (MPSV) derived LTR devoid of negative control region and the dl587rev derived primer binding site, respectively. The vector allows continued expression of the transgene against shut off/ silencing in vivo.
3. pMD.G *(8)*: This is an expression vector for the vesicular stomatitis virus G protein (VSV-G).

2.2. Packaging Cell Lines

Packaging cell lines generally used in gene delivery into HSCs are the ecotropic lines GP+E-86 *(9)* and BOSC23 *(10)*, and the amphotropic cell line PA317 *(11)*. 293 gp cells have been engineered to express MMLV derived Gag and Pol by transfection of pCMVgag-pol into the human adenovirus 5 trans-

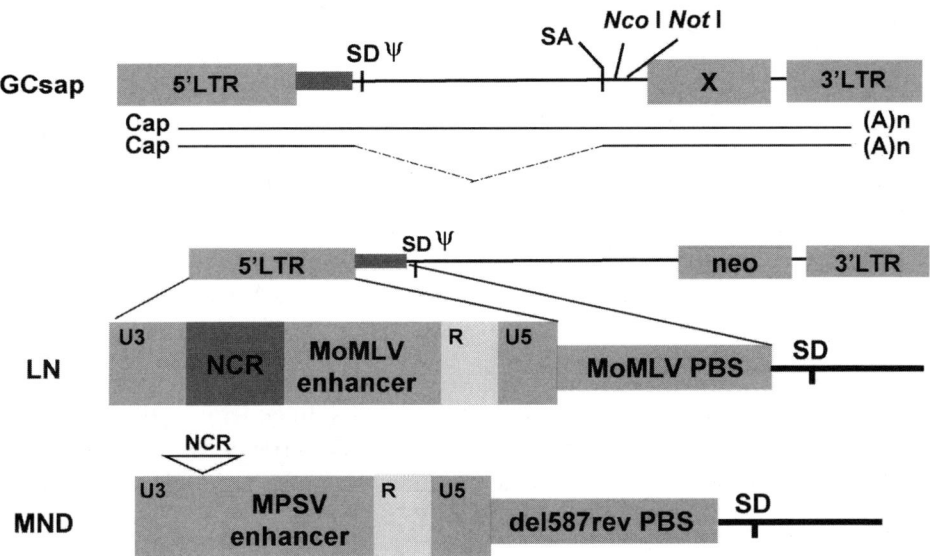

Fig. 1. Structures of the retroviral vectors GCsap and MND. The gene of interest is cloned into the *Nco*I site of the GCsap vector such that the gene translational start site is located precisely at the env translational start site in the wild-type virus. The conventional vector LN is shown to compare the MND vector. LTR, long terminal repeat; SD, SA, splice donor/splice acceptor sequences; NCR, negative control region; PBS, primer binding site; MPSV, myeloproliferative sarcoma virus-derived LTR.

formed embryonic kidney cell line 293 *(12)*. Maintain packaging cell lines and 293gp in Dulbecco's modified Eagle's medium (high glucose DMEM; 4.5 mg/mL) supplemented with 10% heat-inactivated fetal bovine serum (FBS), penicillin G sodium (100 U/mL), streptomycin sulfate (100 µg/mL), and 2 m*M* L-glutamine. All vectors and packaging cells lines are available from the developers.

2.3. Culture Medium, Cytokines, and Monoclonal Antibody.

1. RPMI1640, DMEM, StemPro-34 SFM, and other culture reagents are purchased from Invitrogen (Carlsbad, CA). StemPro-34 SFM is a serum-free medium for the culture of mouse hematopoietic progenitor cells.
2. Mouse and human cytokines of interleukin (IL)-3, IL-6, stem cell factor (SCF), Flt-3/Flk-2 ligand (FL), and Thrombopoietin (TPO) are available from R&D Systems (Minneapolis, MN).

3. Monoclonal Antibodies (MAbs): Biotinylated MAbs against lineage markers used in negative selection are Ly-6G/ Gr-1 (RB6-8C5) for granulocytes, CD11b/ Mac-1 (M1/70) for myeloid and monocytes, CD45/ B220 (RA3-6B2) for B cells, CD4/ L3T4 (GK1.5), CD8a/ Ly-2 (53-6.7) for T cells, and TER-119/ Ly-76 (TER-119) for erythroid cells. For positive selection or FACS sorting, FITC-conjugated MAb for mouse CD34 (49E8), PE conjugated MAb for Sca-1 (E13-161.7), biotin- or APC conjugated MAb for c-KIT for positive selection or FACS sorting, respectively, and Texas Red streptavidin are used. PE conjugated CD45.1 (A20) are used to distinguish donors derived cells (B6-Ly5.1) from recipient's (B6-Ly5.2) post-transplant.

Clone names of MAbs are shown in parentheses. In general, antibodies stated in **Subheading 2.3.3.** are used at 0.5–1 µg per 1 × 10⁶ cells, but the suitable amount of each antibody should be determined by researchers. All antibodies and Texas Red streptavidin are available from BD PharMingen (San Diego, CA).

2.4. Mice, Reagents, Solutions, and Others

1. C57BL/6 (B6-Ly5.1 and B6-Ly5.2) mice 8–12 wk old are purchased from Charles River Japan, Inc. (Yokohama, Japan).
2. MBS Mammalian Transfection Kit for calcium phosphate transfection (Stratagene, La Jolla, CA).
3. Pentobarbital (P-3761) and protamine sulfate (P-4020) are available from Sigma (St. Louis, MO). Lymphosepar II (gravity 1.090 +/− 0.001) is purchased from IBL (Fujioka, Japan).
4. Recombinant human fibronectin fragments CH-296 (RetroNectin, Takara Shuzo, Otsu, Japan).
5. Staining medium (SM) is Ca^{2+} and Mg^{2+} free phosphate-buffered saline (Sigma) with 1% bovine serum albumin (BSA) (Sigma).
6. ACK solution is made by dissolving 8.29 g of NH_4Cl, 1 g of $KHCO_3$, and 37.2 mg of Na_2 EDTA in 800 mL of H_2O and adjusting the pH to 7.2–7.4 with 1 N HCl. Finally, add H_2O to 1 L, filter-sterilize through a 0.2-µm filter, and store at room temperature.
7. Streptavidin-magnetic beads (M-280) and Magnetic Particle Concentrators (MPC) are available from Dynal Biotech Inc (Lake Success, NY).
8. 0.45-µm pore size microfilter (Millex-Ha; Millipore, Bedford, MA).
9. Surgical scissors and forceps.
10. 21 G needles, 1-mL syringes.
11. Irradiation apparatus; HITACHI, Deep X-ray Apparatus, Model MBR-1520R (Abiko, Chiba-ken, Japan).

12. Dual laser flow cytometer; FACS Vantage, Becton Dickinson (San Jose, CA)

3. Methods

3.1. Production of Recombinant Retroviruses for HSC Transduction

It is important to select suitable packaging cell lines as well as retroviral vectors in accordance with the species or types of the target cell population. In the case of mouse HSCs, GP+E-86 cells that express the ecotropic envelope are mainly used as the stable virus producer clone. BOSC23 cells, expressing the ecotropic envelope, are also used for transient virus production. Amphotropic packaging cell lines such as PA317, although they produce recombinant retroviruses to infect mouse cells, are not generally used for gene delivery to HSCs because mouse HSCs express few receptors for amphotropic viruses on their cell surface *(13)*. Recently, pseudotype retroviruses enveloped with the VSV-G protein have been used for transduction into mouse HSCs because of their broad host range, including murine cells. Furthermore, VSV-G retroviruses show greater stability, allowing concentration of the vectors to high titer *(14)*. However, the virus is generally generated by transient transfection because the VSV-G protein is very toxic to cells. A method to produce and concentrate the VSV-G pseudotype retroviruses is described below.

1. D 0: Plate 2×10^6 293 gp cells in 10 mL DMEM/10% FBS in a 100-mm culture dish approx 16 h before transfection.
2. D 1: Transfect 20 μg of retroviral vector (e.g., GCsap) together with 10 μg of pMD.G into 293 gp cells by calcium-phosphate-precipitation using the MBS Mammalian Transfection Kit as described by the supplier. Wash the cells with PBS after a 3 h incubation with the DNA-CaPO$_4$ complex. Add 10 mL of fresh DMEM/10% FBS and incubate for 16 h.
3. D 2: Collect the virus supernatant with a pipet and store it in a 15-mL culture tube at 4°C. Add 10 mL of fresh medium to the dish and continue incubating the culture for an additional 24 h.
4. D 3: Repeat the procedures of d 2.
5. D 4: Collect the virus supernatant and combine with those stored on d 2 and 3 (the total volume is about 30 mL). Filter the virus supernatant through a 0.45 μm pore microfilter, transfer it into a 50-mL centrifuge tube, then centrifuge for 16 h at 6000g at 4°C.
6. D 5: Aspirate the supernatant carefully and resuspend the virus pellet gently in 300 μL of StemPro-34 SFM using a shaker for 48 h at 4°C. (This concentrates the virus 100-fold.)
7. D 7: Store the concentrated virus supernatant at −80°C until used.

3.2. Isolation of Mouse HSCs and Transduction of Retroviral Vectors into Mouse HSCs

3.2.1. Isolation of Mouse HSC (see **Note 2**)

1. Anesthetize mice with an intraperitoneal (ip) injection of 40 mg/kg pentobarbital and disinfect mice with 70% ethyl alcohol.
2. Cut the skin from knees to hip joints and expose femurs in each leg by removing the quadriceps with surgical scissors.
3. Dislocate the hip joints with scissors and remove the muscle behind the femurs.
4. Dislocate the knee joints with scissors and remove the femurs.
5. Make pinholes at both ends of the femur with a 21 G needle (**Fig. 2**).
6. Attach the 21G needle to a 2-mL syringe containing 1.5 mL of culture medium and flush the marrow out into a 15-mL culture tube by pushing the medium in the syringe into the bone.
7. Repeat **steps 5** and **6** and remove the marrow from the other femur.
8. Combine the marrow from both femurs and bring the volume to 6 mL with PBS.
9. Carefully layer the marrow onto 2 mL of Lymphosepar II in a 15-mL tube.
10. Centrifuge at 20°C for 20 min at 400g.
11. Aspirate the upper layer leaving the mononuclear cell (MNC) layer at the interphase.
12. Transfer the cells at the interphase into a 15-mL tube containing 10 mL of SM.
13. Pellet the cells by centrifugation at 4°C for 5 min at 200g. Discard the supernatant, then wash the pellet again with SM. The expected number of MNCs obtained from two femurs is approx $3–4 \times 10^7$.
14. Resuspend the cells in 200 µL of SM (approx 1×10^5/µL) and stain with biotinylated antibodies against lineage markers (Mac-1, Gr-1, B220, CD4, CD8, and TER119) for 30 min on ice.
15. Add 10 mL of SM, then pellet the cells by centrifugation at 4°C for 5 min at 200g. Discard the supernatant.
16. Resuspend the cells in 100 µL of SM and incubate with Streptavidin-magnetic beads (M-280; approx 6 beads/cell) for 30 min on ice.
17. Add 10 mL of SM then pellet the cells by centrifugation at 4°C for 5 min at 200g. Discard the supernatant.
18. Resuspend the cells in 7 mL of SM in a 15-mL polystyrene tube and place the tube in a Dynal MPC for 5 min.
19. Transfer the supernatant containing nonbound cells to another tube as lineage negative cells (negative selection). Approximately $3–4 \times 10^6$ cells

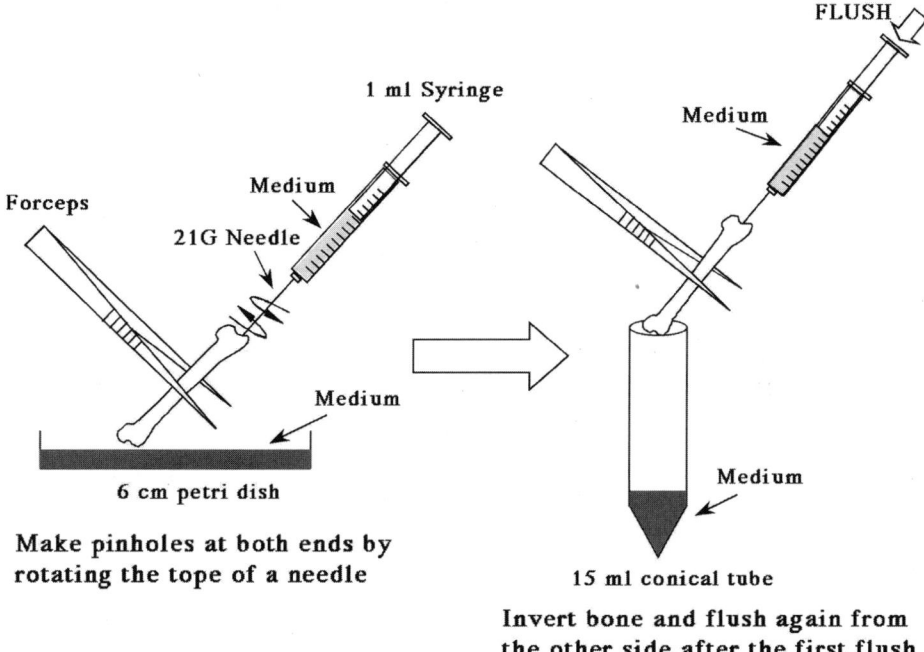

FLUSH

1 ml Syringe

Medium

Medium

Forceps

21G Needle

Medium

Medium

6 cm petri dish

Make pinholes at both ends by rotating the tope of a needle

15 ml conical tube

Medium

Invert bone and flush again from the other side after the first flush

Fig. 2. Collection of bone marrow cells.

(one-tenth of the starting cell number) should be recovered at this step. See **Subheading 3.2.2.1.** for transduction into mouse HSCs.

20. Pellet the nonbound cells by centrifugation at 4°C for 5 min at 200g. Discard the supernatant then resuspend the cells in 100 µL of SM.

21. To isolate purified HSCs as c-KIT positive/lineage negative cells (KL cells), stain lineage negative cells with biotin conjugated anti-c-KIT MAb and isolate using a Dynal MPC (positive selection). Alternatively, stain the cells with FITC- anti-CD34 antibody, PE-Sca-1, APC-c-KIT antibody, and Texas Red streptavidin to isolate CD34$^{-/low}$ c-KIT$^+$ Sca-1$^+$ lineages$^-$ (34-KSL) cells using a dual laser flow cytometer *(2)*. Use Texas Red streptavidin to stain lineage positive cells that have already stained with biotin-conjugated antibodies in the step of negative selection, allowing complete depletion of lineage positive cells by FACS sorting. Because the ratio of 34-KSL cells to total bone marrow cells is one per 1–4 $\times$ 10^4, approx 500–1000 34-KSL cells should to be recovered from two femurs. Both of these procedures will reduce the total number of target cells for transduction into mouse HSCs (*see* **Subheading 3.2.2.2.**).

3.2.2. Transduction into Mouse HSC

3.2.2.1. TRANSDUCTION BY CO-CULTURE WITH RETROVIRUS PRODUCER CLONES (*SEE* NOTE 3).

The co-culture system is suitable if the number of target cells is large (e.g., when using total bone marrow or lineage negative cells).

1. D 0: Plate 2×10^6 BOSC23 cells in 10 mL of DMEM/10% FBS in a 100-mm culture dish 16 h before transfection.
2. D 1:
   a. Transfect 15 µg of retroviral vector (e.g., GCsap) into the BOSC23 cells by calcium phosphate precipitation using the MBS Mammalian Transfection Kit as described by the supplier to produce the recombinant retrovirus.
   b. Plate total bone marrow or lineage negative cells (*see* **Subheading 3.2.1.**) at $1–2 \times 10^6$ cells/mL in RPMI1640 supplemented with 10% FBS on another 100-mm culture dish. Add mouse IL-3, mouse IL-6, and mouse SCF, each to a final concentration of 50 ng/mL. Incubate for 16 h.
3. D 2: transfer the cytokine stimulated nonadherent marrow cells onto the transfected BOSC23 cells with 5 µg/mL of protamine sulfate and culture for 48 h.
4. D 4: Collect the non-adherent cells gently from the culture, centrifuge at 4°C for 5 min at 200g, and resuspend them with fresh culture medium at 1 $\times 10^6$ cells/mL. Use the cells as soon as possible for the subsequent experiments (*see* **Subheading 3.3.**).

3.2.2.2. TRANSDUCTION USING THE VIRUS SUPERNATANT (*SEE* NOTE 4).

Transduction using the virus supernatant is recommended if the number of target cells is small (e.g., KL or 34-KSL cells).

1. D 1: Plate the target cells (5000 to 10^4 of KL or 500–1000 of 34-KSL; *see* **Subheading 3.2.1.**) in 150 µL of StemPro-34 SFM in a well of a 96-well plate and add human TPO to a final concentration of 100 ng/mL and mouse Flt-3 and mouse SCF, each to a final concentration of 10 ng/mL. Put 150 µL of PBS or sterile water in the wells around the well containing the cells. Incubate for 24 h.
2. D 2: Add 50 µL of the concentrated supernatant of VSV-G pseudotype retrovirus (see **Subheading 3.1.**) with 1 µg/mL of CH296 (RetroNectin) and 5 µg/mL of protamine sulfate into the culture, then culture for an additional 24 h.

3. D 3: Collect the transduced cells from the 96-well plate, centrifuge, and resuspend them in 100–200 µL of fresh culture medium. Use the cells as soon as possible for subsequent experiments (*see* **Subheading 3.3.**).

3.3. Transplantation of the Transduced Mouse HSC into Mice

The protocol below describes transplantation of bone marrow cells into B6-Ly5.2 mice. This is an accurate system for measuring transduction efficiency of HSCs. Estimation of transduction efficiency by an in vitro assay using long-term culture initiating cells (LTC-IC) and colony forming unit cells (CFU-C) *(15,16)* do not accurately reflect the transduction efficiency measured by the in vivo method.

1. Irradiate B6-Ly5.2 mice at a dose of 950cGy from a cesium gamma source 4 h before transplantation.
2. Anesthetize mice with an intraperitoneal injection of 40 mg/kg pentobarbital.
3. Resuspend either $1–2 \times 10^6$ of the transduced total bone marrow or lineage negative cells, 5000 to 10^4 of the transduced KL cells, or 100–1000 of the transduced CD34-KSL cells together with 1×10^5 of B6-Ly5.2 bone marrow as rescue cells in 500 µL of PBS (*see* **Note 5**). Transplant these into the irradiated B6-Ly5.2 mouse through the tail vein using a 21 G needle and a 1-cc syringe. No selection procedure to enrich for the transduced cells is performed before transplantation unless otherwise used for special purposes.
4. Collect peripheral blood from the tail vein of recipients at various time points posttransplantation (e.g., 2, 6. 12, 24, 48 wk). To collect peripheral blood cells, restrain the mouse, extend the tail of the mouse with one hand, using the other hand insert a 25 G needle 3–4 mm into the lateral vein and collect dropping blood from the hub of the needle into a microfuge tube.
5. To remove red blood cells from the samples, add 1.0 mL of ACK buffer per 100 µL of blood. Incubate 5 min at 37°C with occasional shaking. Centrifuge the microfuge tube for 1 min at 200*g* and discard the supernatant. Wash the pellet with SM, then resuspend in an appropriate volume of SM for subsequent procedures (e.g., cell counting, FACS analysis, colony-forming unit (CPU) cells assay, or PCR).
6. For FACS analysis, stain cells prepared in **step 5** with PE-conjugated CD45.1. Use the FL-1 (FITC) channel to analyze the transgene expression and determine the chimerism of donor cells in the transgenic mice by calculating the ratio of CD45.1 positive cells to the total cells. To determine

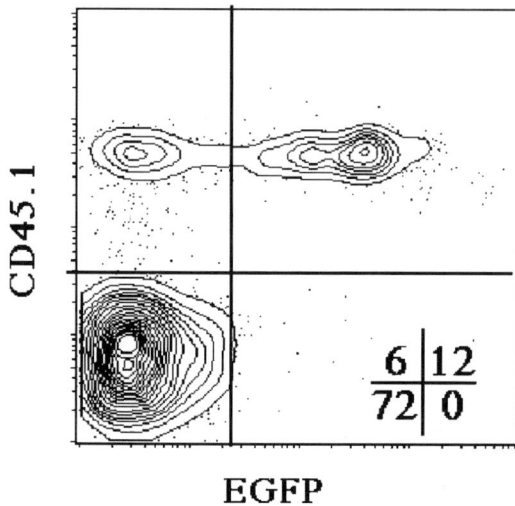

EGFP

Fig. 3. Dual-parameter contour plots by FACS analysis posttransplantation. Values in the right lower box indicate the percentage of cells in each quadrant. X- and Y-axes represent the florescent intensity of enhanced green fluorescent protein (EGFP) as the transgene and CD45.1, respectively. In this experiment, the chimerism of donor cells in transplanted mice and the transduction efficiency are 18% (6% + 12%) and 67% (12%/6% + 12%), respectively

the transduction efficiency, calculate the ratio of transgene expressing cells in total CD45.1 expressing cells (*see* **Fig. 3**).
7. Expression of the transgene in serially transplanted mice strongly confirms successful gene delivery into mouse HSCs. For the secondary or tertiary transplants, harvest bone marrow from the primary or secondary recipient mice at 3–4 mo posttransplant and transplant 2×10^6 bone marrow cells into lethally irradiated B6-Ly5.2 mice. Enrichment for the transduced cells is not necessary unless otherwise used for special purposes.

4. Notes

1. With high expectations built up for HSC gene therapy, human CD34$^+$ cells derived from cord blood have been used as the target cell population in retroviral-mediated gene transfer experiments. As stated in **Subheading 3.1.**, selection of packaging cell lines suitable for transduction into human CD34$^+$ cells is important. Recently, packaging cell lines PG13 *(17)* or RD18 *(18)* that express the viral envelope derived from gibbon ape leukemia virus or feline virus RD114, respectively, have been used in transduc-

tion experiments. Although VSV-G envelope pseudotype retrovirus can also infect human CD34$^+$ cells with high efficiency, the VSV-G envelope is more toxic to CD34$^+$ cells than are other types of the viral envelopes. Regarding the detailed protocol of gene transfer into human CD34$^+$ cells, refer to Chapter 28 by Hiroyuki Miyoshi.

2. When bone marrow is to be isolated, administration of 150 mg/kg of 5-fluorouracil intraperitoneally (ip) into mice 48 h prior to marrow harvest increases the number of HSCs in the bone marrow, but changes surface phenotypes of mouse HSCs *(19)*.

3. Because HSCs are very difficult to transduce with retroviruses, it is important to produce a high titer of retrovirus for transduction of HSCs. In general, the transient expression system using BOSC23 cells generates retroviruses with titers over 10^5 CFU/mL by transfection with retroviral vectors. Because BOSC23 cells are derived from 293T cells that express an SV40 T antigen, however, expression of the ecotropic envelope in BOSC23 cells is unstable and is sometimes lost during long-term culture.

4. The VSV-G pseudotype retroviruses are resistant to the gravity of centrifugation enough to be a pellet while maintaining their infectivity. Resuspending the virus pellet in a small volume of culture medium increases the virus titer up to 10^7 to 10^8 CFU/mL (a 100-fold concentration). Virus with a VSV-G envelope can also be concentrated by ultracentrifugation at 113,000g for 2 h.

5. Transplant experiments using a single CD34-KSL have demonstrated that three out of ten CD34-KSL cells are capable of reconstituting hematopoiesis of irradiated mice completely *(2)*. However, in vitro, gene transfer into CD34-KSL results in differentiation of these cells into mature cells that are no longer capable of reconstituting bone marrow hematopoiesis and decreases the ratio of HSCs to one per 30–50 cells at the end of 2 d in culture. Therefore, more than 50 cells of the transduced CD34-KSL should be transplanted to reconstitute bone marrow hematopoiesis of lethally irradiated mice.

References

1. Till, J. E., McCulloch, E. A., and Siminovitch, L. (1964) A stochastic model of stem cell proliferation, based on the growth of spleen colony-forming cell. *Proc. Natl. Acad. Sci. USA* **51**, 29–36.
2. Osawa, M., Hanada, K., Hamada, H. and Nakauchi H. (1996) Long-term lympho-hematopoietic reconstitution by a single CD34-low/negative hematopoietic stem cell. *Science* **273**, 242–245.
3. Mulligan, R. C. (1993) The basic science of gene therapy. *Science* **260**, 926–932.

4. Hoogerbrugge, P. M., van Beusechem, V. W., Fischer, A., Debree, M., le Deist, F., Perignon, J. L., et al. (1996) Bone marrow gene transfer in three patients with adenosine deaminase deficiency. *Gene Ther.* **3**, 179–183.

5. Wang, L., Robbins, P. B., Carbonaro, D. A., and Kohn, D. B. (1998) High-resolution analysis of cytosine methylation in the 5' long terminal repeat of retroviral vectors. *Hum. Gene Ther.* **9**, 2321–2330.

6. Onodera, M., Nelson, D. M., Yachie, A., Jagadeesh, G. J., Bunnell, B. A., Morgan, R. A., and Blaese, R. M. (1998) Development of improved adenosine deaminase retroviral vectors. *J. Virol.* **72**, 1769–1774.

7. Halene, S., Wang, L., Cooper, R. M., Bockstoce, D. C., Robbins, P. B., and Kohn, D. B. (1999) Improved expression in hematopoietic and lymphoid cells in mice after transplantation of bone marrow transduced with a modified retroviral vector. *Blood* **94**: 3349–3357.

8. Tahara-Hanaoka, S., Sudo, K., Ema, H., Miyoshi, H., and Nakauchi, H. (2002) Lentiviral vector-mediated transduction of murine CD34(-) hematopoietic stem cells. *Exp. Hematol.* **30**, 11–17.

9. Markowitz, D., Goff, S., and Bank, A. (1988) A safe packaging line for gene transfer: separating viral genes on two different plasmids. *J. Virol.* **62**, 1120–1124.

10. Pear, W. S., Nolan, G. P., Scott, M. L., and Baltimore, D. (1993) Production of high-titer helper-free retroviruses by transient transfection. *Proc. Natl. Acad. Sci. USA* **90**, 8392–8396.

11. Miller, A. D. and Buttimore, C. (1986) Redesign of retrovirus packaging cell lines to avoid recombination leading to helper virus production. *Mol. Cell Biol.* **6**, 2895–2902.

12. Burns, J. C., Friedmann, T., Driever, W., Burrascano, M., and Yee, J. K. (1993) Vesicular stomatitis virus G glycoprotein pseudotyped retroviral vectors: concentration to very high titer and efficient gene transfer into mammalian and nonmammalian cells. *Proc. Natl. Acad. Sci. USA* **90**, 8759–8760.

13. Orlic, D., Girard, L. J., Anderson, S. M., Pyle, L. C., Yoder, M. C., Broxmeyer, H. E., and Bodine, D. M. (1998) Identification of human and mouse hematopoietic stem cell populations expressing high levels of mRNA encoding retrovirus receptors. *Blood* **91**, 3247–3254.

14. Yang, Y., Vanin, E. F., Whitt, M. A., Fornerod, M., Zwart, R., Schneiderman, R. D., and Grosveld. G. (1995) Inducible, high-level production of infectious murine leukemia retroviral vector particles pseudotyped with vesicular stomatitis virus G envelope protein. *Hum. Gene Ther.* **6**, 1203–1213.

15. Dexter, T. M., Allen T., and Lajtha L. G. (1997) Conditions controlling the proliferation of haemapoietic stem cells in vitro. *J. Cell Physiol.* **91**, 335–344.

16. Ogawa, M., Matsuzaki, Y., Nishikawa, S., Hayashi, S., Kunisada, T., Sudo, T., et al. (1991) Expression and function of c-kit in hemopoietic progenitor cells. *J. Exp. Med.* **174**, 63–71.

17. Miller, A. D., Garcia, J. V., von Suhr, N., Lynch, C. M., Wilson, C., and Eiden, M. V. (1991) Construction and properties of retrovirus packaging cells based on gibbon ape leukemia virus. *J. Virol.* **65**, 2220–2224.

18. Cosset, F. L., Takeuchi, Y., Battini, J. L., Weiss, R. A., and Collins, M. K. (1995) High-titer packaging cells producing recombinant retroviruses resistant to human serum. *J. Virol.* **69**, 7430–7436.

19. Sato, T., Laver, J. H., and Ogawa, M. (1999) Reversible expression of CD34 by murine hematopoietic stem cells. *Blood* **94**, 2548–2554.

VII

DELIVERY USING ALPHAVIRUSES

35

Delivery and Expression of Heterologous Genes in Mammalian Cells Using Self-Replicating Alphavirus Vectors

Gunilla B. Karlsson and Peter Liljeström

1. Introduction

The RNA genomes of alphaviruses have been exploited to create highly efficient vectors for transient expression of foreign genes in mammalian cells and for use as vehicles for genetic vaccines (*1–6*). For Semliki Forest virus (SFV), a representative alphavirus, three strategies for in vitro and in vivo gene-delivery have been developed. One method relies on the packaging of recombinant vectors into suicidal viral particles and infection of target cells, whereas the other two methods are based on direct transfection of target cells, either by using naked DNA encoding the SFV replicon placed downstream of an RNA polymerase II dependent promoter, or by using in vitro transcribed RNA encoding the SFV replicon. All three approaches result in the delivery of a self-replicating SFV vector into target cells, with expression of foreign genes being driven from a highly efficient viral subgenomic promoter (**Fig. 1**).

The SFV system has been successfully employed to express a variety of complex proteins, including highly glycosylated surface-expressed viral antigens, G protein-coupled receptors and ion channels, families of proteins for which expression in prokaryotic systems is notoriously difficult or yields nonfunctional products. Using the SFV system, surface densities in the range of 5 $\times$ 10^6 receptors per cell can be obtained, facilitating functional studies significantly. The recent adaptation of the SFV technology to mammalian suspension

From: *Methods in Molecular Biology, vol. 246:*
Gene Delivery to Mammalian Cells: Vol. 2: Viral Gene Transfer Techniques
Edited by: W. C. Heiser © Humana Press Inc., Totowa, NJ

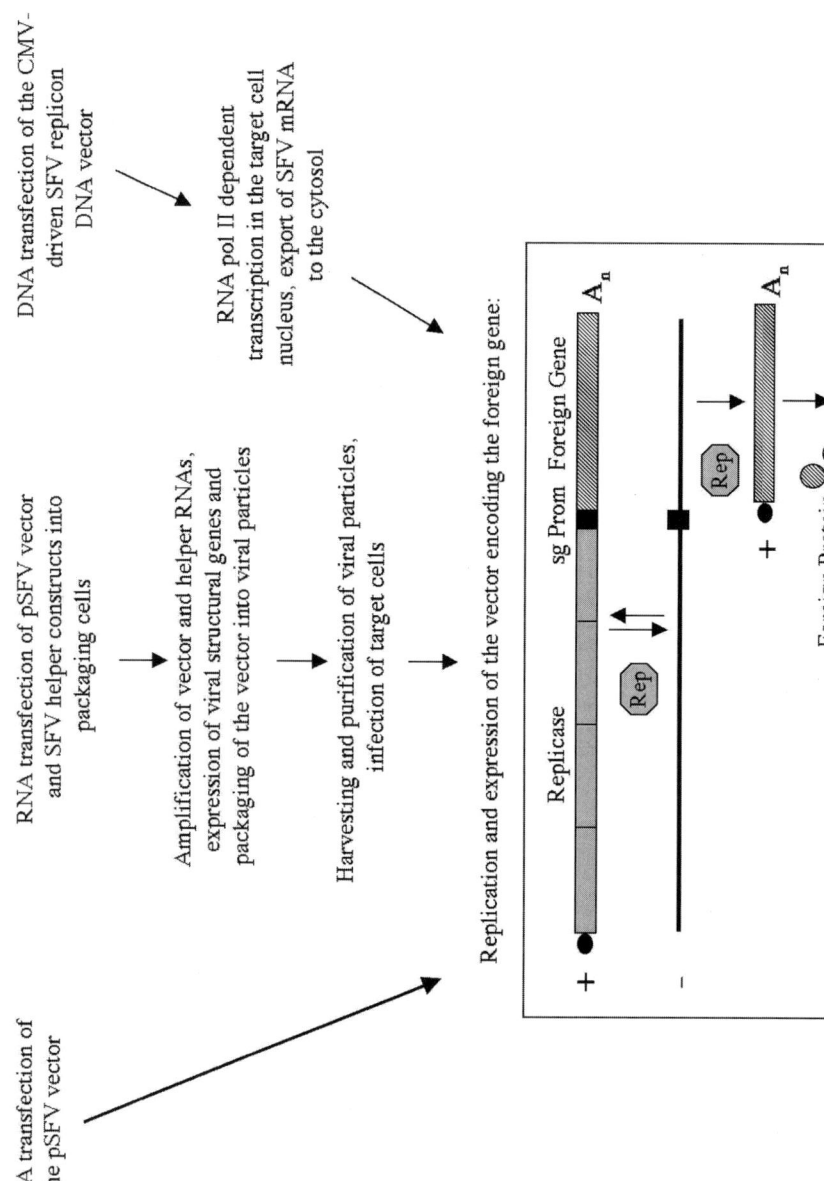

DNA transfection of the CMV-driven SFV replicon DNA vector

RNA pol II dependent transcription in the target cell nucleus, export of SFV mRNA to the cytosol

RNA transfection of pSFV vector and SFV helper constructs into packaging cells

Amplification of vector and helper RNAs, expression of viral structural genes and packaging of the vector into viral particles

Harvesting and purification of viral particles, infection of target cells

Replication and expression of the vector encoding the foreign gene:

RNA transfection of the pSFV vector

Fig. 1. Methods to deliver self-replicating SFV vectors into mammalian cells.

cultures now allows large-scale purification of many otherwise "difficult" proteins *(7–9)*. In addition, the SFV system has been used in several immunization studies, yielding very promising results *(10–15)*.

The alphavirus genome is a capped and polyadenylated positive-strand RNA molecule of 11–12 kb *(16)*. Productive infection can be initiated by transfection of the full-length genomic RNA into cells as well as through infection. SFV has a broad tropism capable of infecting distantly related hosts, ranging from insects and birds to mammals. Furthermore, cells of many different histological origins are susceptible to infection, including muscle, fat, neuronal, lymphoid, and epithelial cells.

Central to the technology built on alphaviruses is the virally encoded replicase, a complex consisting of four nonstructural proteins (nsp) 1–4, which are encoded by the 5' two-thirds of the genome. The replicase proteins, which are synthesized very early after infection, produce a negative strand that acts as a template for the production of new genomic RNAs for packaging. Importantly, a subgenomic promoter is exposed on the negative strand, resulting in the production of an additional RNA species corresponding to the 3' one-third of the viral genome. This transcript encodes the structural proteins of the virus, synthesized as a polyprotein precursor (NH_2-C-p62-6K-E1-COOH), which is co-translationally cleaved by the capsid proteinase and signal peptidase to produce the capsid protein (C) and the envelope proteins p62, 6K and E1. A sequence at the 5' end of the SFV capsid gene functions as a translational enhancer, providing high-level production of the structural proteins *(17)*. In the SFV vectors, the replicase gene and the 5' and 3' sequences necessary for replication are intact, while the structural genes are replaced by the foreign gene-of-interest. As a result, the vector is self-amplifying and foreign genes can be expressed to high levels *(9)*.

A characteristic of alphavirus infection is the shut-off of host-cell protein synthesis, occurring within hours of infection and ultimately leading to apoptotic death in most cell types *(18,19)*. Expression of foreign genes from the SFV vectors also results in target cell apoptosis and consequently protein expression in these systems is transient and cannot (at least in its present form) be used for making constitutively expressing stable cell lines. However, for many applications, vector-induced cell death is not a limitation as high levels of expression is achieved during the initial period of infection. Many cell types remain viable and produce protein for at least 48 h after infection. For genetic vaccinations, the onset of apoptosis in antigen-expressing cells may well contribute positively to the strong immune-response SFV vectors elicit. Additionally, suicidal systems are considered quite safe as bio-hazardous problems otherwise often associated with genetic vaccinations, such as integration of the transgene into the chromosome or induction of tolerance owing to prolonged expression of the antigen, are circumvented.

2. Materials

2.1. Plasmids

The plasmids required for the methods described in this chapter have recently been reviewed *(20)*. Briefly, the pSFV1 and pSFV10 vectors, as well as the Helper-C (S219A) and Helper-S2 split helper constructs used for packaging SFV particles, are pGEM-based and confer ampicillin resistance. They contain a SP6 promoter to drive synthesis of RNA in vitro, and unique restriction endonuclease sites allowing linearization of the plasmids are present immediately downstream of the poly A sequence in the viral 3′ end (*Spe*I for pSFV1 and the two helper constructs, *Nru*I for pSFV10) (**Fig. 2**).

1. pSFV1: pSFV1 encodes a functional replicase gene, including an intact packaging signal. The structural genes encoded by the subgenomic RNA in the wild-type SFV genome have been removed in the vector constructs to allow expression of foreign proteins or antigens in this position (*see* **Note 1**). The subgenomic transcript starts 31 nucleotides upstream of the *Bam HI* restriction endonuclease site (**Fig. 2A**).

2. pSFV10: In the pSFV10 vector, several silent point mutations in the replicase gene have been introduced to remove recognition sites for useful cloning enzymes such as *Rsr*I, *Xho*I, and *Not*I. A versatile poly-linker, which contains these enzyme sites and others, has been introduced allowing insertion of foreign genes in this position. The 3′ SFV sequence between the poly-linker and the poly A sequence has been trimmed from 888 nucleotides (in pSFV1) to 459 (in pSFV10) (**Fig. 2B**).

   The development of the SFV split helper system has been described in detail *(21)*. Briefly, Helper-C (S219A) and Helper-S2 were constructed by deleting 6091 bases of the replicase gene (*Acc*I [308] to *Acc*I [6399]). The resulting RNA molecules cannot be packaged into particles as a region at the end of nsP1 required for this function is removed by the deletion. Both Helper-C (S219A) and Helper-S2 retain the 5′ and 3′ sequences necessary for RNA replication.

3. Helper-C (S219A): An artificial stop codon was introduced immediately downstream of the natural cleavage site between capsid and spike, after the ultimate 3′ residue of the capsid protein (a tryptophan residue). The capsid gene was further modified to abolish its natural protease activity by mutating serine 219 to alanine (agt [S] to gcc [A]) at the core of the enzyme active site (**Fig. 2C**).

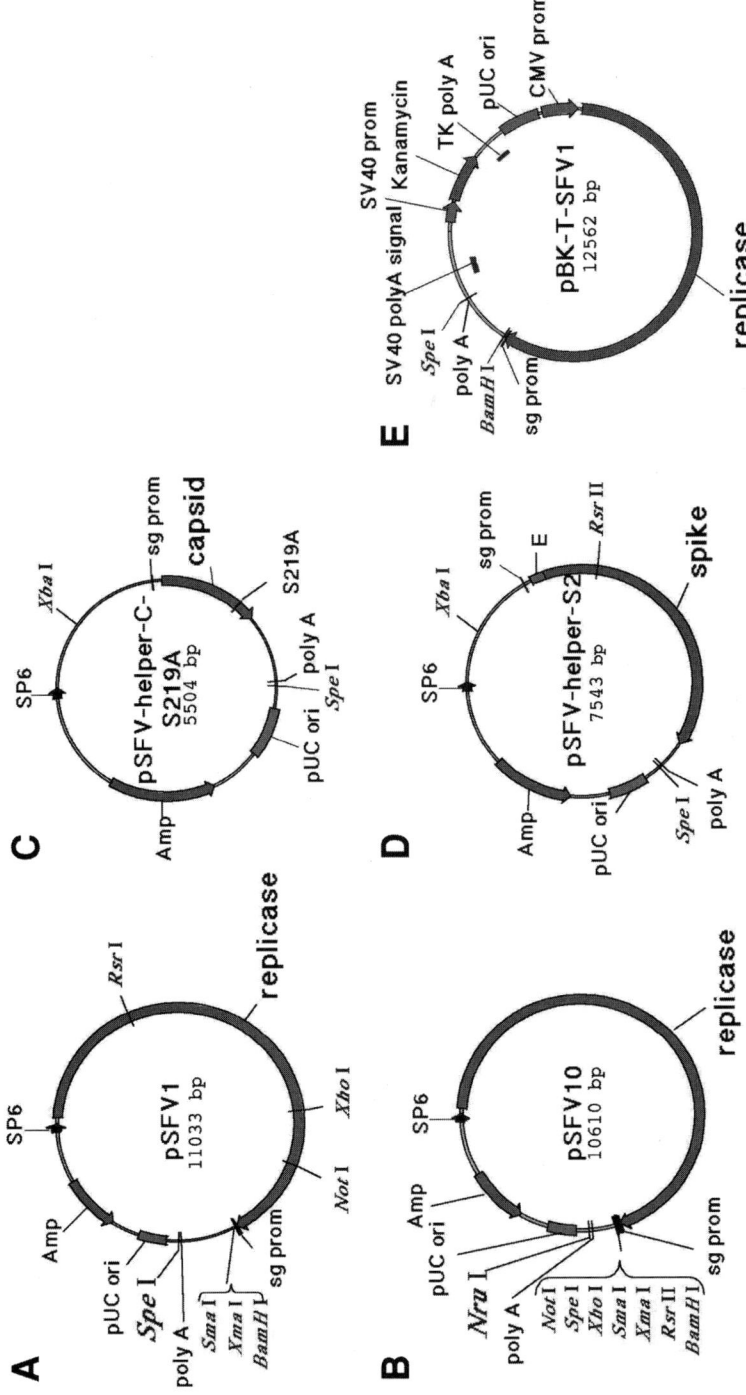

Fig. 2. Molecular constructs. (**A**) pSFV1, (**B**) pSFV10, (**C**) pSFV-helper-C-S219A, (**D**) pSFV-helper-S2, and (**E**) pBK-T-SFV1. Sg prom, the 26S subgenomic promoter; poly A, SFV polyadenylation sequence; Amp, ampicillin resistance gene; SP6, the SP6 prokaryotic promoter; CMV prom, CMV promoter; E, capsid enhancer linked to the 2A autoprotease.

4. Helper-S2: The translational enhancer region corresponding to the first 34 amino acids of the capsid gene was included in the Helper-S2 construct to obtain similar levels of protein from spike Helper construct as the capsid Helper construct *(17)*. A short sequence encoding autoprotease 2A of the Foot and Mouth Disease virus (FMDV) was inserted in frame between the enhancer sequence and the spike protein *(21)*. This allows processing of the spike proteins and generates a p62 protein, which starts from the second amino acid (Ala) of the natural sequence (**Fig. 2D**).

5. pBK-T-SFV1: The SFV sequence was placed downstream of the immediate early cytomegalovirus (CMV) promoter in pBK-CMV (Stratagene). The CMV promoter drives the expression of the SFV replicon, allowing replication of the vector and expression of foreign genes from the viral subgenomic promoter once the RNA polymerase II generated mRNA has been transported into the cytoplasm and the nsp1–4 genes have been expressed (**Fig. 2E**).

6. pBK-T-SFV1-EGFP: The enhanced green fluorescent protein (EGFP) was inserted into pBK-T-SFV1.

7. pBK-T-SFV1-enhancer-EGFP: This plasmid is identical to pBK-T-SFV-EGFP, except that the EGFP gene was inserted in frame with the SFV capsid-enhancer sequence. The enhancer (E), which is described above for pSFV-helper-S2, augments protein expression by about five- to ten-fold as shown in **Figs. 3D** and **E**.

8. pCMV-EGFP: A conventional expression plasmid, in which expression of EGFP is driven from a CMV promoter. This plasmid is used for comparative purposes as shown in **Fig. 3C**.

2.2. Apparatus and Reagents

1. Electroporator and electroporation cuvet (0.2 or 0.4 cm), Bio-Rad (Hercules, CA).

2. SP6 RNA polymerase, ^{35}S-methionine, m7 (5′) ppp (5′) G, and rNTP mix, Amersham BioSciences (Uppsala, Sweden).

3. Rnasin (an RNase inhibitor), Promega (Madison, WI).

4. BHK-21 cells are available from American Type Culture Collection (ATCC, Manassas, VA)

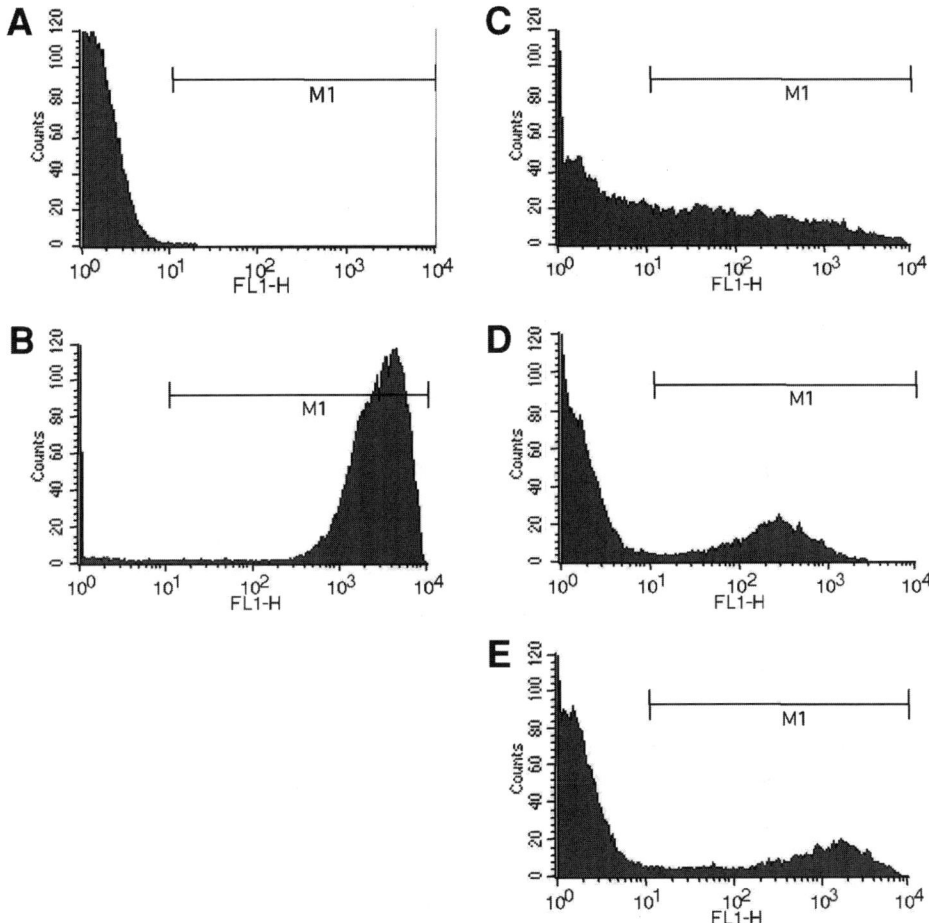

Fig. 3. Expression of EGFP by different SFV vectors. Cells were mock-infected (**A**), infected with recombinant SFV-EGFP particles at MOI = 20 (**B**), or cells were Lipofect-amine-transfected with pBK-EGFP plasmid DNA (**C**), pBK-T-SFV1-EGFP plasmid DNA (**D**), or with pBK-T-SFV1-enhancer-EGFP (**E**). Expression in BHK-21 cells was measured at 12 h postinfection or transfection using flow cytometric analysis. The percentage transfected cells (% inf.) and the mean channel values (MCV) within the M1 gate in the different samples were as follows: (**A**) (mock) 0.1% inf., MCV = 22, (**B**) (infection with recombinant particles) 97.4% inf., MCV = 2928, (**C**) (pBK-EGFP) 42.5% inf., MCV = 498, (**D**) (pBK-T-SFV1-EGFP) 25.0% inf., MCV = 312, (**E**) (pBK-T-SFV1-en-hancer-EGFP) 23.4% inf., MCV = 1269. High expression from almost all the cells in the culture can be achieved by infecting with recombinant SFV-EGFP particles [B]). The transfection efficiency of the conventional CMV promoter-driven pBK-EGFP plasmid (C) is higher than of the pBK-T-SFV1 plasmids (D, E), likely owing to the much smaller size of pBK-EGFP. However, the expression level from the SFV-driven plasmids is more homogenous with all positive cells showing similar levels of EGFP signal. When the translational enhancer is used, the expression level from the pBK-SFV-1 vector increases significantly (compare D and E).

2.3. Solutions and Cell-Culture Medium

1. TBE gel buffer (1X) for agarose gels: 50 mM Tris base, 50 mM H_3BO_4, 2.5 mM ethylenediaminetetraacetic acid (EDTA). pH will be 8.3, do not adjust.
2. 5X TD solution: 20% Ficoll 400, 25 mM EDTA, pH 8.0, 0.05% bromphenol blue, 0.03% Xylene cyanol.
3. 10X SP6 buffer: 400 mM HEPES-KOH, pH 7.4, 60 mM MgAc, 20 mM spermidine-HCl.
4. TNE buffer: 50 mM Tris-HCl, pH 7.4, 100 mM NaCl, 0.5 mM EDTA.
5. PBS without Mg^{+2} and Ca^{+2}, G-MEM, Trypsin-EDTA, and E-MEM can be obtained from Invitrogen (Paisley, UK).
6. BHK-21 medium: G-MEM containing 5% fetal calf serum (FCS), 10% tryptose phosphate broth, 10 mM HEPES, 2 mM L-glutamine, 100 U/mL penicillin (optional), 100 µg/mL streptomycin (optional).
7. Starvation medium: methionine and cysteine-free MEM is available from Sigma (St. Louis, MO). Add 2 mM L-glutamine, 10 mM HEPES.
8. Chase medium: E-MEM containing 2 mM L-glutamine, 10 mM HEPES, 150 µg/mL unlabeled methionine and cysteine.
9. 1X lysis buffer: 1% NP-40 (use 10% stock), 50 mM Tris-HCl, pH 7.6, 150 mM NaCl, 2 mM EDTA
10. Mowiol mounting medium: mix 6 g glycerol and 2.4 g Mowiol (Calbiochem, San Diego, CA) in 6 mL H_2O thoroughly. Incubate at room temperature for 2 h. Add 12 mL 0.2 M Tris-HCl pH 8.5. Incubate in +50°C to dissolve the Mowiol. Clarify by centrifugation at 5000g for 15 min. Aliquot and store at −20°C.

3. Methods

3.1. Preparation of RNA In Vitro for Production of Recombinant SFV Particles

For technical tips relating to the preparation of RNA, see **Note 2** and **Fig. 4**.

1. Linearize 5 µg of the vector plasmid (based on pSFV1 or pSFV10) and 5 µg each of the two split-helper plasmids, by digesting with the appropriate restriction enzymes (*Spe*I for pSFV1 and the split helper constructs, *Nru*I for the pSFV10 vector). Phenol extract and ethanol precipitate the DNA. Resuspend the DNA in H_2O to get a final concentration of 1.5 µg/5 µL.
2. Set up in vitro transcription reactions for each plasmid: 5 µL DNA (1.5 µg), 5 µL 10X SP6 buffer, 5 µL 50 mM dithiothreitol (DTT), 5 µL 10 mM m^{7}G(5′)ppp(5′)G, 5 µL rNTP mix, 23 µL H_2O, 1.5 µL RNasin (50 U), 0.5 µL (30 U) SP6 RNA polymerase.
3. Incubate at 37°C for 60–90 min, then take 1 µL aliquot into 10 µL of H_2O, add 3 µL of DNA gel loading solution (5XTD) and run sample on a 0.5%

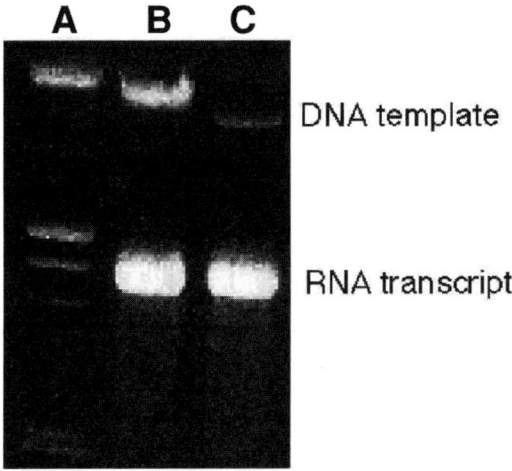

Fig. 4. A typical agarose gel analysis of SFV transcripts, in which 1 μL of a standard in vitro transcription mixture was analyzed on a 0.8% agarose gel (commonly used for DNA restriction fragment analysis). (**A**) Lambda DNA (*Hind*III + *Eco*RI) restriction fragments were used as markers (albeit not as true molecular-weight indicators). (**B**) and (**C**) show transcripts of the same linearized DNA template; in (**B**) the amount of template per transcription mixture is significantly greater than in the sample in (**C**). This shows that if the correct protocol is used, the amount of template per defined amount of units SP6 is saturated, i.e., both transcription reactions yield the same amount of RNA.

agarose gel to check the RNA. The quality of the RNA should be as shown in **Fig. 4.** Use lambda DNA (e.g., *Eco*RI +*Hind*III) cut) as marker. This protocol yields approx 50 μg RNA per construct, which is the amount used for one electroporation.

4. Freeze the rest in aliquots at −80°C

3.2. Transfection of BHK-21 Cells by Electroporation

The transfection efficiency is critical for obtaining high-titer viral stocks (*see* **Note 3**).

1. Grow BHK-21 cells to late log phase in complete BHK medium.
2. Wash cells once with PBS (without Mg^{+2} and Ca^{+2}).
3. For a 75-cm² flask, add 2 mL of trypsin and incubate at 37°C until the cells detach (about 1 min), then briefly pipet cell solution back and forth to ensure that a single-cell suspension is obtained (monitor by microscope). Stop trypsinization by adding 10 mL of BHK-21 medium.
4. Harvest cells by centrifugation for 5 min, 400*g*, resuspend cells in 10–20 mL PBS (without $MgCl_2$ and $CaCl_2$).

5. Harvest cells as in **step 4**, and resuspend in PBS (without $MgCl_2$ and $CaCl_2$) to get 10^7 cells/mL.

6. Transfer 0.8 mL of cell suspension to an Eppendorf tube containing the RNAs to be transfected. For virus packaging, use 50 µL of RNA from each of the three plasmids (vector and the two split helpers). Mix thoroughly by pipetting and transfer the mixture to a 0.4 cm electroporation cuvet.

7. Pulse twice at 850 V/25 µF at room temperature. The time constant after the pulse should be 0.4.

8. Dilute transfected cells 10–20-fold in complete BHK-21 medium and rinse the cuvet with the same medium to collect all cells. Seed cells from one electroporation into one 75-cm² flask or cells from three electroporations into one 225-cm² flask. It takes about 1 h for the cells to reattach to the dish.

9. Incubate the transfected BHK-21 cells in a 5% CO_2 incubator for 24 h at 33°C to allow the cells to assemble and release virus particles (*see* **Note 4**).

10. Collect and clarify the medium containing the recombinant SFV particles by centrifuging at 40,000*g* for 30 min at +4°C to remove remaining cells and cell debris. Aliquot and freeze the supernatant quickly on dry ice or in liquid nitrogen. Store at −80°C.

3.3. Purification and Concentration of SFV Particles

To concentrate and purify recombinant virus from the medium of transfected BHK-21 cells, sediment the particles by ultracentrifugation through a sucrose cushion. See **Note 5** for safety considerations.

1. Transfer the viral supernatant to ultracentrifuge tubes, (two 35-mL Beckman 25 × 89 mm tubes are suitable for a viral supernatant resulting from three electroporations). Add 5 mL 20% sucrose with a pipet through the supernatant down into the bottom of the tube. Fill the tube to the top with viral supernatant or medium. Balance the tubes carefully and spin at 140,000*g* for 90 min at 4°C.

2. Tilt the tube and aspirate the entire medium and sucrose fraction, without touching the bottom of the tube. Add 250–500 µL TNE buffer per tube and resuspend the virus pellet from the bottom of the tube.

3. Filter the concentrated virus stock through a 0.22-µm filter, using a small syringe.

4. Aliquot and freeze the purified virus stock quickly, store at −80°C.

3.4. Infection of Target Cells

To obtain full infection of most cell types, such as BHK-21 cells, a multiplicity of infection (MOI) of 20 may be required (*see* **Fig. 3B**). For example, from a virus stock of 10^9 infectious particles per mL, use 10 µL of virus stock diluted in a total volume of 500 µL to infect 0.5×10^6 cells in a 35-mm well (a 1:50 dilution of the virus). In general the virus needs to be diluted 25–100 times depending on the titer of the stock and infectability of the target cells. The following is a standard protocol for in vitro infection of target cells.

1. Wash 80–100% confluent cells thoroughly with PBS.
2. Thaw recombinant virus preparation quickly (room temperature), then dilute as needed in E-MEM containing 0.2% BSA, 2 mM L-glutamine, and 20 mM HEPES.
3. For a 35-mm plate, apply 500 µL of virus solution (diluted as required from the frozen stock) on the cells and incubate at 37°C for 45–60 min.
4. Remove the virus solution, add 3 mL complete BHK-21 medium (or other suitable medium required for the experiment) and continue incubation as required.

3.5. Titer Determination of Recombinant Virus Particles

Use different dilutions of the virus stock to infect cells. Detect protein expression by immunofluorescence, using a primary antibody with specificity for the heterologous protein expressed by the vector.

1. Grow cells on glass cover slips to about 70% confluency.
2. Infect the cells with different dilutions of the recombinant virus stock diluted in EMEM containing 0.2% bovine serum albumin (BSA), 2 mM glutamine, and 20 mM HEPES. Allow 10–16 h for expression of protein of interest.
3. Rinse cover slips twice with PBS, then fix cells in -20°C methanol for 5 min.
4. Remove methanol and wash cover slips three times (3X) with PBS.
5. Block nonspecific binding by incubating with PBS containing 0.5% gelatin and 0.2% BSA (30 min at room temperature).
6. Replace blocking buffer with same buffer containing primary antibody. Incubate at room temperature for 30 min.
7. Wash 3X with PBS, then bind secondary antibody as in **step 5**.
8. Wash 3X with PBS and 1X with water, drain, and let cover slip air-dry.
9. Mount on glass slide using 10–20 µL Mowiol 4-88 containing 2.5% DABCO (1,4-diazobicyclo-[2.2.2]-octane) (the DABCO will reduce fading of FITC)
10. Count 20 fields of a dilution for which individual positive cells can be clearly identified (usually the 1:10,000 or the 1:100,000 dilution) to obtain a reliable average value per field (ideally count around 20 positive cells per field). Virus titer is determined based on the dilution factor and a microscope specific constant (depending on size of eye field, lens used, and area of the dish), according to the following formula: (Average value counted cells) $\times$ (constant) $\times$ (dilution factor) = infectious units (IU) per mL.

3.6. Gene Delivery by Transfection of SFV Vector Encoding DNA

Follow standard DNA transfection protocols, such as those using Lipofecta-mine. Because the pBK-SFV plasmids are large (approx 12.5 kb without insert), transfection efficiencies are usually lower than what can be obtained with conventional (smaller) expression plasmids (see **Fig. 3C, D**).

3.7. Gene Delivery by Transfection of SFV Vector RNA

Follow the protocols in **Subheadings 3.1.** and **3.2.**, using only the SFV vector and omitting the split helper plasmids.

3.8. Analysis of Protein Expression by Metabolic Labeling of Cells

Protein expression from SFV infected cells (*see* **Subheading 3.4.**), or cells transfected with SFV RNA (*see* **Subheading 3.7.**) or with DNA encoding the SFV replicon (*see* **Subheading 3.6.**) is readily monitored by using ^{35}S-methion-ine/cysteine metabolic labeling followed by analysis by sodium dodecyl sulfate polyacrylamide gel electrophoresis (SDS-PAGE). The ideal times to perform the labeling after infection or transfection are outlined in **Note 6**. The following protocol is given for 35-mm tissue-culture plates (80–100% confluent).

1. Aspirate growth medium and wash cells twice with 3 mL PBS pre-warmed to 37°C, then overlay cells with 2 mL starvation medium and incubate plates at 37°C (5% CO_2) for 30–45 min.
2. Aspirate medium and replace with 500 μL of the same containing 50–100 μCi/mL of ^{35}S-methionine, incubate for appropriate pulse time at 37°C. A pulse time of 10 min is suitable for most purposes.
3. Remove pulse medium and wash cells once with 2 mL of chase medium, then overlay cells with 2 mL of chase medium and incubate for required chase time. To follow maturation, posttranslational modifications and/or surface expression or secretion of a protein, chase times of 0, 0.5, 2, and 4 h are usually suitable.
4. Remove medium and wash cells with 3 mL ice-cold PBS, add 300 μL of lysis buffer and incubate on ice for 10 min.
5. Resuspend cells and transfer solution into an Eppendorf tube, spin at max speed at 4°C in an Eppendorf centrifuge for 5 min to pellet unbroken cells, nuclei, and cellular debris. Transfer supernatant to a fresh tube and store at −80°C.
6. Assay for protein expression by SDS-PAGE and autoradiography.

4. Notes

1. A variety of foreign genes have been expressed using the pSFV1 and pSFV10 vectors. Large genes, such as *LacZ*, are readily expressed without

affecting the viral titers. Vectors containing double subgenomic promoters, expressing two foreign genes from the same vector, have also successfully been constructed. As is often seen in recombinant viral systems, the efficiency of packaging may become compromised as the total length of the inserted cDNA increases, resulting in a decrease in viral titers. This will likely also be the case in the SFV system as it is pushed to its limits.

2. Spermidine is present in the SP6 buffer so the reaction mixture should be set up at room temperature to avoid precipitation of the DNA. For one in vitro transcription reaction using 30 U of SP6 RNA polymerase, a total of 1.5 µg of linear DNA is required. Under the conditions described, the transcription mixture is saturated for DNA and yields about 50 µg of RNA. Although the gel is nondenaturing and thus does not reflect the molecular weight of the RNA produced, it is nevertheless useful for checking the quality and quantity of the RNA. The RNA band should be defined (no smearing) and relatively thick in comparison to the DNA bands (*see* **Fig. 4**). If necessary, RNA samples of known concentration can be run for comparison.

3. By using this protocol a transfection efficiency near 100% can be obtained in BHK-21 cells. The optimal electoporation parameters are likely to vary greatly between different cell types so they have to be determined on a case-by-case basis if other cell types are used. To optimize the electoporation protocol on a new cell type, it is necessary to compare differences in voltage, capitance, time constant of electrical pulse, and the number of pulses given to identify the optimal conditions. If other cell types are to be used, it is important to check their ability to support virus particle formation. This can be achieved by performing an infection with wild-type SFV.

4. Incubation of the BHK-21 cells at 37°C gives, on average, a 10-fold lower titer than incubation at 33°C, probably because the onset of apoptosis is delayed when the cells are cultured at the lower temperature. A second harvest can be performed at 48 h after transfection, if fresh media is placed on the cells after the 24-h harvest time point. The second harvest can yield as high titers as the first one, doubling the total amount of virus that can be obtained from each transfection. The anticipated titer from 10^7 BHK-21 cells is 10^9–10^{10} infectious recombinant particles.

5. As with all recombinant viral system, care should be taken when handling viral supernatants, especially following high-speed ultracentrifugation when aerosol formation can occur. The SFV split-helper system has been designed to minimize the risk of recombination *(21)* and so far the appearance of replication-proficient viruses (RPVs) has not been reported. Nevertheless, all work using these systems should be carried out in biosafety level 2 laminar flowhoods according to standard biosafety level 2 practices.

6. Metabolic labeling experiments of cells expressing foreign proteins from the SFV vectors are easy to analyze, because the vector RNA replication results in a general shut-off of host protein synthesis, reducing the background of labeled host proteins dramatically. The labeling can be done already at 4–6 h after RNA transfection or infection. However, in order to assure minimum labeling of host proteins, it is recommended that 8–9 h has elapsed before adding the radioactivity. If the cells have been infected with a high MOI, achieving full infection of the cell population, the vector-expressed protein will be seen as the major species on an SDS gel without performing immunoprecipitation. After transfection with DNA encoding the SFV replicon (e.g., pBK-T-SFV1), the radioactivity can be added around a similar time point. However, because in most cases it is not possible to achieve 100% transfection of the cell population using methods such as Lipofectamine, it is advisable to perform an immunoprecipitation prior to analysis by SDS gel. Protein expression from SFV vectors in which the foreign gene is proceeded by the translation enhancer (E) is about 10-fold increased. This difference can readily be detected when samples are analyzed by SDS-PAGE or by other means of analysis, such as by flow cytometry, as shown in **Fig. 3D**, **E**.

References

1. Liljeström, P. and Garoff, H. (1991) A new generation of animal cell expression vectors based on the SemLiki Forest virus replicon. *Biotechnology* **9,** 1356–1361.
2. Pushko, P., Parker, M., Ludwig, G. V., Davis, N. L., Johnston, R. E., and Smith, J. F. (1997) Replicon-helper systems from attenuated Venezuelan equine encephalitis virus: expression of heterologous genes *in vitro* and immunization against heterologous pathogens *in vivo*. *Virology* **239,** 389–401.
3. Xiong, C., Levis, R., Shen, P., Schlesinger, S., Rice, C. M., and Huang, H. V. (1989) Sindbis virus: an efficient, broad host range vector for gene expression in animal cells. *Science* **243,** 1188–1191.
4. Frolov, I., Hoffman, T. A., Prágai, B. M., Dryga, S. A., Huang, H. V., Schlesinger, S., and Rice, C. M. (1996) Alphavirus-based expression vectors: strategies and applications. *Proc. Natl. Acad. Sci. USA* **93,** 11371–11377.
5. Liljeström, P. (1994) Alphavirus expression systems. *Curr. Opin. Biotechnol.* **5,** 495–500.
6. Garoff, H. and Li, K. J. (1998) Recent advances in gene expression using alphavirus vectors. *Curr. Opin. Biotechnol.* **9,** 464–469.
7. Blasey, H. D., Lundström K., Tate, S., and Bernard, A. R. (1997) Recombinant protein production using the SemLiki Forest virus expression system. *Cytotechnology* **24,** 65–72
8. Lundström, K., Michel, A., Blasey, H., Bernard, A. R., Hovius, R., Vogel, H., and Surprenant, A. (1997) Expression of ligand-gated ion channels with the SemLiki Forest virus expression system. *J. Rec. Signal Transd. Res.* **17,** 115–126.

9. Lundström, K., Mills, A., Allet, E., Ceszkowski, K., Agudo, G., Chollet, A., and Liljeström, P. (1995) High-level expression of G protein-coupled receptors with the aid of the SemLiki Forest virus expression system. *J. Rec. Signal Transd. Res.* **15,** 23–32.

10. Zhou, X., Berglund, P., Rhodes, G., Parker, S. E., Jondal, M., and Liljeström P. (1994) Self-replicating SemLiki Forest virus RNA as recombinant vaccine. *Vaccine* **12,** 1510–1514.

11. Brand, D., Lemiale, F., Turbica, I., Buzelay, L., Brunet, S., and Barin, F. (1998) Comparative analysis of humoral immune responses to HIV type 1 envelope glycoproteins in mice immunized with a DNA vaccine, recombinant SemLiki Forest *virus* RNA, or recombinant SemLiki Forest virus particles. *AIDS Res. Hum. Retrovirol.* **14,**1369–1377.

12. Berglund, P., Smerdou, C., Fleeton, M. N., Tubulekas, I., and Liljeström, P. (1998) Enhancing immune responses using suicidal DNA vaccines. *Nat. Biotechnol.* **16,** 562–565.

13. Fleeton, M. N., Chen, M., Berglund, P., Rhodes, G., Parker, S. E., Murphy, M., et al. (2001) Self-replicative RNA vaccines elicit protection against influenza A virus, respiratory syncytial virus, and a tickborne encephalitis virus. *J. Infect. Dis.* **183,** 1395–1398

14. Vajdy, M., Gardner, J., Neidleman, J., Cuadra, L., Greer, C., Perri, S., et al. (2001) Human immunodeficiency virus type 1 Gag-specific vaginal immunity and protection after local immunizations with sindbis virus-based replicon particles. *J. Infect. Dis.* **184,** 1613–1616.

15. Harrington, P. R., Yount, B., Johnston, R. E., Davis, N., Moe, C., and Baric, R. S. (2002) Systemic, mucosal, and heterotypic immune induction in mice inoculated with Venezuelan equine encephalitis replicons expressing Norwalk virus-like particles. *J. Virol.* **76,** 730–742.

16. Strauss, J. H. and Strauss, E. G. (1994) The alphaviruses: gene expression, replication, and evolution. *Microbiol. Rev.* **58,** 491–562.

17. Sjöberg, E. M., Suomalainen, M., and Garoff, H. (1994) A significantly improved SemLiki Forest virus expression system based on translation enhancer segments from the viral capsid gene. *Bio/Technology* **12,**1127–1131.

18. Glasgow, G. M., McGee, M. M., Sheahan, B. J., and Atkins, G. J. (1997) Death mechanisms in cultured cells infected by SemLiki Forest virus. *J. Gen. Virol.* **78,**1559–1563.

19. Ying, H., Zaks, T. Z., Wang, R. F., Irvine, K. R., Kammula, U. S., Marincola, F. M., et al. (1999) Cancer therapy using a self-replicating RNA vaccine. *Nat. Med.* **5,** 823–827.

20. Smerdou, C. and Liljeström, P. (2000) Alphavirus vectors: from protein production to gene therapy. *Gene Ther. Reg.* **1,** 33–63.

21. Smerdou, C. and Liljeström, P. (1999) Two-helper RNA system for production of recombinant SemLiki forest virus particles. *J. Virol.* **73,** 1092–1098.

Index